Topics in Applied Physics   Volume 64

Topics in Applied Physics Founded by Helmut K. V. Lotsch

Volumes 1–56 are listed on the back inside cover

Sputtering by Particle Bombardment III

Characteristics of Sputtered Particles, Technical Applications

Edited by R. Behrisch and K. Wittmaack

With Contributions by
R. Behrisch   W. Hauffe   W. O. Hofer
N. Laegreid   E. D. McClanahan   B. U. R. Sundqvist
K. Wittmaack   M. L. Yu

With 190 Figures

Springer-Verlag Berlin Heidelberg GmbH

Dr. *Rainer Behrisch*

Max-Planck-Institut für Plasmaphysik, EURATOM Association
D-W-8046 Garching bei München, Fed. Rep. of Germany

Dr. *Klaus Wittmaack*

GSF-Forschungszentrum für Umwelt und Gesundheit GmbH
D-W-8042 Neuherberg bei München, Fed. Rep. of Germany

ISBN 978-3-662-31104-2   ISBN 978-3-540-46881-3 (eBook)
DOI 10.1007/978-3-540-46881-3

© Springer-Verlag Berlin Heidelberg 1991
Originally published by Springer-Verlag Berlin Heidelberg New York in 1991.
Softcover reprint of the hardcover 1st edition 1991

Typesetting: Macmillan India Ltd., India
54/3140-543210 – Printed on acid-free paper

Preface

This is the third and last volume of a series published in Topics of Applied Physics dealing with all aspects of sputtering, i.e. the release of atoms from the surface of a solid (or liquid) bombarded with energetic particles. The first volume (TAP 47) gives an introduction to the theory of collision cascade sputtering and a summary of sputtering yields measured for amorphous, polycrystalline and monocrystalline elemental solids bombarded primarily with noble-gas and hydrogen ions. The second volume (TAP 52) is devoted to sputtering of multicomponent solids and the influence of chemical effects on sputtering. It also describes the evolution of bombardment-induced surface topographies and sputtering by particles other than ions such as electrons, photons and neutrons. The present volume deals with the characteristics of sputtered particles and various technical applications of sputtering.

The review of the physics and applications of sputtering with chapters written by the experts in the various areas was first planned, about 15 years ago, for publication as one comprehensive book. It soon became clear, however, that at least two books would be needed and now three volumes have finally resulted. Unfortunately, publication of this last volume has taken many years more than originally planned. In fact, the first versions of Chaps. 2 and 7 of this book were first completed almost ten years ago together with drafts on other topics. These manuscripts have been updated. On the other hand, the last decade has seen a rather rapid development in many areas of applied sputtering. This could be included in this book and it is documented by the chapters on sputter depth profiling, ion induced desorption of biomolecules and the production of microstructures by ion beam sputtering. The understanding of the charge and excitation states of sputtered particles has also advanced considerably in recent years. The early review on this topic could, unfortunately, not be brought to an end, because the authors had meanwhile left the field. Another colleague joined the team of authors at a late stage. Two other contributions which were originally planned for inclusion in this volume either were not completed (Sputter Ion Sources) or have already been published elsewhere [E. Taglauer "Surface Cleaning using Sputtering", Appl. Phys. A51, 283 (1990)].

We wish to apologise for all these changes and delays. Furthermore, we are grateful to all authors for their patience and kind collaboration, and to Springer-Verlag for their constant cooperation, which finally made it possible to complete this series on sputtering.

Garching, Neuherberg   *R. Behrisch*
May 1991   *K. Wittmaack*

Contents

Contributors

Behrisch, Rainer

Max-Planck-Institut für Plasmaphysik, EURATOM Association,
D-W-8046 Garching bei München, Fed. Rep. of Germany

Hauffe, Wolfgang

Sektion Physik, Technische Universität Dresden, Mommsenstr. 13,
D-O-8027 Dresden, Fed. Rep. of Germany

Hofer, Wolfgang O.

Projekt Kernfusion, Forschungszentrum Jülich, EURATOM Association,
D-W-5170 Jülich, Fed. Rep. of Germany

Laegreid, Nils

Advanced Materials Section, Pacific Northwest Laboratory,
Battelle Boulevard, P.O. Box 999,
Richland, WA 99352, USA

McClanahan, Edwin Davidson

Advanced Materials Section, Pacific Northwest Laboratory,
Battelle Boulevard, P.O. Box 999,
Richland, WA 99352, USA

Sundqvist, Bo U.R.

Division of Ion Physics, Department of Radiation Sciences,
Uppsala University, Box 535,
S-75121 Uppsala, Sweden

Wittmaack, Klaus

GSF—Forschungszentrum für Umwelt und Gesundheit GmbH,
Ingolstädter Landstr. 1,
D-W-8042 Neuherberg, Fed. Rep. of Germany

Yu, Ming Lun

IBM Thomas J. Watson Research Center, P.O. Box 218,
Yorktown Heights, NY 10598, USA

1. Introduction

Rainer Behrisch and Klaus Wittmaack

With 4 Figures

More than 125 years after the discovery of sputtering, the main features and basic physical processes of this phenomenon have been widely investigated, and most features are reasonably well understood. In the various fields of applications, a remarkable level of sophistication has been achieved, for example, in surface analysis, depth profiling, sputter cleaning, micromachining, and sputter deposition. This book reviews the characteristics of the sputtered-particle flux, as well as the various analytical and practical applications of sputtering.

1.1 Overview

Sputtering is the removal of material from the surface of a solid through the impact of energetic particles. The material released from the bombarded surface consists predominantly of single atoms and a sometimes sizable fraction of homo- or heteronuclear clusters. The ejected particles have a broad energy and angular distribution, the mean energy being on the order of 10 eV. Except for the case of compound targets with appreciable ionic binding, only a small fraction of the sputtered particles is in an excited or charged state.

Generally any energetic radiation composed of ions, neutrals, neutrons, electrons, or photons that causes damage in the bulk of a solid will also cause sputter ejection of surface atoms. The basic process is a collision cascade between the lattice atoms, initiated by the incident radiation in the bulk or in the surface layer of the solid. The same models are used for describing radiation damage and sputtering.

Sputtering was first observed in a gas discharge as the erosion of the cathode [1.1–4] that was bombarded by energetic ions from the plasma. Today two major technologically important aspects of sputtering can be distinguished. On one hand, sputtering has become an indispensable tool in such modern technologies as the deposition of high-quality thin films on almost any substrate, surface and depth microanalysis, surface cleaning and micromachining, and generating ion beams of solid materials. Sputtering has also been successfully applied in the analysis of large organic molecules. On the other hand, in gas discharges, sputtering results in a mostly undesirable erosion of the vessel walls and a contamination of the plasma. Sputtering thus constitutes one of the most critical problems in high-temperature plasma experiments, especially in attempts to build a fusion reactor [1.5–7].

Topics in Applied Physics, Vol. 64
Sputtering by Particle Bombardment III Eds.: R. Behrisch · K. Wittmaack
© Springer-Verlag Berlin Heidelberg 1991

1.2 Ion Bombardment Phenomena

If ions with energies exceeding a few tens of eV impinge on the surface of a solid, several processes are initiated:

1) A small fraction of the incident ions is backscattered in collisions with surface and near-surface atoms. The backscattered particles are mostly neutral atoms with a broad energy distribution. The backscattering yield depends on the energy and mass of the incident ions, as well as on the mass of the target atoms [1.8–13].

2) The major fraction of the incident ions is slowed down in collisions with atoms and electrons of the solid. The energy transferred to target atoms may initiate a collision cascade. Energy and momentum can thus be transported back to the surface and may cause sputtering, i.e., release of surface atoms [1.14, 15].

3) At the end of their range the injected projectiles may be trapped and accumulate in the solid, or, depending on their chemical identity and the properties and temperature of the host material, they may diffuse back to the bombarded surface or into the bulk. Implanted gaseous ions may form bubbles and cause blistering of the surface layer [1.16].

4) Ion impact on the surface of a solid may also give rise to the emission of electrons and photons [1.17–24 and Chap. 3]. The electrons may be ejected by potential emission [1.22], kinetic emission [1.23, 24], or Auger emission from excited sputtered particles [1.17, 18]. Photons may originate from backscattered or sputtered particles, excited in collisions with a surface atom before emission, or directly from excitations in the near-surface region of the solid [1.17, 18].

1.3 Sputtering Mechanisms

As incident ions in collisions with atoms and electrons of the solid slow down, energy in excess of the lattice binding energy (on the order of 10 eV) may be transferred to an atom of the solid. This occurs mostly in direct collisions with target atoms. Energy may, however, also be transferred by local electronic excitation and ionization, which may lead to a modification of the interaction potential and a repulsion. Atoms removed from their original sites are subsequently slowed down in the solid by the same mechanisms as the incident ions. Knockon atoms may also remove other atoms from their lattice sites. Thus a collision cascade develops. Surface or near-surface atoms will be emitted if they receive a momentum in the direction of the vacuum half-space with enough energy to overcome the surface binding. Generally more than 60% of the ejected atoms originate from the first atomic layer, while the remainder originate from the layers underneath [1.25–29]. This process of surface erosion has been investigated throughout the world. In English it is called *sputtering*, in German *Zerstäubung*, in French *pulvérisation*, in Italian *polverizzazione*, in Spanish *pulverizacion*, in Russian *raspilenie*, in Japanese *s(u)pattering(u)*, and in Chinese *jian shi.*

The removal of surface and near-surface atoms by knockon sputtering –, i.e. by direct momentum transfer in a collision cascade – can be called *physical sputtering.* It is a nonthermal process, with a major fraction of the lattice atoms staying cold during the initial spread of the cascade that leads to sputtering. The energy is finally dissipated in displacements of atoms and in lattice vibrations. Most of the atoms removed from their sites fall back to their original locations; i.e., these defects annihilate spontaneously. Atoms coming to rest at a sufficiently large distance from their original location may form stable Frenkel defects. Vacancies abound near the center of the incident particle trajectory, whereas interstitials cluster in the volume around it [1.30–32].

Collision cascade evolution is influenced by the crystal lattice structure through channeling, blocking, and focusing [1.27, 28]. Depending on the energy and mass of the incident ions, three different collision cascade regimes can be distinguished [1.29]:

1) For low-energy and light-ion bombardment, the collision cascades are very dilute and involve only a few atoms. This is the so-called *single-knockon regime.*
2) Under bombardment with ions of medium or high mass at energies exceeding a few keV, the collision cascades get larger and more dense. If each target atom is at rest before it gets hit, we are dealing with the *linear-cascade regime.*
3) With ions of high atomic number, or with molecular ions of sufficiently high energy (above about 100 keV), and with target atoms of high atomic number, the cascades get very dense and each target atom may be knocked on while it is already in motion. This is named the *collision-spike regime*, and nonlinear effects may play a sizable role. Generally a local thermal equilibrium is not established during the evolution of the spike, so it is not justified to refer to this event as a thermal spike.

If the incident ions and the atoms of the solid interact chemically and volatile molecules are formed, the erosion rate may increase much beyond the value corresponding to knockon sputtering. This process has been termed *chemical sputtering* [1.33, 34].

A very special phenomenon may occur if interstitials produced by the collision cascade diffuse to the surface and escape by sublimation. This *radiation-enhanced sublimation* is assumed to be responsible for the large increase in the erosion yield of carbon observed during ion bombardment at temperatures above 1200 °C [1.35, 36].

1.4 Modeling and Microscopic Investigation of the Sputtering Process

The sputtering process and collision cascade development have been described analytically [1.29, 37, 38] and by computer simulation [1.26, 27, 32, 39–47], for both amorphous and crystalline solids. With the analytical approach, average quantities are calculated using Boltzmann transport theory, whereas computer simulation allows tracing of individual particle trajectories in the solid. The results of both approaches agree reasonably well if the basic assumptions are the same [1.40].

A plot of the trajectory of one 20 keV Ge ion injected into amorphous silicon, as obtained with the computer simulation program TRIM.SP [1.41, 47], is shown in Fig. 1.1 (thick line); also shown are the trajectories of knockon target atoms (thin lines) [1.47]. Because of the projection of the trajectories into the drawing plane (normal to the surface) the cascade appears very dense, but we are still in the linear-cascade regime. As a result of this impact, 2078 atoms are knocked from their sites. Out of these, 372 atoms are permanently removed. Three surface atoms are ejected into vacuum; i.e., they are sputtered.

To get more experimental information about the individual cascades and the sputtering process, the surface topography of solids after low-fluence heavy-ion bombardment have been investigated using the transmission electron microscope (TEM) [1.48] and very recently using scanning tunneling microscopy (STM) [1.49]. After 500 keV Bi ion bombardment of Au foils TEM observations sometimes showed large craters with a volume equivalent to as many as about 2000 atoms per incident ion. These craters were attributed to collision spikes of

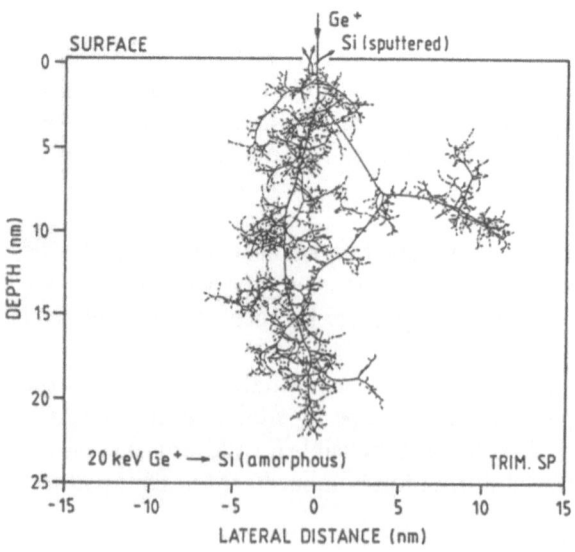

Fig. 1.1. Trajectories of atoms displaced in a collision cascade, as calculated with the computer simulation program TRIM.SP for normal incidence of a single 20 keV Ge ion on amorphous Si. The thick line represents the trajectory of the Ge ion, while the thin lines correspond to the Si trajectories. All trajectories are projected on the drawing plane normal to the surface [1.47]. The surface is located at depth zero, and it corresponds to the location of atoms standing farthest out from the solid.

8 keV Kr$^+$ ⟶ PbS (001)

Fig. 1.2. Scanning tunneling microscopy image of a PbS (100) surface after bombardment with 3×10^{12} Kr ions/cm^2 at 8 keV in a random direction. The dark area represents Pb and S atoms in the center of a crater, while the bright area represents the rim of the crater [1.49]

very high density generated by favorable impacts. In addition, many dark spots due to dislocation loops have been found [1.48]. STM investigations on Si and PbS surfaces after bombardment with Ge, Ar, and Kr ions at energies between 8 keV and 1 MeV showed craters with a dip in the center and a surrounding rim (see Fig. 1.2) [1.49]. These features have been attributed to agglomeration of vacancies in the center of the cascade, causing the dip, and to relaxation of the surrounding interstitial atoms forming the rim [1.30–32, 49].

1.5 The Sputtering Yield

The most important and most widely investigated quantity for describing the sputtering process is the sputtering yield Y, which is defined as the average number of atoms released at the solid's surface per incident particle. Commonly, sputtering is accomplished using energetic ions or neutral atoms. When sputtering is initiated by bombardment with energetic electrons and photons [1.50] or with neutrons [1.51] the same definitions of the sputtering yields can be applied.

Sputtering yields measured under impact of molecular ions are generally quoted as a yield per incident atom at correspondingly lower impact energy. In the case of bombardment with ions of the target material, named *self-sputtering* [1.52, 53], the incident ions are mostly implanted and trapped in the solid. A self-sputtering yield of unity means that on the average one atom is removed per incident ion; i.e., there is no mass change of the sample during the bombardment.

The sputtering yield depends on the mass, the energy, and the impact angle of the incident particles; on the mass and the binding energy of the atoms in the solid and at the surface [1.29]; and on the crystallinity of the sample and the orientation of low-index directions with respect to the direction of the incident particle beam [1.27, 28]. Generally, the sputtering yield can be described by two terms: one describes the density of the energy deposited by the incident particles into nuclear motion of target atoms near the surface, and the other, a material factor, contains the atomic density and the surface binding energy of the solid [1.29, 37, 52].

The sputtering yield features a threshold energy E_{th} of typically 20–50 eV, which is determined by the necessary energy transfer between the incident particles and the atoms of the solid, as well as by the surface binding energy U_0 [1.52, 54, 55]. For bombardment energies larger than E_{th} the sputtering yield increases monotonically with the energy of the incident particles. It reaches a maximum of about 10^{-2}–10^{-1} at a few keV for light incident particles, and of about 1–20 at energies of 5–500 keV for heavy incident particles [1.52]. At higher incident energies the sputtering yield decreases, because the ions penetrate more deeply into the solid and deposit less energy near the surface. The sputtering yield increases with increasing angle of incidence up to a maximum at an angle between 55° and 85° with respect to the surface normal. The decrease at larger angles of incidence is related to larger particle and energy reflection [1.52, 53]. In the linear-cascade regime the knockon sputtering yield generally does not depend on the target temperature [1.52, 56, 57].

Very sensitive measurements involving bombardment of virgin surfaces have shown that the partial sputtering yield of matrix atoms may initially be very low. A stationary yield is observed only after bombardment with an ion fluence corresponding to the removal of about one monolayer [1.58, 59]. This finding can be attributed to the removal of surface contaminants [1.59]. Interpretation along these lines suggests a very shallow depth of origin of sputtered atoms. One might also argue that some bombardment-induced roughening is required to obtain a stable sputtering yield [1.58].

Some time ago analytical calculations showed that the sputtering yield due to the impact of individual ions can exhibit large fluctuations [1.60]. Such fluctuations must exist where the sputtering yield is smaller than unity, as with light-ion bombardment [1.52]. Fluctuations in the sputtering yield can be studied conveniently using computer simulations [1.27, 40–47, 61, 62]. Figure 1.3 shows the probability distribution for sputter ejection of N_s atoms for 10 keV Xe bombardment of amorphous Ni at normal incidence, as calculated with the computer simulation program TRIM.SP [1.41, 47, 62]. The distribution can be described by a negative binominal distribution. The data in Fig. 1.3 suggest that about 10% of the impacts do not cause sputtering at all. The most probable event is the ejection of two to three Ni atoms per ion. In 1% of the ion impacts, 15 Ni atoms are removed, and the probability for removing 27 Ni atoms per incident ion is still about 10^{-4}. These numbers can be compared with a calculated mean sputtering yield of (4.87 ± 0.04) Ni atoms/Xe ion.

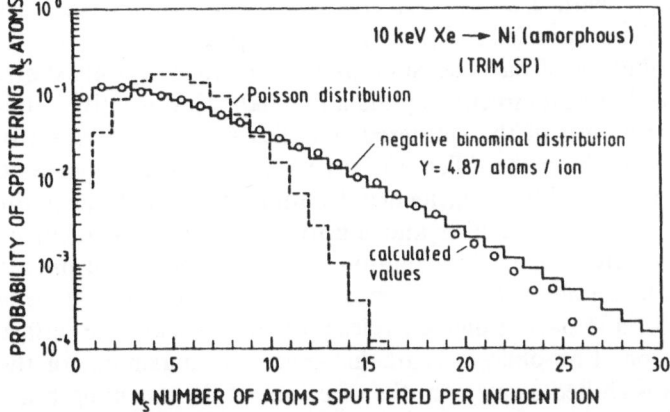

Fig. 1.3. Probability distribution for sputtering N_s atoms per incident projectile for the bombardment of amorphous Ni by 10 keV Xe ions at normal ion incidence [1.62]

Undoubtedly, the statistics of sputtering illustrated by Fig. 1.3 must have an effect on many of the observable quantities, for example, on the cluster size distribution [1.63]. Also the depth resolution in sputter depth profiling will depend to some extent on the fluctuations of the ejection probability. Investigations in this area are needed.

1.6 Distributions of Sputtered Particles

The particles sputtered from the surface of a solid are emitted with a broad distribution in energy E_1 and exit angles (Chap. 2). Most particles emitted from metallic targets are neutral atoms in the ground state, a fraction are excited atoms and ions, and some are clusters, which are again partly excited and partly ionized (Chap. 3).

The distributions of the sputtered particles are described by differential sputtering yields, with $\partial Y/\partial E_1$ giving the energy distribution and $\partial^2 Y/\partial^2\Omega$ the angular distribution of the sputtered particles. Here Y^q denotes the yield of particles sputtered with charge state q, and Y^* the yield of atoms sputtered in an excited state*. Thus $\partial^3 Y^{q*}/\partial E_1 d^2\Omega$ describes the yield of atoms per incident particle having a charge state q, an excitation state *, and an energy E_1, emitted into the solid angle Ω (Chap. 3). Measurements of differential yields allow the extraction of detailed information about the sputtering process and the collision cascade (Chap. 2). Investigations of this kind are, however, still rather incomplete.

The energy distribution of the sputtered particles generally exhibits a maximum between $U_0/2$ and U_0, where U_0 is the surface binding energy [1.29, 64]. At high ejection energies the number of sputtered particles decreases roughly proportional to $(1/E_1)^m$ with $3/2 < m < 2$. Deviations are observed at very oblique angles of incidence and for collision spikes (Chap. 2).

The angular distribution of particles sputtered from single crystals shows maxima in close-packed crystal directions [1.65 and Chap. 2]. For polycrystalline material the distribution of the sputtered particles is a superposition of the emission distributions from the differently oriented crystallites exposed to the incident beam. For normal beam incidence the angular distributions of the particles sputtered from polycrystalline and amorphous solids can often be described to first order by a cosine distribution. For heavy ions and impact energies close to the threshold, a larger fraction of atoms is emitted at large angles, while for light-ion impact at high energies more atoms leave the surface in the normal direction. For oblique beam incidence the maximum of the emission distribution is shifted away from the direction of the incoming beam (Chap. 2). In sputtering of compounds, the different components may be sputtered with slightly different angular distributions [1.66].

1.7 Analytical Applications Based on Sputter Erosion

The controlled removal of material on an atomic scale from the surface of solids by sputtering is the basis of many current analytical techniques that aim at determining a sample's composition as a function of depth (sputter depth profiling) (Chap. 4). Two approaches can be distinguished.

The first relies on elemental analysis of the flux of sputtered particles. The most important of these techniques is Secondary Ion Mass Spectrometry (SIMS), which features high sensitivity and a large dynamic range [1.67]. The general applicability of SIMS to a wide variety of sample materials is, however, hampered by the dependence of the elemental ion yields on the sample composition (a phenomenon commonly called the matrix effect) [1.68, 69]. If the sputtered neutral particles are postionized on passage through a plasma [1.70] or a high-temperature oven [1.71], or by the impact of an electron [1.72, 73] or a laser beam [1.74, 75] matrix effects are generally less pronounced but may still be serious [1.76]. Compared with SIMS these Sputtered Neutral Mass Spectrometry (SNMS) techniques generally have less-favorable sensitivities [1.77] or they are extremely mass selective, like resonant laser ionization [1.74].

For obtaining the lateral distribution of the target composition two techniques have been developed: (a) The sample is bombarded with a broad ion beam, and an ion microscope is used to generate a magnified image of the flux of secondary ions on a suitable screen or detector [1.78], or (b) a microfocus ion beam is scanned across the surface area of interest, and the signal of the detected sputtered ions is displayed synchronously with the ion beam position [1.79].

In the second category of sputter-based depth profiling methods the original as well as the sputter-eroded sample surface are characterized with a surface-sensitive or depth-resolving technique, for example Auger Electron Spectroscopy (AES) [1.80, 81], low-energy Ion Scattering Spectrometry (ISS) [1.82, 83], or Rutherford Backscattering Spectrometry (RBS) in glancing angle geometry [1.84, 85]. Compositional analysis involving these techniques is straightforward and relatively easy, but the detection limits are generally poor compared to SIMS. This is due to the presence of an inherent background, poor mass resolution, and/or an overlap of the mass and depth scales (Chap. 4).

In sputter depth profiling the sputtering process can induce a number of artifacts. One of these is collisional relocation of target atoms [1.86], which takes place in ion impacts preceding the sputter removal of the atom under consideration. As a result the apparent distribution measured as a function of the depth is broadened, compared with the original concentration distribution in the unbombarded sample. Conditions resulting in the least distortion –, i.e., giving the best depth resolution – have to be determined experimentally [1.87].

In addition to collisional relocation due to the ion bombardment, transport of atoms due to defect migration or segregation can limit the actual depth resolution [1.88]. Moreover element-differential (preferential) sputtering [1.89] can have a pronounced effect on the measured depth profiles. In SIMS analysis primary ion species like oxygen or cesium are commonly applied to maximize secondary ion yields [1.90, 91]. These ions can induce chemical effects that affect the migration and segregation of target atoms.

The depth resolution achievable in sputter profiling also depends on how much the surface is roughened during prolonged ion bombardment. Sputtering of polycrystalline solids generally results in rough surfaces. This can be attributed to the sputtering yield of individual grains depending on the orientation of the crystallites with respect to the beam axis [1.92 and Chap. 6]. In sputter depth profiling this effect can be reduced by rotating the sample during the erosion at oblique ion incidence [1.93]. Even on samples like Si or GaAs, for which a significant roughening had not previously been reported, wave-type features with a roughness amplitude on the order of 100 nm can be generated at impact angles between 30° and 60° using bombardment with or in the presence of oxygen (Fig. 1.4) [1.94–96]. This effect is still not well understood. For analysis it is better to perform the sputtering measurements at angles of incidence where roughening does not occur.

Sputtering can also be used to remove large intact molecules from the surface of a sample (Chap. 5). This phenomenon was first observed using low-energy primary ions and low bombardment fluences [1.97, 98]. In an alternative approach heavy primary ions with energies in the 100 MeV region have been used to cause the desorption of very large organic molecules [1.99]. In this regime sputtering yields of more than 1000 intact molecules per incident ion have been measured [1.100]. Using SIMS the ejected molecular species are commonly detected as either protonated or deprotonated ions. Even though the mechanism responsible for the emission of large molecules by fast heavy-particle

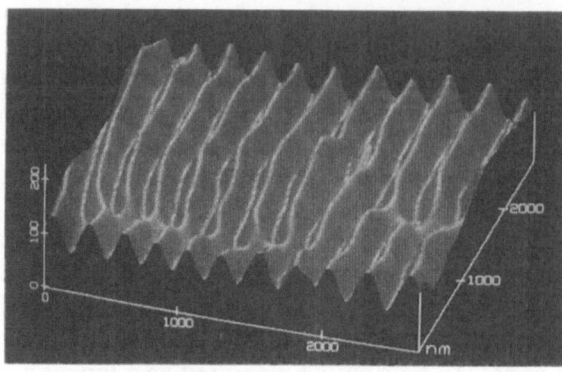

Fig. 1.4. Scanning tunneling microscopy image of a GaAs surface after sputter erosion by bombardment with 10.5 keV O_2 ions at an impact angle of about 37°. Sputtered depth 1.7 nm [1.96].

bombardment is not clear, this technique has already been used successfully for clarifying the previously unknown composition of complex molecules [1.101], and its areas of application may grow rapidly.

1.8 Micromachining by Ion Beam Sputtering

Ion beam sputtering is also widely used to remove material from a surface for surface cleaning and thinning as well as to produce a desired surface topography (Chap. 6).

When most materials undergo ion beam sputtering, their surfaces get rough. This effect can be increased by depositing a small amount of low-sputtering-yield material on the surface (seeding), for example, Mo and W [1.102]. Technically this method is used to prepare surfaces with good bonding to deposited layers. On the other hand, sputtering under conditions where surface roughening is kept small [1.93] is used for a controlled preparation of thin films (ion beam milling) [1.103].

Well-defined microstructures are obtained using a focused ion beam moved across the target. The process is controlled either by moving the target or by deflecting the ion beam. Microstructures have also been obtained with a large ion beam structured on passage through a mask in front of the target. If the mask is located at some distance from the target, the ion beam may be demagnified by ion optical means to obtain microstructures even from a larger mask [1.104].

Finally, an intense ion beam with a sharp edge has been produced using a beam-limiting screen. Such beams have successfully been used for nearly distortion-free well-defined cuts of any material, both normal to a surface or in a sloping fashion (the process is called Ion Beam Slope Cutting, IBSC) [1.105, 106 and Chap. 7]. With two cuts from both sides even thin films for transmission electron microscopy have been produced [1.106].

1.9 Sputter Deposition

Along with analytical applications and micromachining, another technological-ly and commercially important application of sputtering is the controlled deposition of sputtered material on substrates of almost any conceivable com-position and geometry (Chap. 8). To optimize the deposition a detailed know-ledge of the sputtering process as well as of the energy and angular distributions of sputtered particles is needed. Complications arise from the sputter deposition often being carried out at rather high residual gas pressures, in which case beam-induced reactions play an important role [1.107]. Optimum operation conditions for film deposition are usually determined empirically.

Generally, the design of the sputter deposition instrument deserves prime interest. Depending on the field of application one can distinguish between a large variety of concepts [1.108]. Many designs have already been realized in commercially available instrumentation.

Acknowledgements. We are indebted to W. Eckstein for preparing Figs. 1.1 and 1.3, and to I.H. Wilson, S.T. Tsong, and A. Karen for providing the originals of Figs. 1.2 and 1.4.

References

1.1   W.R. Grove: Philos, Mag. **5,** 203 (1853)
1.2   J.P. Gassiot: Philos. Trans. R. Soc. London **148,** 1 (1858)
1.3   J. Plücker: Ann. Phys. (Leipzig) **103,** 88, 90 (1858)
1.4   A. Günterschulze: J. Vac. Sci. Technol. **3,** 360 (1953)
1.5   D. Post, R. Behrisch (eds.): *Physics of Plasma Wall Interactions in Controlled Fusion* (Plenum, New York 1987)
1.6   R. Behrisch, J. Roth, G. Staudenmaier, H. Verbeek: Nucl. Instrum. Methods **18,** 692 (1987) and in *Fundamental and Applied Aspects of Sputtering of Solids,* ed. by E.S. Mashkova (MIR, Moscow (1989)
1.7   G.M. McCracken: Contrib. Fusion Plasma Phys. **29,** 1273 (1987)
1.8   J. Bøttiger, J.A. Davies, P. Sigmund, K.B. Winterbon: Radiat. Eff. **11,** 69 (1971)
1.9   E.S. Mashkova, V.A. Molchanov: Radiat. Eff. **16,** 143 (1972); ibid. **23,** 215 (1974); ibid. **110,** 227 (1989)
1.10 H. Verbeek: In *Material Characterisation using Ion Beams,* ed. by J.P. Thomas, A. Cachard (Plenum, New York 1978) p. 303
1.11 E.S. Mashkova, V.A. Molchanov: *Medium-Energy Ion Scattering from Solids,* Modern Prob-lems in Condensed Matter Sciences, Vol. 11 (North-Holland, Amsterdam 1985)
1.12 W. Eckstein, H. Verbeek: In *Data Compendium for Plasma Surface Interactions,* ed. by R.A. Langley, J. Bohdansky, W. Eckstein, P. Mioduszewskyi, J. Roth, E. Taglauer, E.W. Thomas, H. Verbeek, K.L. Wilson: Nucl. Fusion special issue (IAEA Vienna 1984) p. 12
1.13 R. Behrisch, W. Eckstein: In [1.5], p. 413
1.14 J. Stark: Z. Electrochem. **14,** 752 (1908)
1.15 F. Keywell: Phys. Rev. **97,** 1611 (1955)
1.16 B.M.U. Scherzer: In *Sputtering by Particle Bombardment II,* ed. by R. Behrisch, Topics Appl. Phys., Vol. **52** (Springer, Berlin, Heidelberg 1983) p. 271
1.17 E. Taglauer, W. Heiland (eds.): *Inelastic Particle-Surface Collisions,* Springer Ser. Chem. Phys., Vol. 17 (Springer, Berlin, Heidelberg 1981)
1.18 R. Baragiola: Radiat. Eff. **61,** 47 (1962)
1.19 E.W. Thomas: Prog. Surf. Sci. **10,** 383 (1982)

1.20   E.W. Thomas: In [1.12], p. 94
1.21   K. Ertl, R. Behrisch: In [1.5], p. 515
1.22   H.D. Hagstrum: Phys. Rev. **96**, 336 (1954)
1.23   E.S. Parilis, L.M. Kishinevskii: Sov. Phys.–Solid State **3**, 885 (1960)
1.24   D.B. Medved, Y.E. Strausser: Adv. Electron. Electron Phys. **21**, 101 (1965)
1.25   J.W. Burnett, J.P. Biersack, D.M. Gruen, B. Jörgensen, A.R. Krauss, M.J. Pellin, E.L. Schweitzer, J.T. Yates Jr., C.E. Young: J. Vac. Sci. Technol. A **6**, 2064 (1988)
1.26   P. Sigmund et al.: Nucl. Instrum. Methods B **36**, 110 (1989)
1.27   M.T. Robinson: In *Sputtering by Particle Bombardment I*, ed. by R. Behrisch, Topics Appl. Phys., Vol. 47 (Springer, Berlin, Heidelberg 1981) p. 73
1.28   H. Roosendaal: In [1.27], p. 219
1.29   P. Sigmund: In [1.27], p. 9
1.30   J.A. Brinkmann: J. Appl. Phys. **25**, 961 (1954)
1.31   A. Seeger (ed.): *Moderne Probleme der Metallphysik* (Springer, Berlin, Heidelberg 1965)
1.32   R.S. Averbeck, T. Diaz de la Rubia, R. Benedeck: Nucl. Instrum. Methods B **33**, 693 (1988)
1.33   J. Roth, J. Bohdansky, W. Poschenrieder, M.K. Sinha: J. Nucl. Mater. **63**, 222 (1976)
1.34   J. Roth: In [1.16], p. 91
1.35   J. Roth, J. Bohdansky, K.L. Wilson: J. Nucl. Mater **111 & 112**, 775 (1982)
1.36   J. Roth: In [1.5], p. 389
1.37   P. Sigmund: Phys. Rev. **184**, 383 (1969)
1.38   H.M. Urbassek: Nucl. Instrum. Methods B **36**, 585 (1988)
1.39   D.D. Jackson: In Proc. Symposium on Sputtering (SOS) ed. by P. Varga, G. Betz, F.P. Viehböck (Technische Universität Wien 1980) p. 2
1.40   J. Biersack, W. Eckstein: Appl. Phys. A **34**, 73 (1984)
1.41   W. Eckstein, W. Möller: Nucl. Instrum. Methods B **7**, 272 (1985)
1.42   H.H. Andersen: Nucl. Instrum. Methods B **18**, 321 (1987)
1.43   Y. Yamamura, W. Takeuchi: Nucl. Instrum. Methods B **29**, 461 (1987)
1.44   J.L. Likonen, M. Hautala: Appl. Phys. A **45**, 137 (1988)
1.45   B.J. Garrison, N. Winograd, D.M. Deaven, C.T. Reimann, D.Y. Lo, T.A. Tombrello, D.E. Harrison Jr., M.H. Shapiro: Phys. Rev. B **37**, 7197 (1988)
1.46   D.E. Harrison, Jr.: In CRC Critical Reviews in Solid State and Materials Sciences, Vol. 14, ed. by J.E. Greene, Supplement S1, (1988)
1.47   W. Eckstein: *Computer Simulation of Ion–Solid Interactions*, Springer Ser. Mater. Sci., Vol. 10 (Springer, Berlin, Heidelberg 1991) and private communication (1989)
1.48   K.L. Merkle, W. Jäger: Philos. Mag. A **44**, 741 (1981)
1.49   I.H. Wilson, N.J. Zeng, U. Knipping, I.S.T. Tsong: Phys. Rev. B **38**, 8444 (1988); Appl. Phys. Lett. **53**, 2039 (1988)
1.50   P.D. Townsend: In [1.16], p. 147
1.51   R. Behrisch: In [1.16], p. 179
1.52   H.H. Andersen, H. Bay: In [1.27], p. 135
1.53   W. Eckstein, J. Biersack: Z. Phys. B **63**, 109 (1986)
1.54   R.V. Stuart, G.K. Wehner: J. Appl. Phys. **33**, 2345 (1962)
1.55   Y. Yamamura, J. Bohdansky: Vacuum **35**, 561 (1985)
1.56   K. Besocke, S. Berger, W.O. Hofer, U. Littmark: Radiat. Eff. **66**, 35 (1982)
1.57   J. Bohdansky, H. Lindner, E. Hechtl, A.P. Martinelli, J. Roth: Nucl. Instrum. Methods B **18**, 509 (1987)
1.58   J.S. Colligon: In *Atomic Collision Phenomena in Solids*, ed. by D.W. Palmer, M.W. Thompson, P.D. Townsend (North-Holland, Amsterdam 1970)
1.59   R. Behrisch, J. Roth, J. Bohdansky, A.P. Martinelli, B. Schweer, D. Rusbüld, E. Hinz: J. Nucl. Mater. **93 & 94**, 645 (1980)
1.60   J.E. Westmoreland, P. Sigmund: Radiat. Eff. **6**, 187 (1970)
1.61   D.E. Harrison: Radiat. Eff. **70**, 1 (1983)
1.62   W. Eckstein: Nucl. Instrum. Methods B **33**, 489 (1988)
1.63   G. Staudenmaier: Radiat Eff. **13**, 87 (1972)
1.64   M.W. Thompson: Philos. Mag. **18**, 377 (1968)
1.65   G.K. Wehner: Phys. Rev. **102**, 690 (1956)
1.66   G. Betz, G.K. Wehner: In [1.16], p. 11
1.67   A. Benninghoven, F.G. Rüdenauer, H.W. Werner (eds.): *Secondary Ion Mass Spectrometry* (Wiley, New York 1987)

1.68 V.R. Deline, W. Katz, C.A. Evans, Jr., P. Williams: Appl. Phys. Lett. **33**, 832 (1978)
1.69 Y. Homma, K. Wittmaack: J. Appl. Phys. **65**, 5061 (1989)
1.70 H. Oechsner: In *Thin Film and Depth Profile Analysis*, ed. by H. Oechsner (Springer, Berlin, Heidelberg 1984) p. 63
1.71 G. Blaise: Scanning Electron Microsc. **1**, 31 (1985)
1.72 H. Gnaser, J. Fleischhauer, H.O. Hofer: Appl. Phys. A **37**, 211 (1985)
1.73 D. Lipinski, R. Jede, O. Ganschow, A. Benninghoven: J. Vac. Sci. Technol. A **3**, 2007 (1985)
1.74 G.S. Hurst, M.G. Payne, S.D. Kramer, J.P. Young: Rev. Mod. Phys. **51**, 767 (1979)
1.75 C.H. Becker, K.T. Gillen: Anal. Chem. **56**, 1671 (1984)
1.76 A. Wucher, W. Reuter: J. Vac. Sci. Technol. A **6**, 2316 (1988)
1.77 W. Reuter: In *Secondary Ion Mass Spectrometry SIMS V*, ed. by A. Benninghoven, R.J. Colton, D.S. Simons, H.W. Werner (Springer, Berlin, Heidelberg 1986) p. 94
1.78 R. Castaing, G. Slodzian: J. de Microsc. **1**, 359 (1962)
1.79 H. Liebl: J. Appl. Phys. **38**, 5277 (1967)
1.80 C.C. Chang: In *Characterisation of Solid Surfaces*, ed. by R.F. Kane, G.B. Larrabee (Plenum, New York 1974) p. 509
1.81 D. Briggs, M.P. Seah: *Practical Surface Analysis by Auger and X-Ray Photoelectron Spectroscopy* (Wiley, Chichester 1983)
1.82 T.M. Buck: In *Methods of Surface Analysis*, ed. by A.W. Czanderna (Elsevier, Amsterdam 1975) p. 75
1.83 E. Taglauer, W. Heiland: Appl. Phys. **9**, 261 (1976)
1.84 W.K. Chu, J.W. Mayer, M.-A. Nicolet: *Backscattering Spectrometry* (Academic, New York 1978)
1.85 K. Wittmaack, N. Menzel: Appl. Phys. Lett. **53**, 1708 (1988)
1.86 U. Littmark, W.O. Hofer: Nucl. Instrum. Methods **168**, 329 (1980)
1.87 K. Wittmaack: Vacuum **34**, 119 (1984)
1.88 L.E. Rehn, N.Q. Lam, H. Wiedersich: Nucl. Instrum. Methods B **7/8**, 764 (1985)
1.89 P. Sigmund: Nucl. Instrum. Methods B **18**, 375 (1987)
1.90 C.A. Andersen: Int. J. Mass Spectrom. Ion Phys. **2**, 61, (1969); ibid. **3**, 413 (1970)
1.91 K. Wittmaack: Surface Sci. **112**, 186 (1981); ibid. **126**, 573 (1983)
1.92 G. Carter, B. Navinsek, J.L. Whitton: In [1.16], p. 231
1.93 A. Zalar: Surf. Interf. Anal. **9**, 41 (1986)
1.94 S. Duncan, R. Smith, D.E. Sykes, J.M. Walls: Vacuum **34**, 145 (1984)
1.95 F.A. Stevie, P.M. Kohara, D.S. Simons, P. Chi: J. Vac. Sci. Technol. A **6**, 76 (1988)
1.96 A. Karen, K. Okuno, F. Soeda, A. Ishitani: In *Secondary Ion Mass Spectrometry SIMS VII* ed. by A. Benninghoven, C.A. Evans, K.D. McKeegan, H.A. Storms, H.W. Werner (Wiley, Chichester 1990) p. 139 and private communication
1.97 A. Benninghoven: Z. Phys. **239**, 403 (1970)
1.98 A. Benninghoven, D. Jaspers, W. Sichtermann: Appl. Phys. **11**, 35 (1976)
1.99 D.F. Torgerson, R.P. Skowronski, R.D Macfarlane: Biophys. Res. Commun. **60**, 616 (1974)
1.100 A. Hedin, P. Håkonnson, M. Salehpour, B.M.U. Sundqvist: Phys. Rev. B **35**, 7377 (1987)
1.101 R.D. Macfarlane, D. Uemura, K. Ueda, Y. Hirafa: J. Am. Chem. Soc. **102**, 875 (1980)
1.102 G.K. Wehner, D.J. Hajieck: J. Appl. Phys. **42**, 1145 (1971)
1.103 P.J. Goodhew: *Thin Foil Preparation for Electron Microscopy*, Vol II, Practical Methods in Electron Microscopy, ed. by A.M. Gauert (Elsevier, Amsterdam 1985)
1.104 B.A. Free, G.A. Meadows: J. Vac. Sci. Technol. **15**, 1028 (1978)
1.105 W. Hauffe: Elektronika **26**, 3 (1985)
1.106 W. Hauffe: Proc. VIIIth Europ. Conf. Electron Microscopy (Budapest 1984) p. 105
1.107 H.F. Winters: Topics Curr. Chem. **90**, 69 (1980)
1.108 J.L. Vossen, W. Kern (eds.): *Thin Film Processes* (Academic, New York 1978)

2. Angular, Energy, and Mass Distribution of Sputtered Particles

Wolfgang O. Hofer

With 36 Figures

The particles sputtered from a solid surface during particle bombardment are mostly neutral atoms emitted with an angular distribution that can be more peaked or broader than a cosine distribution. For single crystals, emission peaks are found in close-packed crystal directions. These emission distributions are influenced by the surface topography that develops during sputtering, by correlated collision sequences in the spread of the collision cascade in single crystals, and by the last collisions close to the surface where the atoms are sputtered. The sputtered atoms have a broad energy distribution that peaks at a few eV, about half the surface binding energy. A small fraction of the sputtered particles are clusters containing up to 15 atoms or more. For spike conditions a fraction of the sputtered atoms have thermal (a few thousand kelvin) energies, but also larger clusters containing up to some 1000 atoms are released. On rough, partly oxidized surfaces emission of microparticles (chunks) with up to 10^{12} atoms was found.

2.1 Historical Survey and Outline

Most published work on sputtering is devoted to the measurement of the total sputtering yield Y, i.e., the overall number of atoms ejected from the solid per incoming ion, irrespective of the energy E or direction Ω of emission, and without distinction between charge and molecular state of the sputtered species. The more than 400 references cited by *Andersen* and *Bay* [2.1] are indicative of this situation. Quite often, however, the corresponding differential quantities $dY/d\Omega$ and dY/dE have decided the issue in the various sputtering models proposed in the past, or caused revision thereof. For instance, Stark's momentum-transfer model [2.2] was regarded as less appropriate after *Seeliger* and *Sommermeyer* [2.3] found cosine-shaped *angular distributions* on polycrystalline and liquid metals. *von Hippel's* hot-spot model [2.4], which treated sputtering as evaporation from a small region around the projectile's impact, had thus found a mainstay, which in turn broke down when *Wehner* discovered anisotropic emission from single crystals [2.5]. *Wehner* had found maxima of emission in low-index lattice directions which manifested themselves as a characteristic "spot pattern" on collectors arranged around the crystal to condense the sputtered material. Even higher contrast and more detailed spot patterns

Topics in Applied Physics, Vol. 64

Sputtering by Particle Bombardment III Eds.: R. Behrisch · K. Wittmaack

© Springer-Verlag Berlin Heidelberg 1991

were found at higher energies by *Yurasova* [2.6–8], *Thompson* and *Nelson* [2.9, 10], *Molchanov* [2.11], and *Perovic* [2.12] and several others thereafter, thus disproving that evaporation processes take place at a high enough rate to account for the sputtering yield. Conversely, thermal spike effects in sputtering from high-density collision cascades were first surmised [2.13] and later clearly resolved in energy distribution measurements [2.14], and they were also apparent in the angular distribution [2.15].

The finding of preferential ejection along close-packed lattice rows has prompted a great many investigations of correlated collision sequences in solids [2.16–18]. Attempts to understand single-crystal sputtering on the basis of these focusing collision sequences [2.19–21] tended to mislead thinking about the phenomenon as a whole. Around 1965, doubts arose about the contribution of the anisotropic component to total ejection [2.22, 23] and about the assignment of the "Wehner spots" to focusing collision sequences [2.24–26]. *Lehmann* and *Sigmund* [2.26], in particular, proposed that surface collisions alone might account for anisotropic emission. *Robinson*'s and *Hofer*'s investigations of the angular distribution of particles sputtered from hexagonal metals strongly favored this model [2.25, 27]. Such measurements also showed that most sputtered particles generally are not ejected preferentially, much as the Wehner spots may dominate the visual impression. This supported applying theories based on (random) collision cascades in amorphous matter to crystalline matter [2.28–32]. Recent measurements on high-Z metals, however, have shown such a pronounced anisotropy [2.33–36] that it appears to be necessary to include the focuson mechanism in theoretical descriptions of the sputtering of such crystals.

Early *energy distribution* measurements by *Thompson, Wehner, Kopitzki*, and their coworkers [2.13, 37, 38] clearly showed the epithermal nature of sputtered particles. Subsequent refined dY/dE measurements confirmed the theoretically predicted E^{-2} high-energy tail, thus again providing strong support for random collision cascades in sputtering [2.30–32, 39–42]. Furthermore, they yielded information on the form and strength of surface binding, and indicated thermal components at elevated temperatures [2.30, 14]. Double-differential yield measurements $d^2Y/d\Omega\,dE$ helped to identify focusing collision sequences, as well as other specific ejection mechanisms, and allowed their characteristic energies to be determined [2.43–45]. Even when their contribution to sputtering is small, more knowledge of these mechanisms is important since their role may be much more pronounced in energy dissipation and radiation damage [2.46, 47]. The laser fluorescence technique in sputtering investigations [2.48] has opened an entirely new field: it is now possible to measure individually the energy distributions of particles in different excited and binding states. This gives information not only on elastic energy transfer during the ejection of surface atoms, but also on the concurrent electronic transitions.

For experimental reasons, in the past *mass distributions* of sputtered particles have been determined only for charged particles [2.49–56]. Since metals under clean conditions are sputtered predominantly in the neutral state, results obtained with secondary ions must be generalized carefully. From the few studies available on neutral sputtered particles, it appears, however, that the ejection of agglomerates of atoms, or clusters as they are called, is not at all a negligible effect [2.57–62]. Investigations of cluster emission should yield information on the low-energy, "late" phase of the cascade, where attractive interatomic forces also come into play. The discovery of anisotropic ionic [2.55] and neutral [2.62] cluster emission stressed the importance of collective motion in low-energy cascades. To understand fully the cluster emission phenomenon, however, we need to know the variance of individual collision cascades.

Clusters are important in such practical applications as secondary ion mass spectrometry of solids, production of molecular beams, studies of the stability of metal agglomerates, and catalysis. Intensive research on the characteristics of clusters began about 1980.

While clusters containing 20–50 atoms can be obtained rather routinely by heavy-ion bombardment at about 10 keV, the emission of very much larger agglomerates is observed at energies exceeding 100 keV. *Merkle* and *Jäger* reported microparticles containing as many as 10^3 atoms [2.63, 64]. These microparticles, which are assumed to be emitted by very dense, near-surface subcascades, are in turn small compared with the "chunks" observed in neutron sputtering. The emission of such micrometer-sized particles, which contain 10^{10} atoms and more, caused serious confusion since it cannot be explained by knockon processes alone. It seems to be agreed now that preformed particulates existed on the surface of these samples [2.69b].

This review is restricted to sputtering by elastic collisions, so-called knockon sputtering. Strictly speaking it therefore applies only to metals and, with a few limitations, to semiconductors. The classification of sputtering regimes into single-knockon, linear cascade, and spike sputtering follows that given by *Sigmund* in [2.69a]. Repeated reference will be made to the publications [2.65–74].

2.2 Angular Emission Distribution

Investigations of the angular distribution of sputtered particles have hitherto served as one of the most powerful sources of information on the basic mechanisms of sputtering. This is because the differential yield $dY/d\Omega$ provides more than just the direction function in the momentum distribution. Since the mass of the ejected particles is a roughly known quantity, and the distribution function of the absolute value of the velocity appears not to be very sensitive to the ejection direction, knowledge of $dY/d\Omega$ is a major step toward the ultimate aim, the momentum distribution. This is good because velocity- or energy-resolved angular distribution measurements are far more difficult to obtain [2.338, 339].

Unfortunately this information obtained in angular distribution measurements that seems at first sight rather straightforward has led to many investigations where the necessary experimental care has not been observed. The results of such work have often been even qualitatively wrong and for a time have confused understanding of the sputtering phenomenon as a whole. The controversies over the contribution of anisotropic effects to the total emission, for instance, and in particular the role of focusing collision sequences in the anisotropic component, are consequences of such shortcomings in experimental technique. Thus a short assessment of the minimum requirements for angular differential yield measurements is presented before the results are discussed.

2.2.1 Experimental Techniques and Their Requirements

So far there is only one way of recording simultaneously for every outgoing direction the number of atoms emitted per unit solid angle: the *collector technique*. Here the emitted particles are collected at suitable collector foils placed near the target (Fig. 2.1); subsequently the thickness of the deposit thus formed is determined. Contrary to angular scans with any sort of particle detector [2.75–81] around the target, the large-area collector method has the crucial advantage of yielding the angular dependence of the differential yield at the same ion fluence. Transient effects in sputtering are usually observed up to the fluences necessary for removal of a surface layer equivalent in thickness to the path length of the projectile, and at fluences larger than required for the initiation of surface structures: these are typically 10^{17} and 10^{19} ions/cm^{-2}, respectively. In extreme cases [2.82], the truly stationary state of damage and projectile accumulation may never be reached. A scanning technique is therefore likely to feign angular effects that are actually fluence effects. Most results reviewed in Sect. 2.2 were obtained with collectors, so the following methodological considerations will be restricted to this procedure.

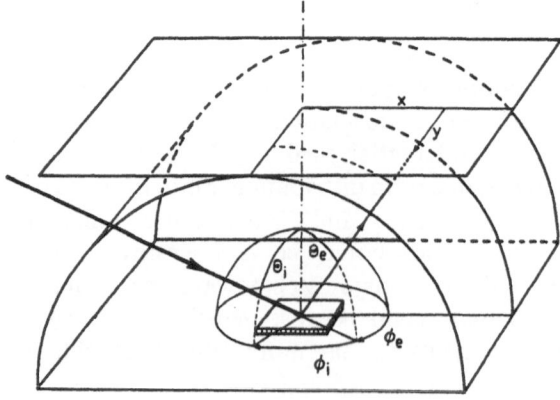

Fig. 2.1. Schematic setup for measuring the angular dependence of the differential sputtering yield $dY/d\Omega$ by the collector technique. Target particles sputtered by the ion beam are collected on a thin foil semicylindrically bent around the beam spot. Distortion-free mapping is achieved only in the $y = 0$ plane. Deposit profiles are usually evaluated on foils flattened after exposure

Measuring the angular distribution with collectors actually requires two steps: true mapping of the emission distribution by the deposit profile and determination of the thickness profile on the collector. In the first step there are *intrinsic* mechanisms that result in deviations of the measured distribution from that of an ideal target surface, such as shadowing by surface structures, shift of preferential ejection due to faceting, reduction of anisotropy by radiation damage, and resputtering of the deposit by backscattered projectiles. Also, there are failures caused by *inadequate transformation* of the emission distribution to a thickness profile on the collector, in particular, geometric distortion on nonspherical collectors and incomplete condensation of the sputtered particles on the collector.

For given bombardment conditions such as the mass, energy, and incident direction of the projectiles, as well as target orientation and temperature, avoiding effects of the first category is difficult. Fortunately, they are usually small for not-too-high projectile energies ($\lesssim 10^5$ eV), moderate fluences ($< 10^{19}$ cm^{-2}), and nonglancing incident/ejection angles. This latter restriction stems from the profound influence of surface topography on the total and differential yields [2.82–87]. Careful electron microscopic control of the target surface should always accompany sputtering investigations, so topographical and radiation damage can be considered in the interpretation.

Inadequate mapping of the emission distribution on the collector, on the other hand, is not a complication we have to put up with. Geometry as well as condensation distortions must and can be avoided by proper layout of the experiment. Note that a posteriori corrections of such failures are practically impossible.

a) Geometric Distortions

As far as true mapping of the emission distribution is concerned, spherical collectors are, of course, the ideal solution [2.36]. Hollow hemispheres, however, are difficult to prepare precisely, to handle during sputtering (especially when they need to be cooled), and to use for accurate profile-thickness determination. For these reasons, spherical collectors were used mostly for localizing and identifying preferential ejection directions [2.25, 88–94] and seldom for determining the actual distribution characteristics [2.36, 93].

Abandoning spherical symmetry entails the problem that the deposit profile is no longer a true image of the emission distribution, since the projection of the solid angle element on the collector now becomes angle-dependent. The measured quantity, the number density of atoms on the collector, dN/dF, then no longer corresponds directly to $dY/d\Omega$, but requires retransformation of the solid angle element $d\Omega$ from the area element dF [2.92]:

$$\frac{dY}{d\Omega}(\theta_e, \phi_e) = T(\theta_e, \phi_e) \frac{dN}{dF} , \tag{2.1}$$

where the transformation function $T(\theta_e, \phi_e)$ reads

$$T(\theta_e, \phi_e) \equiv \frac{dF}{d\Omega} =$$

$$\begin{cases} R_c^2 \cos^{-3}\theta_e \text{ or } R_c^2\left[1 + \left(\frac{x}{R_c}\right)^2 + \left(\frac{y}{R_c}\right)^2\right]^{3/2} & \text{for plane collectors,} \\ R_c^2\left[1 + \left(\frac{y}{R_c}\right)^2\right]^{3/2} & \text{for cylindrical collectors.} \end{cases} \tag{2.2}$$

Here R_c is the radius of the cylinder or the collector–target distance; for x and y see Fig. 2.1.

Cylindrical collectors are certainly a reasonable compromise: the foils can be bent over the target as sketched in Fig. 2.1, thus yielding distortion-free mapping in the azimuth $\phi_e = 0$. They also allow thickness measurements on the flattened foil.

The frequently used plane collector, however, causes insoluble correction problems. *Wehner* [2.88, 89], *Robinson* and *Southern* [2.95], *Schulz* and *Sizmann* [2.92], and *Francken* and *Bonnijns* [2.96] have pointed out the implications of projectional distortions.[1] These warnings were reiterated in [2.73], but plane collectors were still being used up until recently. Therefore, the consequences of neglecting these elementary geometric facts should be summarized again:

1) Deposit maxima do not coincide with the corresponding lattice directions of maximum emission, the deviation depending on the shape and width of the emission distribution to be determined.
2) Equal-intensity preferential emission at different polar angles is reproduced by strongly differing deposit densities, whereby low-intensity ejections may be completely lost (Fig. 2.2).
3) The correction procedure for obtaining undistorted ejection characteristics is a complicated iterative process and requires the whole deposit pattern to be evaluated.

This last requirement is impossible to meet; the reasons, are discussed in the following two sections.

Figure 2.2 reveals two classic examples of deception by plane collectors: first, the high deposit density around the (111)-target normal was erroneously interpreted as $\langle 111 \rangle$-assisted focusing ejection [2.97], while the deposit on the semicylindrical collector shows that this is just the $\cos^3 \theta$ projection transformation (2.2); $\langle 111 \rangle$ preferential ejection from fcc crystals does not exist in spot patterns. Second, the $\langle 100 \rangle$ ejection is not visible at all on the plane collector,

---

[1] There exists the possibility of mounting flat collectors in a special orientation with respect to the ejection direction of interest, e.g., perpendicular to a close-packed lattice direction [2.96]. This alleviates the problem somewhat for this ejection distribution, but at the expense of the total angular distribution in 2π.

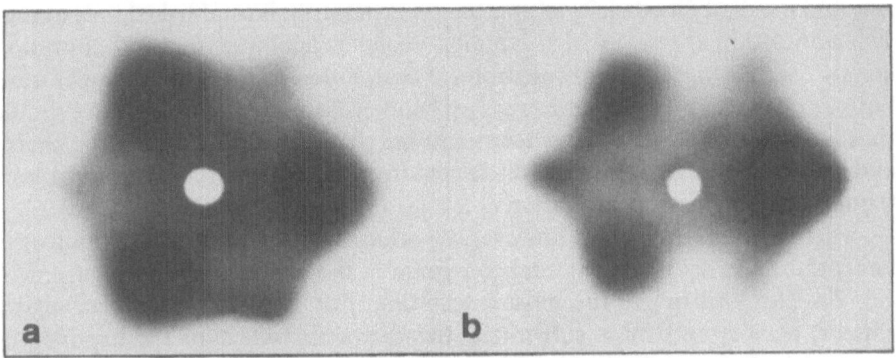

Fig. 2.2a, b. Spot pattern of an aluminium (111) crystal bombarded with 20 keV Ar$^+$ ions. The pattern on the left was obtained on a plane collector, the one on the right, on semicylindrical collectors. Distortion-free mapping of the emission distribution is thus achieved in the (110) azimuth (here the horizontal line). From [2.92]

although it is the second-strongest emission from fcc crystals (Sect. 2.2.3). It is clearly discernible in distortion-free geometry.

Nonetheless, there exist situations where plane geometries cannot be circumvented, such as in the direct imaging of emitted or desorbed particles. As in Electron Stimulated Desorption Ion Angular Distribution (ESDIAD) measurements for desorbed ions [2.69c], sputtered-ion or positionized-neutrals distributions can be imaged with (planar) channel-plate multipliers (Sect. 2.2.3). Fortunately, this detection technique is not affected as much by incident angle dependences as are condensation and film growth. Therefore, rectification of the geometric distortion (2.2) should not be too complicated.

b) Distortion due to Incomplete Condensation

Sputtered atoms arriving at the collector surface must transfer their kinetic energy to the lattice of the substrate to enter the adatom-state, from which growth of the deposit occurs by agglomeration. Because of the rather high energy of sputter-ejected particles, the accommodation problem is delicate, in particular when proportionality between the arriving flux and the film growth is required for true imaging. To substantiate this crucial assumption, *Hofer* [2.98] modified *Volmer* et al.'s [2.99] theory of heterogeneous nucleation. *Hofer* found that the growth rate $I(\dot{n}_i, T_c)$ is

$$I = C_1 \dot{n}_i \exp\left\{-C_2\left[T_c^3 \ln^2\left(\frac{\dot{n}_i}{\dot{n}_\infty}\right)\right]^{-1}\right\} \tag{2.3}$$

where C_1, C_2 are calculable material constants, T_c is the collector temperature, and $\dot{n}_\infty$ is the flux density of atoms reevaporated from the condensate (drops exponentially with T_c). According to (2.3), a linear relation between the growth rate and incident flux density of sputtered particles, $\dot{n}_i$, is established only at high supersaturation $\dot{n}_i \gg \dot{n}_\infty$, which can most easily be achieved at low T_c. For most metals condensing on metal substrates, $I \propto \dot{n}_i$ holds true at room temperature, but this changes rapidly with increasing condensation temperature. Care should thus be exercised at high target temperatures, where not only the kinetic energy and heat of sublimation must be dissipated from the collector, but also heat from the target's temperature radiation.

There are several metals, however, for which cooling to near liquid nitrogen temperatures is necessary in order to remain in the proportionality regime: Mg, Cd, Zn, Hg, and In are the most interesting. For Cd and Zn condensing on copper, plastic, and mica substrates, the distorting effect of the exponential function in (2.3) has been studied experimentally as well as theoretically. As shown in Fig. 2.3, considerable sharpening of the ejection distribution is effected by insufficient condensation. A similar effect was recently reported for sputtered silver on aluminium collectors [2.36]. Therefore a sticking factor of less than 1 by no means leads to a uniformly weaker mass deposit on the entire collector, but to overemphasis of the regions of high emission intensity. In other words, anisotropy appears sharper than it actually is, and random ejection ("background") as well as low-intensity preferential emission is suppressed – the deposit profile no longer represents the angular emission distribution from the target.

Metal collectors should be used instead of the rather common plastic, glass, or mica substrates for several more reasons. Because of the relatively high metal–metal desorption energies U_{des}, not only is the mean residence time of the adatoms longer, but accommodation is also facilitated [2.99]. Moreover, the accommodation conditions do not change during the formation of the first continuous layer of the deposit, an effect which again would cause sharpening of deposit distributions.

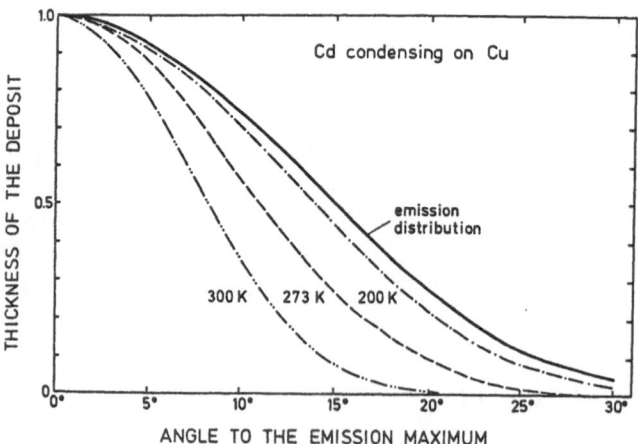

Fig. 2.3. Deviation of deposit distributions from the corresponding emission distribution for cadmium condensing on copper at different substrate temperatures. Calculations are from [2.98]; for experimental evidence in the case of Ag see [2.36]

Accommodation of sputter-ejected particles becomes completely uncontrollable at oblique incidence; the transverse momentum component can lead to an adatom energy distribution not in equilibrium with the substrate, causing enhanced desorption. This is why a posteriori correction of deposit profiles may be greatly in error.

c) Evaluation of the Deposit Profile

Determining the thickness profile of the condensate on the collector is a general problem in thin-film physics and will be discussed only briefly here. The requirements of the method applied for spot-pattern evaluation are rather exacting: $0.1-10^2$ monolayers of covered thickness range, approximately 100 μm lateral resolution, and insensitivity to the chemical state of the deposit. Furthermore, convenience and generality, the method should be nondestructive, applicable over a large range of the periodic table, and not require separation from the substrate. The crucial requirement, however, is its *blindness to the chemical state of the condensate*: during growth of the deposit the flux ratio $\dot{n}_i(\theta_e)/\dot{n}_{O_2}$ of sputtered to residual gas (oxygen) particles changes with the ejection angle θ_e, and so does the oxidation (nitrification) rate. Even before exposure to the atmosphere, some areas of the deposit will thus be strongly oxidized, while emission maxima (spots) will be mainly in the metallic state. Exposure to air before or during the thickness measurement aggravates the situation. This is of no consequence when evaluation is performed by autoradiography [2.15, 19, 100–103], characteristic X-ray excitation [2.98, 104–109], electron or light-ion backscattering [2.33–36, 86, 109–112], or γ-ray spectroscopy [2.22, 23, 113], but renders the previously often-used optical-absorption method unsuitable. The optical extinction coefficient depends on the growth rate of the film and varies drastically with (reactive gas) impurity content. Only a few prominent examples of the wealth of information on this effect are cited here: for Au [2.114], for Cu [2.10], for Zn and Mg [2.98, 115], and for Zr [2.93]. The general observation is that oxidation reduces optical absorption, so that *Wehner*-spots appear much sharper than they actually are.

Curiously, then, both condensation and optical effects lead to an overemphasis of anisotropy – a situation carried to the extreme when plane collectors are used. The visual impression created by the spot pattern is then that ejection from single crystals is predominantly anisotropic. This in turn prompts interpretation of the sputtering process in terms of focusing collision sequences. Understanding of the phenomenon has been significantly hampered by these cumulative artifacts.

2.2.2 Angular Distribution from Polycrystalline and Amorphous Solids

Despite the great interest in single crystals following *Wehner*'s discovery of anisotropic emission [2.5], ejection distributions from polycrystalline solids continue to be the subject of experimental and theoretical investigations [2.80,

106–148]. Among the subjects investigated are the relative contributions of specific ejection mechanisms, the verification of basic assumptions and predictions of theoretical approaches [2.80, 100, 109, 116–124], thin-film production [2.130–132], and the determination of the partial fluxes from multicomponent targets [2.123, 133, 137, 144]. This latter information is badly needed for insight into preferential sputtering and radiation-induced surface segregation.

Theories of sputtering based on collision cascades generally assume *random* media, whereas experimental investigations were most often carried out on *polycrystalline* materials. Since emission from polycrystals is a superposition of emissions from individual single crystal grains, commensurability of the two systems is not a priori given. The main justification for comparing results from theory on amorphous targets and experiments on crystalline targets is that also for single crystals the random component is dominant in the sputtered flux (Sect. 2.2.3). The anisotropic component, on the other hand, is expected to average out for random grain orientations of polycrystals. In practice, however, truly texture-free targets are very difficult to obtain and maintain during irradiation [2.149–151].

For comparisons with random cascade theory, *semiconductor* targets are the best choice, since they amorphize during particle irradiation at temperatures below 300°C [2.73, 152]. Ion-induced amorphization of Si and Ge, known for about 20 years, was only recently used for systematic angular distribution investigations [2.120, 122, 124, 126, 146].

However, the main features of angular distributions from nonsingle-crystal targets upon heavy-ion impact were already apparent at the time the last comprehensive reviews of sputtering were written [2.69, 70], despite the aforementioned reservations concerning experimental technique and target qualification. This information was the basis for the cascade theory of sputtering. In light-ion sputtering, on the other hand, angular distribution measurements only confirmed existing models and theories [2.109, 111, 121, 137].

a) Cascade Sputtering

Linear cascade sputtering has evolved as a kind of reference standard in sputtering. This is due to the wealth of experimental yield data in this regime, and to their rather satisfying description by linear collision cascade theory [2.31, 69a]. *Sigmund*'s theory reproduces total yields quite well and seems to describe angular distributions correctly. Under bombardment conditions where random collision cascade theory applies – i.e., for keV ions of not too low mass – experiment and linear transport theory both give cosine-like distributions:

$$\frac{dY}{d\Omega} \sim \cos^{\nu_e} \theta_e, \qquad 1 \leq \nu_e < 2 , \tag{2.4}$$

where θ_e is the polar angle of the emitted particle. Furthermore, note that (2.4) is independent of the incident angle θ_i, except for glancing angles, where additional effects such as projectile reflection and surface structure influences come into play. More or less pronounced "over-cosine" distributions ($v_e > 1$) are often observed [2.80, 106, 107, 122], a feature enhanced by increasing projectile energy. At energies above 10^2 keV, the angle of incidence θ_i can be increased up to 80° without changing the emission distribution (Fig. 2.4). This enhanced emission near the surface normal indicates that the momentum distribution of recoils near the surface is more outwardly peaked than assumed in the first-order (isotropic) version of the theory [2.139, 145, 153–155].

In the regime of cascade sputtering, the angular distributions are also independent of the azimuth ϕ_e; i.e., the ejection distribution is rotationally symmetric with respect to the surface normal, whatever the direction of the incoming beam. This again is a consequence of a recoil cascade that has "forgotten" the details of the primary projectile–target interaction while slowing down. This effect has already been described rather well in the early review literature [2.69, 70, 72], so further discussion will only summarize characteristic results acquired since:

1) *Scharmann* and coworkers [2.106–108] obtained, through careful measurements of the polycrystalline angular distribution in the energy range 10^2–10^3 keV, the results shown in Fig. 2.4b. Similar results were obtained on Fe, Pt and Au targets, all of them showing independence of the incident angle and emphasis of ejection along the surface normal ($v_e \approx 2$) at projectile energies above 10^2 keV.

2) *Betz* et al. [2.101, 119] found that agreement with the cosine law and independence of the projectile incident direction are the more pronounced the lower the sublimation energy. This is plausible because the approximation to an isotropic recoil spectrum improves with progressive slowing down of the cascade.

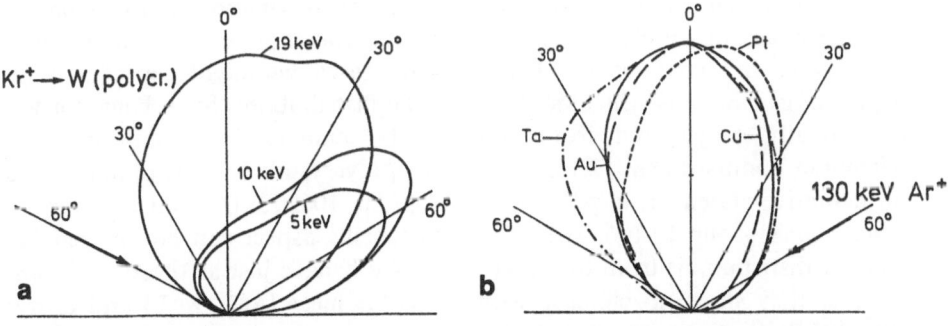

Fig. 2.4a, b. Emission distribution from various polycrystalline targets at oblique projectile incidence. With increasing energy, ejection is determined more and more by random collision cascades, resulting in axially symmetric intensity distributions around the surface normal. Owing to the rather high fluences applied, influences of surface topography cannot be dismissed: (a) Kr^+ from [2.134], (b) Ar^+ from [2.106]

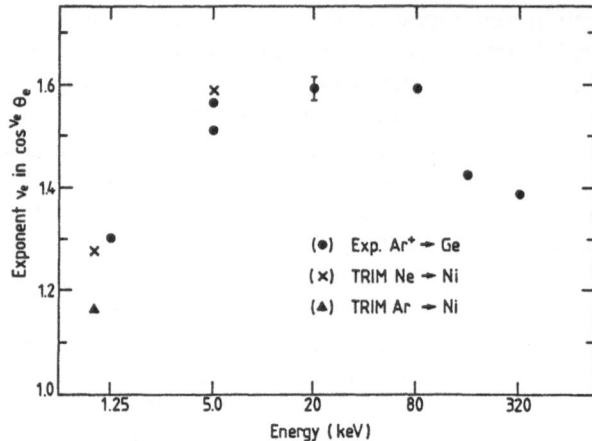

Fig. 2.5. Exponent v_e from fits of (2.4) to experimental Ge angular distributions vs Ar^+ projectile energy. Some data from TRIM.SP computer simulations [2.142] are also shown. From [2.122]

3) *Andersen* and coworkers [2.122] concluded from their own measurements (1.25–350 keV $Ar^+ \rightarrow$ Ge) as well as from analysis of published work that over-cosine distributions are a rather general feature in the cascade regime (Fig. 2.5). They question whether this anisotropy is due to the near-surface recoil flux or scattering by surface atoms. There is presently some dispute about whether these (and earlier) results could have been influenced by surface contamination [2.67, 155]; surface layers may reduce emission at large θ_e.

4) *Besocke* et al. [2.80] studied the temperature dependence of the sputtering of silver and found agreement with the aforementioned results up to temperatures near the melting point, i.e., where the sublimation rate far exceeds the sputtering rate.

Very little information is available on angular distributions in the regime of nonlinear cascades or elastic collision *spikes*. As spikes are often compared with and treated like confined hot high-pressure gases, we might expect a larger degree of isotropy and thus more cosine like distributions. Such behavior was occasionally conjectured from single-crystal emission distributions under heavy-ion bombardment [2.15, 34], but for polycrystals or amorphous targets no definitive statement is possible as yet [2.80, 106, 128]. From the work of *Scharmann*'s group [2.106, 108] no spike-specific aspect can be inferred: the angular distributions from the spike-prone Au behave just as those of Cu and Zr, and they show no changes with projectile mass (Ar^+, Xe^+) and energy (10^2–10^3 keV).

b) Single-Knockon Sputtering: Heavy Ions

In the regime of low-energy ($\leqslant 1$ keV) projectiles the emission distribution depends on the incident angle, and it deviates markedly from the cosine law, giving higher emission at grazing ejection angles. These so-called under-cosine distributions are generally explained as a result of shallow cascades, where too few collisions are available to acquire momentum randomization [2.69, 75, 146, 156]. At perpendicular incidence the ejection of atoms by primary recoils (Fig. 2.6a) is then likely, as can be estimated from the differential collision cross section

$$d\sigma = C_m E^{-m} T^{-1-m} dT \ , \tag{2.5}$$

where C_m is a constant depending on the atomic numbers Z_1 and Z_2 of the projectile and target atom respectively. Here m is the exponent in the interaction potential applied,

$$V(r) \propto r^{-1/m} \ , \tag{2.6}$$

and T is the energy transferred in the collision

$$T = T_m \cos^2 \phi'' \quad \text{with} \tag{2.7}$$

$$T_m = 4 \frac{M_1 M_2}{(M_1 + M_2)^2} E \ , \tag{2.8}$$

ϕ'' being the recoil angle.

According to (2.5) small energy transfers at small scattering angles will prevail, giving appreciable cross sections at low projectile energies E. Integrating $d\sigma$, we obtain the mean-free-path length

$$\lambda = (N\sigma_d)^{-1} \ , \tag{2.9}$$

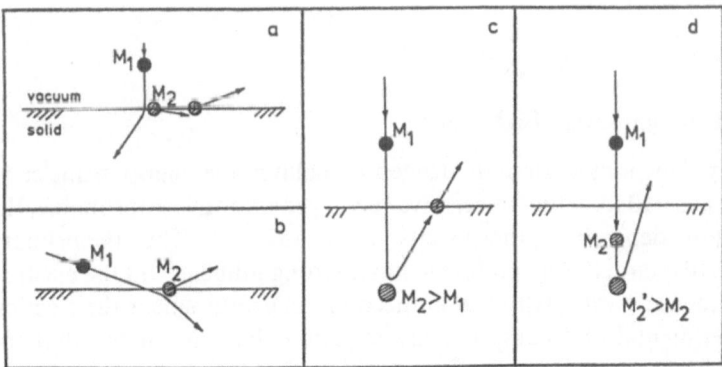

Fig. 2.6a–d. Collision kinematics of low-energy projectile impacts leading to ejection of surface atoms (single-knockon sputtering): Ejection of (**a**) higher-order (secondary) recoils and (**b**) primary recoils; (**c**) ejection by backscattered projectiles; (**d**) ejection of atoms recoiling at heavier constituents in the target

in which

$$\sigma_d = \int\limits_{E_d}^{T_m} d\sigma = \frac{1}{m} C_m (ET_m)^m \left[\left(\frac{T_m}{E_d} \right)^m - 1 \right] \tag{2.10}$$

is the total interaction cross section for displacement collisions (the displacement threshold E_d for many metals is around 20 eV). For projectile energies in the 10^2–10^3 eV regime, λ is on the order of one monolayer distance; surface atoms therefore receive their momentum predominantly at small angles with respect to the surface (Fig. 2.6a). Ejection at high polar angles is correspondingly intensified. This effect is important in single crystal sputtering as well, where it accounts for the polar angle dependence of preferential ejection intensities of crystallographically equivalent lattice directions, [2.27, 2.156].

Although far more pronounced in light-ion sputtering, backscattering processes may also play a role in low-energy heavy-ion sputtering. *Wehner* et al. [2.135] reported a change in isotope abundance in the emission intensities of target atoms at low polar angles. Their explanation is based on backscattering of the lighter target atoms from the heavier ones (Fig. 2.6d). The magnitude of the enrichment of the lighter component is a few percent at most, and it decreases with increasing energy. The authors expected that the effect would become less significant as more random cascade sputtering took over, but this was not confirmed by computer simulations. Using a molecular dynamics computer code, *Shapiro* et al. [2.148] simulated sputtering of Cu by 1 and 5 keV Ar$^+$ ions. Throughout this transition region from single-knockon to cascade sputtering the authors found an enrichment of the lighter isotope, in particular at small polar ejection angles. This is explained by both the 180° scattering possibility of the lighter isotope, and an asymmetric energy sharing between the different isotopes. For instance, the lighter cascade atoms have higher average energies when the projectile is lighter in mass than either target atom, and vice versa.

c) Single-Knockon Sputtering: Light Ions

In sputtering by light ions such as hydrogen or helium, the energy transfer to target atoms is so small (2.7) that successive recoil generation is improbable; the energy is, therefore, deposited rather locally [2.140, 157–162]. Thus the primary event, the projectile–target atom collision, has a strong influence on the ejection process; this influence is especially pronounced at ion energies near the ejection threshold. Experimental and computer investigations have intimated that the direct ejection of surface atoms by backscattered projectiles is the preponderant ejection mechanism for light-ion bombardment (Fig. 2.6b, c). At perpendicular ion incidence, the ejection must then show a preferential component perpendicular to the surface, an effect found in low-energy sputtering of V [2.109], Ni, and W [2.111]. While the usual cosine distributions were observed in the

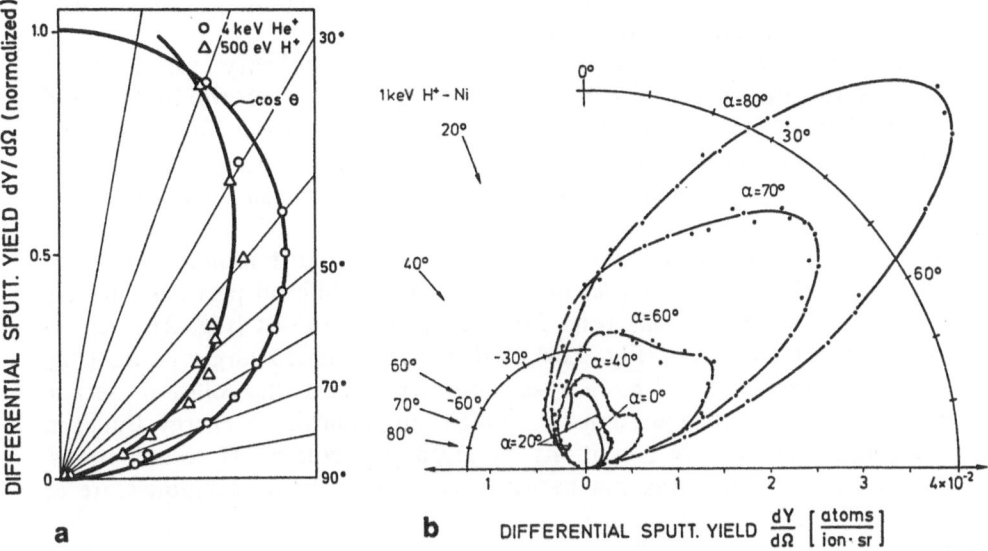

Fig. 2.7a, b. Emission distributions of ions sputtered by low-energy light ions. (a) From polycrystalline vanadium; note over-cosine distribution due to atoms sputtered by backscattered projectiles, see Fig. 2.6c. (Perpendicular incidence, normalized data, from [2.109].) (b) From polycrystalline nickel; note preferential forward emission at grazing incidence due to direct projectile–surface atom collisions, see Fig. 2.6b. Fine structures in the shape of the emission distribution are caused by crystalline effects (e.g., texture) of the target. From [2.111]. Both ejection mechanisms become second-order effects as random collision cascades take over ejection

keVregime, distinct over-cosine distributions at 500 eV H^+ or He^+ irradiation testified to ejection by backscattered ions (Figs. 2.6c, 7a). This behavior is exactly opposite to the aforementioned heavy-ion sputtering angular distributions, where enhanced oblique ejection manifests itself in under-cosine distributions.

Even more pronounced is the influence of primary interaction events with surface atoms at oblique incidence (Fig. 2.6b). Here the direct ejection of surface atoms by projectiles on their way into the solid is the dominant sputtering event in light-ion sputtering. Suspected to be the reason for the strong enhancement of the total yield in the Y vs θ_i dependence [2.163], this effect was clearly identified in angular distribution measurements [2.111] (Fig. 2.7b). In the emission characteristic the ejection of primary surface recoils causes a lobe- or club-shaped emission at polar angles opposite to the projectile incidence direction ("specular" emission), which is very much more pronounced than the preferred forward emission observed in heavy-ion sputtering [2.118, 124, 134]. Particularly interesting is this effect with compound targets such as NbB_2 or TaC, where two kinds of recoils with different masses result in different emission distributions [2.121]. Theoretical treatment of this emission process is not very complicated;

the main problem is the refraction effect of surface binding. For this reason numerical simulation appears suitable [2.140].

Bay and coworkers [2.164, 165] have investigated primary recoil emission by measuring the velocity distribution of sputtered atoms. This will be discussed in Sect. 2.3, but here is a conclusion from *Bay*'s work which holds true equally well for the angular distribution: "The anisotropy of that part of the collision cascade which contributes to the sputtering process is increasing with decreasing maximum transferable energy."

Quite generally, the following can be stated: The contribution of all ejection mechanisms based on a special orientation of the collision partners – namely ejection by primary recoils (Fig. 2.6a), of primary recoils (Fig. 2.6b, c), by backscattered projectiles (Fig. 2.6c), and of backscattered target atoms (Fig. 2.6d) – becomes more and more a second-order effect as the collision cascade develops. This is usually just a matter of high enough projectile energy. But even under optimum cascade conditions for given projectile/target combinations, these ejection mechanisms can be seen, provided sensitive techniques are at hand (Sects. 2.3, 4).

d) Influence of Surface Topography

The development of surface structures on crystalline solids is a major problem in all sputtering investigations where fluences approach 10^{19} ions/cm^2. With poly-crystalline targets, the situation is worse because the stochastic orientation of the grains causes discontinuities in erosion speed at the grain boundaries. These discontinuities are likely to induce surface structures such as ridges and pyramids, in addition to the characteristic structures developing on the individual single-crystalline grains. A polycrystalline surface subjected to intense particle irradiation will therefore be covered with arrays of differently oriented structures, mostly of the facet type. The orientation of the structures within the array (grain surface) often evinces astonishing regularity [*Carter* in 2.69b, 67, 106, 166–168].

Any deviation from a flat target surface must give rise to changes in yield and angular distribution of sputtered particles, irrespective of structure type and orientation. Because of the yield increase caused by oblique incidence,

$$Y(\theta_i) = Y(0)\cos^{-\nu_i}(\theta_i) \quad \text{for} \quad 0 \lesssim \theta_i \lesssim 70°, \quad \tfrac{1}{2} \lesssim \nu_i \lesssim 3 , \qquad (2.11)$$

structured surfaces lead to enhanced emission, but recapture of ejected particles at these ridges, facets, or cones may compensate for this effect. *Littmark* and *Hofer* [2.85] considered this problem theoretically and found the following equations for the angular dependence of the differential yield from regularly faceted surfaces:

$$\frac{dY^{A+B}/d\Omega_e}{dY/d\Omega_e} = \begin{cases} \left(\dfrac{\cos\theta_i^A}{\cos\theta_i}\right)^{\nu_i+1}\left(\dfrac{\cos\theta_e^A}{\cos\theta_e}\right)^{\nu_e-1}, & \text{case I,} \\[2ex] \left(\dfrac{\cos\theta_i^A}{\cos\theta_i}\right)^{\nu_i+1}\left(\dfrac{\cos\theta_e^A}{\cos\theta_e}\right)^{\nu_e}\dfrac{\sin\beta}{\sin(\alpha+\beta)} \\[2ex] \quad + \left(\dfrac{\cos\theta_i^B}{\cos\theta_i}\right)^{\nu_i+1}\left(\dfrac{\cos\theta^B}{\cos\theta_e}\right)^{\nu_e}\dfrac{\sin\alpha}{\sin(\alpha+\beta)}, & \text{case II,} \\[2ex] \left(\dfrac{\cos\theta_i^B}{\cos\theta_i}\right)^{\nu_i+1}\left(\dfrac{\cos\theta^B}{\cos\theta_e}\right)^{\nu_e-1}, & \text{case III,} \end{cases}$$

(2.12)

where α and β are the angles of the facets, (θ_i, ϕ_i) defines the beam incidence direction, and (θ_e, ϕ_e) the emission direction. The cases in (2.12) correspond to exit angles where either no shadowing (case II), or shadowing by the facets occurs. Since any angular emission distribution – whether under-cosine, over-cosine, or just cosine shaped – can be described by a polynomial of (2.4), Eq. (2.12) is general.

Two special evaluations are shown in Fig. 2.8. Clearly, substantial deviations of the observed angular distribution from the distribution obtained from flat

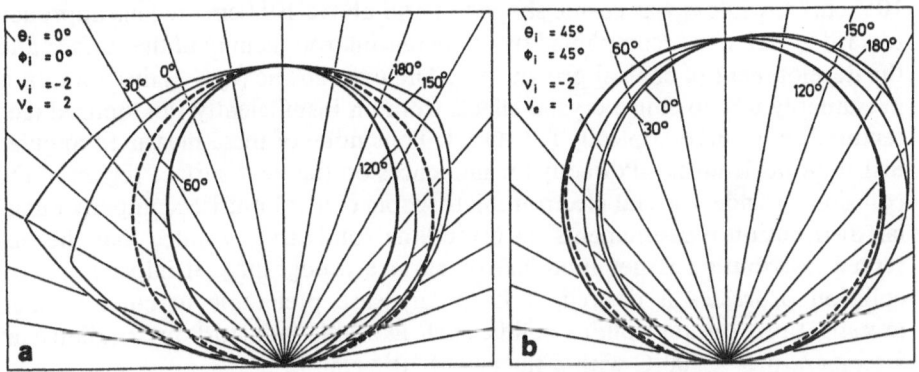

DIFFERENTIAL SPUTT. YIELD dY/dΩ (arb. units)

Fig. 2.8a, b. Calculated angular emission distributions from faceted surfaces. In both cases facet angles of $\alpha = 30°$, $\beta = 60°$ and a $\cos^{-2}\theta_i$ dependence for the total (flat-surface) sputtering yield were assumed. Curve parameter is the azimuthal orientation of the registration plane with respect to the facets. The heavy, full-drawn curve represents the genuine, flat-surface emission distribution; the dashed curves are obtained by averaging over all azimuthal orientations to simulate polycrystalline target conditions: (a) $\Phi_i = \theta_i = 0°$, $\nu_e = 2$ (over-cosine distribution), (b) $\Phi_i = \theta_i = 45°$, $\nu_e = 1$ (cosine distribution). Pronounced deviations from the genuine emission distribution are observed at fixed azimuth and for integrated distributions, the latter barely distinguished from a cosine characteristic. From [2.85]

surfaces ensue from surface structures.[2] This is particularly true at oblique ejection, where shadowing by the structures is most pronounced.

An interesting special case should, however, be mentioned: $v_i = 1$, $v_e = 1$, i.e.,

$$Y(\theta_i) \propto 1/\cos\theta_i \qquad \frac{dY}{d\Omega} \propto \cos\theta_e .$$

Here neither the total yield Y nor the differential yield depends on the surface structure. Under these conditions the faceted surface cannot be distinguished from a perfectly flat one. The assumption $v_i = 1$ does not generally apply, so findings by *Gurmin* et al. [2.134] that the surface structure is not important are an inadmissible generalization of the $v_i = 1$ case.

Consider also the ϕ_e-averaged curves in Fig. 2.8: these distributions were obtained by averaging over the azimuth of the ejected particles to simulate the situation of polycrystalline targets. Certainly this is only a rough approach, because the same type of facet structure has been assumed on the differently oriented grains. Notwithstanding, there appears to be a remarkable randomization effect due to surface structures. The resulting distribution is barely distinguishable from a cosine function, although the "intrinsic" distribution may have anything but a cosine shape. *Hasuyama*'s observation [2.136] may be interpreted here: the angular distribution of particles sputtered from a polycrystalline gold target by 20 keV Ar ions had an over-cosine shape at fluences below 10^{16} cm^{-2}, while it was cosine shaped at and above 10^{17} cm^{-2}. This approach toward $v_e = 1$ was accompanied by an increasing roughening of the surface and the development of conical protrusions. The over-cosine ($v_e > 1$) distribution is presumably due to single-crystal effects from an insufficiently randomized (i.e., textured) but microscopically flat target. Thus none of these measured angular distributions lends itself directly for inferences on the near-surface region of the collision cascade. Careful electron microscopic control must accompany angular distribution measurements on crystalline solids to guarantee that the observed distribution is not modified by surface topography. If information on collision processes is desired, it is in any case preferable to conduct such investigations on amorphous solids, such as semiconductors preirradiated at a temperature $<300°C$ with a fluence of 10^{14} cm^{-2}.

When needle-like cones develop at the surface the situation is different. Relation (2.12) no longer holds. Projectile reflection at the steep slopes of the cones reduces sputtering from the cone mantle, while recapture of particles prevents most ejected particles from leaving the target. The total yield is therefore reduced [2.83, 84], and the angular distribution is distorted to an over-cosine characteristic [2.109]. Under clean experimental conditions, however, structures with such steep slopes are the exception to the rule.

---

[2] This is true of the total yield as well: faceted surfaces in general give higher yields than flat surfaces [2.85].

2.2.3 Angular Distribution from Single Crystals

Preferential ejection of sputtered target atoms in the direction of close-packed lattice rows was first found in the low-energy (10^2 eV) regime [2.5], but was soon verified at high energies too [2.6–12]. It is now established over five orders of magnitude of projectile energy, is observable in backsputtering as well as transmission, and occurs for metals, semiconductors, and insulators under heavy particle and electron bombardment. *Preferential ejection is a general irradiation effect in crystalline solids* when the energy transfer exceeds the surface binding energy.

On the other hand, the appearance of intensity maxima in certain crystallographic directions also proves the persistence of the regular lattice structure of the solid subjected to energetic particle irradiation. This has been impressively demonstrated with semiconductors: only above a certain critical temperature can enhanced ejection along the $\langle 111 \rangle$ direction be observed with silicon and germanium single crystals, while below this temperature ejection distributions typical of random structures are obtained. This specific amorphous/single crystalline transition has been confirmed independently by surface-structure analysis and ion-induced electron emission [2.169–174]. Furthermore, the existence of anisotropic ejection precludes evaporation from being the pertinent sputtering mechanism, since sublimation from single crystals is isotropic [2.175–177].

The most prominent preferential emission directions are listed in Table 2.1 with their ejection mechanisms. Since emission along the most closely packed lattice rows is usually also the most intensive one, it has attracted the greatest interest. It offered, in particular, the prospect of studying the then newly found mechanism of focusing collision sequences [2.16–19 and *Robinson* in 2.69a].

a) Preferential Ejection in Close-Packed Lattice Directions: Models

In 1957 *Silsbee* [2.16] pointed out the possibility of a lattice influence on the energy dissipation by energetic recoils. In particular, he demonstrated by applying the hard-sphere approximation how momentum focusing along $\langle 110 \rangle$ in fcc lattices could be accomplished, provided the interaction potential was

Table 2.1. Lattice directions of most prominent preferential ejection in single-crystal sputtering

fcc	hcp	bcc	Diamond	Mechanisms of anisotropic emission
Lattice structure				
$\langle 110 \rangle$	$\langle 11\bar{2}0 \rangle$	$\langle 111 \rangle$	—	Direct-focusing collision sequences, binary collisions in regular surfaces
$\langle 100 \rangle$	$\langle 0001 \rangle$	$\langle 100 \rangle$	$\langle 111 \rangle$	Assisted-focusing collision sequences, ejection through potential minimum
—	$\langle 20\bar{2}3 \rangle$	—	$\langle 111 \rangle$	Binary collisions in regular surface

sufficiently strong. *Wehner's* findings of 1955 and 1956 were not evident enough for the existence of focusing collision sequences, mainly because of the low (10^2 eV) projectile energies in these experiments. With the confirmation of preferential ejection at higher (10^4 eV) energies, however, this restraint no longer applied, and *focusons* became for years the general interpretation for ejection along close-packed directions.

With Fig. 2.9, we can derive equations that clearly show the gross effect of momentum focusing: the distance of closest approach in a head-on collision of atoms is given by

$$2R = a_{BM} \ln \frac{2A_{BM}}{E_1} ,$$
(2.13)

where A_{BM} and a_{BM} are the constants in the Born-Mayer potential

$$V(r) = A_{BM} e^{-r/a_{BM}}$$
(2.14)

and E_1 is the energy of the first atom moving toward its neighbor in the string. At noncentral impact the respective collision angles are related by

$$\sin \theta_2 = \sin \theta_1 \left[\frac{D}{2R} \cos \theta_1 - \sqrt{1 - \left(\frac{D}{2R} \sin \theta_1 \right)^2} \right] ,$$
(2.15)

which reduces to

$$\frac{\theta_2}{\theta_1} = \frac{D}{2R} - 1$$
(2.16)

at small angles. The ratio $\theta_2/\theta_1 = \Lambda(\theta_1)$ is called the focusing parameter; it is often used just to represent the scattering law, without necessarily implying focusing. In the hard-sphere approximation a simple criterion for momentum focusing – i.e., $\theta_2 < \theta_1$ – follows directly from (2.16):

$$2R > \frac{D}{2} \quad \text{or} \quad D < 4R .$$
(2.17)

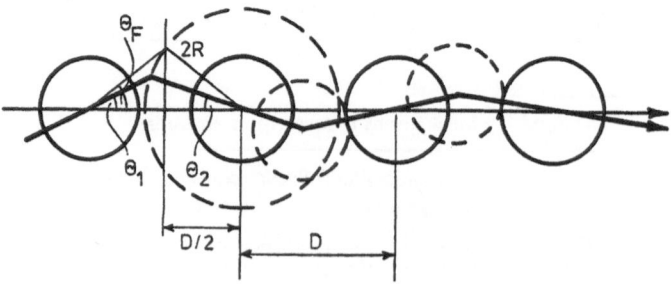

Fig. 2.9. Schematics of a focusing collision sequence in the hard-core approximation

In other words, the impacting atom must not cross the halfway plane between the two atoms. This relation is also obvious from purely geometric considerations (Fig. 2.9). According to (2.13, 17) there is an upper limit above which no correlated collision sequence can transport energy along the lattice row

$$E_f = 2A_{BM}e^{-D/2a_{BM}} \; . \tag{2.18}$$

The higher this so-called focusing energy, the larger the portion of the recoil spectrum that may undergo correlated collision sequences. Hence the contribution of focusons to energy dissipation in general, and anisotropic ejection in particular, was expected to be most pronounced for heavy elements (large A_{BM}) and along the most closely packed lattice rows (small D). In fact, *Silsbee* was the first to suggest comparing the $\langle 110 \rangle$-ejection characteristics from Au and Al crystals to check the significance of focusing collision sequences; in such open lattices as Al this effect was not expected to play a role.

The whole concept of focusing collision sequences was modified considerably by *Leibfried* and coworkers [2.17, 18, 178, 179]. A correct treatment of the collision problem under realistic potentials resulted in significantly lower focusing energies, crowdion-like mass transport in the regime $E_f/2 < E < E_f$, and nonzero acceptance angles for focusing sequences at $E_1 = E_f$.[3] On the basis of these findings, *Andersen* and *Sigmund* [2.181] suggested a formula for the focusing energy which, for copper at least, agrees quite well with *Lehmann* and *Leibfried*'s results [2.18]:

$$E_f = 2A_{BM}\exp\left[-\frac{D}{4a}\left(1 + \sqrt{1 + \frac{8a_{BM}}{D}}\right)\right] \; . \tag{2.19}$$

All focusing energies given in this chapter were calculated from this formula, but focusing energies derived from experiment and theory deviate by up to a factor of four, so these values are rather an order-of-magnitude statement [*Robinson* in 2.69a].

Apart from the focusing energy, the decisive quantity determining the range of focusons, as well as their importance in sputtering, is the focus on energy loss. This loss results from vibrational displacements in the lattice chain itself [2.19, 181], potential barriers established by neighboring lattice rows [2.17, 18, 178, 179], and, when the velocity is comparable with the speed of sound, collective interaction with the lattice. The first two mechanisms are well understood, but very little is known about phonon excitation by collision sequences. Because of the lack of theoretical treatment, an expression first given by *Lidiard* [2.182] for the energy loss of channeling particles is often extrapolated to low energies to give a rough estimate:

$$\Delta E \sim \frac{U_{latt}^2}{E}\,\text{eV/impact} \; . \tag{2.20}$$

---

[3] The latter two effects are prohibited in the hard-core approximation, as is apparent from (2.16).

According to this relation, the energy loss rises with decreasing energy, thus defining a limited focuson range even for perfectly aligned sequences. The potential barrier U_{latt} erected by the neighboring unrelaxed lattice again depends on the atomic number, giving lower energy loss values for Al and Mg than for the rather dense lattices of Ag and Au. Thus, as regards the energy losses, focusons may still have appreciable ranges in low-Z metals. The question is, however, whether the focusing energy is large enough for a major part of the collision spectrum to focus along certain lattice directions. We will deal with this in the context of preferential ejection from Al and Mg, but it is mentioned here that the (somewhat oversimplified) computer calculations of *Nelson* and *v. Jan* [2.73, 183] showed that the focuson contribution to anisotropic emission can be neglected only under conditions where

$$E_f/\Delta E^\Sigma < 10 \ ; \tag{2.21}$$

ΔE^Σ denotes the sum of all energy losses to the surrounding lattice. This inequality is barely satisfied for any close-packed lattice row in fcc, bcc, and hcp crystals (Table 2.2). So preferential ejection due to focusing collision sequences should be a common phenomenon for all metal crystals.

Lehmann and *Sigmund* have suggested quite a different mechanism for preferential ejection [2.26]. Several inconsistencies in the focuson model of sputtering led them to assume that preferential ejection may just be a consequence of the predominant low-energy part of the recoil spectrum and a regularly ordered surface lattice. According to (2.7), only nearly head-on collisions can lead to emission of a surface atom when the subsurface atom energy is on the order of the binding energy U_s (Fig. 2.10). In other words, the existence of

Table 2.2. Focusing energy E_f, its fraction of the surface binding energy U_s, and the energy loss to the neighboring lattice ΔE^Σ, respectively, for a selected group of metal crystals. Where experimental values are available, they are given first

Direct focusing

	Mg	Al	Cu	Zn	Ag	Au
Lattice direction	$\langle 11\bar{2}0 \rangle$	$\langle 110 \rangle$	$\langle 110 \rangle$	$\langle 11\bar{2}0 \rangle$	$\langle 110 \rangle$	$\langle 110 \rangle$
E_f [eV]	3.5	7.5	$50 \pm 10; 36$	16.5	20	$167 \pm 25; 42$
E_f/U_s	2	2	14	12	7	42
$E_f/\Delta E^\Sigma$	65	42	26	—	—	100

Assisted focusing

	Mg	Al	Cu	Zn	Ag	Au
Lattice direction	$\langle 0001 \rangle$	$\langle 001 \rangle$	$\langle 001 \rangle$	$\langle 0001 \rangle$	$\langle 001 \rangle$	$\langle 001 \rangle$
E_f [eV]	29	16	$320 \pm 50; 65$	165	40	500 ± 100
$E_f/\Delta E^\Sigma$	10	—	46	28	—	23

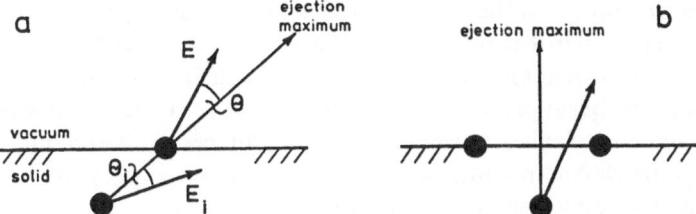

Fig. 2.10. Anisotropic ejection according to the surface ejection mechanism of *Lehmann* and *Sigmund* [2.26]: (a) direct ejection in close-packed directions by head-on collisions of low-energy recoils, (b) ejection of subsurface atoms through potential minima set up by surface atoms

a positive surface binding energy results in a selection of small impact parameters, the discrimination of large impact parameters (i.e., large ejection angles) being the more complete the nearer E_i is to U_s. According to *Lehmann* and *Sigmund* the differential yield is given by

$$Y(\theta)\,d\theta = \int g(E_i)dE_i\,d\Omega_i \tag{2.22}$$

which with (2.16) becomes

$$Y(\theta)\,d\Omega = d\Omega_i \int\limits_{E > U_s}^{\infty} \Lambda^{-2} g(E_i)dE_i \ . \tag{2.23}$$

The focusing parameter Λ, (2.16), is well described by [2.184]

$$\Lambda(E_i) = \sqrt{E_i/E_f} \ , \tag{2.24}$$

and the recoil spectrum

$$g(E_i) \propto E_i^{-2} \tag{2.25}$$

is in good agreement with both theory and experiment. Thus straightforward integration of (2.23) yields for the half-width of the emission distribution

$$\theta_{1/2} = \frac{1}{2}\left(1 + \sqrt{\frac{E_f}{U_s}}\right)^{-1} \ . \tag{2.26}$$

Evaluation of this equation for Cu gives, in view of the simplifying assumptions made, quite good agreement with experiment. *Schulz* [2.91, 185] carried out the integration somewhat more accurately and again found for Au reasonable conformity with measured data. However, experimental half-widths are generally somewhat smaller; i.e., anisotropic ejection is more pronounced in reality.

In the derivation of (2.24), an isotropic momentum distribution was the basis. Certainly, the regular lattice structure does exert an influence on energy transport, so it is not surprising that slightly peaked momentum distributions for the subsurface atoms make for better agreement between experiment and theory [2.186]. The advantage of the simple "surface model" (as opposed to the focuson "bulk" model) is that it can explain preferential ejection under conditions where the focuson mechanism must fail. The first such condition is the $\langle 20\bar{2}3 \rangle$ ejection from the basal plane of hcp lattices; in fact (2.26) describes well the ejection characteristics of this latter binary collision direction. The second is the low-energy bombardment regime, where the penetration ranges are too short for generating focusons toward the surface.

In the regime of *single-knockon sputtering*, specific surface collisions of the sort in Fig. 2.6a, b may also generate anisotropy. In fact *Henschke*'s model for explaining the Wehner spots was such a "surface billard" model [2.187]. Computer simulations by *van Veen* [2.77–79], *Yurasova* and coworkers [2.189–193], as well as from experimental results discussed below, show that this mechanism is significant only in the near-threshold sputtering regime.

b) Preferential Ejection in Close-Packed Lattice Directions: Experiments with Cubic Metals

For the reasons discussed in Sect. 2.2.1, the experimental situation concerning reliable emission distributions is far less promising than would be expected from the large number of publications on this subject. Undistorted ejection distributions are available over a useful range of bombardment conditions only for Au [2.33–35, 77, 91, 114, 185]; Mg, Al [2.27, 98]; and Zn [2.27]. Limited data are available for Cu [2.75, 77, 95, 156] and Ag [2.36, 75, 77, 95, 156, 177]. There is plenty of qualitative information, however, on preferential ejection along different lattice directions in fcc, bcc, hcp, and diamond lattices. This holds true particularly in the *low-energy regime* ($< 10^3$ eV), where the general features of anisotropic ejection can be summarized as follows:

1) Preferential ejection is observable down to projectile energies less than 10^2 eV, but its distribution appears here to be rather broad [2.34, 35, 77, 78] (Fig. 2.11a).
2) In the near-threshold regime, preferential ejection may also be due to specific projectile/surface atom collisions, such as the emission by scattered projectiles. In the case of Au, where this effect is expected to be particularly pronounced, *Szymczak* and *Wittmaack* [2.34] found ejection intensities as high as the $\langle 110 \rangle$ emission (Fig. 2.11b).
3) With increasing penetration depth of the projectile (i.e., increasing energy, decreasing mass) preferential ejection of crystallographically equivalent lattice directions shifts from high to low polar angles [2.75, 77, 156, 194].

The last-mentioned effect is obvious from (2.9, 10) and had in fact already been interpreted this way by *Koedam* [2.156]. The effect is observable in the *medium-energy regime* (10^3–10^5 eV) too, although it applies here mainly to assisted-focusing ejection [2.10], Sect. 2.2.3f. Emission distributions along the most

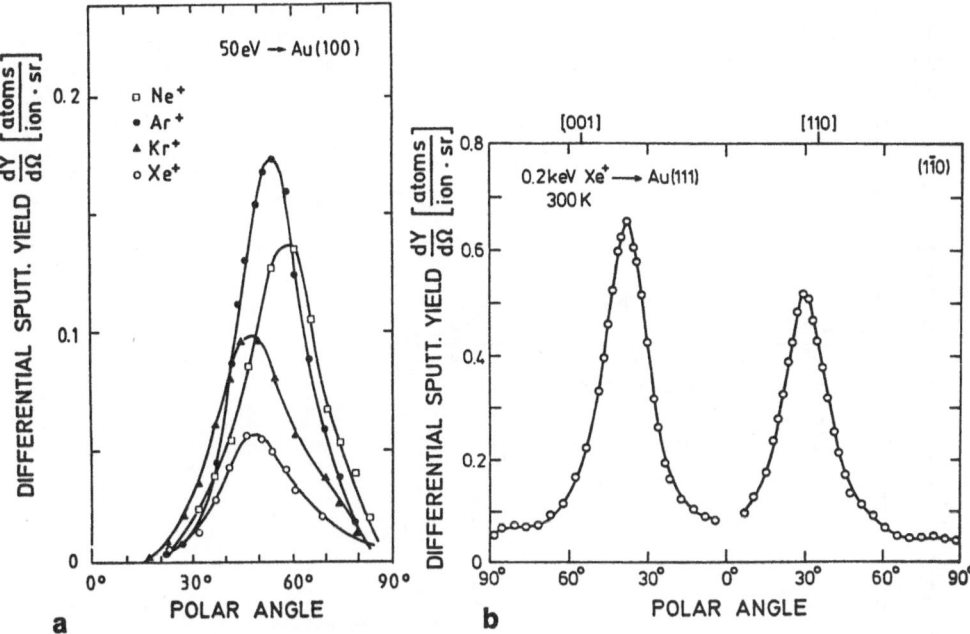

Fig. 2.11a, b. Angular distributions from single-crystal gold surfaces in the single-knockon regime: (a) ⟨110⟩ emission from a Au(100) surface; [2.78]. (b) full poloidal scan in the (1̄10) azimuth for the emission from a Au(111) surface, from [2.35]

closely packed directions are rather sharp and are ideal for systematic investigations. These have been carried out most extensively by *Szymczak* and *Wittmaack* [2.33–35], who investigated preferred emission from a Au(111) surface by He⁺, Ne⁺, and Xe⁺ ions in the energy range from 200 eV to 270 keV. Thus this work covered the energy regime difficult-to-access from single-knockon sputtering (Fig. 2.11) to that where collision cascades determine sputtering (Fig. 2.12). The ⟨110⟩ emission characteristic sharpens with increasing energy in this transition regime and then stays constant. The projectile mass has little effect in the cascade regime on the shape of the genuine ⟨110⟩ emission. The background emission, however, increases with the mass and energy of the projectile. For this reason, the researchers had to deconvolute the measured distributions to obtain the true anisotropic component (Fig. 2.13). In previous investigations, deconvolution was performed only in special cases [2.27, 36, 91, 114, 115], mostly because only very small angular regimes lend themselves to a background fit of the kind of (2.4). Accurate background-component determination requires knowledge of the total emission distribution, not just in one azimuthal plane. This has been performed so far only by *Linders* and coworkers [2.36]. Their measurements have, furthermore, quantified for the Ag ⟨110⟩ emission what *Schulz* and *Sizmann* [2.185] had observed for the Au ⟨110⟩ emission, namely, that the ⟨110⟩ emission is not rotationally symmetric. It is somewhat broader in constant-polar-angle scans than in constant-azimuth scans. The Ag ⟨110⟩

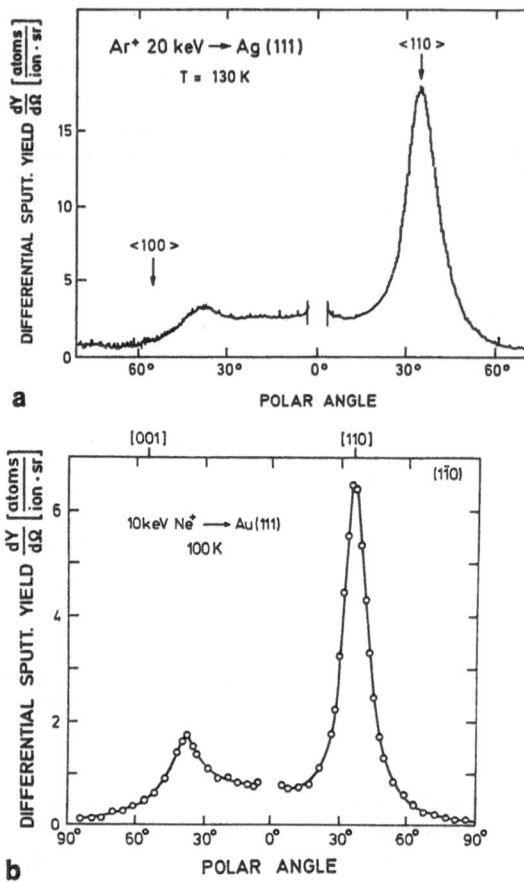

Fig. 2.12a, b. Angular distributions from (111) surfaces bombarded by inert gas ions in the cascade-sputtering energy regime: (a) ⟨110⟩ and shifted ⟨100⟩ emission from Ag, from [2.36]; (b) ⟨110⟩ and shifted ⟨100⟩ emission from Au, from [2.35].

emission, shown in Fig. 2.12a, for instance, is about 4° broader in the plane orthogonal to the plotted polar angle scan [azimuth: (1$\bar{1}$0)].

Szymczak and *Wittmaack* never found a bombardment condition where the anisotropic emission prevailed over the background. Even under low-energy, low-temperature, light-ion bombardment, the background component was the dominant emission component (Fig. 2.13). Their background-subtraction procedure also resulted in a marked sharpening of the ⟨110⟩ emission characteristics.[4] The authors were unable to reconcile this highly collimated emission with the surface ejection mechanism, and thus concluded that the ⟨110⟩ emission was largely due to emission by focusing collision sequences.

The *temperature dependence* of preferential emission is somewhat more pronounced than that of total or random emission. This expresses itself not so much in the partial ("spot") yields (Fig. 2.13) [2.91], as in the angular distribution (Fig. 2.14). The broadening of the ⟨110⟩ emission with increasing thermal

---

[4] The *as-measured* (deposit) profile shapes compare to those obtained earlier in the same laboratory [2.91, 194].

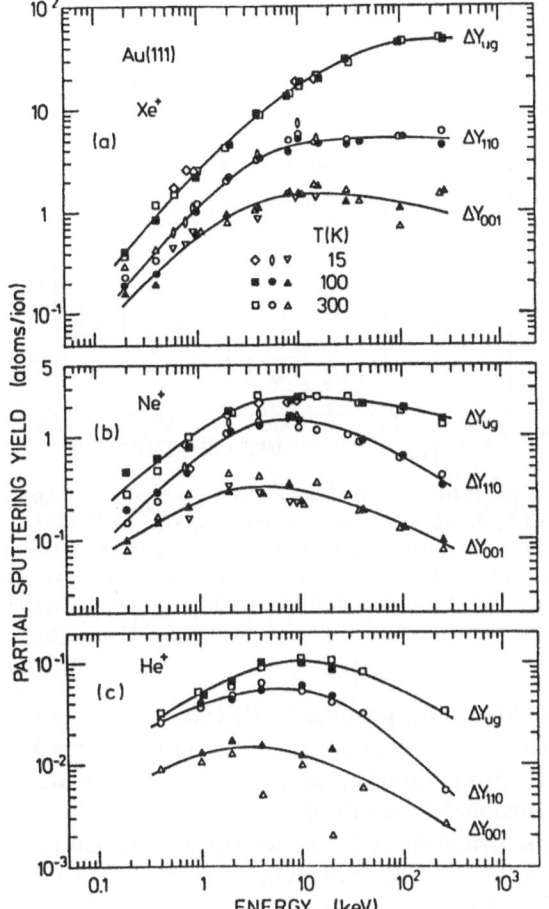

Fig. 2.13. Partial yields ΔY for preferred $\langle 110 \rangle$, $\langle 100 \rangle$ and background emission from a Au(111) surface under (**a**) Xe^+, (**b**) Ne^+, and (**c**) He^+ bombardment. The data points also show that the target temperature has little influence on the respective yields. Deconvolution of the emission distributions was performed by fitting cosine functions (2.4). From [2.35]

displacement of atoms along the $\langle 110 \rangle$ string has been investigated both theoretically and experimentally by *Nelson* et al. [2.19] on the basis of the focuson model. They calculated the mean squared angular deviation

$$\langle \theta^2 \rangle \equiv \frac{\int \theta^2 n(\theta)\, d\Omega}{\int n(\theta)\, d\Omega} \tag{2.27}$$

as a function of temperature. Here $n(\theta)$ is the probability distribution for anisotropic ejection as a function of the ejection angle θ with respect to the lattice row. Experimental data available so far agree with theory [2.91, 114, 185, 195]. *Schulz* and *Sizmann* [2.185] claim, however, that equally good agreement can be obtained within the surface ejection model. They point out further the large degree of arbitrariness in evaluating $\langle \theta^2 \rangle$ owing to uncertainties in $n(\theta)$ (Fig. 2.14). A meaningful test of the models with the mean squared angular

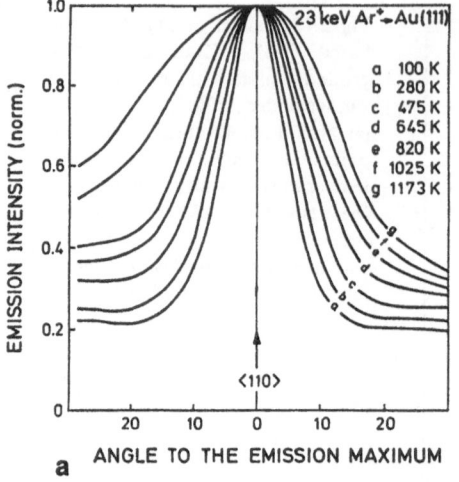
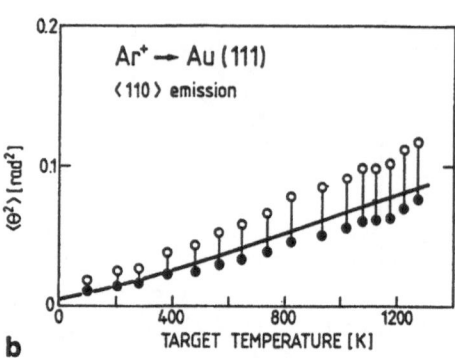

Fig. 2.14a, b. Temperature dependence of the $\langle 110 \rangle$ emission from a (111) gold crystal. (a) Experimental intensity distribution as measured by *Schulz* and *Sizmann* in the (110) azimuth [2.185], (b) Mean squared angular deviation. *Nelson* et al.'s theoretical result based on ejection by focusing collision sequences is shown as a solid line [2.19]. The results of *Schulz* and *Sizmann* [2.185] agree quite well with theory. The different symbols refer to different distribution functions in the averaging procedure

deviation requires accurate knowledge of the genuine $\langle 110 \rangle$ emission characteristic $n(\theta)$. Especially at high temperatures, where the distributions $n(\theta)$ are rather broad, their separation from random emission is practically impossible without knowledge of the full, three-dimensional distribution.

There is another reason why information from the temperature dependence of the ejection characteristic is hardly suited for assessing the respective ejection mechanisms: relaxation of surface atoms — the atoms that are actually ejected! – and their anharmonic vibration exert too strong an influence on the ejection distribution to be treated as a minor disturbance [2.196]. Conceivably this last collision is the dominating process, while the two competing momentum transport processes – collision cascade and correlated collision sequence – manifest themselves mainly in the momentum distribution of the *sub*-surface atoms. This notion comes closer to the surface-ejection model in the version of *Lenskjaer* et al. [2.186].

A remark concerning thermal versus collisional emission is appropriate here. In the introduction to this section anisotropic emission from single crystals was mentioned as evidence against sputtering being a thermal process in a hot microspot [2.2–4]. This is true only if *sublimation* from single crystals is isotropic. To the best of my knowledge, the angular distribution of subliming atoms on metal or semiconductor surfaces has not been investigated. *Cooper* and *Comas* [2.175] reported a structureless deposit on a plane collector when a Ag(100) crystal was heated to 815 °C, while sputtering the same surface at room temperature produced the well-known spot pattern. Similar observations were made in other laboratories [2.176] but remained unpublished. Figure 2.15 shows results from a recent systematic investigation by *Chen* et al. [2.177]. Under distortion-free geometry [2.80, 81], the emission from a Ag(111) surface was measured with and without 1 keV Ar⁺ bombardment. For comparison, the distribution at room-temperature bombardment is also shown. At 680 °C the ⟨110⟩ emission is broadened so much that its maximum can be distinguished from background only with difficulty; azimuthal scans at constant $\theta_e = 35°$, however, still clearly show the threefold modulation characteristic of the (111) surface. Conversely, plain sublimation conforms almost perfectly to a cosine distribution.

The observation of an isotropic emission in sublimation – devoid of any information on the crystal surface – contrasts with anisotropic desorption of molecules in, for example, the ESDIAD patterns [2.68]. More importantly in this context, the contrast between isotropic thermal and anisotropic collisional emission is not in keeping with the notion of ion bombardment as an all-destructive process: sputtered particles obviously carry more information on the emitting crystal than the more gently removed thermal particles. The facts are, first, that the ion-induced damage is effectively annealed in metals during the lifetime of the collision cascade, and, second, that thermal emission takes place from adatom positions too remote from the regular potential field of the bulk crystal.

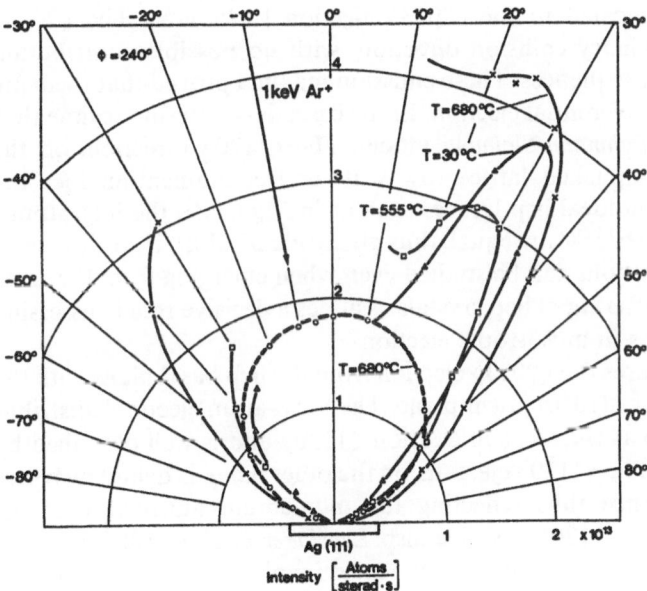

Fig. 2.15. Sublimation and sputtering from a Ag(111) surface. Angular distribution in the (110) azimuth. From [2.177]

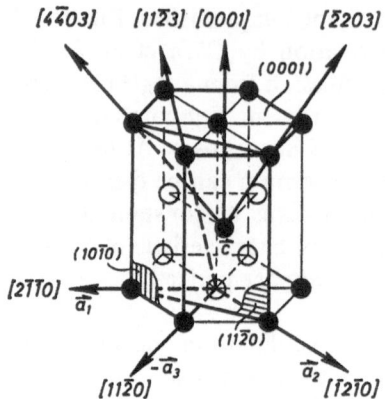

Fig. 2.16. Hexagonal close-packed lattice structure with most prominent ejection directions. While focusing collision sequences may contribute to emission along the $\langle 11\bar{2}0 \rangle$ lattice rows, this is virtually impossible for the pair-structured $\langle 20\bar{2}3 \rangle$ direction. Ejection along $\langle 0001 \rangle$ is very pronounced for most hcp crystals; it is a typical assisted-focusing emission

c) Preferential Ejection in Close-Packed Lattice Directions: Hexagonal Metals

The highly symmetric fcc and bcc lattices hardly lend themselves to a distinction between the two ejection mechanisms, primarily because there is no pure two-atom ejection direction available. Unraveling the extra contribution of focusons with more than two collisions is difficult because the momentum distribution of the recoils has to be known accurately.

The situation is different for *hcp lattices* (Fig. 2.16). *Perovic* [2.90], and later others [2.25, 197], found for the (0001) basal plane six distinct ejection maxima,[5] which were finally allocated to the $\langle 20\bar{2}3 \rangle$ ejection [2.25, 115]. The $\langle 20\bar{2}3 \rangle$ direction is an ideal binary collision direction, with no possible contribution from focusing collision sequences. These emission maxima proved that there are other mechanisms of preferential ejection. In the cascade-sputtering regime they are evidence for the *Lehmann-Sigmund* model. The $\langle 11\bar{2}0 \rangle$ direction, on the other hand, is a close-packed lattice row with perfect momentum focusing qualities. In addition, in ideal hcp lattices such as in Mg or Zr, the interatomic distances $D^{\langle 20\bar{2}3 \rangle}$ and $D^{\langle 11\bar{2}0 \rangle}$ are equal, thus giving identical E_f/U_s ratios. Since these two ejection directions can be studied even when emerging from the same crystal surface, the sputtering of hcp crystals assumes a decisive role in assessing the dominant processes in anisotropic ejection.

Figure 2.17a compares the $\langle 20\bar{2}3 \rangle$ ejection from the hcp basal plane with the $\langle 11\bar{2}0 \rangle$ ejection from the (10$\bar{1}$0) prism plane. The "two-atom ejection" distribution is nearly as sharp as the focuson ejection $\langle 11\bar{2}0 \rangle$ and is well described by (2.26) with $E_f/U_s = 2$. The $\langle 11\bar{2}0 \rangle$ ejection, on the other hand, is definitely better collimated, the difference thus reflecting the momentum anisotropy of the subsurface atoms in the $\langle 11\bar{2}0 \rangle$ row; see also *Lenskjaer* et al. [2.186]. A rough estimate of the contribution of collision sequences involving more than the two

---

[5] On plane collectors, anisotropy can only be inferred from hexagonal-shaped deposits. The patterns show no ejection maxima [2.23, 195].

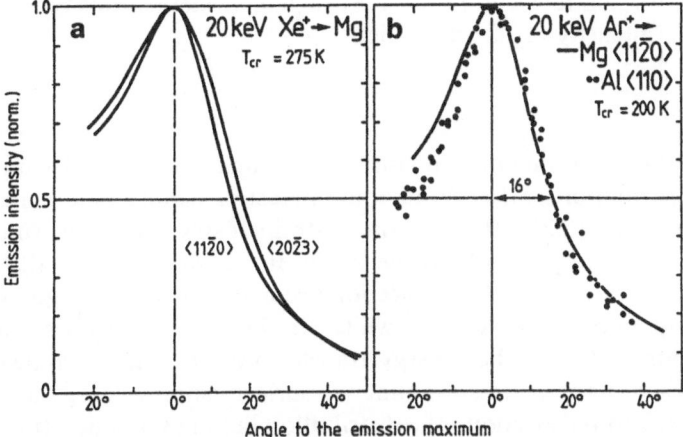

Fig. 2.17a, b. Comparison of preferred emission from low-Z single crystals. (a) $\langle 11\bar{2}0 \rangle$ and $\langle 20\bar{2}3 \rangle$ emission distributions from hcp Mg – for $\langle 20\bar{2}3 \rangle$ only the Lehmann-Sigmund mechanism (Fig. 2.10) can cause preferred emission, while for $\langle 11\bar{2}0 \rangle$ focusons can contribute also. (b) $\langle 11\bar{2}0 \rangle$ emission from hcp Mg and $\langle 110 \rangle$ emission from fcc Al. Both lattice rows are closest-packed and allow focusing collision sequences. From [2.27]

topmost atoms was obtained by sputtering a crystal surface from which $\langle 20\bar{2}3 \rangle$ and $\langle 11\bar{2}0 \rangle$ emerge at the same polar angle; the upper limit was 25% for the anisotropic part in the $\langle 11\bar{2}0 \rangle$ emission [2.27].

Since preferential ejection is usually a minor part of all the sputtered particles, the contribution of focusons to sputtering is small under the usual bombardment conditions. Only under special channeling conditions, where focuson generation near the surface is artificially enhanced, is this sometimes not true (Sect. 2.2.3e) [2.188, 197–200].

The behavior of preferential ejection with respect to temperature, projectile mass, and energy is essentially identical for $\langle 11\bar{2}0 \rangle$ and $\langle 20\bar{2}3 \rangle$, which is to be expected from the surface ejection mechanism. This is, however, not a strong argument against focuson ejection, because the crucial final (surface) collision may be the reason for the similarities (Sect. 2.2.3b). The similarity also indicates again the difficulties encountered when only close-packed lattice *rows* are at hand for investigating ejection mechanisms.

At *high projectile energies* ($\gtrsim 10^5$ eV), no particularly new features appear in the emission distribution. The increasing low-polar-angle ejection results in the disappearance of most emission at oblique angles when the 10^5 eV level is reached, leaving only a few ejection directions near the surface normal accessible to investigations on anisotropy. Furthermore, an increasing contribution from collision spikes can result in even more prevalent isotropic emission [2.15], so the heavier projectile masses have to be excluded as bombardment species. Thus, this energy regime appears generally not very well suited to preferential ejection studies, unless one is up to transmission-type experiments [2.10], or is mainly interested in assisted-focusing ejection. At such high energies this type of ejection compares in intensity to direct ejection.

d) Silsbee's Proposal

In the course of an attempt at making a quantitative comparison of the Au and Al $\langle 110 \rangle$ ejection distributions, *Silsbee* found an interesting feature that seems to favor the surface model [2.27]: the ejection width increases with increasing transferable energy from the projectile to the target atoms. According to (2.7) this means, for instance, that the Al $\langle 1\bar{1}0 \rangle$ ejection becomes sharper when going in the series of inert gases from Ne$^+$ to Xe$^+$, while Au $\langle 1\bar{1}0 \rangle$ ejection shows just the opposite behavior [2.27]. Higher energy transfer near the surface allows atoms at larger impact angles θ_1 to surmount the surface binding energy and thus contribute to preferential ejection – the distribution becomes broader. If we disregard this fact and compare the Al and Au ejection distributions obtained at the same projectile mass, we get for Xe bombardment the incomprehensible situation of $\theta_{1/2}^{Al} \approx \theta_{1/2}^{Au}$. Since E_f/U_s is on the order of 2 and 10 in Al and Au, respectively, both models predict sharper ejection distributions for Au. The widths of the ejection characteristics come into better perspective, however, when we compare at the same (mean or maximum) transferred energy. We get

$$\theta_{1/2}^{Al} = 1.8 \, \theta_{1/2}^{Au} \, , \tag{2.28}$$

which is reasonable in the framework of both theories.

Correcting the measured distributions for background emission by fitting a cosine distribution has the effect of merely changing the numerical factor in (2.28) to 1.6. The procedure is unreliable when used for only one specific azimuth. Furthermore it raises the fundamental question as to whether the "background" includes more than the random part of the collision cascade. For instance, as long as it is not clear what part of the background the surface-defocused focusons constitute, simple subtraction procedures introduce ambiguity rather than clarity. Double-differential measurements might be helpful here.

Thus (2.28) is the result of the experiment suggested by *Silsbee* 30 years ago [2.16]. Several attempts in different laboratories had failed, mainly because strong oxidation of the aluminum deposit hindered evaluation of the deposit profile [2.91]. A typical example of an Al ejection distribution is given in Fig. 2.17b. It also shows no significant difference between Al $\langle 110 \rangle$ and Mg $\langle 11\bar{2}0 \rangle$ ejection, proving the latter to be truly representative of an ejection distribution along a close-packed lattice row.

The question arises, however, whether low-Z metals such as Al and Mg are suitable for studying focuson effects, since the rather weak repulsive potential gives focusing energies on the order of U_s. In fact, this was why the Al-Au comparison was suggested. However, $E_f < U_s$ not only renders the focuson model ineffective, it also results in far too broad an ejection distribution in the surface model. Thus the sharp $\langle 110 \rangle$ and $\langle 11\bar{2}0 \rangle$ ejection distributions of Al and

Mg, respectively, call for reconsideration of the potential parameters assumed so far in calculations of E_f [2.180]. Further arguments based on solid-state and radiation-damage quantities have resulted in $E_f^{Al} \approx 10$ eV [2.27], Both models, then, describe the experimentally observed ejection distribution correctly.

There is one more important reason focuson effects in energy dissipation in low-Z metals cannot be precluded: although it is true that the "small core size, compared to the cell size" [2.16] gives low focusing energies, it also results in lower focuson energy losses. With the data in Table 2.2, the key quantity $E_f/\Delta E^\Sigma$ is thus not at all in the regime where focusons are expected to be negligible [2.183]. In fact, $E_f/\Delta E^\Sigma$ is even larger for Al than for Cu. To take a somewhat extreme view, we might therefore say that in Al focusons are generated in the most prominent regime of the collision spectrum, while in heavier elements they are transformed into heat by collective (phonon) lattice interactions in this energy regime (2.20). Phonon generation in Al, by contrast, becomes dominant only at energies lower than U_s, i.e., at energies no longer of interest for sputtering. Unfortunately, interaction in the energy regime $U_s < E < E_f$ is poorly understood, and the interatomic potentials are only vaguely known. Therefore any statement concerning this collision cascade regime, which, of course, is most important for ejection, can hardly be more than semiquantitative.

e) Suppression of the Random Cascade

Under special bombardment conditions the focuson component can be more clearly discernible: in sputtering by electrons and in sputtering under certain channeling conditions. In both cases, the random cascade is reduced – at least near the surface – while excitation of focusing collision chains is promoted.

Sputtering by electrons has long been recognized as a way to excite focusons at defined energies. Unlike with ion bombardment, the energy transfer by electrons is rather localized and small, allowing controlled generation of recoil atoms in an undisturbed lattice. In a transmission-sputtering experiment, therefore, the range of focusons traveling toward the "downstream" surface can be varied with the electron energy. Relativistic electrons are necessary to transfer to the atoms an energy equivalent to the surface binding energy (250 keV for Au with $U_s = 3.9$ eV).

The first angle-resolved electron-sputtering experiment on metals was performed on Au(111) single-crystal foils in a high-voltage electron microscope [2.105]. Plane carbon films were used as collectors and electron-induced characteristic X-ray generation was applied to evaluate the deposit. Figure 2.18 is taken from this work. While the intensity distribution along the polar angle is influenced largely by the electron-atom interaction cross section, the increasing influence of the regular lattice can best be seen in the azimuthal distribution. Again, specifying the onset of collision chains involving more than two atoms is difficult (note that the direct ejection of a surface atom by an electron causes no anisotropy; the *Lehmann-Sigmund* mechanism involves two atoms and requires

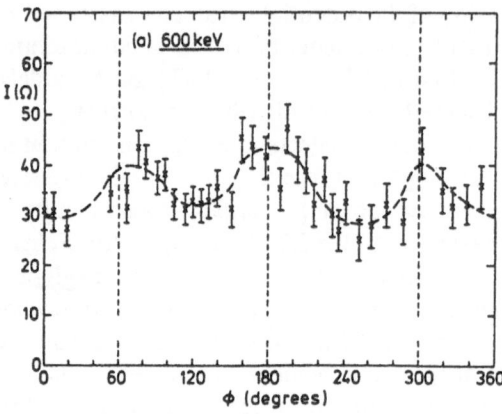

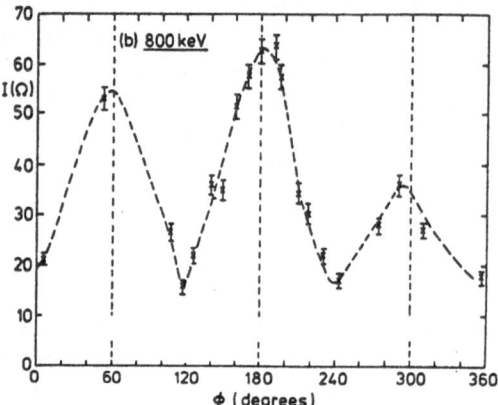

Fig. 2.18a, b. Preferential ⟨110⟩ ejection in transmission electron sputtering of a (111) gold crystal. The anisotropic structure in azimuthal intensity scans becomes more pronounced as the electron beam energy is raised from 500 to 1100 keV. Below 600 keV ejection is accomplished only by electron/surface atom collisions ($T_m = 5$ eV) and thus shows no anisotropy. Lattice effects appear only for ejection of surface atoms by recoiling subsurface atoms. From [2.105]

displacement of a *subsurface* atom). In principle, however, this should be possible, provided the ejection distribution can be measured over a sufficiently large energy interval.

Electron sputtering is hardly representative of ion-bombardment conditions, because no collision cascade is generated in the solid. But the reduction of isotropic ejection thus achieved makes deconvolution of the recorded emission distribution much less subject to ambiguity, while at the same time variation of the collision sequence length should allow unraveling of the focuson characteristics. The situation resembles investigations on Frenkel-pair generation by electrons and neutrons, respectively. The small and controllable energy transfer in electron–atom collisions allows us to probe energy transport processes precisely, while the same experiment with neutrons is complicated by cascades generated around the neutron/nucleus primary interaction [2.46, 198]. Electron-sputtering experiments are very promising, but have unfortunately not been pursued since work by *Cherns* et al. [2.105].

Bombardment along lattice channels also reduces the energy deposited near the surface by the random collision cascade. The result is a general reduction of the sputtering yield. But, under special geometric conditions, focusons may be generated in the crystal which then transport energy to the surface and cause ejection of atoms along this lattice direction. Under such conditions, the aniso-tropic component may become the dominant ejection type and accordingly stand out more clearly in the angular distribution. This explanation was given by *Elich* and *Roosendaal* [2.196] when they found enhanced ejection along $\langle 1\bar{1}0 \rangle$ for bombardment of a Cu (100) crystal under a slightly higher incident angle that corresponds to the $\langle 110 \rangle$ channel.

Different laboratories repeated this experiment. The results were different, too, but this might be due to the slight differences in the experimental conditions *Roosendaal* in [2.69a, 188, 200, 201].

f) Ejection Through Symmetric Arrangements of Atoms

Preferential ejection in single-crystal sputtering is observed not only along the most closely packed lattice rows, but also in certain highly symmetric crystal directions where neighboring atoms form "rings" around the collision direction (Fig. 2.19). Typical examples are the $\langle 100 \rangle$ ejection from fcc and bcc crystals and $\langle 0001 \rangle$ from hcp crystals. Two models were again proposed for the ejection mechanism.

According to *Nelson* and *Thompson* [2.10], momentum becomes collimated along such lattice directions because of the focusing action of surrounding ring atoms. These may be regarded as electrostatic lenses in the small-angle approx-imation. In these so-called *assisted*-focusing directions, the correlated collision sequence is thus established by the neighboring lattice. Conversely, for the

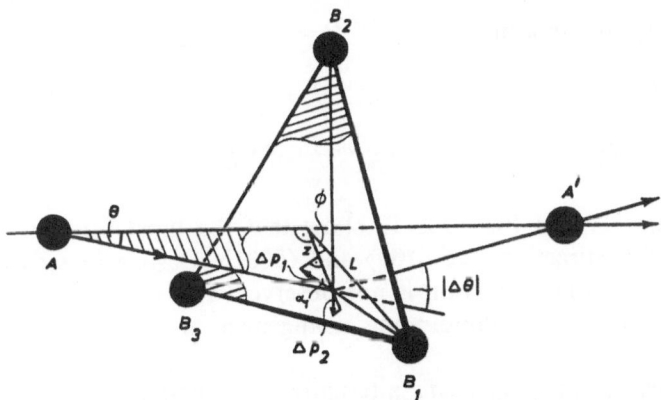

Fig. 2.19. Focusing of recoil trajectories (A, A') by symmetric "rings' of neighboring atoms (B_1, B_2, B_3) as demonstrated on the $\langle 0001 \rangle$ emission direction of hcp crystals. The effectiveness of this assisted focusing is obvious when compared with similar configurations in fcc structures: the focusing power of the ring is larger compared with $\langle 100 \rangle$ and $\langle 111 \rangle$ emission since the radius of the ring, L, is smaller and the interatomic distance in the string $AA' = c_0$ is smaller, respectively. From [2.27]

direct-focusing sequence, the neighboring lattice hardly affects focusing and mostly causes energy loss. Theoretical treatments [2.10, 27, 71, 178] considered the collision of the recoiling atom with the ring atoms in the momentum approximation, and the collisions within the lattice chain in the hard-core approximation. For collision directions with n-fold symmetry this yields

$$E_f^{(n)} = c^{(n)} \frac{\pi}{2La_{BM}} A_{BM} \frac{c_0}{2} \left(\frac{L}{2a_{BM}} - 1 \right) e^{-L/a_{BM}} \qquad (2.29)$$

with $c^{(4)} = 2$ corresponding to, for instance, $\langle 100 \rangle$ emission from fcc crystals [2.178], and $c^{(3)} = 3/2$ corresponding to $\langle 111 \rangle$ emission from fcc crystals or $\langle 0001 \rangle$ emission from hcp crystals [2.27]. Here A_{BM} and a_{BM} are the Born-Mayer parameters (2.14) and L is the radius of the "ring" (Fig. 2.19).

Perhaps more important than the focusing energy of the collision chain is the *focusing power* of the ring, since this quantity is meaningful even if collision sequences cannot be established, and preferential ejection must again be ascribed to some surface collision mechanism. Expressed as a focal length [2.10], the inverse focusing power of the symmetric atom arrangement is

$$f^{(n)} = c_1^{(n)} \frac{2a_{BM}}{(L - 2a_{BM})} \frac{E}{A_{BM}} \frac{2La_{BM}}{\pi} e^{L/a_{BM}} , \qquad (2.30)$$

where $c_1^{(4)} = 1/2$ and $c_1^{(3)} = 1/3$. The ring's influence on the penetrating particle's trajectory thus decreases with increasing ring radius and particle energy.

Evaluating (2.30) allows interesting comparisons for different crystal directions and structures, for instance,

$\langle 100 \rangle$ versus $\langle 111 \rangle$ focusing in Al: $\qquad f^{\langle 100 \rangle} = 3.5 f^{\langle 111 \rangle}$

$\langle 111 \rangle$ versus $\langle 0001 \rangle$: $\qquad\qquad\qquad f_{Mg}^{\langle 0001 \rangle} = 2.8 f_{Al}^{\langle 111 \rangle} \qquad (2.31)$

$\langle 0001 \rangle$ focusing in Mg and Zn: $\qquad f_{Mg}^{\langle 0001 \rangle} = 5.8 f_{Zn}^{\langle 0001 \rangle}$

These data and Table 2.2 and Figs. 2.2, 19 show that the most prominent assisted-focusing ejection directions are $\langle 100 \rangle$ and $\langle 0001 \rangle$. Despite its strong focusing power, preferred $\langle 111 \rangle$ ejection is not observed with fcc lattices. As discussed below, this is due to the double-ring arrangement around the $\langle 111 \rangle$ lattice row.

According to *Lehmann* and *Sigmund*'s alternative explanation [2.26], the symmetric ring of atoms not only exerts focusing forces on particles penetrating its center region, it also constitutes a potential barrier for low-energy recoil atoms. This barrier has a strong minimum at the center of the ring, roughly given by

$$E_{min} = n A_{BM} e^{-L/a_{BM}} , \qquad (2.32)$$

through which escape of near-surface recoils should be easiest. In other words, cascade atoms arriving at the surface "see" a surface barrier with large spatial variations in barrier height. Preferential ejection along the ring center is then just a consequence of a larger proportion of recoil atoms being able to overcome this local potential minimum.

Experimental evidence for assisted-focusing ejection is far scarcer than for ejection along close-packed directions. Usually only qualitative descriptions of its characteristics are possible [2.10, 27, 202–205], such as

1) Assisted-focusing ejection is a medium- to high-energy effect; it does not exist for projectile energies in the 10^2 eV regime.
2) Ejection intensities may well compare to direct ejection at high bombardment energies ($\sim 10^5$ eV), but the distribution is usually not as well collimated [2.202].
3) The position of the ejection maximum seems to be very sensitive to defects in the ring structure [2.95, 203].

There is an observation that shows preferential ejection along such lattice rows is more than just the penetration of cascade atoms through a relative potential hole in the surface barrier: all medium-to-high-Z hcp metals evince *pronounced* ⟨0001⟩ *ejection* [2.25] (Fig. 2.20), *but there is no equivalent preferred* ⟨111⟩ *ejection* from fcc crystals. In the surface collision model, ejection along ⟨111⟩ and ⟨0001⟩ should be identical, because the difference between these two directions comes into play only with the third atomic layer (layer sequence ABCABC . . . and ABAB . . . in fcc and hcp structures, respectively). This third layer, however, plays no role in a model which ascribes preferential ejection to particle release through potential barrier minima. The pronounced ⟨0001⟩ ejection, together with the nonexistence of ⟨111⟩ ejection, therefore proves the necessity of at least one more collision within the sequence (just as in Fig. 2.19).

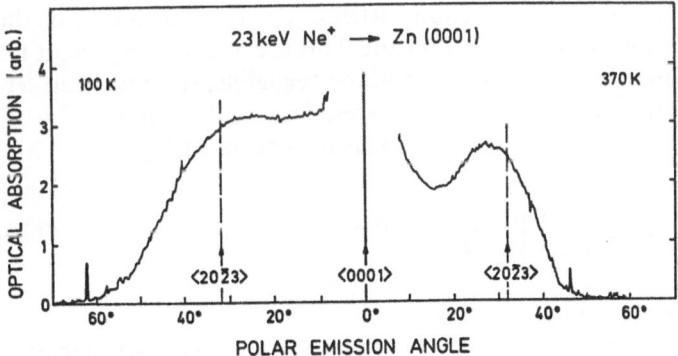

Fig. 2.20. Deposit density profile of zinc atoms sputtered from the (0001) basal plane. At low crystal temperatures (*left*) the central ⟨0001⟩ emission completely covers the ⟨20$\bar{2}$3⟩ ejection. Reduced thermal vibration reduces both the focusing power and the transparency of the three-atom ring around the ⟨0001⟩ lattice direction. From [2.27]

In other words, a true assisted-focusing collision must be established. At the same time, this finding disproves collision sequences where *two* rings have to be penetrated before an atom can collide with its next neighbor in the chain.

Investigations of the emission from the (0001) surface of zinc [2.27, 115] found the following to be typical of assisted-focusing ejection:

1) The $\langle 0001 \rangle$ ejection is temperature dependent; its intensity (and probably width) increases with decreasing temperature. Reduced thermal vibration thus leads not only to better momentum focusing in the lattice row, but also to higher focusing ring transparency.
2) Penetration of atoms through a ring of surface atoms may result in destruction of the ring, i.e., ejection of ring atoms as well. For $\langle 0001 \rangle$ ejection, the energy transfer to the ring atoms is given by [2.27, 206]

$$\Delta E = 3E \sin^2 \frac{\bar{\theta}}{2}\left(1 - \frac{3}{4}\sin^2\frac{\bar{\theta}}{2}\right) , \tag{2.33}$$

where the ejection angle θ_e is

$$\sin^2\frac{\bar{\theta}}{2} = \frac{L/2a_{BM} - 1}{(L/4a_{BM} + (E/2A_{BM})e^{L/a_{BM}})^2 - (E/A_{BM})e^{L/a_{BM}}} \tag{2.34}$$

with

$$\tan\theta_e = 2\cot\frac{\bar{\theta}}{2} . \tag{2.35}$$

Evaluating (2.33–35) shows that dissolution of the ring around $\langle 0001 \rangle$ extends over a rather broad energy regime (Zn: $10 - 400$ eV) and, furthermore, that the ejection angles are so oblique that successive collisions with surface atoms will strongly influence the ejection distribution. Deconvolution of the emission distribution then is extremely difficult. This effect is less important for crystals with high binding energies, but it is never negligible, since the transferred energy rises with increasing interatomic forces. This can be seen in (2.33–34) but is even more evident in the momentum approximation [2.27]

$$\Delta E = 3\frac{A_{BM}^2}{E}\left[\frac{L}{a_{BM}}K_0\left(\frac{L}{a_{BM}}\right)\right] \sim \frac{A_{BM}^2}{E}e^{-L/a_{BM}} . \tag{2.36}$$

Figure 2.21 is an evaluation of (2.33–36).

Much more work is necessary for a better understanding of assisted-focusing ejection, both in theory and in experiment [2.207–209]. *Szymczak* and *Wittmaack*'s statement summarizes the confusing situation 30 years after the first publications on assisted-focusing emission. In their thorough study on anisotropic emission from Au (111) they found "no evidence whatsoever for the

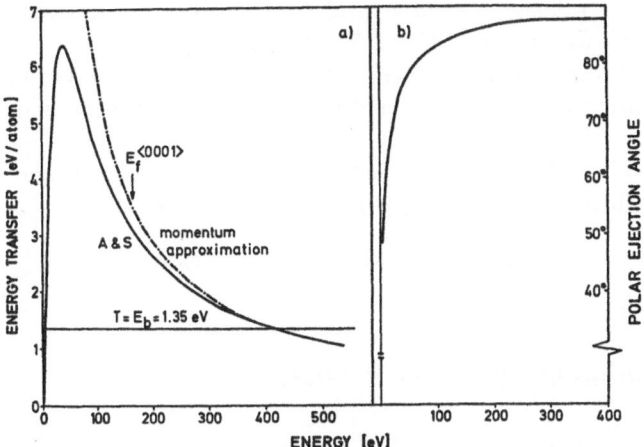

Fig. 2.21. (a) Energy transfer per ring atom in $\langle 0001 \rangle$ assisted-focusing emission from zinc. Over nearly the whole recoil spectrum the transferred energy is larger than the sublimation energy, causing destruction of the ring in the surface. These ring atoms, in turn, are ejected at rather oblique angles to the surface (b) and suffer further collisions with surface atoms. The calculations are based mainly on a theory by *Andersen* and *Sigmund*; for details see [2.27]

existence of assisted focusons" [2.34]. The understanding of direct ejection, in contrast, has reached such a level that only double-differential yield measurements are expected to lead to major further improvement.

g) Angular Distributions of Ions and Clusters

Measuring the angular distribution of sputtered ions or clusters requires energy and mass filters to separate the particles of interest from the total flux. Such spectrometers are too bulky to be scanned around the target. For this reason, general practice is to rotate the target so that the particular ejection directions are scanned over the fixed entrance aperture [2.55, 62, 210–214]. In a strict sense, such measurements do not yield angular distributions, since the projectile impact conditions are varied as well.[6] Nevertheless, some important features of angular distributions have been identified in this way:

1) Sputtered *ions* show emission maxima, too, and they coincide with those of sputtered neutrals (total flux) [2.210–214].
2) In close-packed directions the percentage of fast ions is larger than in random directions [2.210].
3) Discriminating low-energy particles increases anisotropy (e.g., the FWHM decreases) [2.55, 62, 214].

---

[6] Only when the rotation axis coincides with the beam direction do such measurements give undistorted angular distributions [2.214].

4) Also charged and neutral *clusters* show preferred emission in low-index lattice directions [2.55, 62, 214] (see Fig. 2.35 below).
5) The energy spectra of clusters shift to lower energies with increasing number of atoms in the cluster and at emission directions outside the emission maxima [2.51, 55, 56, 62, 216, 217].
6) Unlike sputtered (Ni^+, Ni_2^+) ions, sputtered (Cu, Cu_2, Cu_3) neutrals show less anisotropy for clusters than for atoms [2.62, 214] (Fig. 2.35).

Several of these features will be discussed in the following sections.

2.3 Energy Distribution of Sputtered Particles

2.3.1 Influence of Surface Binding Forces

The energy distribution of ejected neutral atoms and ions affords vital information about the sputtering phenomenon as a whole. Thermal components, single-crystal effects, and the E^{-n} high-energy tail ($n \simeq 2$) have been identified this way. Transformation of the measured "external" spectrum to the *recoil spectrum* in the cascade, however, is seriously hampered by lack of knowledge about surface binding forces. Depending on the type of surface barrier potential, the intersection of the cascade by the surface either causes a small modification of the ejection spectrum or dominates the whole low-energy part ($\lesssim 50$ eV). If surface binding is understood to be the energy U_s that has to be exceeded by the recoiling atom, irrespective of its ejection angle, the external spectrum differs from the internal one by just U_s. Since the theoretically predicted spectrum decreases monotonically with increasing energy while the experimentally determined external one evinces a distinct maximum, the model of a spherically symmetric surface barrier is generally considered inappropriate – at least as far as the total emission is concerned.[7] The work-function-type barrier is preferred, because it correctly reproduces the energy spectrum shape. Treatment of the ejection processes on an atomic scale indicates, however, that either model is an oversimplification [*Robinson* in 2.69a, 208, 209].

Briefly, the pronounced maximum in the emission spectrum for a continuously falling recoil spectrum is generally explained by a kind of refraction effect of recoils moving toward the surface. As in light diffraction, the momentum component will be reduced by an amount equivalent to the binding force

$$p_i \sin \theta_i = p \sin \theta ,$$

$$\frac{p_i^2}{2M} \cos^2 \theta_i = \frac{p^2}{2M} \cos^2 \theta + U_s , \tag{2.37}$$

---

[7] As distinguished from ejection into a narrow cone around a special lattice direction (Sect. 2.2.3).

the latter just expressing energy conservation

$$E_i = E + U_s \; , \tag{2.38}$$

where p_i or p is the momentum, and θ_i or θ the angle with respect to the normal for a target atom inside or outside (ejected state) the surface, respectively.

Equations (2.37, 38) give the "refraction" relation

$$\frac{\sin \theta}{\sin \theta_i} = \sqrt{1 + \frac{U_s}{E}} \; . \tag{2.39}$$

For an isotropic momentum distribution within the collision cascade, the spectral flux density of atoms coming from $(\theta_i, \theta_i + d\theta_i)$ and arriving at the surface in the energy interval $(E_i, E_i + dE_i)$ is by definition $N(E_i) dE_i \sin \theta_i \, d\theta_i$. This quantity is conserved during the transition. The transformation relation between the observable emission distribution and the recoil spectrum is thus

$$N(E_i, \theta_i) \, dE_i = \frac{\sin \theta \; d\theta}{\sin \theta_i \; d\theta_i} N(E, \theta) dE$$

$$= \cos^{-1} \theta \sqrt{\left(1 + \frac{U_s}{E}\right)\left(\cos^2 \theta + \frac{U_s}{E}\right)} N(E) dE \; , \tag{2.40a}$$

which reduces for near-perpendicular ejection to the oft-cited form

$$N(E_i) dE_i = \left(1 + \frac{U_s}{E}\right) N(E) dE \; . \tag{2.40b}$$

It is obvious from (2.40) that the external spectrum is identical to the internal spectrum for $E \gg U_s$. For low energies, however, the external spectrum is governed by the refraction effect, which causes a narrow cone of recoil trajectories within the solid to spread out into the whole half-space above the surface, thus diluting the particle flux accepted by the energy spectrometer and consequently causing the recorded flux to approach zero.

Figure 2.22 shows an application of this plane-barrier transformation to obtain the recoil distribution. In agreement with theory and earlier investigations by *Thompson* and coworkers [2.29–32], the E^{-n} tail with $n \approx 2$ is rather well reproduced, despite the low projectile energies used here [2.273]. Several other such cases will be presented after a discussion of experimental techniques.

2.3.2 Experimental Techniques

a) Analysis of Sputtered Ions

Measuring the energy distribution of sputtered ("secondary") ions is far easier than analyzing neutrals: none of the problems associated with postionization

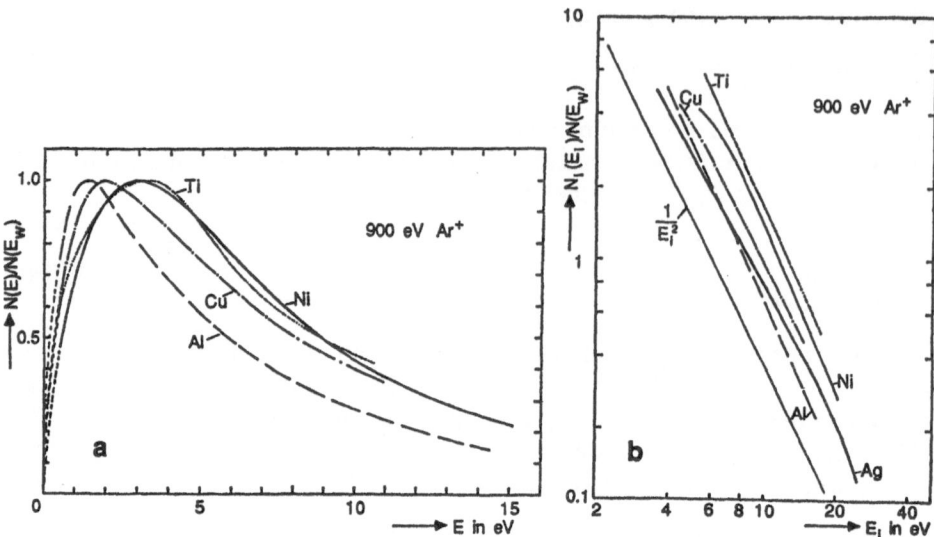

Fig. 2.22. Energy spectra of sputtered particles (**a**) and their spectral distribution before emission (**b**). Equation (2.40b) was used for evaluating the "internal" or recoil spectra. From [2.41]

and/or particle detection arise, and straightforward energy dispersive elements such as electrostatic condenser analyzers can be used. Hence many more measurements have been published for this minority group of sputtered species than for the predominant neutrals [2.216–226]. However, such spectra are questionable as far as information on the collision cascade is concerned, not so much because secondary ions constitute only a negligible fraction of sputtered particles, but rather due to the *strong influence of charge-exchange processes at the passage of recoils through the solid – vacuum interface.* These processes are most pronounced in the low-energy regime, thus again causing a shift of the most probable energy. The general understanding is that the probability of ion neutralization decreases with increasing velocity, thus shifting the most probable energy toward higher energies [2.224]. It appears hopeless for the time being to account quantitatively for this as well as the surface-binding effect to obtain the internal recoil spectrum. As with the binding force, on the other hand, the spectrum is relatively unaffected at higher energies ($\gtrsim 50$ eV).Therefore, structures in that part of the ion spectrum can be well correlated with processes within the cascade. Moreover, there are several cases where one must use ion spectra, because no other information is available. The most prominent examples of unavailable information are the energy distributions of cluster particles and of different components of an alloy. The great number of energy spectra of secondary ions thus should not be rejected out of hand.

Recently devised spectrometer systems allow polar angle scans at a fixed primary beam/target orientation [2.223, 226]. Double differential measurements $d^2 Y/dE\, d\Omega$ can thus be performed by analogy with the neutral particle analyzer at the University of Sussex [2.39, 40]. Owing to the high transmission achievable with secondary ion spectrometers, energy spectra can be obtained at considerably lower ion fluences [2.214], so avoiding all the problems associated with beam-induced surface topography.

b) Analysis of Neutrals

Most particles sputtered from clean metal surfaces are ejected in the electrically neutral state. Energy analysis of these particles can be performed by time-of-flight (TOF) methods, Doppler-shift measurements, or postionization followed by conventional charged-particle energy analysis.

The first and most extensive investigations of sputtered neutrals used the *TOF technique* [2.13, 14, 37–40, 227–237], particularly the pioneering work of *Thompson* and coworkers. They used a high-speed spinning rotor for time-resolved registration of particles sputtered by a chopped ion beam. The particle density condensed at the rotor rim reflects the velocity distribution of the sputtered flux. Many of the results discussed below were obtained with *Thompson*'s TOF spectrometer. In interpreting these results, however, two complications should be kept in mind. Rather high projectile fluences are required, causing changes in the surface topography, and, probably more important, the sputtered flux contains not only atoms but also clusters of target atoms. Both these complications can be dealt with, but require separate investigations.

Several authors present and discuss their results in TOF plots rather than energy or velocity distributions. One should not be misled by such presentations, however, when the relative intensities of the different contributing mechanisms are to be considered: converting the TOF scales to an energy distribution not only affects the abscissa

$$E = \tfrac{1}{2}Kt^{-2} \, , \tag{2.41}$$

it strongly changes the ordinate as well,

$$N(E) = N(t)\frac{dt}{dE} = Kt^3 N(t) \, , \tag{2.42}$$

where the constant $K = M_2 L^2$ for a given mass M_2 of analyzed particles depends on the length L of the drift distance. Hence high-energy events such as ejection by focusing collision sequences will appear much more distinctly in an $N(t)$ diagram than in an $N(E)$ plot, where the t^3 conversion factor will mask such events.

An instrument that overcomes the cluster problem in TOF measurements is shown schematically in Fig. 2.23: an rf quadrupole mass filter was incorporated in the drift path to enable separate measurements of atom, cluster, and molecule distributions. This is important because the energy distribution of clusters is

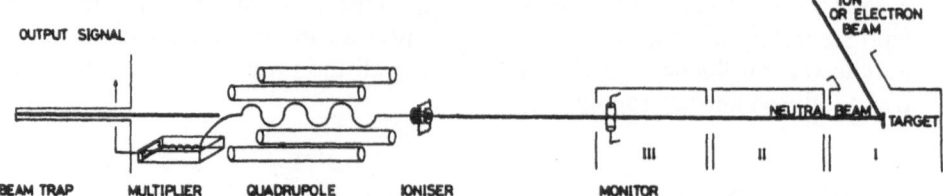

Fig. 2.23. Schematic setup of the time-of-flight spectrometer used at the FOM Institute, Amsterdam, for velocity distribution measurements of neutral sputtered particles. To increase the duty cycle of the instrument, the primary ion beam is pulsed at random intervals and the correlation technique is applied. The drift tube comprises an rf-quadrupole mass spectrometer and, in front of it, an electron impact ionizer. A small fraction of the neutral flux is ionized, mass separated, and detected by a particle counter. From [2.235]

strongly peaked at low energies, thus blurring the distribution of atomic particles in this energy regime. Furthermore, the mass filter allows measurement of different components in an alloy target. On the other hand, it is a severe potential error source owing to the energy dependence of the spectrometer's transmission. Quadrupole transmission is generally highest in the 10–20 eV range and falls off at either end rather sharply, depending on the electronic resolution and mass to be analyzed. This instrumental function must be carefully determined to allow accurate unfolding of the spectra.

The most promising method for energy analysis of neutral particles in the 100 meV to 10 keV regime uses the *Doppler shift* of the characteristic radiation [2.48, 238–252]. The method was first attempted by *Stuart* et al. [2.230] but only recently achieved a breakthrough when tunable high-power lasers became available. It is based on the shift of the excitation frequency, Δv, of atoms with velocity v and angle θ with respect to the photon beam:

$$\Delta v = \frac{v}{c} v \cos \theta \ . \tag{2.43}$$

The velocity distribution is obtained by measuring the fluorescence generated by the photon beam as a function of the laser detuning. Figure 2.24 shows the schematics of the instrument, which is capable of analyzing the energy distribution of sputtered atoms in the energy regime between 10^{-1} and 10^3 eV. By contrast, with mechanical velocity selectors [2.236, 237] the energy range is typically 1–100 eV. However, the main advantage of the Doppler-shift laser spectrometer (DSLS) lies in the possibility of distinguishing between ground-state and electronically excited particles [2.242, 246, 252]. With this method, for the first time energy distributions of excited metastable neutral particles could be determined (Fig. 2.25). These investigations showed that with increasing excitation level the energy spectra shift toward higher energies. This is in keeping with the conception of enhanced de-excitation and neutralization with

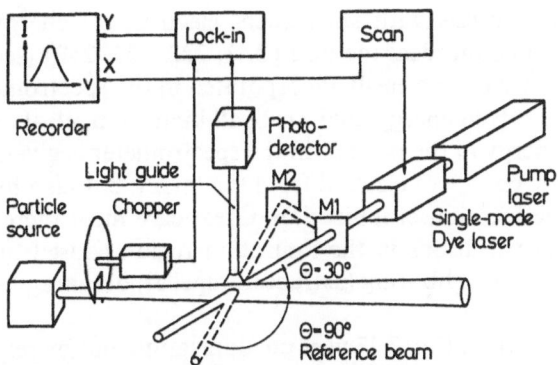

Fig. 2.24. Principle of the Doppler-shift laser spectrometer at the Technical University of Vienna. The beam of a narrow-bandwidth tunable dye laser is crossed with the sputtered beam both at right angles and at 30°. In the former case, resonance absorption of the photons occurs at exactly the excitation frequency, while at oblique intersection of the beams the laser must be tuned to the Doppler-shifted frequency. The frequency shift necessary to obtain resonance absorption is taken as a measure of the beam velocity. The number of excited atoms is measured by the intensity of the fluorescence light; only at low particle densities is the beam chopped and the lock-in technique used. From [2.238]

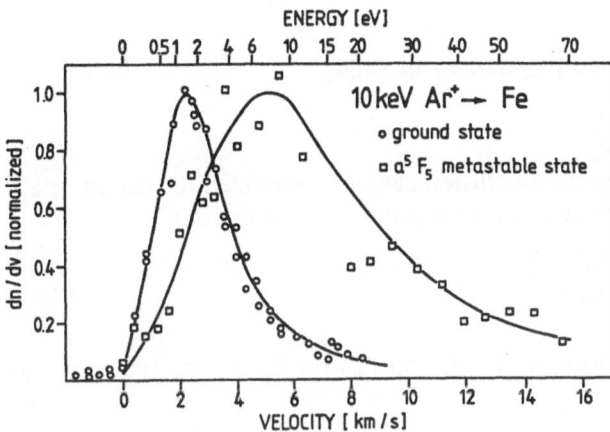

Fig. 2.25. Velocity distribution of Fe atoms sputtered in the ground state and in the metastable state a^5F_5, which lies 0.86 eV above the ground state. The distribution of the latter is much broader and shifted toward higher velocities (average velocities: $v_{gst} = 9.1$ km/s, $v_{mst} = 19.6$ km/s). From [2.246]

increasing dwell time of the exiting particles in the surface region: slow atoms are preferentially de-excited and appear in the ground-state flux (Sect. 4.2a and Chap. 3).

Such measurements have also revealed a dramatic decrease of ground-state neutrals when the surface of the sputtered target is covered with reactive gases. Oxygen coverage of chromium, for instance, reduces the Cr ground-state flux by more than a factor of 50 and causes a deficiency of low-energy particles, which are presumed to be ejected as molecules [2.242]. These findings also have a great bearing on surface analytical applications of sputtering techniques.

Postionization of sputtered neutrals with subsequent electrostatic energy analysis appears at first sight to be an ideal method [2.41, 222, 253–259] (see Fig. 2.32 below). Since a mass filter can be easily incorporated in the spectrometer, such systems allow mass-resolved energy analysis of all kinds of neutral as well as charged particles. The energy of the particle in the spectrometer is given, however, not only by the kinetic energy of ejection from the target, but also by the electrostatic potential at the location of ionization. Great care must therefore be exercised with potential variations in the ionization region caused by electrostatic field penetration from neighboring (extractor, repeller) diaphragms as well as by space charges.

Also the TOF instrument shown in Fig. 2.23 uses postionization, but merely to render the particles susceptible to mass analysis and detection. Information on the particle energy is deduced from the flight time and thus is much less influenced by potential variations in the ionizer [2.233–235, 261]. In both cases, the mass filter should be considered: its transmission is energy-dependent; the measured data therefore require unfolding.

As to terminology, note that all energy (velocity) distribution measurements are performed at a spectrometer acceptance $\ll 2\pi$. To a certain degree they are, therefore, angle resolved. For truly double-differential results $d^2 Y/dE\,d\Omega$, however, an acceptance on the order of 10^{-3} sr or less is required.

2.3.3 Information on Collision Phenomena in Solids

a) Linear Cascades

Theoretical treatments of random-collision cascades predicted the energy spectrum of the recoils to be of the form (first-order approximation)

$$N(E_i, \theta_i)\,dE_i\,d\Omega_i \propto \frac{\cos\theta_i}{E_i^{2-2m}}\,dE_i\,d\Omega_i \;, \tag{2.44}$$

which gives with the transformation relation (2.40b) the energy spectrum for ejected atoms

$$N(E)\,dE \propto \frac{E}{(E + U_s)^{3-2m}}\,dE \;. \tag{2.45}$$

This is often referred to as the Thompson formula. For the most probable energy we obtain

$$E_p = \frac{U_s}{2 - 2m} \;. \tag{2.46}$$

Since $m = 0$ to $m = 1/4$ gives a reasonable description of the interaction potential for the low-energy part of the recoil cascade, the spectrum falls off approximately with E^{-2} at $E \gg U_s$, and the most probable energy occurs at a fixed ratio of the surface binding energy ($\approx U_s/2$). This behavior had indeed already been found in the first experimental investigations and has been confirmed ever since where linear collision cascades apply. Figure 2.26 shows an example where polycrystalline gold was bombarded with 20 keV Ar$^+$ ions at various temperatures.

b) Collision Spikes

Figure 2.26 not only shows good agreement between linear cascade theory and experiment at the high-energy tail of the spectrum; it also reveals a remarkable feature at the low-energy end, namely a "thermal" component when the target was kept at elevated temperature. Sputtering from *thermal spikes* had already been anticipated from yield and angular distribution measurements long before, but this interpretation was not conclusive. A spike, by definition, is a limited volume in the solid where most atoms are temporarily in motion [2.262–270]. Such a high-density gas-like situation for the collision cascade is conceivable if energy dissipation to the surrounding lattice is sufficiently slow. The recoil energy is then confined for a time of approximately 10^{-11} s to the cascade volume. Average energies per atom on the order of the sublimation energy are then obtained, causing enhanced atomic emission. At the same time, this emission bears no resemblance to the original lattice. Such a situation can be met at high-energy heavy-particle irradiation [*Sigmund, Andersen* and *Bay* in 2.65a]. The effect of an increasing spike contribution with increasing projectile mass can also clearly be seen in the energy spectra, where the most probable energy shifts from the theoretically expected $U_s/2$ to lower values [2.268].

However, spike conditions probably cannot be generated by heating the target to temperatures near the melting point [2.14, 80, 265–268]. Along with

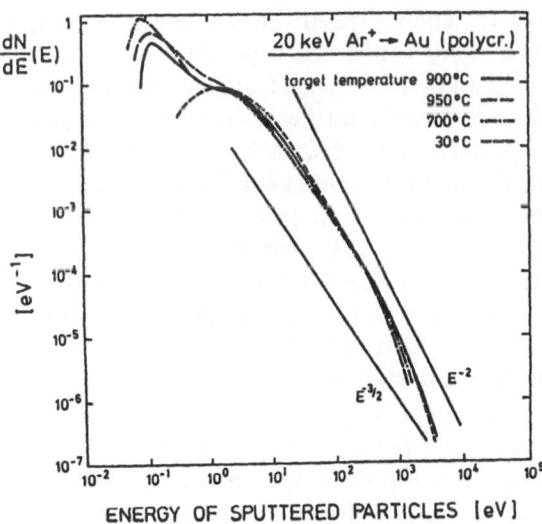

Fig. 2.26. Temperature dependence of sputtered-particle energy distributions. A pronounced low-energy peak appears at elevated temperatures and is ascribed to thermal spikes. The high-energy part of the spectrum is barely influenced by the target temperature and conforms well to linear random-collision cascades. This investigation [2.14] as well as those of Figs. 2.30 and 31, were performed with the TOF spectrometer described in detail in [2.29, 39]

ENERGY OF SPUTTERED PARTICLES [eV]

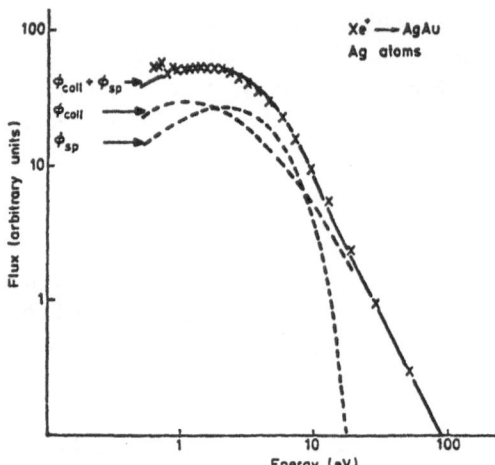

Fig. 2.27. Energy spectrum of silver atoms sputtered from a AgAu alloy by 6 keV Xe^+ ions. Deconvolution of the measured distribution reveals two components, a random-collision component according to E^{-2} and a contribution from "thermal" spikes. From [2.234]

the counterarguments relating to angular distribution measurements [2.80] and recent work on this subject [2.265], we note that no change of the energy spectrum of sputtered Cr and Ca could be detected by *Husinsky* et al. [2.241], nor did they see a change of the yield. This controversy with *Chapman* et al. [2.14] is still not resolved.

Although there is little doubt about the existence of the spike effect, its relative contribution compared with the linear cascade is still disputed. The largest controversies, however, have been caused by attempts to expand the spike model to plasma-like states to explain *inelastic* effects such as secondary-ion or photon emission. This concept of a local thermodynamic equilibrium (LTE) has received some support from *Szymonski* et al. [2.230] with the instrument shown in Fig. 2.23, where about equal contributions from spikes and linear cascades have been deduced from the energy spectra shown in Fig. 2.27. Spike temperatures as high as 20 000 K were concluded from these spectra; these temperatures are 4–5 times larger than those deduced from the results shown in Fig. 2.26. Such high spike temperatures would, furthermore, result in an exceedingly high ion fraction in the sputtered flux. This has never been reported. Moreover, a 50% or even prevailing spike component from gold at an ion energy as low as 6 keV is not in keeping with most earlier work: in Au self-sputtering, projectile energies up to 50 keV were necessary to produce a substantial deviation from the linear cascade [2.271]. Also, in *Thompson*'s earliest energy distribution measurements [2.13], at projectile energies seven times as high, the spike amounts to only 12% of the total emission. More work on unfolding the spectrum from instrumental functions is required before the instrument's powerful characteristics can be fully utilized for quantitative work.

The spikes discussed above are better referred to as elastic collision spikes, rather than thermal spikes, because it is doubtful whether energy equilibrium can be attained between the atoms in the spike. More importantly, the electrons in the cascade volume never assume a distribution that corresponds to the spike "temperature". This is evident from investigations on ion-induced electron emission [2.269, 61], where even from extreme high-density cascades (up to 10 eV/atom) no thermionic emission was observed. There is no physical basis for LTE in metals.

c) Single-Knockon Emission

As with elastic collision spikes, the energy distribution also deviates from Thompson's formula (2.45) for *light-ion sputtering* [2.270]. Ejection of surface atoms upon hydrogen or helium-ion impact occurs mostly as primary recoil (Fig. 2.6b, c). These collision mechanisms have been discussed in Sect. 2.2, hence we restrict ourselves to presenting supplementary information.

Indications of the profound influence of projectile/surface-atom collisions on ejection can be seen in a "hardening" of the energy spectra; i.e., the spectra are broadened and shifted toward higher energies with increasing projectile energy (Fig. 2.28). Ejection is accomplished predominantly by backscattered projectiles. The most probable energy is no longer directly correlated to the binding energy; if it is used under such conditions, it is merely a fitting para-meter.

Even more pronounced – but still not as dramatic as in angular distribu-tions – is the influence of single-collision ejection for *oblique* incidence (Fig. 2.29). This allows the projectile to eject surface atoms already at its first nuclear collision while *entering* the solid (mechanism in Fig. 2.6b). Velocity spectra obtained under such conditions bear no resemblance to collision cascade emis-sion [2.260, 270].

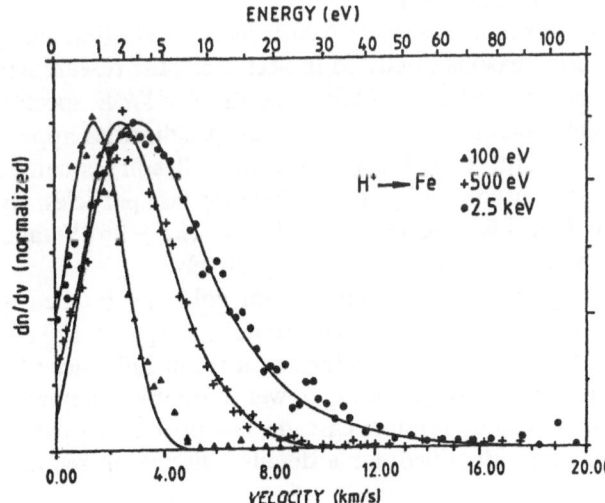

Fig. 2.28. Velocity distribution of atoms sputtered from poly-crystalline iron by perpendicu-larly incident H^+ ions. The average velocities are 1.8, 4.0, and 6.2 km/s for 100, 500, and 2500 eV bombardment, respect-ively. From [2.243]

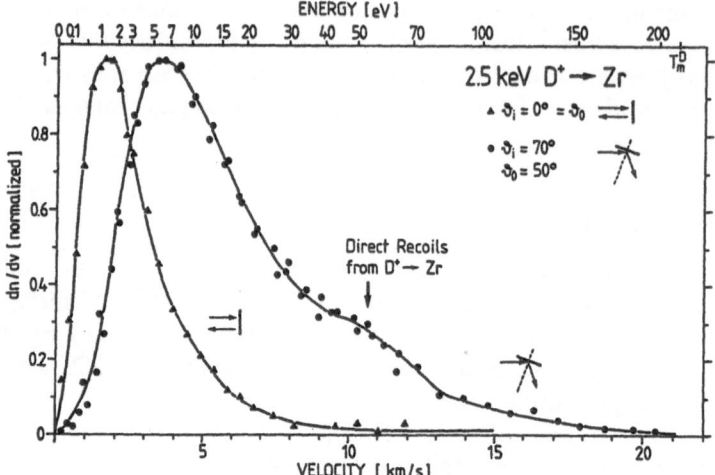

Fig. 2.29. Velocity distribution of atoms sputtered from polycrystalline zirconium by 2.5 keV D^+ ions at perpendicular and oblique incidence. Because of direct ejection of surface atoms by obliquely incident D^+ projectiles (mechanism in Fig. 2.6b) the spectrum is shifted toward higher velocities. From [2.260]. (See also Fig. 2.7, where the equivalent angular distributions are shown)

d) Single Crystals

Much more information can be gained from angle-resolved energy distributions, i.e., from the determination of $d^2 Y/(dE \, d\Omega)$. Especially for single crystals, double differential measurements permit looking for preferred momentum-transfer processes. These are, for instance, direct and de-/reflected recoils [2.43, 227], or assisted- and direct-focusing collision sequences [2.44, 270]. Also, the influence of point defects on ejection can be studied and the effective surface binding energy determined [2.30, 228, 268].

The time-of-flight spectra shown in Fig. 2.30a, b correspond to ejection along ⟨110⟩ and ⟨100⟩ from Au. For reasons discussed in Sect. 2.3.2, the researchers prefer to present their results in terms of dY/dt instead of dY/dE spectra. Especially for energetic events such as focusing collisions, peculiarities appear more distinctly in TOF plots. Still, emission by focusing collision sequences along ⟨110⟩ appears only as a discontinuity in the slope for fast particles. By unfolding the spectra, a focusing energy of $E_f^{\langle 110 \rangle} = 167 \pm 25 \, \text{eV}$ and $50 \pm 10 \, \text{eV}$ has been deduced for gold and copper, respectively.

Note that the ⟨100⟩ assisted-focusing ejection from gold can be clearly resolved (Fig. 2.30b). The peak is observed at an ejection energy of 170 eV, irrespective of the bombardment conditions – as long as it is not influenced by the direct recoil. This is a beautiful example of the power of double differential measurements: even near the prevailing random cascade, low-probability events can be detected and energetically identified; for a detailed discussion see also [2.272].

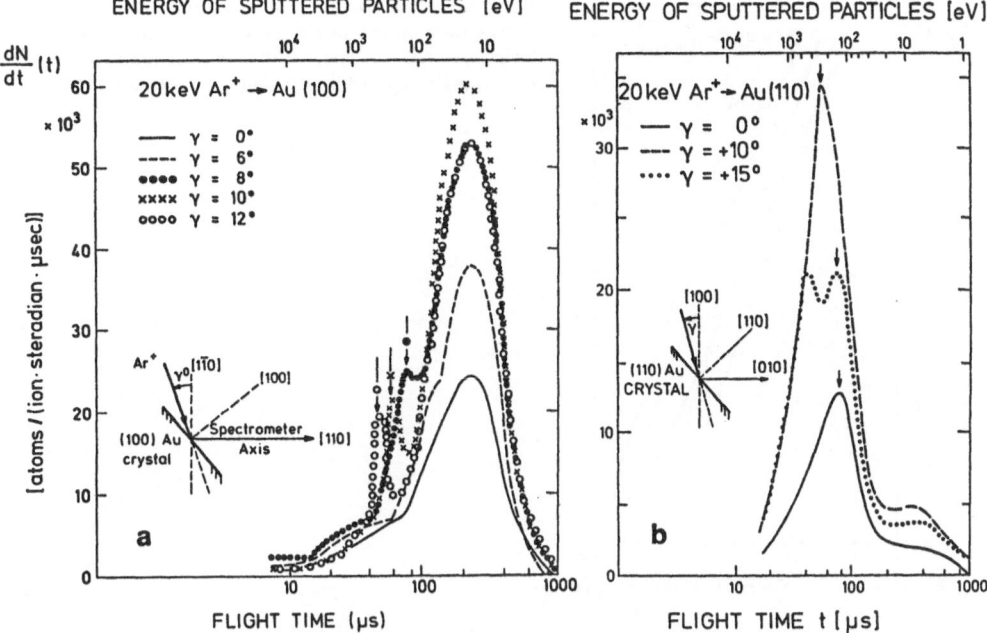

Fig. 2.30a, b. Time-of-flight spectra of particles sputtered from single gold crystals. (a) Ejection from the (100) surface. The spectra show clearly resolved high-energy peaks, which could be ascribed to primary recoils ejected from the surface. (b) Ejection from the (110) surface in the ⟨100⟩ direction. There is a distinct difference to random spectra, not only in the better resolved direct-recoil peak, but also in the appearance of a second maximum, which was interpreted as assisted-focusing ejection. From [2.43]

e) Alloy Targets

A problem not yet solved in sputtering is preferential ejection from alloys. Deviation of total and differential yields from stoichiometry is a common observation but still little understood. Recent findings of a significant shift in the energy distribution of an alloy component both with respect to its respective elemental target (Fig. 2.31) and with respect to other components [2.48, 254] may help to clarify the phenomenon. *Oechsner* and coworkers used (2.46) to find the surface binding energy U_s by determining the power m from the slope of the recoil spectrum. They correlated the data obtained in this way with surface composition investigations by Auger electron spectrometry. For the time being, it is not clear why the spectrum falls off more steeply for the alloy than for the pure elements. Perhaps the rather large mass difference within the target inhibits the population at the high-energy end of the distribution. This finding is difficult to interpret because the experiment is in the transition region between single-knockon and cascade sputtering. *Urbassek*'s expansion of transport theory to low energy is presently applicable only in the equal-mass case [2.273]. It

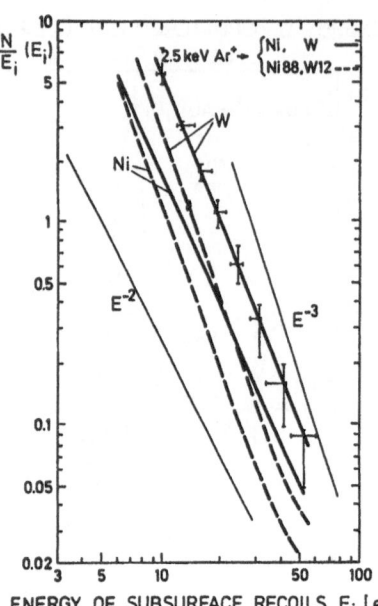

Fig. 2.31. Recoil energy spectra as determined from the energy distribution of ejected atoms. Transformation from "external" to "internal" distribution functions was performed using (2.40) as in Fig. 2.22. From [2.254]

describes well the experimental results for elemental targets shown in Fig. 2.22, but composite ones are presently out of range for a rigorous analytical treatment.

Also, the effective surface binding energy may change in multicomponent targets. Deconvoluting the energy spectrum of Fig. 2.27 and comparing it with those of pure Ag and Au, *Szymonski* et al. [2.234] deduced the following reductions of U_s in a Ag-40 Au alloy:

Ag 3.1 eV → 2.1 eV

Au 3.8 eV → 3.3 eV .

The total and partial yields change accordingly.

Determining the energy distribution of various constituents of a multicomponent solid is a common practice in secondary ion mass spectrometry (SIMS) [2.56, 224, 226, 257, 274, 275]. Hence examining the large number of spectra of *sputtered ions* is useful for information on the physical processes involved. In a rather phenomenological way the energy distribution of sputtered positive ions can be described by

$$N^+(E)dE = R^+(E) N(E)dE \propto R^+(E) \frac{E}{(E + U_s)^3} dE , \qquad (2.47)$$

where (2.45) has been used in the $m \to 0$ limit (low recoil energies). Here $R^+(E)$ is the charged fraction of sputtered species and is often [2.274–278] assumed to follow a power law

$$R^+(E) \propto E^\beta, \qquad \beta > 0 \ . \tag{2.48}$$

This gives for the most probable energy E_p^+

$$E_p^+ = U_s \frac{\beta + 1}{2 - \beta} > E_p \ , \tag{2.49}$$

with its aforementioned shift toward larger energies as compared with the distribution of the neutrals. Only for $\beta = 0$ does the ionic spectrum directly reflect the distribution of neutral particles. Occasionally it has been argued that such a $\beta = 0$ situation can be met with surfaces covered with reactive gases [2.226]. However, this procedure changes not only the surface binding forces but also the composition of the subsurface region, owing to implantation of the sorbed atoms [2.258, 279–281]. Finally, an observation of *MacDonald* and coworkers shows the need for caution with gas-covered surfaces: there are transient low-energy peaks at contaminated surfaces, which only gradually disappear with dynamic surface cleaning during irradiation [2.218, 219].

The intricate charge exchange processes accompanying the exit of particles from the surface makes it difficult to calculate $R^+(E)$ from first principles [2.282]. Without detailed knowledge of the ionization/neutralization function, however, the information extractable from energy distributions of sputtered ions is rather limited. Most basic details of the collision cascade must still be studied with the aid of neutral particles. There is, on the other hand, one phenomenon that has been investigated almost exclusively by measuring charged particles, namely the ejection of clusters and molecules. This is dealt with next.

2.4 Mass Distribution of Sputtered Particles

The investigation of the mass distribution of sputtered particles is, for reasons mentioned in the previous section, a domain of secondary ion mass spectrometry (SIMS). Hence the following is based largely on results obtained with the charged fraction of emitted particles. For example, the impression that most sputtered species are monatomic stems from secondary ion mass spectra. Especially with quadrupole mass filters the atomic ions predominate; sputtered agglomerates, so-called *clusters*, are regarded as a minority. In recent years, however, evidence that this contribution to the total emission can by no means be neglected has increased. It has been recognized that instrumental factors strongly influence the apparent cluster (size) distribution, and that many clusters decay into stable fragments after a lifetime of several microseconds. As with energy distributions (Sects. 2.3.2a, 2.3.3e), however, the characteristics of sputtered ions are to be transferred with care to neutral particles. This is particularly true of clusters, since, in addition to the *charge-changing processes* during passage through the selvage of the solid, *stability conditions* must also be met.

2.4.1 Instrumental

The principle of the instrumental setup for investigations of the sputtered-particle mass distribution has remained the same from the initial experiments [2.49, 50, 57, 283, 284] to now [2.59, 62, 216, 253–259, 285–297]; a modern version of such an instrument is shown and explained in Fig. 2.32. An intriguing modification has been developed by *Oechsner* and his coworkers [2.59, 60]; here an rf-heated plasma provides both the sputtering particles and the ionizing electrons.

As-measured abundance distributions may show strong discrimination effects against larger masses. This depends both on the choice of the spectrometer and the setting of the instrument parameters. The main discrimination effects are as follows:

1) The *transmission* of the mass filter decreases with increasing mass; this is particularly pronounced with rf-quadrupole filters.
2) The *detection probability* decreases with mass. Decreasing velocity decreases ion-to-electron conversion efficiency in the detector.
3) Many clusters are unstable and *disintegrate* during the first microseconds after emission [2.291, 294]. Although this is a long time compared with the emission event, it is short compared with the flight time in quadrupole instruments. Disintegration may be further promoted by postionization.

Some numbers may illustrate the latter two items. With quadrupole mass filters, typical pass energies of the particles to be analyzed are $10 - 10^2$ eV; this allows only moderate ion extraction from the target and results in a flight time to the detector of some 10^{-4} s. Magnetic spectrometers, by contrast, allow pass

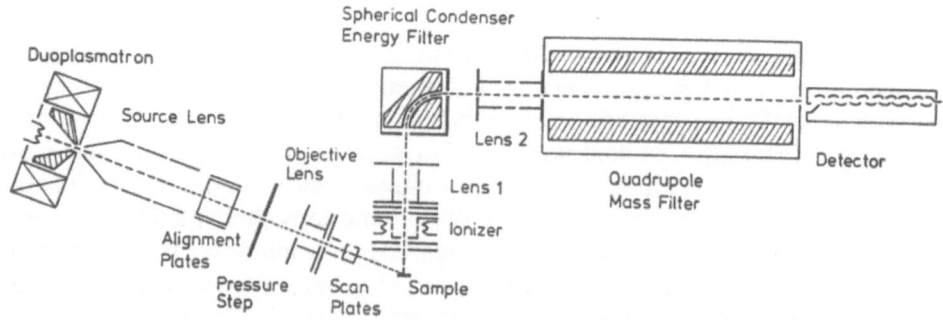

Fig. 2.32. Schematic layout of an instrument that measures the mass distribution of sputtered particles. It consists of an ion beam generating/focusing system and an energy/mass spectrometer. An electron impact ionizer in front of the spectrometer ionizes sputtered neutrals. From [2.256]

energies of $10^3 - 10^4$ eV. This results in efficient charged-cluster extraction and reduces the flight time into the 10^{-6} s regime. Such mass distributions come nearer to those characteristic of the emission. Moreover, as ion detection is based on ion-to-electron conversion, a process virtually zero below a threshold velocity of 5×10^6 cm/s, quantitative cluster registration constitutes a serious problem. Because the impact energy is essentially given by the detector potential (for a channeltron multiplier typically 3 keV), registration would be limited to about 700 atomic mass units. The fact that much larger clusters are often recorded [2.53, 54, 291–294] can be explained only by secondary effects such as ion desorption. The detection efficiency via such processes is unknown but indubitably far below unity. Ion–electron converters operated at high voltages are a partial solution. [2.290, 292].

These effects may account for the general observation that mass distributions obtained with rf-quadrupole mass filters show fewer heavy clusters than with magnetic mass filters. The fact that the intensity of *neutral* clusters falls off with cluster size n faster than charged ones [2.59, 60, 62] may also have its origin partly in stability: neutrals cannot be extracted from the target and may, furthermore, become destabilized by the ensuing ionization process.

All in all, many more clusters are emitted than recorded.

2.4.2 Cluster Emission

a) Cluster Stability

Most early cluster-emission studies were performed on silver: *Honig* [2.49] first observed dimer clusters in the mass spectrum of positive ions; *Krohn* [2.50] found Ag clusters containing up to five atoms and noted, furthermore, that in special cases cluster intensities may even be stronger than atomic emission. *Hortig* and *Müller* [2.53] reported negative clusters containing up to 60 Ag atoms when they bombarded polycrystalline silver targets with 15 keV Kr^+ ions; simultaneous Cs deposition was performed here to facilitate negative-ion formation. This record in cluster size was only recently broken by *Katakuse* et al. [2.291], who found Ag_n^+ clusters up to $n = 200$ (> 20000 amu!) from polycrystalline silver bombarded with 10 keV Xe^+. In addition, strong oscillatory behavior in the abundance distribution was observed for clusters up to Ag_{30}^-, the odd numbers being significantly more intense than the even ones. Referring to *Dörnenburg* et al. [2.284], *Hortig* and *Müller* gave an interpretation in terms of binding-electron parity. Concurrently, *Blaise* and *Slodzian* [2.52] explained in the same scheme the oscillations they found for positive ion clusters of Cr and Cu. Figure 2.33 gives an example of this effect from a more recent investigation. Owing to fragmentation of less stable clusters during their transit time through the spectrometer (approx. 10^{-4} s), the mass distributions reflect the clusters' stability distribution rather than their genuine emission distribution.

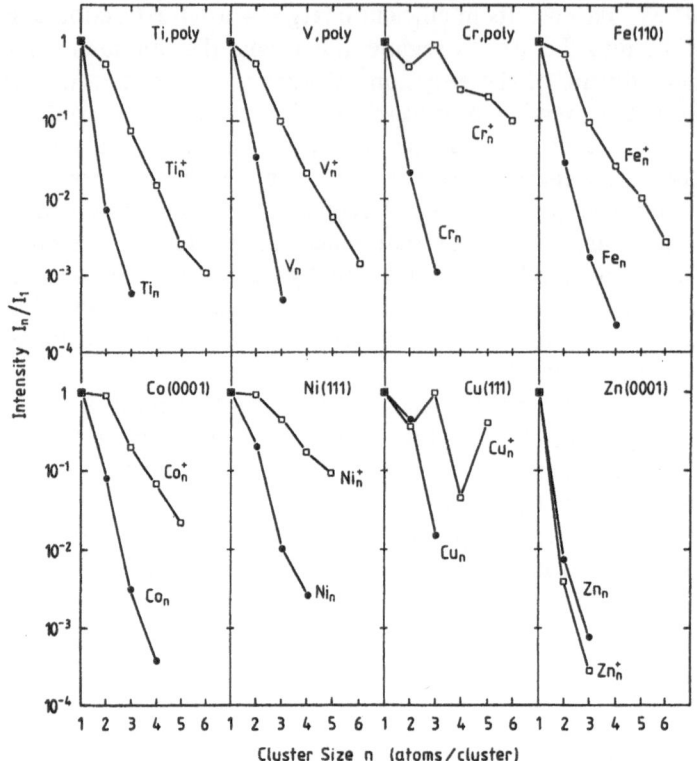

Fig. 2.33. Abundance distribution of positively charged and neutral clusters sputtered from 3d transition metals by 5 keV Ar^+ impact. The measurements were carried out with the instrument sketched in Fig. 2.32. The data were taken with constant energy band width at the respective most probable energies; owing to the differences in the energy distribution of atomic and cluster/molecular particles (cf. Sect. 2.4.2b), the atoms' intensity I_1, to which all cluster intensities I_n are normalized, is suitable only for approximative intensity comparisons. Note the significantly steeper intensity decrease for neutral clusters as well as the absence of odd-even oscillations in their abundance distribution. From [2.311]

A straightforward consequence of cluster stability is the abnormally small abundance of zinc clusters (Fig. 2.33); their binding energy is so small – the cohesive energy of the Zn-dimer, for example, is an order of magnitude smaller than that of Cu – that collisional emission of such agglomerates is unlikely and their resistance to fragmentation is weak. This case of a low-clustering but high-yield target has a bearing on the conflicting models of cluster emission in sputtering; it urges caution when ratios of cluster yield to total yield are related for different elements (Sect. 2.4.3b, c) [2.59, 289, 302].

Abundance distributions of sputtered molecules and clusters are now quite generally interpreted in terms of the agglomerates' stability. For not too large metal clusters, stability is strongly influenced by the number of valence electrons and their orbitals' parity; even parity yields higher stability. For neutral Cr, Cu, and Ag clusters[8] for instance, this would mean that clusters containing an even number of atoms are more stable than odd ones, while for singly charged ions the odd clusters are more tightly bound (and thus more abundant in the sputtered flux). For clusters formed by atoms with two s-electrons[9] in the outer shell, on the other hand, there is no parity alternation with varying number n of constituents, and hence no intensity oscillation (see the Ni distribution in Fig. 2.33). This *parity-rule* model agrees with the experimental situation not only for sputtered positive [2.285, 291, 294, 311] and negative [2.53, 291] clusters, but also for clusters from high-frequency sparks [2.284] clusters from laser-generated plasmas in solids [2.299, 300], and for field-ionized clusters from liquid metals [2.345]. It also agrees with the molecular distribution in sublimation from graphite [2.298]; this latter accordance with a thermodynamic equilibrium process emphasizes again the importance of molecular stability.

Stability with respect to dissociation or fragmentation, however, is only one aspect of tight-binding molecular orbitals. Stability with respect to ionization, i.e., the ionization potential and the functional dependence of the ionization cross section, is the other. In sputtered cluster research, this consequence of an alternating parity of valence electrons appears to have been generally overlooked. The recent finding, that sputtered neutral Cr and Cu clusters do not follow odd-even intensity alternations, Fig. 2.33, therefore met interpretive complications [2.311]. In research with clusters in supersonic nozzle beams, on the other hand, this fundamental difference in neutral and charged cluster distributions is well known. There, it is the (laser) ionization process which changes a monotonic spectrum of neutral clusters, of Cu for instance, into an alternating one for the positively charged species [2.346]. This is interpreted as an effect of near-threshold ionization of Cu clusters of alternating electronic structure. At even parity, spin compensation is achieved for the $4s\sigma$ orbitals, to the effect of high ionization potentials and small electron affinities for even-number neutral Cu-clusters. The influence of electron affinity is readily observable in the oscillating flux distribution of negatively charged clusters; this distribution, in turn, shows resemblance to that obtained in sputtering [2.347]; it is expected that all monovalent metals, the alkali and inert metals in particular, will show a similar behavior.

It is interesting to note that the oscillations observed in the field-ionized flux from liquid-metal tips are similar to those registered for the charged sputtered flux, i.e., to those measured in SIMS [2.345]. This similarity between methods, where the charged particles originate in the very surface, is another demonstration of the sensitivity of surface ionization to the ionization potential; (electron beam) post-ionization well above the threshold is apparently much less selective – a fact that is the origin of the advantage of SNMS (sputtered neutral mass spectrometry) over SIMS in surface analysis.

---

[8] Electronic configurations of the atoms: Cr: $(Ar)3d^5 4s^1$, Cu: $(Ar)3d^{10} 4s^1$, Ag: $(Kr)4d^{10} 5s^1$.

[9] Ni: $(Ar)3d^8 4s^2$.

Joyes [2.286] and *Leleyter* and *Joyes* [2.301, 302] have made theoretical attempts to apply quantum-chemical methods, such as the tight-binding (Hückel) approximation, to calculate molecular stability, ionization potentials for Cu_n^+, and electron affinities for Ag_n^-. For the low-Z elements Li, Be, and C, the CNDO method was also applied. The qualitative agreement with experimental results is good, but the treatment has to be considerably refined to give quantitative information for general systems. Such calculations are being carried out [2.303–305].

Now there is great theoretical interest in a more quantitative understanding of the "magic number" clusters: in addition to the aforementioned odd–even abundance alternations, clusters of Na, Cu, Ag, and Au that contain 2, 8, 18, 20, 34 or 40 *s*-electrons show pronounced intensity and stability [2.291, 294, 306]. This effect is explained by a one-electron shell model in which the free *s*-electrons are bound in a spherically symmetric potential well.

As regards stability, it thus transpires from present investigations that for clusters of group-IV elements the *atomic configuration* plays the decisive role, while for metal clusters it is the *electronic configuration*. For these clusters, the importance of atomic configuration not only seems to be minor, rather we should visualize metal clusters generated by sputtering as highly heated agglomerates. Vibrational and rotational temperatures between 1000 K and 1500 K have been measured for alkali metal dimers by laser spectroscopy; detailed decay studies have revealed that clusters lose excess energy by evaporating single atoms. These processes seem to come to an end about 10^{-4} s after emission. Obviously, mass distributions measured at this time are governed by stability criteria rather than by collisional interactions at the emission event.

Apart from stability considerations, there should be no difference in principle between ejection of electrically charged and neutral clusters; i.e., the ejection mechanism is presumed to be the same, while the charge state is determined when the particle escapes from the surface. This view agrees with the few early investigations of sputtered neutral clusters: *Woodyard* [2.58] found up to 20% Cu atoms in dimers but did not detect any trimers [Ar^+ 2 keV → Cu (100)]; similarly *Gerhard* and *Oechsner* [2.59] found an intensity ratio of $Cu_2/Cu_1 \approx 10^{-1}$, while Cu_3/Cu_1 was two orders of magnitude lower. Many sputtered atoms are thus found to be bound in clusters. This emerged also from preliminary measurements of *Giber* and *Hofer* [2.308].

How large then is the fraction of atoms sputtered in a bound state? This key question in sputtering still cannot be answered conclusively. Until very recently, no information was available on neutral cluster distributions. All knowledge of the fraction of atoms ejected in a bound state had to be taken from charged clusters. This inference may, however, be misleading, as one realizes after inspection of Fig. 2.33: the number of atoms bound in charged clusters may well amount to 50% of the atomic ion intensity, whereas neutral particles – the vast majority in the sputtered flux – show fewer clusters and, thus, a smaller bound-atom fraction. We have no knowledge, however, of the genuine emission distribution, neither for charged nor for neutral clusters. The frequency distribution of clusters after ejection and before modification towards its stability spectrum via elimination of less stable (metastable) agglomerates has, as yet, eluded registration.

b) Energy and Angular Distributions

Since cluster stability thus exerts such a decisive influence on the abundance distribution of sputtered particles, information on the collision cascade from such measurements is limited. Angular and energy distributions of the different species are therefore more informative than mass distributions. Especially the finding that both the most probable energy and the width of the energy spectrum decrease with increasing cluster size [2.51, 54, 56, 216] indicates the restrictions in relative energy and momentum at the ejection process (Fig. 2.34). The (probably nonbinary) collision must be sufficiently soft in order not to separate the cluster constituents at or after ejection.

The importance of the relative momentum of cluster components is further emphasized by angular-distribution studies that have found an anisotropic component for the clusters as well. *Staudenmaier* [2.55] reported preferential ejection of W_2^+ and W_3^+ clusters along $\langle 111 \rangle$ of a tungsten (110) single crystal when bombarded with 150 keV Ar$^+$ ions. Unfortunately, in these measurements the ejection angle could not be varied without changing the angle of incidence, too. The results are therefore slightly falsified by varying fractions of channeled projectiles (and thus deposited energy), but the effect has been confirmed by

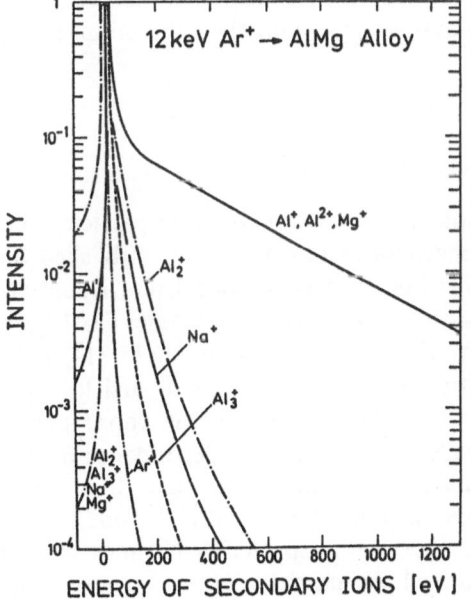

Fig. 2.34. Energy distribution of atomic and cluster ions sputtered from a MgAl alloy. The curves are not corrected for changes in spectrometer acceptance with particle energy. In reality the energy distributions will drop off less steeply, but their shift toward lower energies with increasing cluster size is well demonstrated by these early measurements. From [2.56]

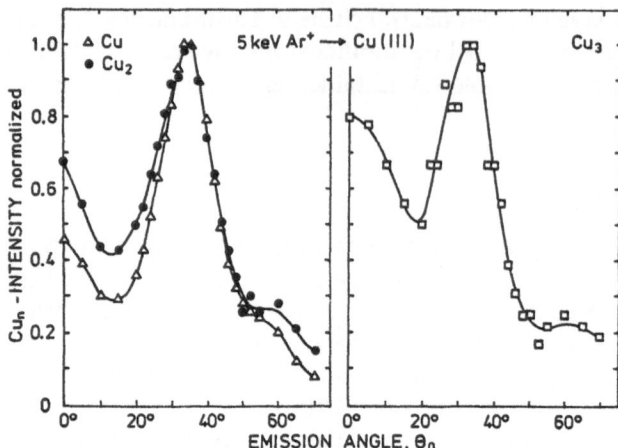

Fig. 2.35. Anisotropic $\langle 110 \rangle$ emission of neutral Cu atoms, dimers, and trimers from a Cu(111) surface when rotated around an axis perpendicular to [110] and [111]. Particles of 2 eV emission energy were selected. From [2.62]

Bernheim on aluminium single crystals [2.309], and by *Holland* et al. for Ni_2^+ ejection along $\langle 100 \rangle$, [2.214].

With the aid of the spectrometer shown in Fig. 2.32, the anisotropic emission of *neutral* clusters has also been proved [2.62, 311] (Fig. 2.35). As in [2.55], the target had to be rotated under the ion beam to scan over the $\langle 110 \rangle$ emission. In a strict sense, then, these results too are not angular distributions. Under the bombardment conditions, used by *Hofer* and *Gnaser* [2.62], however, the incidence-angle dependence of the sputtering yield is well known [2.310]. From these data the authors could show that there is no major distortion of the $\langle 110 \rangle$ distribution owing to varying channeled fractions. Evidently, preferred ejection along close-packed lattice directions also occurs for clusters. This is understandable in view of the collimated momentum transport along these directions in the single crystal.

2.4.3 Mechanisms of Cluster Emission

Cluster emission is one of the least understood fields in energetic-particle/solid interactions. Even the basic ejection mechanism has not been established on a generally accepted basis. Before *Honig*'s first observation of clusters in sputtering [2.49], clusters were found in sparks between graphite electrodes [2.313], in sublimation of carbon, and in evaporation from liquid germanium [2.314, 315]. For germanium the researchers explicitly stated that these bound particles must have been emitted from the surface rather than formed in the vapor phase; agglomeration in the vapor would result in a far stronger dependence on the vapor pressure than observed. By substituting "sputtering yield" for "vapor pressure", similar considerations can be made for collisional emission.

For one-and-a-half decades after its discovery in sputtering, cluster emission was perceived as the ejection of the whole particle. Now, however, the frequently discussed model assumes cluster formation as statistical agglomeration processes of independently sputtered atoms [2.59, 287, 316]. But even the promoters of this agglomeration mechanism seem to agree that the comparatively high intensity of higher-order clusters ($n \geqslant 3$) cannot be explained by pure reaction kinetics. With reference to the charged Cu clusters in Fig. 2.33, for instance, the statistical probability of formation of agglomerates containing up to 5 Cu atoms would be much smaller than the measured 10^{-1}; there must thus be a correlation of the atoms' momenta in space and time.

a) Direct Emission of Clusters

In the direct emission model, clusters are assumed to be ejected "as such" from the surface. This perception is not restricted to collisional emission. With appropriate modifications, it applies also to sublimation in electric sparks, plain sublimation (C, Ge), laser-generated plasmas in metals, and pulsed-laser stimulated field desorption [2.284, 300, 315, 317].

Before *Staudenmaier*'s work [2.54, 55] little attention was paid to the actual collision mechanisms that would accomplish the emission of bound particles. He first showed that cluster emission is not related to surface topography – at least not as far as scanning electron microscopy can resolve surface structures. Within this limit, cluster emission is not the direct emission of preformed, loosely bound surface aggregates. *Staudenmaier* further showed that the injected gas particles would not cause cluster emission by way of bubble rupture: there is no relevant difference between metal ion and inert gas ion bombardment [2.54]. He also discussed the stringent conditions for collisions that may lead to bound-particle emission. To avoid separating the knockon atoms, the time interval for momentum transfer must be on the order of 10^{-14} s or smaller. In addition, the difference in the momenta must be very small with respect to amount and direction. With these restrictions, it is suggestive indeed to perceive the whole process as a plural or collective interaction event [2.61]. A strong indication of momentum alignment's importance is, of course, the appearance of preferred cluster emission along close-packed lattice directions (Fig. 2.35).

Furthermore, these conditions show that cluster emission is an effect of *individual* collision cascades. Especially in this field more knowledge of collision cascade fluctuations is badly needed.

For the special case of diatomic molecules and clusters (dimers), rather detailed calculations exist on the dynamics of and after emission [2.320–324].

b) Agglomeration After Ejection

An alternative to direct cluster emission is the assumption that individually ejected atoms associate above the surface [2.59, 316, 318, 319]. This model, referred to as the atomic (re)combination model [2.318], or the statistical cluster-formation model, has mostly been applied to dimers. *Oechsner* and coworkers tend to ascribe the whole dimer yield to (re)combination above the surface, while *Bitenskii* and *Parilis* [2.320] find 10% – 30% due to a specific direct emission process [2.322]. The latter authors note in addition that "these mechanisms . . . are not the primary mechanisms for the formation of clusters containing more particles", rather they "are governed by collective effects of the transport".

The problem with the agglomeration-in-vacuum model is basically that the sputtered flux is a directed one. Contrary to a vapor in equilibrium with its containment, sputtered particles are ejected into the half-space above the target's surface with no possibility of reencounter; they may interact with each other only for a short time and distance after ejection. The probability for larger clusters ($n > 2$) of forming by purely statistical combinations is far too small compared with experimental evidence (Fig. 2.33).

Kelly has revisited the problems of multimer emission in the light of different phases of collision-cascade evolution [2.321]. Within this scheme, cluster emission is expected to occur primarily in the "slow collisional" ($10^{-14} - 10^{-12}$ s after the projectile's impact) and "prompt thermal" (up to 10^{-10} s) ejection phases. Furthermore, *Kelly* states that: "The use of the term recombination is . . . not to be taken too literally. The atoms which appear as a molecule were initially bonded in the solid, even if not as nearest neighbours, and never went through a state of being non-interacting. Had such a stage occurred, a subsequent linking up would be impossible." Most workers in the field of cluster emission would agree. There is not much physical justification for a (re)combination-in-vaccum model. Consequently, *Haring* et al. [2.326] have recently modified the model of *Können* et al. [2.316] in that the cluster atoms are assumed to be bound already *before* the passage through the surface, i.e., before the ejection.

c) **Discussion and Summary**

It has been argued that the agglomeration-in-vacuum model is supported by the higher power dependence of the multimer yield on the total sputtering yield Y. *Oechsner* and *Gerhard* [2.59, 318, 319], for instance, observed the partial dimer yield to be proportional to Y^2 with a variety of metals. *Winograd* et al. [2.215] found in computer simulations some agreement of the dimer yield behavior with second-order rate processes. *Wittmaack* [2.289] raised objections to the experimental basis of this argument. *Gnaser* and *Hofer* [2.311] found for Zn the lowest cluster yields among the 3d transition metals, even though Zn has by far the highest total sputtering yield. This result would not agree with either cluster-emission model, if it were not resolved by the dominance of stability conditions (Sect. 2.4.2a).

An understanding of cluster emission must take the statistical fluctuations of the cascades into account *Sigmund* [2.68I], p. 121, and in [2.68VI], p. 1; *Eckstein* in 2.325e, p. 33. Average quantities – such as the yield Y – have little significance when the variance of the distribution is of the same order. Both experimental and theoretical information now available indicates that an appreciable fraction of the yield is due to a few collision cascades that develop near the surface. Within 10^{-12} s a large number of particles are thus enabled to overcome the binding to the solid. The effect of such shallow cascades is visible in the transmission electron microscope as well as in field-ion microscopy. We can plausibly assume that the material removed from the observed depleted zones or even microcraters has been emitted with a high degree of momentum correlation [2.61–68, 325–336].

An analytical treatment of cluster emission appears to be very complicated because, apart from tackling the fluctuation problem, we have to trace the development of the collision cascade down to very low energies, where attractive forces as well as plural interactions have to be taken into account. In addition, nonlinear effects come more into play. Thus computer simulation appears more likely to deliver information on those collision processes that lead to cluster ejection. In fact, molecular dynamics simulation studies of 600 eV Ar bombardment of Cu single crystals give rather detailed answers to the basic emission processes [2.215, 331, 333]. Although these bombardment conditions are hardly representative of random-collision cascades, several conclusions are important: most dimer constituents have not been nearest neighbors in the solid; clusters are formed in the interaction region ($\lesssim 4$ Å) of the surface. With increasing cluster size the cluster components represent more and more the combination of atoms in the solid. Moreover, agglomeration does not take place in the vacuum but in and near the surface, where, in interactions with neighboring atoms, the relative kinetic energy E_{kin} can be reduced. Finally, multimer emission happens at the very-low-energy end of the collision sequences or, in terms of the lifetime of the cascade: multimer emission is a "late" event.

Closer inspection of the agglomeration-in-vacuum model shows, furthermore, that fulfillment of the condition for cluster binding, that is,

$$E_{\text{rel kin}} < \sum_{i,k} V_{ik}(r_{ik}) , \tag{2.50}$$

where $V_{i,k}$ is the mutual interaction potential (e.g., of the Morse type) of the cluster components (i, k), essentially rules out cluster formation from atoms that have not initially been nearest or next-nearest neighbors. This is because at larger separation, $r_{ik} > a_0$, $V(r_{ik})$ is close to zero, while the relative kinetic energy E_{rel}, the positive term in (2.50), must increase in order to accomplish particle attraction well within the surface interaction region. The distinction between the two cluster formation models is then more a semantic one.

To summarize, cluster emission appears to be an event at the very low-energy stage of the cascade, where nonbinary interactions favor simultaneous ($\Delta t \leqslant 10^{-13}$ s) knockon processes near and in the surface, and where soft collisions one egress from the surface reduce the relative kinetic energy of the particles to be bound in the cluster/molecule. These cluster components need not stem from nearest-neighbor sites in the lattice, although this becomes a more stringent condition with increasing cluster size. In all probability, nonlinear effects in high-density cascades near the surface will promote cluster emission.

2.4.4 Applications

The high cluster-ejection probability in sputtering can be used in applications and fundamental studies. Since clusters contain the composition characteristics of the microregion in the solid from which they originate, mass analysis of clusters can offer quantitative information on the stoichiometry and homogeneity of the target [2.290]. Clusters can be postionized more efficiently than sputtered atoms, making this method particularly appealing. Charged clusters can, furthermore, provide particles that are otherwise difficult to generate. *Thum* and *Hofer* [2.269], for instance, used sputtered clusters of more than 20 different metal elements and alloys to study secondary electron emission by composite particles; such heavy clusters as V_{14}^+, Nb_{11}^+, and $V_3Nb_4^+$ were obtained by bombarding the respective targets with 8 keV Ar^+ and Xe^+ ions.

Clusters may eventually be studied for their own sake, because they constitute the gradual transition from atomic to molecular to solid states [2.286, 303]. To what extent *sputtered* clusters can be used for this purpose will depend on the efficiency of dissipation of internal energy. Evaporative cooling [2.294, 304] appears to be a promising mechanism, but for now, clusters formed by agglomeration in supersonic nozzles are a closer approach to the ground state than collisional generation from the solid state [2.303].

Cluster research is presently greatly stimulated by the potential use of clusters as catalysts [2.303, 342].

2.4.5 Microparticle and "Chunk" Emission

An extreme case of nonlinearity and collective emission has been reported by *Merkle* and *Jäger* [2.63, 64]. In electronmicroscopy investigations of defect structures created by heavy-ion bombarded gold, *Merkle* observed (in addition to collapsed depleted zones in the bulk) surface craters of about 50 Å diameter. About 10^3 atoms have to be removed to form such a crater. The number of these craters increases sharply with energy and is particularly pronounced for molecular-ion (Bi_2^+) bombardment (Fig. 2.36). At an energy of 100 keV the material removed from these craters alone – there are presumably many more of submicroscopic size – reaches the yield value calculated from linear cascades. This is the same energy range where *Bay* et al. [2.334] have measured enhanced Au self-sputtering yields, which they have attributed to collision spikes. *Merkle* follows this explanation but points out that it is not so much the average (linear) cascade size that determines high-energy ($\geqslant 100$ keV) sputtering, but the individual *subcascade*, as long as it is generated near enough to the surface. These subcascades are the zones that constitute the spike, and their distribution in space must

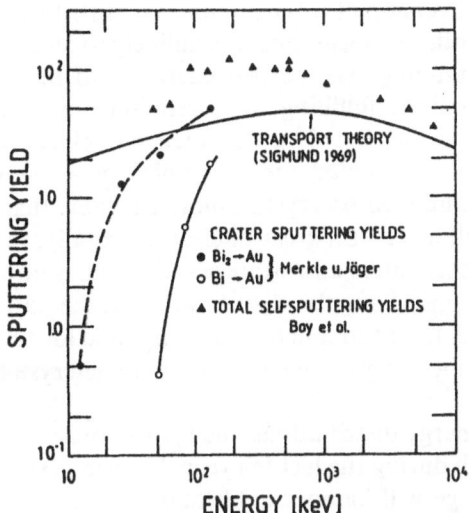

Fig. 2.36. Energy dependence of the total (▲) and partial (○, ●) sputtering yield of polycrystalline gold due to Au$^+$, Bi$^+$, and Bi$_2^+$ bombardment, respectively. From [2.330]

be known for a more accurate calculation of sputtering yields. This problem is thus again coupled to the fluctuations in energy deposition in individual cascades.

Microparticle emission had already been observed in the first erosion investigations of cathodes in glow discharges. *Plücker* [2.335] reported luminous objects leaving the cathode without any influence of electric or magnetic fields on their trajectory. *Hittorf* [2.336] explained them in terms of cone formation: after prolonged bombardment of Al cathodes he observed needle-like or hair-like structures which, owing to their reduced thermal contact to the bulk of the cathode, start melting at their tips and thereby emit macroscopic particles. Such an explanation can also be found in recent work on exfoliation and blistering, [*Scherzer* in 2.69b].

Kaminsky and coworkers found that micrometer-sized particles (called chunks) containing up to 10^{10} atoms were emitted in neutron sputtering, see [2.69b]. Chunks caused the sputtering yields to exceed the physically reasonable values ($\lesssim 10^{-5}$ atoms/neutron) by orders of magnitude. This high erosion triggered intense research in several laboratories. The final consensus seems to be that chunk emission in neutron sputtering is due to loosely bound particles on "technical surfaces" [2.65b].

2.5 Summary and Outlook

After more than 30 years of research on angular distributions from single crystals we still do not fully understand the phenomenon. Preferred ejection into assisted-focusing directions, for instance, appears even more uncertain today than it did ten years ago, when the first version of this overview was distributed [2.337]. I said then that the standard collector technique was unlikely to lead to a major improvement of our understanding. We needed energy- and mass-resolved angular distributions. A technique fulfilling these requirements has since been developed [2.338]. Although the first data and their modeling by computer simulation [2.339, 340] have not modified our notion of single-crystal sputtering, further results – preferably obtained on crystals more tractable than the Rh(111) target used in [2.339] – will meet with great interest. Generally, we can say that with a new generation of computers and strategies, the time is now ripe for *statistically acceptable* simulations of angular distributions. In the past, physically sound trajectory calculations took too much computing time for the high statistical accuracy required for meaningful angle-resolved single-crystal sputtering simulation [2.344].

In research on sputtered-particle energy distributions, the laser fluorescence technique has achieved a breakthrough during the last ten years [*Bay* in 2.312]. The comparatively small dynamic range will hamper application of this technique in investigations of specific single-crystal emission processes, but in research on electronic excitation states of sputtered species the laser technique has just opened up a wide, unknown field. I should, furthermore, like to point out again the still unsettled question of the existence of a partial thermal component in the energy spectra and its dependence on the lattice temperature. This is not a merely academic issue, as it may provide further insight into the nonlinear effects regime.

The fraction of clusters in the flux of sputtered particles is still a largely unknown quantity, and fragmentation processes, intrinsic and experimental ones, will probably keep it in that state for a while. With this uncertainty, we can only try not to underestimate the process of bound-particle ejection and not to take measured mass distributions directly as emission distributions. Here again, the expectations placed on computer simulations are high, in particular for identifying the collisions leading to ejection. Accurate modeling of the atomic interaction forces and their variation when the atoms pass through the surface and become entities on their own is a precondition to successful computer calculations of the mass distribution of sputtered particles.

Acknowledgements. The work on this review article was begun more than ten years before its publication. For stimulating comments on the 1979 version [2.337], I am very grateful to U. Littmark, M.W. Thompson, H.E. Roosendaal, and V.E. Yurasova. Because of postponements beyond the author's control, it became necessary to update the manuscript twice. I very much appreciate valuable discussions of these versions with H.L. Bay, R. Behrisch, and M.M. Ferguson. I would also like to thank B. Goebbels and G. Herstix for their help in carrying out the innumerable modifications to the article.

My deepest thanks go to my wife Ingeborg for her understanding and patience during the preparation of this article. She died suddenly in the course of the last modifications, leaving me with a deep regret for having spent so many evenings and weekends on this work.

References

2.1   H.H. Andersen, H.L. Bay: In [2.69a], pp. 145–218
2.2   J. Stark: Z. Elektrochem. **14**, 752 (1908); ibid. **15**, 509 (1909)
2.3   R. Seeliger, K. Sommermeyer: Z. Phys. **93**, 692 (1935)
2.4   A.v. Hippel: Ann. Physik, **81**, 1043 (1926); ibid. **86**, 1006 (1928)
2.5   G.K. Wehner: J. Appl. Phys. **26**, 1056 (1955); Phys. Rev. **102**, 690 (1956)
2.6   V.E. Yurasova: Zh. Tekh. Fiz. **28**, 1966 (1958) [Engl. transl.: Sov. Phys. – Tech. Phys. **3**, 1806 (1958)]
2.7   V.E. Yurasova, N.V. Pleshivtsev, I.V. Orfanov: Zh. Eksp. Teor. Fiz. **37**, 966 (1959) [Engl. transl.: Sov. Phys. – JETP **37**, 689 (1960)]
2.8   V.E. Yurasova, I.G. Sirotenko: Zh. Eksp. Teor. Fiz. **41**, 1359 (1961) [Engl. transl.: Sov. Phys. – JETP **14**, 968 (1962)]
2.9   M.W. Thompson: Philos. Mag. **4**, 139 (1959)
2.10   R.S. Nelson, M.W. Thompson: Proc. R. Soc. London A **259**, 458 (1961)
2.11   V.A. Molchanov, V.G. Tel'kovskii, V.M. Chicherov: Dokl. Akad. Nauk. SSSR **138**, 824 (1961) [Engl. transl.: Sov. Phys. – Dokl. **6**, 486 (1961)]
2.12   B. Perovic: Bull. Inst. Nucl. Sci. "Boris Kidrich" **11**, 37 (1961)
2.13   M.W. Thompson, R.S. Nelson: Philos. Mag. **7**, 2015 (1962)
2.14   G.E. Chapman, B.W. Farmery, M.W. Thompson, I.H. Wilson: Radiat. Eff. **13**, 121 (1972)
2.15   R.S. Nelson: Radiat. Eff. **7**, 263 (1971)
2.16   R.H. Silsbee: J. Appl. Phys. **28**, 1246 (1957)
2.17   G. Leibfried: J. Appl. Phys. **30**, 1388 (1959); ibid. **31**, 117 (1960)
2.18   Chr. Lehmann, G. Leibfried: Z. Phys. **162**, 203 (1961)
2.19   R.S. Nelson, M.W. Thompson, H. Montgomery: Philos. Mag. **7**, 1385 (1962)
2.20   Chr. Lehmann: Nucl. Instrum. Methods **38**, 263 (1965)
2.21   R. Frére: Phys. Status Solidi **3**, 1252, 1441, 1453 (1963)
2.22   N.Th. Olson, H.P. Smith: AIAA J. **4**, 916 (1966); Phys. Rev. **157**, 241 (1967) T.B. Higgins, N.Th. Olson, H.P. Smith: J. Appl. Phys. **39**, 4849 (1968)
2.23   R.R. Hasiguti, R. Hanada, S. Yamaguchi: J. Phys. Soc. Jpn. **18**, Suppl. III, 164 (1963)
2.24   D.E. Harrison Jr., J.P. Johnson III, N.S. Levy: Appl. Phys. Lett. **8**, 33 (1966)
2.25   M.T. Robinson, A.L. Southern: J. Appl. Phys. **39**, 3463 (1968)
2.26   Ch. Lehmann, P. Sigmund: Phys. Status Solidi **16**, 507 (1966)
2.27   W.O. Hofer: Ref. 65, p. 263 W.O. Hofer: Thesis Universität München (1972), also Report GSF **P44** (1972)
2.28   W. Brandt, R. Laubert: Nucl. Instr. Meth. **47**, 201 (1967)
2.29   M.W. Thompson: *Defects and Radiation Damage in Metals*, (Cambridge Univ. Press, Cambridge 1969)
2.30   M.W. Thompson: Phil. Mag. **18**, 377 (1968)
2.31   P. Sigmund: Phys. Rev. **184**, 383 (1969)
2.32   M.T. Robinson: Phil. Mag. **12**, 145 (1965) and **17**, 639 (1968)
2.33   W. Szymczak, K. Wittmaack: in [2.66], p. 236
2.34   W. Szymczak, K. Wittmaack: Nucl. Instrum. Methods **170**, 341 (1980)
2.35   W. Szymczak: Thesis, Universität München (1985), also Report GSF 18/85
2.36   J. Linders, H. Niedrig, M. Sternberg: Nucl. Instrum. Methods Phys. Res. B **2**, 649 (1984)
2.37   R.V. Stuart, G.K. Wehner: J. Appl. Phys. **35**, 1819 (1964)
2.38   H. Beuscher, K. Kopitzki: Z. Phys. **184**, 382 (1965); both Kopitzki and Wehner [2.37] have short communications in Proc. 6th Int. Conf. Ionis. Phen. Gases, Vol. 2, Paris (1963), ed. by P. Hubert, E. Crémieu-Alcan
2.39   M.W. Thompson, B.W. Farmery, P.A. Newson: Philos. Mag. **18**, 361 (1968)
2.40   B.W. Farmery, M.W. Thompson: Philos. Mag. **18**, 415 (1968)

2.41 H. Oechsner: Z. Phys. **238**, 433 (1970) and in [2.42], p. 31
2.42 Proc. 8th Int. Conf. Phenomena in Ionized Gases, Vienna (1967), Contributed papers (IAEA Vienna, Austria 1967)
2.43 I.H. Reid, B.W. Farmery, M.W. Thompson: In Proc. 7th Int. Conf. Atomic Collisions in Solids, Vol. 2, ed. by J.V. Bulgakov, A.F. Tulinov (Moscow State University Publishing House, Moscow 1980) p. 8, also published in Radiat. Eff. **46**, 163 (1980)
2.44 I.H. Reid, M.W. Thompson, B.W. Farmery: Philos. Mag. A **42**, 151 (1980)
2.45 I.H. Reid, B.W. Farmery, M.W. Thompson: In [2.66], p. 280
2.46 P. Jung: Phys. Rev. B **23**, 664 (1981), and references therein
2.47 M.L. Jenkins, M. Wilkens: Philos. Mag. **34**, 1155 (1976)
2.48 A. Elbern, E. Hintz, B. Schweer: J. Nucl. Mater. **76 & 77**, 143 (1978)
2.49 R.E. Honig: J. Appl. Phys. **29**, 549 (1958)
2.50 V.E. Krohn: J. Appl. Phys. **33**, 3523 (1962)
2.51 Z. Jurela, B. Perovic: Can. J. Phys. **46**, 773 (1968) and in [2.65], p. 161
2.52 G. Blaise, G. Slodzian: C.R. Acad. Sci. B **266**, 1525 (1968)
2.53 G. Hortig, M. Müller: Z. Phys. **221**, 119 (1969)
2.54 G. Staudenmaier: Radiat. Eff. **13**, 87 (1972)
2.55 G. Staudenmaier: In [2.65], p. 181; Thesis, Universität München (1973)
2.56 R.F.K. Herzog, W.P. Poschenrieder, F.G. Satkiewicz: In [2.65], p. 173; 15th Ann. Conf. Mass Spectr. and Allied Topics, Denver (1967)
2.57 J.R. Woodyard, C.B. Cooper: J. Appl. Phys. **35**, 1107 (1964);
      C.B. Cooper, R.J. Woodyard: Phys. Lett. **79A**, 124 (1980)
2.58 J.R. Woodyard, 15th Ann. Conf. Mass Spectr. Allied Topics, Denver (1967), p. 79
2.59 W. Gerhard, H. Oechsner: Z. Phys. B **22**, 41 (1975)
2.60 H. Oechsner, E. Stumpe: Appl. Phys. **14**, 43 (1977)
2.61 W.O. Hofer: Nucl. Instrum. Methods **170**, 275 (1980)
2.62 W.O. Hofer, H. Gnaser: Ref. [2.312], p. 605
2.63 K.L. Merkle: 35th Ann. Proc. Electron Microscopy Soc. Amer., Boston, MA (1977), ed. by G.W. Bailey, p. 36
2.64 K.L. Merkle, W. Jäger: Philos. Mag. A **44**, 741 (1981)
2.65 R. Behrisch, W. Heiland, W. Poschenrieder, P. Staib, H. Verbeek (eds.): *Ion Surface Interaction, Sputtering and Related Phenomena* (Gordon and Breach, London 1973); also published as volume 18/19 of Radiat. Eff.
2.66 P. Varga, G. Betz, F.P. Viehböck (eds.): *Symposium on Sputtering*, Perchtoldsdorf/Vienna, 1980 (IAP, Technische Universität Wien, Austria 1980)
2.67 G. Kiriakidis, G. Carter, J.L. Whitton (eds.): Proc. NATO Advanced Study Institute on Erosion and Growth of Solids Stimulated by Atom and Ion Beams, Heraklion, Crete (1985), (Nijhoff, Dordrecht, Holland 1986)
2.68 Workshop Series on Inelastic Ion-Surface Collisions:
      I, ed. by N.H. Tolk et al., (Academic, New York 1977)
      II, ed. by R. Kelly, Surf. Sci. **90**, 205–687 (1979)
      III, Inelastic Particle–Surface Collisions, ed. by E. Taglauer, W. Heiland, Springer Ser. Chem. Phys., Vol. 17 (Springer, Berlin, Heidelberg 1981)
      IV, ed. by P. Sigmund, Phys. Scr. (1983) T6 (1983)
      V, ed. by P. Williams, Nucl. Instrum. Methods B **14**, (1986)
      VI, D.M. Gruen et al.: Nucl. Instrum. Methods B **27** (1987)
2.69 R. Behrisch: *Festkörperzerstäubung durch Ionenbeschuß*, Ergeb. Exakt. Naturwiss. **35**, 295 (1964)
2.69a R. Behrisch (ed.): *Sputtering by Particle Bombardment I*, Topics Appl. Phys., Vol. 47 (Springer, Berlin, Heidelberg, 1981)
2.69b R. Behrisch (ed.): *Sputtering by Particle Bombardment II*, Topics Appl. Phys. Vol. 52 (1983)
2.70 M. Kaminsky: *Atomic and Ionic Impact Phenomena on Metal Surfaces* (Springer, Berlin, Heidelberg, 1965)
2.71 G. Leibfried: *Bestrahlungseffekte in Festkörpern* (Teubner, Stuttgart 1965)

2.72 G. Carter, J.S. Colligon: *Ion Bombardment of Solids* (Heinemann, London 1968)
2.73 R.S. Nelson: *The Observation of Atomic Collisions in Crystalline Solids*, Defects in Crystalline Solids, Vol. 1 (North-Holland, Amsterdam 1968)
2.74 Ch. Lehmann: *Interaction of Radiation with Solids and Elementary Defect Production*, Defects in Crystalline Solids, Vol. 10 (North-Holland, Amsterdam 1977)
2.75 C.H. Weijsenfeld: Thesis, Rijksuniversiteit Utrecht, Holland (1966)
2.76 J.N. Smith, Jr.: J. Nucl. Mater. **78**, 117 (1978)
2.77 A. van Veen: Thesis, Rijksuniversiteit Utrecht, Holland (1979)
2.78 A. van Veen, J.M. Fluit: Nucl. Instrum. Methods **170**, 341 (1980)
2.79 A. van Veen, A.G.J. de Wit, J.M. Fluit: Ref. 2.66, p. 226
2.80 K. Besocke, S. Berger, W.O. Hofer, U. Littmark: Radiat. Eff. **66**, 35 (1982)
2.81 S. Berger, K. Besocke, W.O. Hofer: Symp. Surface Science, ed. by P. Braun et al., Obertraun, Austria (1983) p. 111
2.82 J.S. Colligon, M.H. Patel: Radiat. Eff. **32**, 193 (1977)
2.83 J.F. Ziegler, J.J. Cuomo, J. Roth: Appl. Phys. Lett. **30**, 268 (1977)
2.84 D.M. Mattox, D.J. Sharp: J. Nucl. Mater. **80**, 115 (1979)
2.85 U. Littmark, W.O. Hofer: J. Mater. Sci. **13**, 2577 (1978)
2.86 J.L. Whitton, W.O. Hofer, U. Littmark, E. Braun, B. Emmoth: Appl. Phys. Lett. **36**, 531 (1980), and Ref. 2.66, p. 632
2.87 V. Alexander, H.-J. Lippold, H. Niedrig: Ref. 2.66 p. 622
2.88 G.K. Wehner: Phys. Rev. **102**, 690 (1955)
2.89 G.S. Anderson, G.K. Wehner: J. Appl. Phys. **31**, 2305 (1960); ibid. **33**, 2016 (1962)
2.90 B. Perovic: Proc. 5th Int. Conf. Ion. Phen. Gases, Vol. 2, ed. by H. Maecker (North-Holland, Amsterdam 1962) p. 1172
2.91 F. Schulz: Thesis, Technische Universität München (1967)
2.92 F. Schulz, R. Sizmann: Philos. Mag. **18**, 269 (1968)
2.93 R. Becerra-Acevedo: Thesis, Technische Universität München (1985); see also Ref. [2.137]
2.94 R.L. Cunningham, J.Ng-Yelim: In *Atomic Collision Phenomena in Solids*, ed. by D.W. Palmer, M.W. Thompson, P.D. Townsend (North-Holland, Amsterdam 1970) p. 290
2.95 M.T. Robinson, A.L. Southern: J. Appl. Phys. **38**, 2969 (1967)
2.96 L. Francken, E. Bonnijns: Phys. Lett. **24A**, 764 (1967)
2.97 R.S. Nelson, M.W. Thompson: Philos. Mag. **7**, 1425 (1962)
2.98 W.O. Hofer: Radiat. Eff. **21**, 141 (1974)
2.99 Volmer's theory of condensation and its various improvements is described in J.P. Hirth, G.M. Pound: *Condensation and Evaporation* (Pergamon, Oxford 1963)
2.100 V.M. Agranovich, O.I. Kapusta, S.Ya. Lebedev, L.P. Semonov: Fiz. Tverd. Tela **11**, 2816 (1969) [Engl. transl.: Sov. Phys.–Solid State **11**, 2280 (1970)
2.101 G. Betz, R. Dobrozemsky, F.P. Viehböck: Ned. Tijdschr. Vacuumtech. **8**, 203 (1970)
2.102 L.T. Chadderton, A. Johansen, L. Sarholt-Kristensen, S. Steenstrup, T. Wohlenberg: Radiat. Eff. **13**, 75 (1972)
2.103 S.Ya. Lebedev, G.V. Lysova: Radiat. Eff. **35**, 109 (1978)
2.104 W.O. Hofer: Thin Solid Films **29**, 223 (1975)
2.105 D. Cherns, M.W. Finnis, M.D. Matthews: Philos. Mag. **35**, 693 (1977)
2.106 K. Rödelsperger, A. Scharmann: Z. Phys. B **28**, 37 (1977); Nucl. Instrum. Methods **132**, 355 (1976)
2.107 G. Bräuer, D. Hasselkamp, W. Krüger, A. Scharmann: Nucl. Instrum. Methods Phys. Res. B **12**, 458 (1985)
2.108 W. Krüger, A. Scharmann, H. Afridi, G. Bräuer: Nucl. Instrum. Methods **168**, 411 (1980)
2.109 W.O. Hofer, H.L. Bay, P.J. Martin: J. Nucl. Mater. **76** & **77**, 156 (1978)
2.110 P. Erlenwein, H. Niedrig: Beitr. elektromikr. Direktabbildung Oberfl. **10**, 247 (1979)
2.111 H.L. Bay, J. Bohdansky, W.O. Hofer, J. Roth: Appl. Phys. **21**, 327 (1980)
2.112 V. Orlinov, G. Mladenov, I. Petrov, M. Braun, B. Emmoth: Vacuum **32**, 747 (1982)
2.113 R.G. Musket, H.P. Smith, Jr.: J. Appl. Phys. **39**, 3579 (1968)
2.114 G.E. Chapman, J.C. Kelly: Aust. J. Phys. **20**, 283 (1967)

2.115 W.O. Hofer, R. Sizmann: In *Atomic Collision Phenomena in Solids*, ed. by D.W. Palmer, M.W. Thompson, P.D. Townsend (North-Holland, Amsterdam 1970) p. 298
2.116 H. Tsuge, S. Esho: J. Appl. Phys. **52**, 4391 (1981)
2.117 H. Patterson, D.H. Tomlin: Proc. R. Soc. London **265**, 474 (1961)
2.118 E. Forman, F.P. Viehböck, H. Wotke: Can. J. Phys. **46**, 753 (1968)
2.119 G. Betz, R. Dobrozemsky, F.P. Viehböck: Int. J. Mass Spectrom. Ion Phys. **6**, 451 (1971)
2.120 S.T. Kang, R. Shimizu, T. Okutani: Jpn. J. Appl. Phys. **18**, 1717 (1979)
2.121 J. Roth, J. Bohdansky, W. Eckstein: Nucl. Instrum. Methods Phys. Res. **218**, 751 (1983)
2.122 H.H. Andersen, B. Stenum, T. Sørensen, H.J. Whitlow: Nucl. Instrum. Methods Phys. Res. B **6**, 459 (1985)
2.123 H.H. Andersen, B. Stenum, T. Sørensen, H.J. Whitlow: Nucl. Instrum. Methods Phys. Res. B **2**, 601, 623 (1984)
2.124 T. Okutani, M. Shikata, S. Ishimura, R. Shimizu: J. Appl. Phys. **51**, 2884 (1980)
2.125 H.J. Kang, Y. Matsuda, R. Shimizu: Surf. Sci. **127**, L179 (1983)
2.126 S. Ichimura, H. Shimizu, H. Murakami, Y. Ishida: J. Nucl. Mater. **128 & 129**, 601 (1984)
2.127 R.G. Allas, A.R. Knudson, J.M. Lambert, P.A. Treado, G.W. Reynolds: Nucl. Instrum. Methods **194**, 615 (1982)
2.128 F.R: Vozzo, G.W. Reynolds: Nucl. Instrum. Methods **209/210**, 555 (1983)
2.129 P.R. Malmberg, R.G. Allas, J.M. Lambert, P.A. Treado, G.W. Reynolds: Nucl. Instrum. Methods Phys. Res. B **2**, 679 (1984)
2.130 J.L. Vossen: J. Vac. Sci. Technol. **11**, 875 (1974)
2.131 M. Mannami, K. Kimura, A. Kyoshima: Nucl. Instrum. Methods **185**, 533 (1981)
2.132 T. Motohiro, Y. Taga, K. Nakajima: Surf. Sci. **118**, 66 (1982)
2.133 M.F. Dumke, T.A. Tombrello, R.A. Weller, R.M. Housley, E.H. Cirlin: Surf. Sci. **124**, 407 (1983)
2.134 B.M. Gurmin, Yu.A. Ryzhov, I.I. Shkrarban: Bull. Acad. Sci. USSR, Phys. Ser. (USA) **33**, 752 (1968)
2.135 G.K. Wehner: Appl. Phys. Lett. **30**, 185 (1977)
       R.R. Olson, M.E. King, G.K. Wehner: J. Appl. Phys. **50**, 3677 (1979)
2.136 H. Hasuyama, Y. Kanda: To be published
2.137 R. Becerra-Acevedo, J. Bohdansky, W. Eckstein, J. Roth: Nucl. Instrum. Methods Phys. Res. B **2**, 631 (1984)
2.138 H.E. Roosendaal, J.B. Sanders: Ref. [2.66], p. 302
2.139 H.E. Roosendaal, U. Littmark, J.B. Sanders: Phys. Rev. B **26**, 5261 (1982)
2.140 J.P. Biersack, W. Eckstein: Appl. Phys. A **34**, 73 (1984)
2.141 J.P. Biersack, L.G. Haggmark: Nucl. Instrum. Methods **174**, 257 (1980)
2.142 J.P. Biersack, W. Eckstein: Appl. Phys. A **34**, 73 (1984)
2.143 W. Eckstein, J.P. Biersack: Nucl. Instrum. Methods Phys. Res. B **2**, 550 (1984)
2.144 W. Eckstein, W. Möller: Nucl. Instrum. Methods Phys. Res. B **7/8**, 727 (1985)
2.145 M. Hautala, H.J. Whitlow: Nucl. Instrum. Methods Phys. Res. B **6**, 466 (1985)
2.146 S.T. Kang, R. Shimizu, T. Okutani: Jpn. J. Appl. Phys. **18**, 1717 (1979)
2.147 T.J. Hoffman, H.L. Dodds, M.T. Robinson, D.K. Holmes: Nucl. Sci. Eng. **68**, 204 (1978)
2.148 M.H. Shapiro, P.K. Haff, T.A. Tombrello, D.E. Harrison: Nucl. Instrum. Methods Phys. Res. B **12**, 137 (1985)
2.149 G.N. van Wyk, H.J. Smith: Radiat. Eff. **38**, 245 (1978)
2.150 G.N. van Wyk, H.J. Smith: Nucl. Instrum. Methods **170**, 433 (1980)
2.151 M. Marinov, D. Dobrev: Thin Solid Films **42**, 265 (1977)
2.152 B. Lang, A. Taoufik: Appl. Phys. A **39**, 95 (1986)
2.153 U. Littmark, P. Sigmund: J. Phys. D **8**, 241 (1975)
2.154 M. Urbassek: Private communication
2.155 M. Szymonski, W. Huang, J. Onsgaard: Nucl. Instrum. Methods Phys. Res. B **14**, 263 (1986), and Ref. [2.67], p. 440
2.156 M. Koedam: Thesis, Rijksuniversiteit Utrecht, Holland (1961)
2.157 R. Behrisch, R. Weißmann: Phys. Lett. **30A**, 506 (1969)

2.158 R. Weißmann, R. Behrisch: Ref. [2.65], p. 55
2.159 R. Weißmann, P. Sigmund: Ref. [2.65], p. 47
2.160 U. Littmark, G. Maderlechner: 8th SPIG, Dubrovnik, Jugoslavia (1976) p. 139
2.161 R. Behrisch, G. Maderlechner, B.M.U. Scherzer, M.T. Robinson: Appl. Phys. **18**, 391 (1979)
2.162 U. Littmark, S. Fedder: Nucl. Instrum. Methods **194**, 607 (1982)
2.163 H.L. Bay, J. Bohdansky: Appl. Phys. **19**, 421 (1979)
2.164 H.L. Bay, W. Berres: Nucl. Instrum. Methods Phys. Res. B **2**, 606 (1984)
2.165 H.L. Bay, B. Schweer, P. Bogen, E. Hintz: J. Nucl. Mater. **111** & **112**, 732 (1982)
2.166 N. Hermanne: Ref. [2.65], p. 161
2.167 G.K. Wehner, D.J. Hajicek: J. Appl. Phys. **42**, 1145 (1971)
2.168 G.K. Wehner: Appl. Phys. Lett. **43**, 366 (1983); J. Vac. Sci. Technol. A **3**, 1821 (1985)
2.169 G.S. Anderson, G.K. Wehner: Surf. Sci. **2**, 367 (1964)
2.170 R.J. MacDonald: Radiat. Eff. **3**, 131 (1970)
2.171 G. Holmen: D. Sci. Thesis, Chalmers Tekniska Högskola Göteborg, Sweden (1974)
2.172 G. Holmen: Radiat. Eff. **24**, 7 (1975)
2.173 H. Sommerfeldt, E.S. Mashkova, V.A. Molchanov: Phys. Lett. **38A**, 237 (1972)
2.174 H. Sommerfeldt, E.S. Mashkova, V.A. Molchanov: Radiat. Eff. **9**, 267 (1971)
2.175 C.B. Cooper, J. Comas: J. Appl. Phys. **36**, 2891 (1965)
2.176 M.W. Thompson: Unpublished
2.177 G.P. Chen, K. Besocke, S. Berger, W.O. Hofer: To be published
2.178 P.H. Dederichs, G. Leibfried: Z. Phys. **170**, 320 (1962)
2.179 G. Duesing, G. Leibfried: Phys. Status Solidi **9**, 463 (1965)
2.180 H.H. Andersen, P. Sigmund: Risø Report No. 103 (1965)
2.181 J. Sanders, J.M. Fluit: Physica **30**, 129 (1964)
2.182 A.B. Lidiard: In *Atomic Collision Cascades in Radiation Damage*, ed. by M.W. Thompson, Harwell (1964)
2.183 R.v. Jan, R.S. Nelson: Philos. Mag. **17**, 1017 (1968) and Ref. [2.42] p. 34
2.184 J.B. Gibson, A.N. Goland, M. Milgram, G.H. Vineyard: Phys. Rev. **120**, 1229 (1960)
2.185 F. Schulz, R. Sizmann: Ref. [2.42], p. 35
2.186 T. Lenskjaer, F. Nyholm, S.D. Pedersen, N.B. Petersen: Phys. Lett. **47A**, 63 (1974)
2.187 E.B. Henschke: J. Appl. Phys. **28**, 411 (1957)
2.188 V.E. Yurasova: 8th SPIG Dubrovnik, Jugoslavia (1976) p. 493
2.189 V.E. Yurasova: V.A. Eltekov: Vacuum **32**, 399 (1982)
2.190 V.A. Eltekov, O.A. Popova, V.E. Yurasova: Radiat. Eff. **83**, 39 (1984)
2.191 M.V. Kuvakin, A.V. Lusnikov, Kh. A. Motavekh: Fiz. Tverd, Tela. **21**, 2870 (1979) [Engl. Transl.: Sov. Phys.–Solid State **21**, 1657 (1979)]
2.192 V.I. Shulga: Radiat. Eff. **70**, 65 (1983)
2.193 V.I. Shulga: Radiat. Eff. **82**, 169 (1984)
2.194 W. Sczymczak: private communication
2.195 K.T. Rie: Report Jül-304-RW, Jülich (1965)
2.196 Proc. Conference Series on *Vibrations at Surfaces*
        VAS I, ed. by H. Ibach, S. Lehwald (Jül-Conf-26, Jülich 1978)
        VAS II, ed. by R. Caudano et al. (Plenum, New York 1982)
        VAS III, ed. by C.R. Brundle, H. Morawitz (Elsevier, Amsterdam 1983)
        VAS IV, ed. by D.A. King, N.V. Richardson, S. Holloway: J. Electron Spectrosc. Relat. Phenom. **38** (1986)
2.197 V.A. Molchanov, V. Soshka, M.A. Faruk: Zh. Tekh. Fiz. **33**, 766 (1963) [Engl. transl.: Sov. Phys.–Tech. Phys. **8**, 573 (1963)]
2.198 J. Gittus: *Irradiation Effects in Crystalline Solids* (Applied Sci., London 1978)
2.199 J.J. Elich, H.E. Roosendaal: Phys. Lett. **33A**, 235 (1970)
2.200 M.V. Kuvakin, L.B. Shelyakin, R.D.G. Schul'tse, V.E. Yurasova: Fiz. Tverd. Tela **18**, 927 (1976) [Engl. transl.: Sov. Phys.–Solid State **18**, 532 (1976)]
2.201 W. Hauffe: Habilitationsschrift, Technische Universität Dresden (1977)

2.202 G. Endzheets, V.A. Molchanov, V.G. Tel'kovskii, M.A. Faruk: Zh. Tekh. Fiz. **32**, 1052 (1962) [Engl. transl.: Sov. Phys.–Tech. Phys. **7**, 752 (1963)]

2.203 M.T. Robinson: J. Appl. Phys. **40**, 4982 (1969)

2.204 S.A. Drentje: Thesis, Universiteit Groningen, Holland (1968)

2.205 V.E. Dubinskii, S.Ya. Lebedev: Fiz. Tverd. Tela **12**, 1906, 2295 (1970) [Engl. transl.: Sov. Phys.–Solid State **12**, 1516, 1834 (1971)]

2.206 H.H. Andersen, P. Sigmund: Mat.-Fys. Medd. Dan. Vidensk. Selsk. **34**, No. 15 (1966)

2.207 S.A. Kurkin, D.D. Odintsov: Fiz. Tverd. Tela **13**, 1495 (1970) [Engl. transl.: Sov. Phys.–Solid State **13**, 1252 (1971)]

2.208 M.V. Kuvakin, A.V. Lusnikov, Kh. A. Motavekh: Fiz. Tverd. Tela **21**, 2870 (1979) [Engl. transl.: Sov. Phys.–Solid State **21**, 1657 (1979)]

2.209 D. Jackson: Can. J. Phys. **53**, 1513 (1975)

2.210 V.E. Yurasova, V.M. Buchanov, N.N. Rimskiy: Ref. [2.42], p. 32

2.211 V.E. Yurasova, A.A. Sysoev, G.A. Samsonov, V.M. Bukhanov, L.B. Shelyakin: Radiat. Eff. **20**, 89 (1973)

2.212 V.M. Bukhanov, Kh. A. Motavekh, V.E. Yurasova: Fiz. Tverd. Tela **18**, 1760 (1976) [Engl. transl.: Sov. Phys.–Solid State **18**, 1025 (1976)]

2.213 S.P. Holland, B.J. Garrison, N. Winograd: Phys. Rev. Lett. **43**, 220 (1979)

2.214 S.P. Holland, B.J. Garrison, N. Winograd: Phys. Rev. Lett. **44**, 756 (1980)

2.215 N. Winograd, D.E. Harrison, B.J. Garrison: Surf. Sci. **78**, 467 (1978)

2.216 F. Dennis, R.J. MacDonald: Radiat. Eff. **13**, 243 (1972)

2.217 Z. Jurela, B. Perovic: Ref. [2.42], p. 30

2.218 A. Bayly, R.J. MacDonald: Radiat. Eff. **34**, 169 (1977)

2.219 K. Snowdon, R.J. MacDonald: Int. J. Mass Spectrom. Ion Phys. **28**, 233 (1978); ibid. **29**, 101 (1979)

2.220 K. Snowdon: Radiat. Eff. **38**, 141 (1978)

2.221 G. Blaise, G. Slodzian: Rev. de Phys. Appl. **8**, 105 (1973)

2.222 Th.R. Lundquist: J. Vac. Sci. Technol. **15**, 684 (1978)

2.223 K. Komori, J. Okano: Int. J. Mass Spectrom. Ion Phys. **27**, 379 (1978)

2.224 K. Wittmaack: Ref. [2.68], Vol. I p. 153

2.225 K. Wittmaack: Vacuum **32**, 65 (1982)

2.226 J.N. Smith: J. Nucl. Mater. **78**, 117 (1978)

2.227 I.H. Reid, B.W. Farmery, M.W. Thompson: Nucl. Instrum. Methods **132**, 317 (1976)

2.228 M.W. Thompson, I.H. Reid, B.W. Farmery: Philos. Mag. **38**, 727 (1978)

2.229 E. Hulpke, Ch. Schlier: Z. Phys. **207**, 294 (1967)

2.230 R.V. Stuart, G.K. Wehner, G.S. Anderson: J. Appl. Phys. **40**, 803 (1969)

2.231 C.E. Young, P.M. Dehner, R.B. Cohen, L.G. Pobo, S. Wexler: J. Chem. Phys. **64**, 306 (1976)

2.232 R.A. Weller, T.A. Tombrello: Radiat. Eff. **37**, 83 (1978)

2.233 M. Szymonski, A.E. deVries: Phys. Lett. **63A**, 359 (1977)

2.234 M. Szymonski, R.S. Bhattacharya, H. Overeijnder, A.E. deVries: J. Phys. D **11**, 751 (1978)

2.235 O. Overeijnder, M. Szymonski, A. Haring, A.E., de Vries: Radiat. Eff. **36**, 63 (1978)

2.236 J. Politiek, J. Kistemaker: Radiat. Eff. **2**, 129 (1969)

2.237 P. Hucks, G. Stöcklin, E. Vietzke, K. Vogelbruch: J. Nucl. Mater. **76 & 77**, 136 (1978)

2.238 D. Hammer, E. Benes, P. Blum, W. Husinsky: Rev. Sci. Instrum. **47**, 1178 (1976)

2.239 W. Husinsky, R. Bruckmüller, P. Blum, F. Viehböck, D. Hammer, E. Benes: J. Appl. Phys. **48**, 4734 (1977)

2.240 D. Hammer: Vacuum **28**, 107 (1978)

2.241 W. Husinsky, G. Betz, I. Girgis: J. Vac. Sci. Technol. A **2**, 698 (1984); ibid. B **3**, 1543 (1985)

2.242 W. Husinsky, G. Betz, I. Girgis, F. Viehböck: J. Nucl. Mater. **128 & 129**, 577 (1984)

2.243 B. Schweer, H.L. Bay: In Proc. IV Int. Conf. on Solid Surfaces and IIIrd Eur. Conf. on Surface Science, ed. by D.A. Degras, M. Costa (Société Francaise du Vide, Paris 1980) p. 1349

2.244 B. Schweer: Thesis, Universität Bochum, Germany, also published as Report JüL-1876 (1983)

2.245 H.L. Bay, W. Berres, E. Hintz: Nucl. Instrum. Methods **194**, 555 (1982)

2.246 B. Schweer, H.L. Bay: Appl. Phys. A **29**, 53 (1982)

2.247 H.L. Bay, B. Schweer: In Symp. on Surf. Sci., Obertraun, Austria, 1985, ed. by G. Betz et al. (TU, Vienna 1985) p. 147, also published as Report JüL-2032 (1985)

2.248 R.B. Wright, M.J. Pellin, D.M. Gruen, C.E. Young: In Proc. 8th Int. Conf. Atomic Coll. Solids, ed. by D.P. Jackson, J.E. Robinson, D.A. Thompson (North-Holland, Amsterdam 1980) p. 295

2.249 M.L. Yu, D. Grischkowsky, A.C. Blatant: Phys. Rev. Lett. **48**, 427 (1982)

2.250 M.J. Pellin, C.E. Young, M.H. Mendelsohn, D.M. Gruen, R.B. Wright, A.B. Dewald: J. Nucl. Mater. **111 & 112**, 738 (1982)

2.251 M.J. Pellin, D.M. Gruen, C.E. Young, M.D. Wiggins: Nucl. Instrum. Methods Phys. Res. **218**, 771 (1983)

2.252 E. Dullni: Appl. Phys. A **38**, 131 (1985)

2.253 F. Bernhardt, H. Oechsner, E. Stumpe: Nucl. Instrum. Methods **132**, 329 (1976)

2.254 H. Oechsner, J. Bartella: In Proc. 7th Int. Conf. Atomic Coll. Solids, ed. by J.V. Bulgakov, A.F. Tulinov (Moscow State University Publishing House, Moscow 1980) p. 55

2.255 H. Oechsner: In *Physics of Ionized Gases*, SPIG 76, ed. by B. Navinsek (I. Stefan Institute, Ljubljana, Yugoslavia 1976) p. 461

2.256 H. Gnaser, J. Fleischhauer, W.O. Hofer: Appl. Phys. A **37**, 211 (1985)

2.257 H. Gnaser, M. Saidoh, J.v. Seggern, W.O. Hofer: Nucl. Instrum. Methods Phys. Res. B **15**, 169 (1986)

2.258 M. Saidoh, H. Gnaser, W.O. Hofer: Appl. Phys. A **40**, 197 (1986)

2.259 D. Lipinsky, R. Jede, O. Ganschow, A. Benninghoven: J. Vac. Sci. Technol. A **3**, 2007 (1985)

2.260 H.L. Bay, W. Berres: Nucl. Instrum. Methods Phys. Res. B **2**, 606 (1984)

2.261 G.K. Cowell, H.P. Smith: J. Appl. Phys. **43**, 412 (1972)

2.262 F. Seitz, J.S. Koehler: In *Solid State Physics*, Vol. 2, ed. by F. Seitz, D. Turnbull (Academic, New York 1956) p. 305

2.263 P. Sigmund: Appl. Phys. Lett. **25**, 169 (1974)

2.264 P. Sigmund: Ref. [2.68], Vol. I, p. 121

2.265 W.O. Hofer, K. Besocke, B. Stritzker: Appl. Phys. A **30**, 83 (1983)

2.266 P. Sigmund, M. Szymonski: Appl. Phys. A **33**, 141 (1984)

2.267 R.S. Nelson: Philos. Mag. **11**, 291 (1965)

2.268 S. Ahmad, B.W. Farmery, M.W. Thompson: Philos. Mag. A **44**, 1387 (1981)

2.269 F. Thum, W.O. Hofer: Surf. Sci. **90**, 331 (1979)

2.270 S. Ahmad, M.W. Thompson: Philos. Mag. A **50**, 299 (1984)

2.271 H.L. Bay, H.H. Andersen, W.O. Hofer, O. Nielsen: Nucl. Instrum. Methods **132**, 301 (1976); Appl. Phys. **3**, 289 (1976)

2.272 M.W. Thompson: Phys. Rep. **69**, 335 (1981)

2.273 M. Urbassek: Nucl. Instrum. Methods Phys. Res. B **4** 356 (1984)

2.274 A.R. Krauss, D.M. Gruen: Nucl. Instrum. Methods **149**, 547 (1978)

2.275 P. Dawson: Surf. Sci. **65**, 41 (1977); Rev. Sci. Instrum. **48**, 159 (1977)

2.276 Z. Sroubek: Surf. Sci. **44**, 47 (1974)

2.277 J.M. Schroeer, T.N. Rhodin, R.C. Bradley: Surf. Sci. **34**, 571 (1973); J.M. Schroeer: Surf. Sci. **35**, 485 (1973)

2.278 W.H. Gries: Int. J. Mass Spectrom. Ion Phys. **17**, 77 (1975)

2.279 R.A. Moline, G.W. Reutlinger, J.C. North: In *Atomic Collisions in Solids*, ed. by S. Datz, B.R. Appleton, C.D. Moak (Plenum, New York 1974) p. 159

2.280 W. Wach, K. Wittmaack: Nucl. Instrum. Methods **149**, 259 (1978)

2.281 U. Littmark, W.O. Hofer: Nucl. Instrum. Methods **168**, 329 (1980)

2.282 B.J. Garrison: Surf. Sci. **167**, L225 (1986)

2.283 R.C. Bradley: J. Appl. Phys. **30**, 1 (1959)

2.284 E. Dörnenburg, H. Hintenberger, J. Franzen: Z. Naturforsch. **16a**, 532 (1961)

2.285 H. Rodriguez-Murcia, H.E. Beske: Report Jül-1292 (1976); Adv. Mass Spectrom. **7**, 593 (1978)

2.286 P. Joyes: J. Phys. Chem. Solids **32**, 1269 (1971)

2.287 F. Honda, G.M. Lancaster, Y. Fukuda, J.W. Rabalais: J.Chem. Phys. **69**, 4931 (1978)

2.288 R. Buhl, A. Preisinger: Surf. Sci. **47**, 344 (1975)

2.289 K. Wittmaack: Phys. Lett. **69A,** 322 (1979)

2.290 J. Schou, W.O. Hofer: Appl. Surf. Sci. **10,** 383 (1982)

2.291 I. Katakuse, T. Ichihara, Y. Fujita, T. Matsuo, T. Sakurai, H. Matsuda: Int. J. Mass Spectrom. Ion Proc. **67,** 229 (1985); **69,** 109 (1986)

2.292 P. Fayet, L. Wöste: Surf. Sci. **156,** 134 (1985)

2.293 F.M. Devienne, J.-C. Roustan: Org. Mass Spectrom. **17,** 173 (1982)

2.294 W. Begemann, K.H. Meiwes-Broer, H.O. Lutz: Phys. Rev. Lett. **56,** 2248 (1986); Ref [2.342], p. 183

2.295 S. Kato, M. Mohri, T. Yamashina: Surf. Sci. **123,** L717 (1982)

2.296 N. Kh. Dzhemaliev, R.I. Kurbanov: Izv. Akad. Nauk SSSR, Ser. Fiz. **43,** 606 (1979)

2.297 M.L. Yu: Appl. Surf. Sci. **11/12,** 196 (1982); Phys. Rev B **24,** 5625 (1981)

2.298 W.A. Chupka, M.G. Inghram: J. Chem. Phys. **21,** 1313 (1953)

2.299 N. Fürstenau, F. Hillenkamp, R. Nitsche: Int. J. Mass Spectrom. Ion Phys. **31,** 85 (1979)

2.300 N. Fürstenau, F. Hillenkamp: Int. J. Mass Spectrom. Ion Phys. **37,** 135 (1981); Ref. [2.66], p. 707

2.301 M. Leleyter, P. Joyes: J. Phys. B **7,** 516 (1974)

2.302 M. Leleyter, P. Joyes: Ref. [2.65], p. 185

2.303 Conference series on *Small Particles and Inorganic Clusters*
    a) 1st Conf., J. de Phys. **38,** C-2 (1977)
    b) 2nd Conf., Surf. Sci. **106** (1981)
    c) 3rd Conf. Surf. Sci. **156** (1985)
    d) 4th Conf., Z. Phys. D **12** (1989)

2.304 C.E. Klots: J. Chem. Phys. **83,** 5854 (1985)

2.305 R. Tang, J. Callaway: Phys. Lett. **111A,** 313 (1985)
    K. Lee, J. Callaway, K. Kwong, R. Tang, A. Ziegler: Phys. Rev. B **31,** 1769 (1985)

2.306 W.D. Knight, K. Clemenger, A.W. de Heer, A.W. Saudners, M.Y. Chou, M.L. Cohen: Phys. Rev. Lett. **52,** 2141 (1984)

2.307 P. Fayet, J.P. Wolf, L. Wöste: Phys. Rev. B **33,** 6792 (1986)

2.308 J. Giber, W.O. Hofer: Ref. [2.66], p. 697

2.309 M. Bernheim: Unpublished

2.310 J.J. Ph. Elich, H.E. Roosendaal, D. Onderdelinden: Radiat. Eff. **14,** 93 (1972)

2.311 H. Gnaser, W.O. Hofer: Appl. Phys. A**48,** 261 (1989)

2.312 G. Betz, W. Husinsky, F.P. Viehböck (eds.): Proc. Int. Symp. on Sputtering, June 1986, Spitz/D., Austria, published in Nucl. Instrum. Methods Phys. Res. B **18** (1987)

2.313 J. Mattauch, H. Ewald, O. Hahn, F. Strassmann: Z. Phys. **120,** 598 (1943)

2.314 R.E. Honig: J. Chem. Phys. **22,** 126 (1954)

2.315 R.E. Honig: J. Chem. Phys. **21,** 573 (1953)

2.316 G.P. Können, A. Tip, A.E. de Vries: Radiat. Eff. **21,** 269 (1974); ibid. **26,** 23 (1975); for corrections see also Ref. [2.326]

2.317 T.T. Tsong: Appl. Phys. Lett. **45,** 1149 (1984)

2.318 H.E. Oechsner: "Formation of Sputtered Molecules" in *The Physics of Ionized Gases,* SPIG 84, ed. by M.M. Popovic, P. Krstic (World Scientific Singapore 1985) p. 571

2.319 H.E. Oechsner, W. Gerhard: Surf. Sci. **44,** 480 (1974)

2.320 I.S. Bitenskii, E.S. Parilis: Zh. Tekh. Fiz. **48,** 1941 (1978) [Engl. transl.: Sov. Phys. – Tech. Phys. **23,** 1104 (1978)]

2.321 R. Kelly: Radiat. Eff. **80,** 273 (1984)

2.322 P. Joyes: J. Phys. B **4,** L15 (1971)

2.323 M. Urbassek: Ref. [2.312] p. 587

2.324 K. Snowdon: Nucl. Instrum. Methods Phys. Res. B **9,** 132 (1985)

2.325 Proc. Int. Conf. Ser. on Atomic Collisions in Solids,
    a) 8th Conf., ed. by D.P. Jackson et al., Nucl. Instrum. Methods **170** (1980)
    b) 9th Conf., ed. by J. Remillieux et al., Nucl. Instrum. Methods **194** (1982)
    c) 10th Conf., ed. by H.E. Roosendaal et al., Nucl. Instrum. Methods B **2** (1984)
    d) 11th Conf., ed. by T.M. Buck et al., Nucl. Instrum. Methods B **13** (1986)
    e) 12th Conf., ed. by F. Fujimoto et al., Nucl. Instrum. Methods B **33** (1988)

2.326 R.A. Haring, H.E. Roosendaal, P.C. Zalm: Nucl. Instrum. Methods Phys. Res. B **28**, 205 (1987)
2.327 M.I. Current, D.N. Seidman: Ref. [2.325a], p. 377
2.328 J. Aidelberg, D.N. Seidman: Ref. [2.325a], p. 413
2.329 D. Pramanik, D.N. Seidman: J. Appl. Phys. **54**, 6352 (1983)
2.330 W. Jäger, K.L. Merkle: In 9th Int. Cong. Electron Microscopy, Toronto, Canada 1978, Vol. 1, p. 378
2.331 B.J. Garrison, N. Winograd, D.E. Harrison: J. Chem. Phys. **69**, 1440 (1978)
2.332 Y. Yamamura, Y. Kitazoe: Radiat. Eff. **39**, 251 (1978)
2.333 D.E. Harrison, C.B. Delaplain: J. Appl. Phys. **47**, 2252 (1976)
2.334 H.L. Bay, H.H. Andersen, W.O. Hofer, O. Nielsen: Nucl. Instrum. Methods **133**, 301 (1976)
2.335 J. Plücker: Ann. Phys. Chem. (Leipzig) **104**, 113 (1858)
2.336 W. Hittorf: Ann. Phys. Chem. (Leipzig) **21**, 90 (1884)
2.337 W.O. Hofer: Habilitationsschrift, Technische Universität Wien, Austria (1979)
2.338 J.P. Baxter, G.A. Schick, J. Singh, P.H. Kobrin, N. Winograd: J. Vac. Sci. Technol. A **4**, 1218 (1986)
2.339 N. Winograd, P.H. Kobrin, G.A. Schick, J. Singh, J.P. Baxter, B.J. Garrison: Surf. Sci. **176**, L 817 (1986)
2.340 B.J. Garrison, C.T. Reimann, N. Winograd, D.E. Harrison, Jr.: Phys. Rev. B **36**, 3516 (1987)
2.341 N. Kh. Dzhemilev, U. Kh. Rasulev, S.V. Verkhoturov: Nucl. Instrum. Methods Phys. Res. B **29**, 531 (1987)
2.342 F. Träger, G. zu Putlitz (eds.): Proc. Int. Symp. Metal Clusters, Heidelberg 1986, Z. Phys. D **3** (1986)
2.343 Y. Yamamura, W. Takeuchi: Nucl. Instrum. Methods Phys. Res. B **29**, 461 (1987)
2.344 H.H. Andersen: Ref. [2.312], p. 321
2.345 N.D. Bhaskar, R.P. Frueholz, C.M. Klimcak, R.A. Cook: Phys. Rev. B **36**, 4418 (1987)
2.346 D.E. Powers, S.G. Hansen, M.E. Geusic, D.L. Michalopoulos, R.E. Smalley: J. Chem. Phys. **78**, 2866 (1983)
2.347 L.-S. Zheng, C.M. Karner, P.J. Prucat, S.H. Yang, C.L. Pettiette, M.J. Craycraft, R.E. Smalley: J. Chem. Phys. **85**, 1681 (1986)

Additional References with Titles

Sect. 2.2 Angular Emission Distribution

Buckard, R., Hasselkamp, D., Scharmann, A., Schartner, K.-H., Seibel, H.-W.: Reemission of sputtered material from collecting foils at high impact energies. Radiat. Eff. Defects Solids **109**, 301–308 (1989)

Dodonov, A.I., Fedorovich, S.D., Krylova, E.A., Mashkova, E.S., Molchanov, V.A.: Spatial distribution of sputtered particles from polycrystals under ion bombardment. Radiat. Eff. **107**, 15–21 (1988)

Eckstein, W., Hou, M.: Angular ejection distributions in low energy single crystal sputtering: The relation to the interatomic potential. Nucl. Instrum. Methods B **31**, 386–392 (1988)

Vicanek, M., Urbassek, H.M.: Energy and angular distributions of sputtered particles: A comparison between analytical theory and computer simulation results. Nucl. Instrum. Methods B **30**, 507–513 (1988)

Waldeer, K.T., Urbassek, H.M.: Collision cascade evolution in media with energy independent scattering: Spatial, energy, and angular distribution. Appl. Phys. A **45**, 207–215 (1988)

Waldeer, K.T., Urbassek, H.M.: On the angular distribution of sputtered particles. Nucl. Instrum. Methods B **18**, 518–524 (1987)

Sect. 2.3 Energy Distribution of Sputtered Particles

Brizzolara, R.A., Cooper, C.B., Olson, Th.K.: Energy distributions of neutral atoms sputtered by very low energy heavy ions. Nucl. Instrum. Methods B **35**, 36–42 (1988)

Dembowski, J., Oechsner, H., Yamamura, Y., Urbassek, M.: Energy distributions of neutral atoms sputtered from Cu, V and Nb under different bombardment and ejection angles. Nucl. Instrum. Methods B **18**, 464–470 (1987)

Likonen, J., Hautala, M.: Sputtering of Cu atoms by Ar ions. Appl. Phys. A **45**, 137–150 (1988)

Schorn, R.P., Zaki Ewiss, M.A., Hintz, E.: Velocity distributions of copper and lithium atoms sputtered from a Cu/Li alloy measured with laser induced fluorescence. Appl. Phys. A **46**, 291–297 (1988)

Szymoński, M., Postawa, Z.: Theoretical energy distributions of atoms sputtered from elastic collision spikes by monomer and dimer ion bombardment. Appl. Phys. A **50**, 269–272 (1990)

Mertens, Ph., Bogen, P.: Velocity distribution of hydrogen atoms sputtered from metal hydrides. J. Nucl. Mater., **128 & 129**, 551–554 (1984)

Bogen, P., Döbele, H.F., Mertens, Ph.: Measurement of velocity distributions of sputtered carbon atoms using laser-induced fluorescence in the vacuum UV. J. Nucl. Mater. **145–147**, 434–437 (1987)

Sect. 2.4 Mass Distribution of Sputtered Particles

Andersen, H.H.: Formation and stability of sputtered clusters. Vacuum **39**, 1095–1099 (1989)

Bitensky, I.S., Parilis, E.S.: Shock wave mechanism for cluster emission and organic molecule desorption under heavy ion bombardment. Nucl. Instrum. Methods B **21**, 26–36 (1987)

de Jonge, R., Benoist, K.W., Majoor, J.W.F., de Vries, A.E., Snowdon, K.J.: Velocity dependent vibrational and rotational energy distributions of sputtered sulfur molecules. Nucl. Instrum. Methods B **28**, 214–226 (1987)

de Jonge, R., Baller, T., Tenner, M.G., de Vries, A.E., Snowdon, K.J.: Internal energy distribution of sputtered sulfur, molecules. Nucl. Instrum. Methods B **17**, 213–226 (1986)

de Jonge, R., Baller, T., Tenner, M.G., de Vries, A.E., Snowdon, K.J.: Rotational and translational energy distributions of sputtered S_2 molecules. Europhys. Lett., **2**, 449–453 (1986)

Dzhemilev, N.Kh., Rasulev, U.Kh., Verkhoturov, S.V.: The fragmentation of sputtered cluster ions and their contribution to secondary ion mass spectra. Nucl. Instr. Methods B **29**, 531–536 (1987)

Dzhemilev, N.Kh., Verkhoturov, S.V., Veriovkin, I.V.: Study of the lifetime and the most probable energies of excited cluster ions. Nucl. Instrum. Methods B**51**, 219–225 (1990)

Gnaser, H., Oechsner, H.: Emission-angle integrated yields of neutral clusters in sub-keV-energy sputtering. Nucl. Instrum. Methods B in press

Haring, R.A., Roosendaal, H.E., Zalm, P.C.: On the energy and angular distribution of sputtered polyatomic molecules. Nucl. Instrum. Methods B **28**, 205–213 (1987)

Oechsner, H.: Secondary neutral mass spectrometric investigations on the formation of sputter-generated molecules by atomic combination. Int. J. Mass Spec. Ion Proc. **103**, 31–43 (1990)

Pellin, M.J., Husinsky, W., Calaway, W.F., Burnett, J.W., Schweitzer, E.L., Young, C.E., Jørgensen, B., Gruen, D.M.: Ion and neutral atomic cluster sputtering yields of molybdenum. J. Vac. Sci. Technol. B **5**, 1477 (1987)

Snowdon, K.J., Haring, R.A.: Diatomic molecule sputtering in the independent binary collision approximation. Nucl. Instrum. Methods B **18**, 596–599 (1987)

Urbassek, H.M.: Sputtered cluster mass distributions, thermodynamic equilibrium and critical phenomena. Nucl. Instrum. Methods B**31**, 541–550 (1988)

3. Charged and Excited States of Sputtered Atoms

Ming L. Yu

With 29 Figures

An important application of sputtering is in materials analysis. Secondary ion mass spectrometry, which is based on detecting the ionized fraction of the sputtered atoms, is now widely used for the determination of major, minor, and trace constituents at surfaces, films, and interfaces, yielding excellent relative (ppb level) and absolute (10^{-20} g) detection sensitivities. In a complementary technique, the optical emission from particles sputtered in excited states is measured. These developments have given much impetus to study sputtered-atom ionization and excitation.

Inelastic phenomena in ion-surface interactions is a challenging field. Electronic transitions can occur by way of the tunneling of valence electrons, the production of core-holes, and the breaking of chemical bonds. These dynamic charge transfer processes between the escaping sputtered atom and the solid surface are further complicated by the strong electronic and structural perturbations from the ion bombardment. The first two sections of this chapter summarize the recent experimental and theoretical developments that have shed much light on the ionization mechanisms. The last section summarizes our present understanding of the excitation process.

3.1 Physical Background of the Phenomena

During sputtering, momentum transferred from the primary ion to the target atoms causes surface atoms to leave the solid [3.1, 2], subjecting electrons of these atoms to strong perturbations. Two major factors affect the electronic states of the sputtered atoms: First, the atomic collisions that lead to sputtering can create electronic excitations. Second, the valence electrons of the sputtered atoms are subjected to a transition from occupying orbitals in the solid to occupying eigenstates of free atoms in a time frame of 10^{-13}–10^{-14} seconds. Electronic excitations can result from this strong time-dependent perturbation. A fraction of the sputtered atoms may be ionized, either positively or negatively, and/or excited above their respective ground states. They are usually called secondary particles to distinguish them from the impinging primary ions inducing the sputtering. These secondary species can originate from collision cascades and from recoil sputtering. The secondary ions can readily be identified by mass spectrometry. Atoms and ions sputtered in the excited states can often be

Topics in Applied Physics, Vol. 64
Sputtering by Particle Bombardment III Eds.: R. Behrisch · K. Wittmaack
© Springer-Verlag Berlin Heidelberg 1991

inferred from observation of the photons emitted during the deexcitation of these species in front of the sample surface. Two important analytical techniques have evolved from the study of these sputtered particles: The first is secondary ion mass spectrometry (SIMS) [3.3]. The second is surface composition by analysis of neutral and ion impact radiation (SCANIIR) [3.4]. Both are discussed in detail in Chap. 4.

The differential sputtering yield $\partial^3 Y_i^{q,*}/\partial E_1 \partial^2 \Omega$ (per incident primary ion) of element i in excited ($*$) charged state (q) from a given target depends on the incident ion energy E_0, the mass of the incident ion M_1, the mass of the target atom M_2, the emission energy E_1, and the emission angle Ω. The angle Ω is defined by the azimuthal angle ψ with respect to a fixed direction on the sample surface, and the polar angle θ measured with respect to the normal of the surface. This chapter focuses on the probability $P_i^{q,*}(E_i', \Omega')$ that an atom, sputtered originally with energy E_i' in the direction Ω', in the immediate vicinity of the surface, would escape in the excited ($*$) charged state (q). Since the potential between the sputtered atom/ion and the surface alters the particle trajectory, the final energy E_1 and direction Ω of the particle differ from E_1' and Ω'. Mathematically the differential sputtering yield $\partial^3 Y_i^{q,*}/\partial E_1 \partial^2 \Omega$ is given by the following equation:

$$\frac{\partial^3 Y_i^{q,*}}{\partial E_1 \partial^2 \Omega} = P_i^{q,*}(E_1', \Omega') \frac{\partial^3 Y_i}{\partial E_1' \partial^2 \Omega'} \cdot \frac{\partial(E_1', \Omega')}{\partial(E_1, \Omega)} , \tag{3.1}$$

where $\partial^3 Y_i/\partial E_1' \partial^2 \Omega'$ is the differential sputtering yield of i, irrespective of the electronic state, in the immediate vicinity of the surface. The Jacobian [3.5] $\partial(E_1', \Omega')/\partial(E_1, \Omega)$ corrects for the transformation of the phase space from (E_1', Ω') to (E_1, Ω).

The emissions of secondary ions and secondary excited species are both very sensitive to the chemical states of the sample surfaces. Metal surfaces, after reacting with electronegative elements, can have positive secondary ion [3.6] and excited atom [3.7] yields two to three orders of magnitude higher than those from clean surfaces. In SIMS and SCANIIR analyses, O_2^+ bombardment or Ar^+ bombardment in an oxygen ambient is widely used to oxidize metallic samples during sputtering. The enhanced positive secondary ion and excited atom yields provide better detection sensitivities. Electropositive elements like cesium, on the other hand, are introduced to the sample during sputtering to enhance the negative secondary ion yields. These chemical effects have had a major impact on the practical applications of these analytical techniques. They also provide much insight into secondary ion and excited atom formation.

The historical development of this field is reflected in the proceedings of the International Conferences on Secondary Ion Mass Spectrometry [3.8–12], the International Workshops on Inelastic Ion Surface Collisions [3.13–18], and book [3.19] and review articles on ionization [3.20–30] and excitation [3.31–36]. The last few years have seen much progress as these phenomena became testing grounds for theories on the dynamic electronic interaction between moving atoms and solid surfaces [3.28–30].

3.2 Experimental Studies on Secondary Ion Emission

Most secondary ions formed during sputtering are singly charged, either positively or negatively. Multiply charged secondary ions are formed with significant abundance only in certain systems (Sect. 3.2.4b). Secondary ions are also emitted in the electronically excited states. Their yields are, however, usually small compared to those of the ground-state ions (Sect. 3.4). Hence we assume that the ion yields measured by mass spectrometry are those of the ground-state ions. Surface crystallinity effects in secondary ion emission [3.37–41] are usually attributed to channeling and structural aspects of sputtering [Ref. 3.1, Chaps. 3, 5]. Crystallinity effects on ionization have not been systematically studied.

The high sensitivity of the ionization probability to the surface chemical state is one of the most important phenomena in this field. A comparative list (Table 3.1) of energy- and angular-averaged values of P^+ for clean and oxidized metal and semiconductor surfaces has been reported by *Benninghoven* [3.42, 43]. Although the absolute values in Table 3.1 may not be completely accurate because of difficulties in measuring the transmission of the instrument, P^+ values from oxidized surfaces are typically up to three orders of magnitude higher than the corresponding values for pure metals. Recently it became

Table 3.1. Secondary ion yields from clean and oxygen-covered surfaces under 3 keV Ar^+ bombardment [3.43]

Metal	P^+ for clean surface	P^+ for oxygen-covered surface
Al	0.007	0.7
Ba	0.0002	0.03
Cr	0.0012	1.2
Cu	0.0003	0.007
Fe	0.0015	0.35
Ge	0.0044	0.02
Mg	0.01	0.9
Mn	0.0006	0.3
Mo	0.00065	0.4
Nb	0.0006	0.05
Ni	0.0006	0.045
Si	0.0084	0.58
Sr	0.0002	0.16
Ta	0.00007	0.02
Ti	0.0013	0.4
V	0.001	0.3
W	0.00009	0.035

apparent that the physical mechanisms for secondary ion emission from chemical compounds and from metals can be quite different [3.28–30]. Local chemical bonds break when metal ions are sputtered from compounds like oxides, while delocalized metallic bonds break in the sputtering of ions from metals. For convenience, we subdivide the experimental data into three categories:

1) Secondary ion emission from clean metals and semiconductor surfaces where the valence-band electrons are important.
2) Secondary ion emission from chemical systems where strong local bonding dominates.
3) Secondary ion emission from systems where inner-shell excitations generated by energetic collisions cause ionization.

3.2.1 Experimental Methods

The basic experimental setup (Fig. 3.1) consists of a primary ion source, a sample mounted on a manipulator, secondary ion extraction optics, an energy filter, and a mass spectrometer. The components are assembled in a vacuum chamber that should maintain an ultrahigh vacuum during sputtering, since secondary ion yields are very sensitive to residual gas contaminations. Both Auger and photoemission spectroscopies have been used for in situ characterization of the sample surfaces. Several systems for fundamental studies have been reported [3.44–51]. Chapter 4 gives more specific details.

The secondary ion intensity I_i^q detected by the mass spectrometer is given by the following integral:

$$I_i^q = \iiint_{\Delta E, \Delta \Omega} \frac{\partial^3 Y_i^q}{\partial E_1 \partial^2 \Omega} \eta(E_1, \Omega) dE_1 d^2\Omega \ , \tag{3.2}$$

where ΔE is the energy-band pass, $\Delta \Omega$ is the acceptance angle, and $\eta(E_1, \Omega)$ is the product of the instrument's transmission and detection efficiencies. In

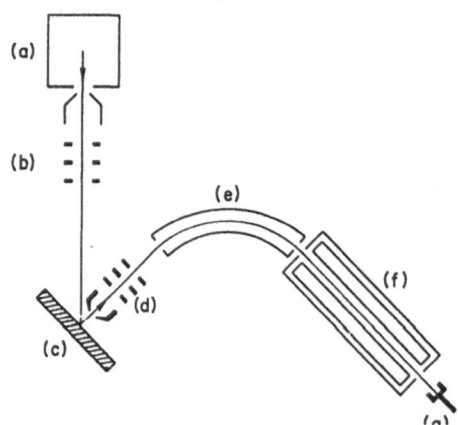

Fig. 3.1. Schematic diagram of the experimental arrangement for secondary ion detection: (a) primary ion source, (b) primary ion beam ion optics, (c) sample, (d) secondary ion extraction optics, (e) secondary ion energy filter, (f) quadrupole mass spectrometer, and (g) detector

angular- and energy-resolved measurements of the ion yield, the resolutions ΔE and $\Delta\Omega$ are typically about 1 eV and 10° respectively [3.51]. Absolute values of the differential secondary ion yields are very hard to measure, because calibrating η accurately is difficult [3.52]. Hence relative ion yields measured at a fixed energy and direction are more frequently reported. Measurements of ion energy distributions are usually not accurate unless the proper ion optics corrections are made [3.53, 54].

To obtain the ionization probability, the differential sputtering yield at the right-hand side of (3.1) has to be known. This quantity is not easily determined experimentally, so it is frequently approximated by the Thompson formula [Ref. 3.1, Eq. (2.3.16b)]:

$$\frac{\partial^3 Y}{\partial E' \partial^2 \Omega'} \propto \frac{E_1'}{(E_1' + U_0)^3} |\cos \theta'| \, , \tag{3.3}$$

where U_0 is the surface-binding energy. Trajectory evolution during ionization has been studied very little: (E_1', Ω') and (E_1, Ω) are usually assumed to be the same, and the value of the Jacobian in (3.1) is set to unity. This can introduce error for low-energy (< 10 eV) trajectories where the kinetic energy is comparable to the binding energy. Corrections using image and Morse potentials have been applied [3.55–57].

There have been some attempts to measure the angular- and energy-averaged ionization probabilities using simultaneous secondary neutral mass spectrometry measurements [3.58], or Faraday cup measurements without a mass spectrometer for the determination of secondary ion yields [3.59]. In the latter case, it was necessary to assume that only one major secondary ion species was emitted.

Since ion bombardment introduces radiation damage on the sample surface, the primary ion fluence is an important parameter in fundamental SIMS studies. Usually, the experiment is performed in either the high-fluence or low-fluence limit. In the high-fluence limit, the sample surface is appreciably affected by the primary ion bombardment. Frequently, a well-focused ion beam is raster scanned across the sample area under examination. To avoid the unevenness at the edge of the sputtered area, only the secondary ions emitted from the central region are recorded. A significant amount of material is removed by sputtering, and a dynamic equilibrium state is finally reached. This procedure is usually referred to as "dynamic mode" sputtering. If an inert-gas ion beam is used, there will be radiation damage and inert gas implantation, and the sample will have a composition dictated by the preferential sputtering characteristics of the target. The surface chemical composition can be altered during sputtering by using reactive primary ions like O_2^+ and Cs^+. Theoretical arguments suggest that the steady-state surface concentration of the reactive species is independent of the beam-current density, but varies with the energy and the angle of the primary ion beam [3.60, 61]. The reactive species can also be introduced by adsorption during sputtering; oxygen jets and cesium effusion sources are commonly used. The steady-state surface concentration can be adjusted by changing the partial pressure of the effusion source [3.60].

The advantage of the high-fluence mode in the fundamental studies of secondary ion formation is the large secondary ion intensity that can be attained. Systems with low secondary ion yields are usually studied in this way. In addition, the high-fluence mode must be used for depth profiling elemental concentrations in SIMS analytical applications. The disadvantage of this mode is that characterizing the sample surface condition is difficult. Ion implantation and radiation damage modify the sample's composition and structure. Emission angles are not well defined, because of surface roughening by the ion bombardment. The different elements of a composite target usually have depth- and chemical-state distributions difficult to delineate even by Auger and photoemission spectroscopies.

In the low-fluence limit, the primary ion fluence is reduced sufficiently so that sputtering practically always occurs on previously undisturbed microscopic regions of the sample. *Benninghoven* introduced this procedure as the "static mode" to eliminate the complications in the "dynamic mode" [3.42]. Typically, noble-gas ions with energy around 1 keV are used for sputtering. Ideally, the secondary ion signal should not change with bombardment time when the experiment is performed in the low-fluence limit.

The advantage of the low-fluence mode is that specifically prepared and characterized surfaces can be used for controlled experiments, without the complications from radiation damage. Many critical issues in secondary ion emission were studied in this mode. The disadvantage is that the secondary ion signals are low because of the low primary ion fluence. Hence this technique is applied quite selectively. Measuring the total sputter yield to determine $P^{\pm}$ in these experiments is still impossible, because the amount of material removed is too small.

3.2.2 Secondary Ion Emission from Metal and Semiconductor Surfaces

The valence electrons in these materials form energy bands, possess high mobility, and are characterized by the Fermi energy ε_F, which basically separates the unoccupied and occupied electronic states. By definition, the work function Φ is the difference between ε_F and the vacuum level. The electrons responsible for the metallic bonding in metals are delocalized spatially. Secondary ion yields from clean metals are usually low (Table 3.1). Much useful information on the ionization mechanism was obtained from the response of the ionization probability $P^{\pm}$ to changes in the physical properties of the metal surface. An important example is the enhancement of the negative secondary ion yields [3.62] by adsorbed alkali atoms – e.g., cesium – on metal surfaces. Here we discuss the various parameters affecting the ionization probabilities. To simplify the notations, Y and $Y^{\pm}$ will be used to denote differential sputtering yields also. Integration over a particular energy or angular range will be stated specifically when appropriate.

a) Dependence of the Secondary Ion Yields on Work Function, Ionization Potential, and Electron Affinity

For the sputtered atom, the important energy parameter is the ionization potential I for positive ionization, and the electron affinity A for negative ionization. For the metallic substrate, the relevant energy parameter is the work function Φ. The quantity $I - \Phi$ is the minimum amount of energy required to ionize positively an atom at infinity and deposit the electron on the solid surface, a situation analogous to the sputtering of a positive ion. Similarly, $\Phi - A$ is the corresponding energy required to transfer an electron from the solid to an atom at infinity to form a negative ion. Hence the formation probability of secondary ions should have a functional dependence on these quantities. The work function Φ can be modified by the adsorption of atoms and molecules. Many electropositive and electronegative adsorbates on metallic surfaces generate dipole moments. Since the electrostatic dipole field is long ranged, a uniform change in Φ is usually assumed in the first approximation, unless the adsorbed layer is grossly inhomogeneous.

Yu and *Lang* demonstrated the importance of the relation between I and Φ [3.63]. In their experiment, Cs$^+$ was sputtered from Au, Al, and Si surfaces covered with a minute ($< 10^{-1}$ monolayer) amount of Cs. Submonolayers of Li were deposited on the surface to change the work function. The data are shown in Fig. 3.2: as long as the work function Φ is higher than the ionization potential I of Cs (3.9 eV), the Cs$^+$ yield is independent of Φ. The Cs$^+$ yield first decreases slowly when Φ is lower than 3.9 eV. For the metallic substrates, it decreases very

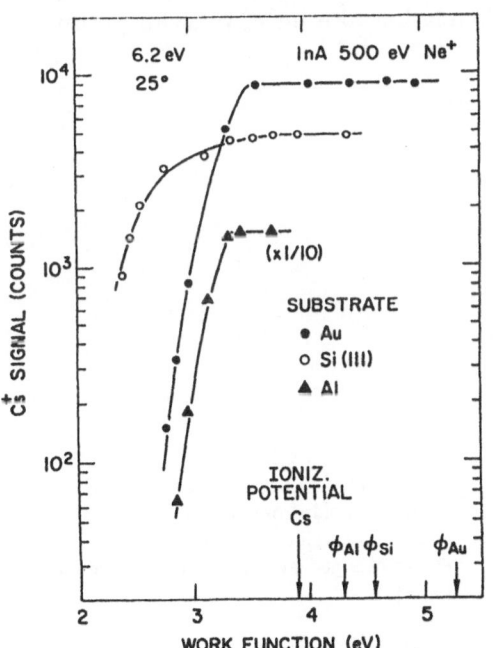

Fig. 3.2. Work-function Φ dependences of the sputtered Cs$^+$ yield for three different substrates during low-fluence sputtering: Φ was changed by depositing Li submonolayers [3.63]

rapidly once $I - \Phi > 0.4$ eV. The work-function dependence is more gradual with the Si substrate.

For most elements, the ionization potentials are so high that $I < \Phi$ rarely occurs. According to the limited data available, lowering the work function of the sample surface by depositing alkali adatoms always suppresses the production of positive secondary ion yields for both metallic [3.63] and semiconducting substrates [3.64]. This holds even for the ionization of sputtered adatoms in a situation where the work-function changes $\Delta\Phi$ are induced by the adatoms themselves [3.65]. In all these cases, we find a simple exponential relation between the positive secondary ion yields and Φ, i.e.,

$$Y^+ \propto \exp(\Delta\Phi/\varepsilon_p) , \qquad (3.4)$$

where the parameter ε_p is specific to the system, and can be a function of the emission energy and angle. An example is shown in Fig. 3.3 where the sputtered lithium ion yield from a lithium-covered (submonolayer) Si(111) surface is plotted against the work-function change induced by depositing Cs [3.63]. This relation was also observed in the emission of Si$^+$ from oxidized Si(100) [3.64].

On the other hand, negative secondary ion yields increase rapidly with decreasing work function [3.19, 66]. The negative secondary ion yields frequently vary exponentially with the change in work-function $\Delta\Phi$ induced by submonolayers of alkali atoms (e.g., Cs, Li):

$$Y^- \propto \exp(-\Delta\Phi/\varepsilon_n) . \qquad (3.5)$$

This relation can hold for several orders of magnitude of change in the ion yields. Again ε_n is system specific. No relation with the ε_p value for positive-ion

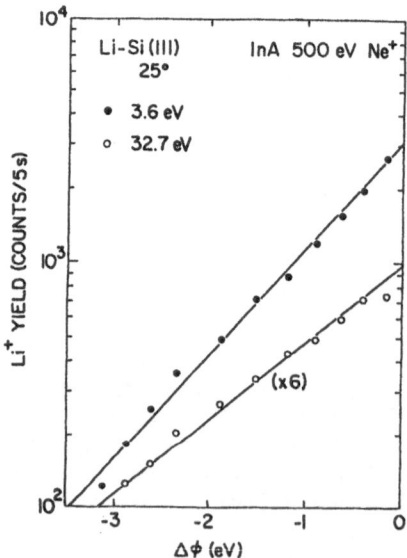

Fig. 3.3. Work-function dependences of the sputtered Li$^+$ yields from an Si(111) substrate at two emission energies during low-fluence sputtering: Φ was changed by depositing Cs submonolayers [3.63]

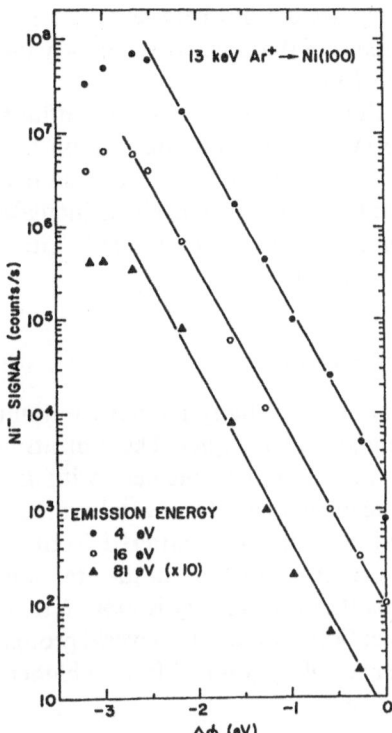

Fig. 3.4. Work-function dependences of the sputtered Ni⁻ yields from an Ni(100) surface at three different emission energies during high-fluence sputtering: Φ was changed by depositing Cs [3.70]

emission has been established. This dependence of the negative-ion yield on $\Delta\Phi$ was observed in static-mode sputtering experiments for H⁻, D⁻, Mo⁻, and O⁻ emission from adsorbate-covered Mo(100) [3.66]; O⁻ from oxygenated Ti [3.67], V and Nb [3.68], and Si(111) [3.69]; and Si⁻ from Si(111) [3.69]. Qualitatively similar behaviors were observed in high-fluence sputtering experiments. These include Cu⁻ from Cu, Ni⁻ from Ni(100), Ag⁻ from Ag(100), Cr⁻ from Cr, Ir⁻ from Ir, H⁻ from contaminated CuNi alloy, and negative phosphorus ions from NiP alloy [3.70]. An example is shown in Fig. 3.4: the Ni⁻ yields at various emission energies are plotted against the work-function changes induced by depositing Cs on the Ni(100) target during high-fluence 13 keV Ar⁺ bombardment [3.70].

Deviations from this simple exponential dependence usually occur at higher alkali coverages, where $\Delta\Phi$ approaches its maximum value. In the high-fluence-mode sputtering of Au⁻ and Cu⁻ from AuCu alloy, and O⁻ from contaminated V, the exponential relation was not observed [3.70], but the enhancement of the ion yield with decreasing Φ was persistent. The exponential relation between the O⁻ yield and $\Delta\Phi$, which was observed with an Li adatom layer on oxygenated Si surfaces, was not obeyed when the more electropositive Cs adatoms were used [3.69], perhaps because the chemistry between the alkali atoms and the surface may cause the exponential relation to break down. Negative secondary ion emission from metallic alloys is not sensitive to alloy composition [3.71]. Also, when there is no strong surface chemistry (e.g., oxidation) between the alkali adatoms and the sample surface, the Φ dependence of the negative secondary ion yield is not sensitive to the alkali species used [3.69].

The work function of a solid surface is much larger than the electron affinity of the sputtered atom for most elements. No systematic study has established a relation between $\Phi - A$ and negative secondary ion formation. The limited data available suggest that the relative negative secondary ion yields increase exponentially with the electron affinities of the elements [3.62, 71]. These experiments were done under high-fluence conditions.

b) Dependences on the Angle and Energy of Emission

The range of E_1 accessible to the experimentalist is usually limited to about 150 eV, because the ion yield decreases rapidly at high energies. The ionization probabilities P^+ of positive secondary ions were found to increase with E_1. Power-law dependences of P^+ on E_1 have been reported by several workers [3.72–74]. *Garrett* et al. [3.75] reported that P^+ for Al^+ sputtered from Al depends on the emission angle through the normal component of the emission velocity $v_\perp = v \cos \theta$. The important message is that the velocity is more fundamental than the energy in determining the ionization probability. Several groups [3.76–79] have reported exponential dependences of P^+ on $v_\perp^{-1}$ for the higher-energy part of the ion energy spectrum:

$$P^+ \propto \exp(- v_0/v_\perp) \tag{3.6}$$

with $v_0 \sim 10^7$ cm/s. At low emission velocities, deviations from this simple relation were observed [3.80]. The above results were all obtained by high-fluence sputtering. Using low-fluence sputtering, *Vasile* [3.56, 81] found that the exponential dependence on $v_\perp^{-1}$ for Au^+, Cu^+, Cr^+, Ag^+, and Zr^+ sputtered from clean elemental surfaces was obeyed down to 3–4 eV emission energy, provided an appropriate image-potential correction to the emission velocity was made (Fig. 3.5). This implies that the atom velocity at the instant of ionization is the parameter that determines the ionization probability. The atom velocity is higher than the measured ion velocity, because the secondary ion has to overcome the image potential on its outward trajectory. This illustrates the importance of trajectory correction in (3.1). The functional dependence on $v_\perp$ is similar to *Hagstrum*'s finding on the neutralization of ions at surfaces [3.82], and is related to the fact that the transition probability for electron transfer depends on the amount of time the ion spends in the interaction region above the surface. The emission velocity also affects the work-function dependence. As shown for the sputtering of Li^+ [3.63], Si^+ [3.64], and Cs^+ [3.65], changing the work function has less effect on the yields of positive secondary ions as the ion velocity increases.

Earlier *Yu* observed similar dependences of the negative secondary ion ionization probability P^- on the angle and velocity of emission [3.83]. He

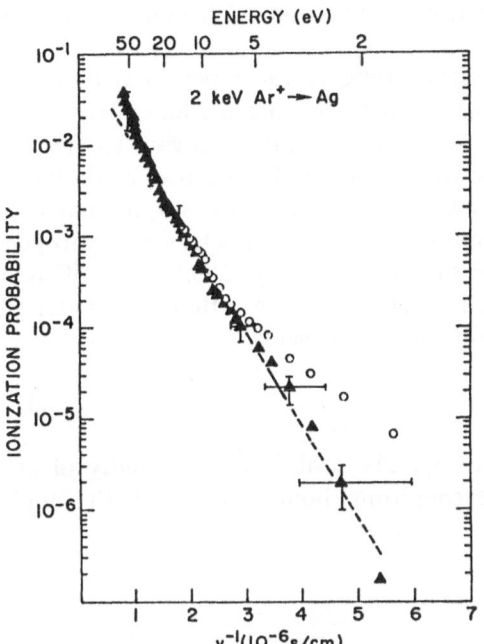

Fig. 3.5. Ionization probability as a function of the inverse velocity for the sputtering of Ag^+ from clean Ag. The open circles are the uncorrected data. The solid triangles are the data corrected with a 1.4 eV image potential [3.56]

studied the work-function dependence of the O^- yield sputtered in the low-fluence mode from oxygen-covered vanadium and niobium surfaces at different emission angles. He found that the parameter ε_n in (3.5) is related to the normal component of the emission velocity, as shown in Fig. 3.6: ε_n increases linearly with $v_\perp$ at high $v_\perp$. The deviation of ε_n from the linear behavior at low $v_\perp$ is believed to be due to trajectory modification by the surface-binding energy, as

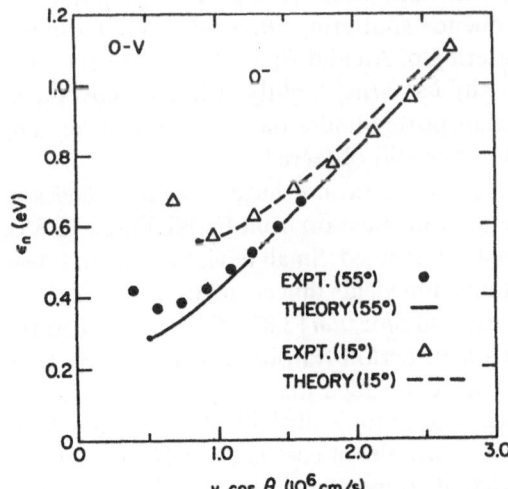

Fig. 3.6. Comparison between the experimental [3.83] and theoretical [3.57] values of the parameter ε_n in (3.5), as a function of the normal component of the emission velocity for O^- sputtered from oxygenated vanadium surfaces at two different emission angles

described in relation to (3.1). Comparison with theoretical calculations will be discussed in Sect. 3.3.1c.

Results from high-fluence sputtering experiments were reported by *Bernheim* and *LeBourse* [3.70]. They found that for negative metallic ions (Au⁻, Ag⁻, Cr⁻, Cu⁻, Ni⁻, Ir⁻) sputtered from metal surfaces, the emission energy has either no or a very weak effect on the work-function dependences as shown in Fig. 3.4. Strong dependence on the emission energy was, however, observed for the emission of negative secondary ions of electronegative elements hydrogen, oxygen, and phosphorus from their samples. The two completely different dependences on emission energy is still unexplained. The authors also reported that for Ag⁻, Ir⁻, and Au⁻ sputtered from the elements,

$$P^- \propto E^{-2} \exp(-\gamma E^{1/2}) \,, \tag{3.7}$$

with $\gamma \simeq 0.17$, 0.19, and 0.22 respectively. The ionization probability of O⁻ varied as v^n with an exponent n decreasing from about 1.5 to zero with sample work functions.

c) Effect of Alloying

Of the few studies made so far on the effect of alloying on the ionization probability of an element, an interesting case was reported by *Cuomo* et al. [3.84]. They studied the sputtering of binary alloys $X_a X_b$ where the component X_a is an electropositive element (Cs or a rare earth), and the component X_b is Cu, Ag, Au, Pt, or Pd. They observed large negative secondary ion yields of the component X_b during sputtering, which correlate with the large electronegativity differences between X_a and X_b. With Sm-Au alloys, they also observed an enhanced Sm⁺ emission (normalized to unit Sm concentration) due to the presence of Au. They proposed that qualitatively there is a charge transfer from X_a to X_b in the alloy, promoting the formation of X_b^- during sputtering. Alloying can happen during high-fluence sputtering. *Bernheim* and *LeBourse* [3.70] reported that during the sputtering of Au and Ag by Ar⁺ in the presence of Cs vapor, the work function actually *increased* slightly at low Cs coverages, suggesting that some Cs atoms are transported under the top atomic layer. The enhancement of Au⁻ and Ag⁻ yields was still observed.

The situation for alloys where X_a and X_b do not have large differences in electronegativity is less clear. Secondary ion emission from Fe-Ni, Fe-Cr [3.85], Cu-Ni, and Cu-Al [3.86] alloys have been studied. Small deviations from linear concentration dependences of the atomic ion yields that correlate with the phase of the alloys have been reported. *Blaise* and *Slodzian* [3.87–90] have studied the secondary ion emission from dilute fourth-period transition-metal alloys: Ti to Cu in Al, Fe, Ni, and Cu matrices. They concluded that the positive secondary ion yield of the solute element increases monotonically with the electronic local density of states at the Fermi level. The prominent case is Cr in Ni, where they related the large Cr⁺ yield with the virtual bound state formed by the Cr atom at the Fermi level. They proposed an autoionization model to explain their results (Sect. 3.3.1b).

Working with a better vacuum, *Yu* and *Reuter* [3.91] obtained contradictory findings. They studied the secondary ion emission from oxygen-free high-concentration Ni alloys: Fe-Ni, Cu-Ni, Pd-Ni, Al-Ni, and Si-Ni in an ultrahigh vacuum using 15 keV Ar^+. They found little or no correlation between the secondary ion yields and the bulk and local density of states near the Fermi level.

d) Secondary Ion Emission from Semiconductors

Secondary ion emission from silicon has been studied extensively. The secondary ion yields increase rapidly with the primary ion energy. It is believed that the energetic collisions induced by the primary ion create Si $2p$ core holes. Subsequent decay of these core holes causes the ionization of the sputtered atoms (Sect. 3.2.4).

Fundamental studies on the secondary ion emission from germanium and group-III–V semiconductors are very scarce. *Croydon* et al. [3.92] performed low-fluence sputtering experiments on (2×4) As-stabilized GaAs(100) surfaces grown by molecular-beam epitaxy. They found that the Ga^+ yield was over three orders of magnitude higher than that of As^+. This large difference is not surprising, because the ionization potential of Ga (6.0 eV) is significantly lower than that of As (9.81 eV). They also noticed that both the positive-ion yields (including molecular ions) and the work function were always higher from this surface than from other differently reconstructed GaAs(100) surfaces. *Sroubek* [3.93] also studied the Ga^+/As^+ ratio under high-fluence sputtering conditions. He found that under normal incidence Ar^+ bombardment, the ratio was increased by an order of magnitude when the primary ion energy was raised from 400 to 2000 eV. Using He^+, on the other hand, gave a small but opposite effect. The yield ratio in this case can be enhanced by simply going to glancing incidence. Similar effects were observed in InAs [3.94], where they were explained by postulating that the ionization process in GaAs and InAs is largely determined by the electronic excitations created by the primary ion in the substrate.

3.2.3 Chemical Effects in Secondary Ion Emission

The reaction of metals with electronegative elements like halogens [3.95–97], oxygen [3.6], and nitrogen [3.98, 99] can result in large enhancement of both positive and negative [3.100, 101] secondary ion yields relative to those from clean metals. Figure 3.7 shows the effect of nitridation on the emission of Si^+ secondary ions from Si(100) surfaces during low-fluence sputtering by 500 eV Ar^+ [3.99]. Nitrogen was cumulatively adsorbed on Si(100) at 1000 °C to form Si_3N_4. Nitridation of the silicon surface up to a monolayer

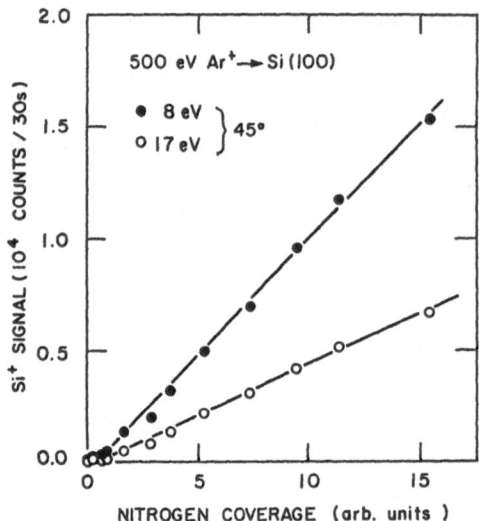

Fig. 3.7. Si$^+$ secondary ion yield as a function of nitrogen coverage on Si(100) for two different emission energies during low-fluence sputtering. The maximum coverage was estimated by X-ray photoemission to be about 7×10^{14}/cm^2 [3.99]

($\simeq 7 \times 10^{14}$/cm^2) enhanced the Si$^+$ yield by at least 150 times. The enhancement effect on positive-ion yields induced by oxidation has been studied most extensively and will be referred to frequently in the following discussions. The key issue is the relationship between the chemical bond and the ionization mechanism of the sputtered atom.

a) Relation to Global Physical Quantities

Oxidation usually proceeds through chemisorption to the formation of surface oxides. Some global properties of the metallic surface change correspondingly: the work function of the surface is usually modified during chemisorption. The valence band on the surface is altered until an oxide band gap is formed.

There is no simple correlation between the secondary ion yields and changes in these properties. Adsorbed oxygen can induce an increase or decrease in the work function, depending on whether the adsorbed oxygen atoms reside above (e.g., Ni, Si, W) or beneath [e.g., Mg, Nb(110), Mo(100)] the first layer of the metal atoms. In all these cases, oxidation enhancement of the positive secondary ion yields was observed [3.22, 102]. *Martin* et al. [3.103] have proposed that the formation of oxide band gaps can reduce or prohibit the reneutralization of sputtered positive secondary ions by electron tunneling. For instance, the valence band edge of SiO$_2$, which has a band gap of 9 eV, lies about 10.2 eV below the vacuum level. Hence the ionization potential of Si (8.1 eV) falls within the oxide band gap where no electronic state is available.

In contrast to this model, there have been observations that the formation of a large oxide band gap is not a necessary condition for the oxygen enhancement of positive secondary ion yields. *Yu* and coworkers [3.99, 104] and *Sander* et al.

[3.58] have used in situ X-ray photoemission and Auger electron spectroscopy to monitor the oxide formation on Si. Both groups found that silicon suboxides $SiO_x(x < 2)$, which do not have a large band gap, are very effective in promoting Si^+ formation. *Yu* et al. [3.104] also used in situ ultraviolet photoemission to track the changes in the valence band, again with no dramatic increase in the Si^+ yield from SiO_2 band-gap formation. In fact, complete conversion from silicon suboxides to SiO_2 increased the Si^+ yield by only a factor of two to four [3.58, 99, 104]. This is small compared with the overall oxidation-induced enhancement of Si^+ yield, which exceeds two orders of magnitude.

b) Relation to Local Chemical Bonds

The idea that ionization is related to the breaking of local chemical bonds was first proposed by *Slodzian* [3.105]. If the bond is partially ionic, like that between a metal atom and an electronegative element, the metal atom is likely to preserve its charge after the bond is broken. This concept is generally accepted as an important mechanism in secondary ion emission [3.23, 106, 107]. In this model, the positive secondary ion yield (per incident ion) Y_M^+ of element M is the sum of the contribution from different bonding configurations (i) involving M [3.99, 108]:

$$Y_M^+ = \sum_i f_i P_i^+ Y_i , \tag{3.8}$$

where f_i, P_i^+, and Y_i are the atomic fraction, ionization probability, and partial sputtering coefficient of the M atoms bonded in the ith configuration. In the first approximation, all P^+ and Y_i values are constants specific to the chemical configuration and are independent of the concentration. The linearity between Y_M^+ and f_i is the unique feature of this localized interaction picture. Unambiguous verification of (3.8) was provided by low-fluence sputtering experiments on the following chemical systems: O-Be [3.109], N-Si(100), O-Si(100), and O-Ge(111) [3.99], where the yields of Be^+, Si^+, and Ge^+ were found to be linearly related to the amount of compounds formed on the surface.

Figure 3.7 shows a simple verification that the interaction for Si^+ formation is localized. Here the Si^+ yields with emission energies 8 eV and 17 eV, and a 45° emission angle, are plotted against the nitrogen coverage. The linear increase of the Si^+ yield with increasing nitrogen coverage indicates that the Si^+ emission is directly proportional to the number of Si_3N_4 molecules formed during nitridation. Expressing it algebraically,

$$Y_{Si}^+ = f_0 P_0^+ Y_0 + f_1 P_1^+ Y_1 . \tag{3.9}$$

The subscripts 0 and 1 represent unreacted Si and Si_3N_4 respectively. Since the Si^+ yield from a clean Si surface ($f_0 = 1, f_1 = 0$) is over 150 times less than that from a fully nitrided surface ($f_0 = 0, f_1 = 1$), $P_0^+ Y_0$ is much less than $P_1^+ Y_1$.

Thus $Y_{Si}^+ \simeq f_1 P_1^+ Y_1$ where f_1 is proportional to the nitrogen coverage. Hence, to a good approximation, the data are consistent with (3.9).

Equation (3.8) was also verified in systems [O-Si(100), O-Ge(111)] where more than one compound was found on the surface [3.99]. Using X-ray photoemission to estimate the amount of suboxides and dioxides formed, *Mann* and *Yu* [3.99] found that dioxides of Si and Ge are more effective emitters of Si$^+$ and Ge$^+$ respectively than their suboxide counterparts. Figure 3.8 shows that the Si$^+$ yields at 8 and 16 eV are proportional to the amount of oxides (normalized to the amount of Si) on the Si surface, and the yields are larger for SiO$_2$. In the above analysis, ion yield measurements only give the products $P^+ Y$, but not P^+ and Y separately; Y should be determined from single (oxide) phase surfaces under low-fluence conditions. Values from high-fluence sputtering experiments can only be used as a guide, because ion bombardment probably modifies the chemical composition.

Secondary ion yields as a function of the oxygen partial pressure during high-fluence-mode sputtering are most frequently reported [3.21, 22]. These data are difficult to analyze, because the oxygen-coverage dependences of the sticking coefficients are rarely known [3.110]. More informative is the dependence of the ionization probabilities on the surface oxygen concentration c_0. However, the experimental results diverge widely. *Deline* et al. [3.111] reported power-law dependences for many elements. *Oechsner* and *Sroubek* also reported a power-law-like dependence for Ta$^+$ from oxidized Ta [3.112]. For the sputtering of Si$^+$ from oxygenated Si, a linear dependence at very low coverages (< 3 at %) [3.113], a power law [3.111, 113, 114], and an exponential dependence [3.58, 115] have all been reported.

These observations can be understood in terms of the local interaction scheme. The averaged ionization probability $\tilde{P}^+$ is defined as the ratio of angle- and energy-integrated secondary ion yield Y_M^+ to the sputtering yield $\tilde{Y}$, as

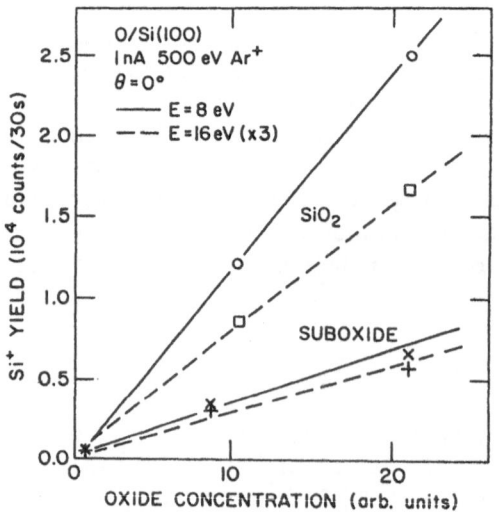

Fig. 3.8. Si$^+$ secondary ion yield as a function of the concentration of silicon suboxides and dioxide formed, as expressed by the silicon content, for two emission energies. The oxide concentrations were determined by X-ray photoemission [3.99]

determined by erosion experiments. The latter is an averaged value given by the following:

$$\tilde{Y} = \sum_i f_i Y_i \ . \tag{3.10}$$

Hence, without special attention, the measured ionization probability P^+ will be a weighted average given by

$$\tilde{P}^+ = \frac{\sum_i f_i P_i^+ Y_i}{\sum_i f_i Y_i} \ . \tag{3.11}$$

Since the product $P_i^+ Y_i$ is site dependent, $\tilde{P}^+$ depends critically on the distribution of the f_i. *Oechsner* and *Sroubek* [3.112] have modeled the case where the oxygen atoms are randomly distributed, and all Y_i are the same. They found that P^+ has a power-law-like dependence on c_0,

$$P^+ \propto [c_0(K - 1) + 1]^N \ . \tag{3.12}$$

The assumption is that the ion-yield-enhancement factor K per oxygen neighbor is a constant. Here N is the maximum number of nearest oxygen neighbors permissible. *Oechsner* and *Sroubek* have fitted (3.12) to the Ta^+ data obtained from the sputtering of oxidized Ta with K and N as free parameters [3.112].

In conventional SIMS experiments with high-fluence-mode sputtering, the distribution of compounds formed (hence f_i) depends on the experimental conditions. *Reuter* demonstrated this with X-ray photoemission monitoring the surface oxides formed during 10 keV O_2^+ bombardment of Si and Ge [3.116]. He found that sites of higher oxidation states are preferentially formed at smaller angles of incidence (i.e., closer to normal). Diverging data on the oxygen-concentration dependence of Si^+ yield sputtered from oxygenated silicon can be explained by a distribution of sites with different oxygen-coordination numbers.

Changing the electronegative species can strongly affect the secondary yield. Recently *Reuter* and *Clabes* [3.97] measured the positive secondary ion yields from 13 elements during 10 keV O_2^+ and F_2^+ high-fluence bombardment. They found that with F_2^+ the averaged ionization probability $\tilde{P}^+$ is at least an order of magnitude higher for Ni, Ag, Pd, and Ir; the same order of magnitude for Ge, Cu, Cr, Fe, Zr, Mo, and W; and an order of magnitude lower for Si than for O_2^+. Without knowing the distribution of the compounds formed, comparing quantitatively the effects of fluorine and oxygen on the ionization probabilities P_i^+ is not possible. A determination of the relationship between the electronegativity of the electronegative element and the ionization probability of the sputtered metal atom would be very useful.

The emission of negative secondary ions is mostly from the electronegative species. Negative metallic secondary ions have also been observed. However, both detailed experimental and theoretical studies are scarce. The O^-/O^+ ion

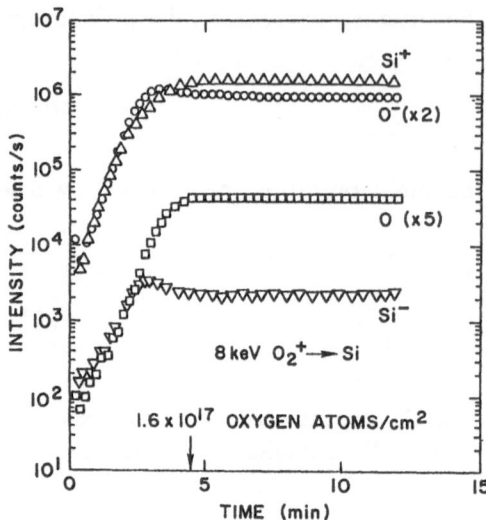

Fig. 3.9. Si$^{\pm}$, O$^{\pm}$ secondary ion yields as a function of oxygen fluence [3.101]

yield ratio has been reported to depend on the surface chemisorption sites on both W(100) [3.117] and Mo(100) [3.118]. *Mann* and *Yu* have reported that the sputtered O$^-$ yields from oxidized Si(100) surfaces are proportional to the surface oxygen coverage [3.99].

Metallic and semiconducting elements also give off negative secondary ions. *Williams* and *Evans* [3.100] have observed an interesting oxidation enhancement of Si$^-$, As$^-$, P$^-$, Ga$^-$, Cu$^-$, and Au$^-$ secondary ion yields over that observed for clean, unoxidized surfaces. *Wittmaack* [3.101] measured the Si$^{\pm}$ and O$^{\pm}$ yields during 8 keV O$_2^+$ bombardment of Si. Figure 3.9 shows his data as a function of the O$_2^+$ fluence. The ion yields are scaled to show the close correlations of the Si$^+$ and Si$^-$ yields with the O$^-$ and O$^+$ yields respectively, at low oxygen fluences ($< 1 \times 10^{17}$ at/cm^2). This is consistent with the various ions originating from similar sites initially. The ion yields saturating at different oxygen fluences suggests that there was more than one type of site, and that each type had its own ionization probabilities for the different secondary ions. The negative ion yields, after enhancement, are usually still more than an order of magnitude smaller than the corresponding positive secondary ion yields. No energy distribution measurement has been reported.

c) Dependences on Sputtered-Atom Properties

There have been many studies of positive secondary ion emission from oxidized multielement metallic targets [3.119–128]; P$^+$ was found frequently to follow a Boltzmann-like dependence on the ionization potential I (Fig. 3.10):

$$P^+ = \frac{Z_+}{Z_0} \exp\left(-\frac{I - \delta I}{k_B T_p}\right) .$$

$$(3.13)$$

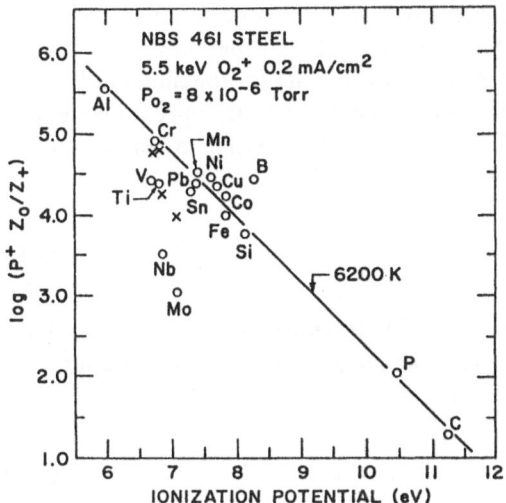

Fig. 3.10. Plot of $\log(P^+ Z_0/Z_1)$ versus ionization potential for the various elements sputtered from an oxygen-saturated NBS 461 steel sample. Crosses are the result from V, Ti, Nb, and Mo after applying molecular-ion correction [3.126]

Here Z_+ and Z_0 can be either partition functions [3.124] or multiplicities of the lowest-lying group of states [3.125] of the positive ion and the neutral atom respectively; δI is a fitting parameter, and k_B is the Boltzmann constant. Also, T_p has the dimension of temperature and ranges from 5000 K to 10 000 K, and is a function of secondary ion energy. *Morgan* and *Werner* [3.126] observed that T_p derived for the secondary ion data from oxidized NBS 461 steel alloy is 6200 K for energies between 0 and 20 eV, and 8400 K for ions between 40 and 60 eV. This Boltzmann-like relation has prompted speculation that a local thermal equilibrium plasma of temperature T_p exists at the sputtering site [3.124], but this has not been verified experimentally. There have been no equivalent studies of chemical enhancement on negative secondary ion emission as a function of the electron affinity.

Detailed studies of the emission energy (velocity) dependence of the ionization probability on well-defined chemical systems are very scarce. *Krauss* and *Gruen* [3.74] have reported that P^+ for Be sputtered from oxidized Be (BeO) approximates a power-law dependence on E_1 with the exponent ≈ 2. They also noticed that P^+ converges to a finite value at low emission energies. *Mann* and *Yu* [3.99] also reported that the yield ratio $Y_R \equiv (P^+ Y)_d/(P^+ Y)_s$ for the emission of Si^+ from silicon dioxide (d) and suboxide (s) depends on the emission energy of Si^+; Y_R was found to be 3.5, 2.7, and 2.6 for Si^+ emitted with 8, 16, and 21 eV, respectively. They reported that the data are consistent with an exponential dependence of P^+ on the inverse of the emission velocity:

$$P^+ \propto \exp(- v_0/v) \ . \tag{3.14}$$

In contrast to (3.6), P^+ does not correlate with the normal component of v [3.99].

The emission of positive secondary ions has an isotope effect. Since the chemical bonding is not expected to vary with the isotopic mass, this effect is probably the result of the difference in emission velocity for the different isotopes emitted with the same energy. The isotope effect can be represented by the isotope fractionation F_{ij}, which is defined by the relative difference of the ionization probabilities of isotope i and j [3.129],

$$F_{ij} \equiv \frac{P_j^+ - P_i^+}{P_j^+} \ . \tag{3.15}$$

Most reported data on F_{ij} are consistent with an emission velocity dependence as expressed by (3.14). This velocity dependence gives in the first approximation the following expression for F_{ij} [3.129]:

$$F_{ij} = -\frac{v_0(M_i - M_j)}{2vM_j} \ . \tag{3.16}$$

The linear dependence of F_{ij} on the isotopic mass difference $(M_i - M_j)$ was verified by *Lorin* et al. [3.130] for many inorganic compounds.

Lorin et al. [3.130] and *Gnaser* and *Hutcheon* [3.129] also observed the linear dependence of F_{ij} on the inverse of the emission velocity (Fig. 3.11). Some deviations at low velocities have been observed [3.131]. *Shimizu* and *Hart* [3.132] and *Schwarz* [3.133] reported more complex functional dependence of F_{ij} on E_1, including oscillating variations. The reason is unclear. The isotope effect also depends on the matrix. For a particular element, F_{ij} depends on the chemical compound [3.130]. This can be understood by noting that the parameter v_0 should change with the chemical bond.

Isotope effects on the emission of negative secondary ions have also been reported by *Yu* [3.134] for the emission of negative hydrogen/deuterium, and O^{16}/O^{18} secondary ions from chemisorption layers on W(100). The lighter

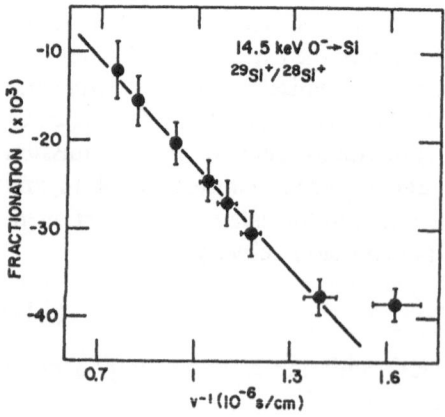

Fig. 3.11. Fractionation as a function of the inverse velocity for Si^+ isotopes [3.129]

isotopes gave the higher yields. This agrees qualitatively with the isotope effect of positive secondary ions.

3.2.4 Secondary Ion Formation by Energetic Collisions and Electronic Excitations

In the previous two sections, the mechanisms of secondary ion formation discussed are related to the chemical bonding and the electronic state of the surface. The primary function of the incident ion is to provide the energy for sputtering the surface atom. Hypothetically the ionization step is completely decoupled from the mechanics of sputtering; the ionization probabilities should be independent of the energy of the primary ions E_0. *Wittmaack* [3.135] and *Tsipinyuk* and *Veksler* [3.136] indeed found only a weak dependence of the ionization probabilities on E_0 for secondary ions emitted from many metallic targets. There are, however, many known cases in secondary ion emission where ionization cannot be decoupled from sputtering.

When a primary ion of energy E_0 strikes a target, the first few atomic collisions can be quite energetic. High-energy electronic excitations – for example, core holes – and low-energy excitations can result from these energetic collisions. These excitations can affect the charge state of the sputtered atoms.

a) Ion Formation in Direct-Recoil Sputtering

When a keV ion beam is directed onto a target surface, surface atoms can be sputtered off by an energetic direct recoil [3.137]. The energy E_r of the recoiled atom in the binary-collision approximation is given by

$$E_r = E_0 \left[\frac{4M_2/M_1}{1 + (M_2/M_1)^2} \right] \cos^2\theta_r , \qquad (3.17)$$

where θ_r, the recoil angle, is the angle between the direction of incidence of the primary ion and the recoiling surface atom.

Eckstein and coworkers measured the charged-state fractions of direct recoiled atoms from Ni [3.138–140] and deuterium-covered Ni targets [3.141]. They measured the ion yields directly, and used a calibrated N_2 stripping cell to detect sputtered neutral atoms. The ionization probability is given by the ratio of the ion yield to the neutral yield. They measured ionization probabilities that ranged from 10^{-2} to 10^{-5}. *Rabalais* and coworkers [3.142–147] used a pulsed ($\simeq 50$ ns) keV primary beam of noble-gas ions at grazing incidence and detected the recoiled ions and neutral atoms with an electron multiplier placed at the end of a flight tube. The mass spectra of the sputtered atoms were recovered from the time-of-flight data. They studied sputtering from compounds and chemisorbed layers on solid surfaces.

Similar to cascade sputtering from metal surfaces, an exponential dependence of the ionization probability on the inverse of the normal component of the

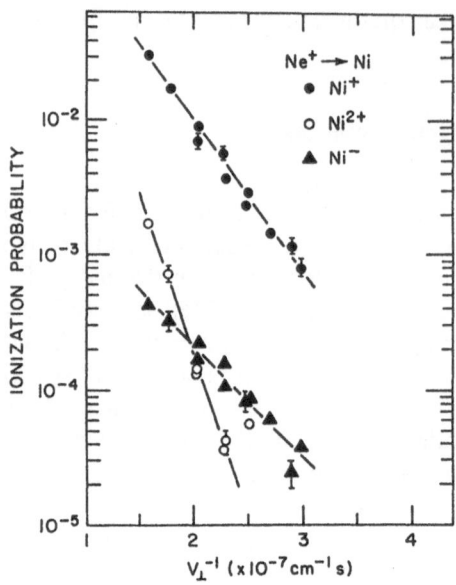

Fig. 3.12. The ionization probabilities of Ni direct recoils as a function of the inverse of the normal velocity component $v_\perp$. The Ne$^+$ incident angle was 70°; $v_\perp$ was changed by varying the Ne$^+$ incident energy and the emission angle (50°, 60°, 70°) [3.140]

emission velocity v (3.6) has been reported. This was observed for Ni$^+$, Ni^{2+}, Ni$^-$ [3.140], and D$^+$ and D$^-$ [3.141] from Ni surfaces. Figure 3.12 shows the results for the Ni ion yields. Some deviations from (3.6) were observed for Ti [3.140], although oxygen contamination was the suspected cause. Chemical effects were also observed. For example, during the direct-recoil sputtering of hydroxylated magnesium [3.147], a higher hydrogen-to-oxygen ratio on the surface enhances O$^+$ but suppresses O$^-$ emission.

A dependence on the primary ion was also observed. Both D$^+$ and D$^-$ were emitted with higher probability at equivalent $v_\perp$ under Ar$^+$ than under Ne$^+$ bombardment. On the other hand, Ne$^+$ bombardment was more effective in causing ionization in recoiled atoms than Ar$^+$ for Ni and Ti, perhaps because it is more efficient in core-hole formation than Ar$^+$ in these metals [3.140]. The results from direct-recoil sputtering indicate that both the energetic primary ion/target atom collisions and the chemical state of the target surface are important for the formation of ionized recoil atoms.

b) Ion Formation by Inner-Shell Excitations

Energetic collisions between the primary ion and target atoms (PT) as well as between target atoms (TT) in collision cascades can also cause secondary ion formation. This effect is particularly evident in the positive secondary ion emission from the light elements Al [3.88, 148–150], Mg [3.88, 148], and Si [3.148, 151] under usual cascade-sputtering conditions. The secondary ion yields from these elements increase rapidly with the primary ion energy, and they have unusually high multiply-charged positive secondary ion yields under noble-gas ion bombardment. For example, the Si^{2+}/Si$^+$ ratio for Si during

15 keV Ar^+ bombardment [3.151] is about 10^{-3}, whereas the Ta^{2+}/Ta^+ yield for Ta is 10^{-5} under identical sputtering conditions [3.135]. Another important observation is that ion-bombardment-induced Auger electron emission is fairly strong in these metals [3.152, 153]. Careful identification of the sharp (atomic-like) features on the Auger electron spectra showed that a significant portion of the electron yield comes from the decay of sputtered atoms and ions with inner-shell excitations [3.154, 155].

It is becoming evident that for the light elements Mg, Al, and Si, inner-shell excitations produced by energetic collisions can generate a significant amount of singly ionized positive secondary ions. This is in addition to the surface-electron-exchange mechanisms discussed above (Sects. 3.2.2, 3). The Auger electrons emitted from neutral sputtered atoms are a direct indication of this collisional mechanism. Ion-beam-induced Auger electrons from the following transitions

$$Al^0(2p^5 3s^2 3p^2) \quad \begin{cases} \rightarrow Al^+(2p^6 3s^2) \\ \rightarrow Al^+(2p^6 3s 3p) \\ \rightarrow Al^+(2p^6 3p^2) \end{cases}$$

$$Si^0(2p^5 3s^2 3p^3) \quad \begin{cases} \rightarrow Si^+(2p^6 3s^2 2p) \\ \rightarrow Si^+(2p^6 3s 2p^2) \end{cases}$$

have been identified [3.154, 156]. The sputtered atoms would be left in the singly ionized state after the transition. *Hennequin* et al. [3.157–159] recently proposed dividing the measured secondary ion yield into the collisional contribution that involves core electrons and the contribution of surface mechanisms that involve valence electrons. They assumed that these two contributions are independent. The surface contribution is linear with the concentration of the element in the target, aside from corrections to the effect of work-function changes. The collisional contribution was assumed to be from TT collisions (see below), and has then a quadratic dependence on the elemental concentration. Working with Fe-Al [3.160] and Ag-Mg [3.157] alloys, they estimated experimentally that the inner-shell excitation mechanism contributed 30% of the Al^+ and 65% of the Mg^+ from their respective elements, when they were sputtered by 10 keV Ar^+. Although no such quantitative estimate has been made for Si^+ sputtering from silicon, data has been reported that is consistent with the collisional mechanism. *Wittmaack* [3.135] has observed that the Si^+ yield depends strongly on the primary ion energy, increasing by tenfold when the energy of the bombarding Ar^+ (on Si) was increased from 5 to 15 keV. *Yu* and *Reuter* [3.91] also reported that the Si^+ yield from Si-Ni alloys is proportional to the square of the Si concentration when subjected to 15 keV Ar^+ bombardment.

The emission of doubly ionized positive secondary ions from these three light elements has been studied most extensively. The ionization probabilities for these species have been estimated [3.22] to be about 10^{-4} for 6 keV Ar^+

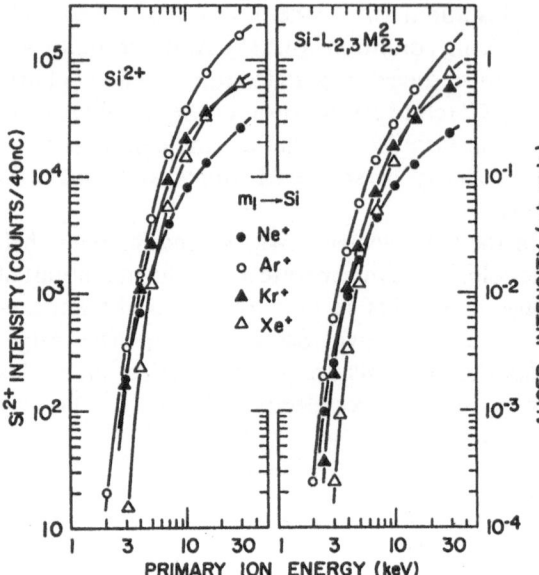

Fig. 3.13. Primary ion energy dependences of the Si^{2+} and Si (LMM) atomic Auger yields for four different projectiles [3.163]

bombardment. These ion yields increase rapidly with the primary ion energy [3.148, 161], and the correlation with the ion-induced emission of atomic-like *L*-level Auger electrons is strong [3.162, 163]. Figure 3.13 shows the data reported by *Wittmaack* [3.163]. The dependences of the Si^{2+} and the atomic-like LMM Auger yields on E_0 are practically identical. This suggests that these doubly ionized positive secondary ions are formed by the Auger decay of singly ionized sputtered atoms with a $2p$ core hole. The idea is reinforced by the observation of respective Auger electrons emitted from the sputtered ions [3.154, 156]. The Auger electron yields scale with the maximum energy that can be transferred in a collision between the primary ion and the target atom [3.156, 164]. This energy is γE_0, where $\gamma = 4M_1 M_2 (M_1 + M_2)^{-2}$ [Ref. 3.1, Eq. (2.2.2)]. The core holes clearly originate in the collision cascades induced by the impact of the primary ions.

The strongest evidence that the core hole results from TT collisions came from the observation that the multiply charged ion intensity varies roughly with the square of the concentration in the alloy [3.150], i.e., with the number of TT neighbors. Another related observation is that the multiply charged secondary ion yields frequently decrease when the metal surface is oxidized, even though the singly charged ion yield is enhanced significantly [3.148, 150, 151] by the chemical effects. During oxidation, the oxygen atoms reduce the number of TT neighbors, and hence the number of symmetric collisions. This is similar to the effect of alloying. The present understanding is that TT symmetric collisions are particularly effective in creating core holes in sputtered atoms [3.165]. Core-hole decay results in the Auger electron emission and the ionization of the sputtered atom.

Triply charged positive secondary ion yields have also been measured. They are usually more than an order of magnitude smaller than the corresponding doubly charged ion yields [3.150, 151]. A study of Al alloys has found that the ratio Al^{2+}/Al^{3+} does not depend on the Al concentration nor on the primary ion energy [3.160]. Similar independence from the primary ion energy has also been reported for the ratio Si^{2+}/Si^{3+} [3.166]. These triply charged secondary ions most likely originate from similar inner-shell excitations. Very few studies have been made on the corresponding negative secondary ion emission. *Wittmaack* reported for Xe^+ bombardment that the Si^- yield has the same dependence on E_0 as Si^+ yield [3.166]. He suggested that the ionization of Si^+ and Si^- is governed by the same process. However Si^+ and Si^- ions have different energy distributions. The mechanism for Si^- production is still not clear.

The energy distributions of the multiply charged secondary ions are usually similar to those of singly charged ions [3.88], with the most probable energies around several eV. High-energy components in the few hundred eV region have been observed [3.157], and the data are shown in Fig. 3.14. This is not surprising in view of the high-energy collisions required to create the inner-shell excitations. The observations are in qualitative agreement with the recent theory of *Veksler* [3.167].

Secondary ions created through inner-shell excitations may also result from the energetic collision of the primary ions with target atoms. During the bombardment of Si by Ar^+ *Wittmaack* [3.161, 164] observed an additional increase in the Si (LMM) atomic-like Auger electron yield and Si^+ ion yields when E_0 exceeded 4 keV. This energy threshold is related to the minimum energy required to generate the Si $2p$ core hole during an Ar-Si collision.

Inner-shell excitation may also cause the emission of adjacent atoms. *Wittmaack* studied the sputtering of H^+ from hydrogenated amorphous silicon by noble-gas ions [3.168]. He found that while the H^+ yield increased by three orders of magnitude when the primary ion energy was increased from 3 to 30 keV, the ratio of H^+ yield to Si^{2+} yield remained quite constant. He

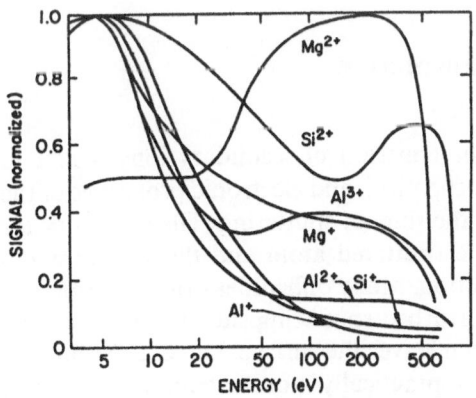

Fig. 3.14. Energy distributions of secondary ions emitted during the sputtering of Al, Mg, and Si by 10 keV Ar^+ at 35° [3.157]

suggested that SiH$^+$ molecules with Si $2p$ core holes were first sputtered; H$^+$ ions were ejected as the molecules disintegrate during the Auger de-excitation process. *Williams* [3.169] has studied the sputtering of F$^+$ from fluorinated silicon by Ar$^+$ and O$_2^+$. He observed a strong correlation between the F$^+$ yield and the Si^{2+} yield, and concluded that Si $2p$ core holes were produced in the Si target. During the interatomic (Si-F) Auger de-excitation, F$^+$ was desorbed [3.170].

c) Ion Formation by Electronic Excitations

The primary ion also dissipates its energy into the electrons of the target. We have already seen that such electronic excitations affect the secondary ion emission from GaAs and InAs (Sect. 3.2.2d). These secondary electrons can also excite the bonds of surface atoms, causing the ejection of secondary ions through electronic sputtering [Ref. 3.2, Chap. 4] or electron-stimulated desorption [3.171]. Positive ions of electronegative species are often emitted by this process. The major difference from cascade sputtering is that the atom (ion) is ejected by the repulsive potential of the excited state, and not by the momentum transfer from an atomic collision. This is outside the scope of this chapter, so we provide only a few examples.

Schultz et al. have studied the emission of F$^+$ from LiF during noble-gas ion bombardment [3.172]. They concluded that for He$^+$ and Ne$^+$ bombardment, F$^+$ is ejected during the Auger neutralization of the primary ions. The F$^+$ yield is independent of the primary ion energy E_0, for E_0 less than 600 eV.

Blauner and *Weller* [3.173, 174] have studied the sputtering of O$^+$ from oxidized V and Al by 25–275 keV noble-gas ions. They found that the O$^+$ yields were proportional to the electronic stopping power for Ne$^+$, Ar$^+$, and Kr$^+$ bombardment. The authors proposed that the secondary electrons generated by the primary ions caused the desorption of O$^+$. This explanation is plausible, because electron-stimulated desorption of O$^+$ has been observed from the oxides of these elements [3.171].

3.3 Models for Secondary Ion Formation

The common approach to explain the emission of secondary ions is first to assume a description of the structural, chemical, and electronic properties of the sputtering site on the solid surface at the time of sputtering. Then a theory for dynamic electron transfer between the sputtered atom and the surface is developed on the basis of such an assumption. Many difficulties originate from our lack of knowledge about the highly perturbed sputtering site during the crucial 10^{-13} s during which the sputtered atom leaves the surface. Direct experimental information of this transient state is practically nonexistent. Hence many models, each with a different assumption about the physical state of the sputtering site, have been proposed. Because of the vast variety of material systems and experimental conditions, certain models are likely to be more suitable for particular situations. The mechanisms for negative-ion formation can be very different from the mechanism for positive-ion formation, even for the same atomic species. Hence we must recognize the boundary conditions of the individual models before applying them to given experimental situations. Both classical and quantum-mechanical models were proposed during the early phase of development in this field. They include the local thermal equilibrium model of *Andersen* and *Hinthorne* [3.121], and the quantum-mechanical models of *Schroeer* et al. [3.175], *Sroubek* [3.176], *Cini* [3.177], *Antal* [3.178], and *Prival* [3.179]. Here we concentrate on more recent developments.

3.3.1 Secondary Ion Emission from Metals and Semiconductors

On metallic surfaces, the electronic interaction between a sputtered atom and the substrate is not localized to one or two electron orbitals, as in a molecule. When an atom is sputtered from a metallic surface, breaking of the metallic bond involves a large number of electrons in the valence band. Also during the sputtering of a positive or negative ion, the electron or hole excitations generated in the metal would be heavily screened by other electrons. The electronic excitation is unlikely to stay localized at the sputtering site, because of the high electron mobility in the metallic substrate. The models presented here reflect this many-electron character. We include certain aspects of secondary ion emission from semiconductors, where they can be discussed in the same context as the metals. Semiconductors with their covalent bonds and band gaps deserve special attention, which has been lacking so far.

a) Computer Simulations

In the simulation of metallic systems, it is usually assumed that all the atoms in the substrate are identical and the interatomic potentials are relatively isotropic. The ionization process is then calculated from first principles [3.180–182].

The usual procedure is as follows: A cluster of atoms (j) is set up to simulate the metallic substrate. A particular atom (a), which can be the same or a different kind, is chosen to be the sputtered atom. The cluster is held together by an interatomic potential $U(r)$, where r is the relative position between two atoms. Usually $U(r)$ is chosen to be the Morse potential $U_M(r) \sim \exp(-2\alpha r) - 2\exp(-\alpha r)$ for r near and beyond the equilibrium distance, whereas the repulsive Born-Mayer potential $U_{BM}(r) \sim \exp(-\beta r)$ is used at small r. Sputtering is simulated by introducing a fast incoming primary atom, or the sputtered atom is allowed to desorb from the surface with preassigned initial conditions. Since the motion of the atomic nuclei is much slower than the velocity of the electrons, the classical trajectory approximation can be used. The evolution of the atomic positions $r_j(t)$ is determined by classical mechanics.

To calculate the ionization probability P^+, the atoms can be assumed to have only a single electronic level each (i.e., only ionization is allowed) or several electronic levels each (i.e., both ionization and excitation are allowed). The Hamiltonian for the electron of the individual atom is taken to be

$$H = H_a + V(r(t)) ,\tag{3.18}$$

where H_a is the atomic Hamiltonian with atomic eigenfunctions Φ_j. Here $V(r)$ is the electronic-interaction potential. The molecular wave functions $\Psi_k(t)$ of the cluster can be expanded in terms of the Φ_j's:

$$\Psi_k(t) = \sum_j c_{jk}(t)\Phi_j .\tag{3.19}$$

The matrix elements $\langle\Phi_i|V|\Phi_j\rangle$ as a function of r_{ij} are either assigned or calculated with an assumed expression for V – e.g., $\sim \exp[-\gamma(r_{ij} - r_{ij}^0)] - r_{ij}^0$ is the equilibrium bond length. Note that all the potentials can be adjusted through the parameters α, β of the Morse and Born-Mayer potentials, and γ. System evolution is calculated with the time-dependent Schrödinger equation $H\Psi_k = i\hbar\partial\Psi_k/\partial t$ or with the Heisenberg representation, using the classical trajectories $r_j(t)$. The ionization probability P^+ of the sputtered atom (a) is given by

$$P^+ = 1 - \sum_k |c_{ak}(t = \infty)|^2 .\tag{3.20}$$

In principle, this treatment can be extended to the formation of negative ions.

Sroubek et al. studied the sputtering of a one-level atom from a cluster of five two-level atoms [3.180]. They calculated one collision sequence for each emission velocity. Interestingly, they found an exponential relation between P^+ and the ionization potential I of the sputtered atom. They also found that P^+ is virtually independent of velocity, which they have attributed to the inclusion of substrate excitations.

Garrison et al. [3.181] studied extensively the effects on P^+ of changing the number of atoms in the cluster (up to 18), the number of energy levels per atom (up to three), the strength of the interaction potential V, and the velocity v of the sputtered atom. An adequate treatment of the ionization process required a cluster of ten atoms. They obtained a dependence of P^+ on v with or without excited levels in the atoms. The velocity dependence ranges from a power law v^n dependence where n is between 2 and 4 at small v, to an exponential dependence on v^{-1} as in (3.6), when the interaction is large. Also observed were some interesting oscillatory variations of P^+ with v. *Olson* and *Garrison* [3.182] obtained even stronger oscillations in the calculations for the Na on W system and attributed these oscillations to a quantum interference effect during nearly resonant electron transfer. This interference was observed [3.183] in the neutralization of noble-gas ions on surfaces when the ionization potential I is nearly equal in energy to a core level of the surface atom, but not in secondary ion emission. These oscillations may be an artifact due to the small number of substrate electronic levels included in the calculations. For a real metal with a continuum of electronic states, such interference effects would be averaged out.

Goldberg et al. [3.184] attempted to develop a more self-consistent calculation with molecular-orbital states. They followed the molecular states of the system and its energies along the sputtered-particle trajectory, so that there was no explicit assumption on the time dependence of either the interaction potentials or the energy positions. They calculated the sputtering of an Al atom from a linear cluster and found an exponential dependence of the ionization probability on the inverse of the emission velocity for $E_1 > 15$ eV. At lower energies, the ionization probability levels off with slightly oscillatory behavior.

The computer simulation approach is still rudimentary. *Olson* and *Garrison* [3.185, 186] have made some progress in deriving more appropriate functions to simulate realistic systems. Simulation, however, provides a universal way to calculate secondary ion yields with combined information on both the momentum transfer and the ionization process.

b) Autoionization of Sputtered Atoms

Blaise and *Slodzian* [3.87–90] postulated that sputtered atoms can be excited into autoionizing states by perturbation on the outer electronic shells when they cross the metal-vacuum interface. These autoionizing states are precursors for the formation of positive secondary ions. For example, Cu has an autoionizing state $3d^9 4s5s$, which is 0.1 eV above the ionization energy of the ground state $3d^{10} 4s$ [3.88]. The autoionizing state can decay through an Auger transition: $4s \rightarrow 3d$, with the ejection of the $5s$ electron to form the $Cu^+(3d^{10})$. In the one-electron approximation, the two electronic levels $4s$ and $5s$ with energies ε_1 and ε_2 can be populated by the two electrons from the valence band, provided that $\varepsilon_F > \varepsilon_M = (\varepsilon_1 + \varepsilon_2)/2$, the mean energy of the electrons in the autoionizing state. Then the probability P^*_{auto} of forming the autoionizing state is a self-convolution of the density of states $\varrho(\varepsilon)$ of the valence band:

$$P^*_{auto} \propto \int_0^{\varepsilon_F} \int_0^{\varepsilon_F} \varrho(\varepsilon_1')\varrho(\varepsilon_2')\delta(\varepsilon_1' + \varepsilon_2' - \varepsilon_1 - \varepsilon_2)\,d\varepsilon_1'\,d\varepsilon_2' \ . \tag{3.21}$$

The δ function simply states the energy-conservation condition. For most elements they have studied $\varepsilon_F - \varepsilon_M \sim 1$ eV. Basically, ε_1 and ε_2 are populated by electrons from a narrow band just below ε_F, and P^*_{auto} would simply be proportional to $\varrho^2(\varepsilon_F)$. The effects of broadening and shifting atomic levels close to a metal surface have not been considered.

This model is consistent with the secondary ion data from dilute transition-metal alloys obtained by *Blaise* and *Slodzian*, but is not consistent with other transition-metal alloy data (Sect. 3.2.2c). The energy condition $\varepsilon_F > \varepsilon_M$ implies that secondary ion formation is favored by a higher ε_F (i.e., a smaller work function), which is in contrast to experimental observations (Sect. 3.2.2a). The applicability of this autoionization model is very much in doubt.

c) Electron-Tunneling Model

When an atom is sputtered from a metal, charge transfer can happen between its atomic level and the delocalized states of the valence band of the metal. The electron-tunneling model describes the electronic transition as a resonant electron-transfer process (i.e., tunneling) between a sputtered atom and the valence band. This many-electron description of the transition is consistent with the delocalized nature of the metallic bond. Conceptually, this model is equivalent to the crossing of the atomic level of the sputtered atom with many electronic levels of the solid [3.187]. The electron-tunneling model was first proposed qualitatively by *van der Weg* and *Rol* [3.188] in 1965. More recently *Yu* [3.66] and *Yu* and *Lang* [3.63] used the concept to interpret the experimentally observed work-function dependences of the ionization probabilities. This is the most comprehensive secondary ion emission model developed so far.

In its simplest form, the metal is represented by a nearly-free-electron valence band with constant density of states and a work function Φ (Fig. 3.15). This is a reasonable assumption for an s electron of the sputtered atom interacting with the sp band of the metal substrate. The electron can tunnel between ε_a, the energy level of the sputtered atom, and the electronic levels of the same energy in the valence band. All the electronic excitations resulting from electron tunneling are assumed to dissipate rapidly and do not affect the tunneling probability. The tunneling probability is determined by the magnitude of the hopping matrix element V_{ak} between the atomic state $|a\rangle$ and the metal state $|k\rangle$. Here V_{ak} is given by

$$V_{ak}(z) = \langle k|V|a\rangle , \tag{3.22}$$

where V is the interaction potential. Since the electron has a finite lifetime in state $|a\rangle$, the atomic level is broadened in energy according to the uncertainty

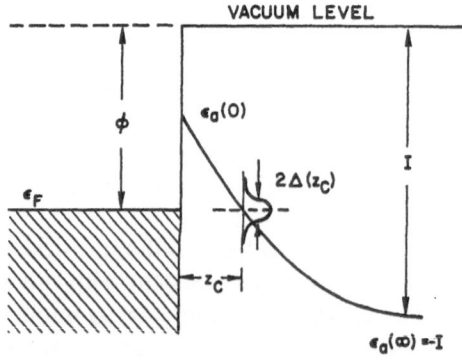

Fig. 3.15. Schematic energy diagram of a sputtered atom leaving a metal surface. The Fermi level ε_F lies below the vacuum level by the work function Φ. The atomic level ε_a is a function of the distance from the surface; ε_a equals ε_F at the crossing point z_c. Here $2\Delta(z)$ of the atomic level is a measure of the tunneling probability

principle. The half width $\Delta(z)$ as a function of the distance z normal to the surface is given in atomic units by

$$\Delta(z) = \pi \sum_k |V_{ak}(z)|^2 \, \delta(\varepsilon_k - \varepsilon_a(z)) \ . \tag{3.23}$$

The electron-tunneling model readily explains the observed relation between I and Φ in the Cs^+ experiment (Fig. 3.2 and Sect. 3.2.2a); Cs atoms are known to be chemisorbed essentially as Cs^+ on most metal surfaces at small coverages, with the empty $6s$ level lying above the Fermi level of most metals. Here $\varepsilon_a(z)$ varies with the image potential. If the work function Φ of the surface is larger than the ionization potential I of Cs (3.9 eV), the Cs level of the sputtered Cs atom will always face empty states of the metal, and little neutralization by electron tunneling can occur; i.e., $P^+ = 1$. When Φ is lowered by the Li overlayer such that $\Phi < I$, the Cs level would have to cross the Fermi level (at $z = z_c$) as the atom escapes, making the neutralization by electron tunneling energetically possible. This causes a rapid decrease in the Cs^+ yield when $\Phi < I$. However, the "onset" of this decrease for metallic substrates (~ 3.5 eV) is shifted by ~ 0.4 eV from the ionization potential of Cs (3.9 eV). The reason is that when Φ is only slightly smaller than I, the crossing point z_c is too far away from the surface for electron tunneling to be effective in the neutralization.

This experiment is presently the most direct verification of the electron-tunneling model. The case with the Si(111) substrate is complicated by the existence of the energy gap (~ 1.1 eV). Aside from the gap states, the role of the Fermi level should be replaced by the valence-band edge. The further shift of the "onset" by 0.8 eV is qualitatively in agreement with the tunneling picture. A detailed theoretical analysis for Si is not yet available.

For a sputtered atom leaving the surface with a classical trajectory $r(t)$, the occupation probability P of the atomic level has been obtained by *Blandin* et al. [3.189], *Brako* and *Newns* [3.190], and *Lang* [3.57]:

$$P = P_1 + P_2,$$

$$P_1 = n_a(0) \exp\left[-2 \int_0^\infty \Delta(r(t)) \, dt \right], \tag{3.24}$$

$$P_2 = \frac{1}{\pi} \int_{-\infty}^{\varepsilon_F} d\varepsilon \left| \int_0^\infty dt \, [\Delta(r(t))]^{1/2} \exp\left\{ i\varepsilon\tau + \int_t^\infty [i\varepsilon_a(r(t')) - \Delta(r(t'))] \, dt' \right\} \right|^2$$

in atomic units. If the atomic level is the ionization level, then P^+ equals $1 - P$. If the atomic level is the electron affinity level, P^- is equal to P. The first term P_1 is proportional to the initial occupation $n_a(0)$ of the atomic level and hence carries the memory of the initial state of the sputtered atom. The second term includes the effect of quantum interference as encountered in simulations (Sect. 3.3.1a), except that now all levels of the valence-band continuum are included. To calculate P we need to know how Δ and ε_a vary along the trajectory. For simplicity, it is common to assume that the metal surface has no lateral inhomogeneity and all quantities depend only on z, and $z = 0$ at $t = 0$. Since the wave function $|k\rangle$ decays exponentially with z, it is a good approximation to use a simple exponential form for Δ:

$$\Delta(z) = \Delta_0 e^{-\gamma z} , \tag{3.25}$$

where γ^{-1} is a characteristic decay length and ε_a varies with z according to the image potential and the chemical force of the system.

According to the first term P_1 of (3.24), memory of the initial occupation $n_a(0)$, including any excitations or excess charge on the sputtered atom initially, is erased very rapidly as the particle moves away from the surface. *Norskov* and *Lundqvist* [3.191] argued that because of the large Δ on the surface, the lifetime of these excitations is too short to contribute to P. This becomes evident if we insert reasonable numbers. With $z = v_\perp t$, P_1 equals $n_a(0) \exp(-2\Delta_0/\gamma v_\perp)$. If $\Delta_0 = 2 \text{ eV}$, $\gamma^{-1} = 10^{-8} \text{ cm}$, and $v_\perp = 5 \times 10^5 \text{ cm/s}$, then P_1 equals 10^{-53}, which is outside the range of detectability. In the electron-tunneling model, the bonding information comes indirectly through $\Delta(z)$ and $\varepsilon_a(z)$. Hence any valence-electron excitation that the sputtered atom has at the instant of sputtering cannot have a large effect on the secondary ion yield. For all practical purposes, the probability to fill the atomic level is given by the second term P_2, which will be discussed below. Possible exceptions are nonmetals, e.g., semiconductors, where no or very few final states for electron tunneling are available, because of the presence of band gaps and filled valence bands. In such cases, the electronic excitations may not be able to deexcite completely within the time scale of sputtering.

In general, P can be calculated by numerical integration of (3.24). Analytic expressions for P have been derived for a few limiting cases. According to *Blandin* et al. [3.189], for a constant ε_a and a large and constant $v_\perp$,

$$P \simeq \frac{1}{2} + \frac{\varepsilon_F - \varepsilon_a}{\gamma v_\perp} . \tag{3.26}$$

When $v_\perp$ is small,

$$P \simeq \begin{cases} 0 \\ 1 \end{cases} \pm \frac{2}{\pi} \exp\left(\frac{-\pi|\varepsilon_F - \varepsilon_a|}{\gamma v_\perp}\right) \quad \text{for} \quad \begin{cases} \varepsilon_F < \varepsilon_a \\ \varepsilon_F > \varepsilon_a . \end{cases} \tag{3.27}$$

However, if ε_a varies rapidly with z and crosses ε_F at the crossing distance z_c with a large slope, P has an exponential dependence on both $\Delta(z_c)$ and $v_\perp$ [3.57]:

$$P \simeq \exp\left[\frac{-2\Delta(z_c)}{\hbar\gamma v_\perp(z_c)}\right] . \tag{3.28}$$

An a priori calculation for the emission of Cs^+ from an Al substrate using the above expression fits the experimental data very well (Fig. 3.16) [3.63]. The

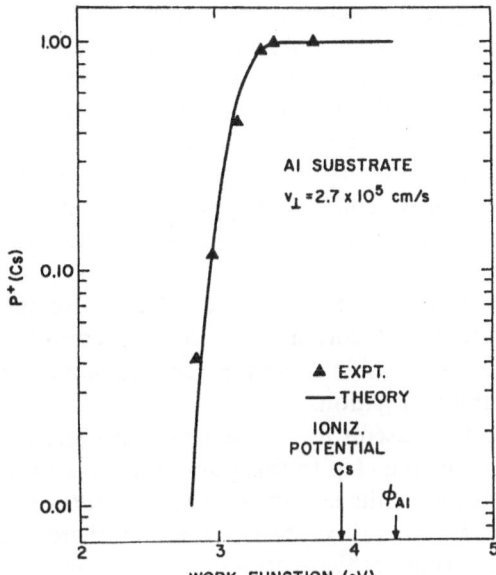

crossing distance z_c is on the order of several Å. The theory also successfully explained the work-function and emission velocity dependences in the sputtering of O^- [3.57].

Here we should emphasize that the exact functional dependence of P on the emission velocity v varies according to the way ε_a and Δ change with z. *Blaise* and *Nourtier* [3.22] have shown that if (3.25) is replaced by a linear decay, or $\varepsilon_a(z)$ or $\Delta(z)$ has a discontinuity in its derivatives, P would have a more power-law-like v dependence.

The ionization probability depends on the normal component $v_\perp$ of the escape velocity because P is related to the amount of time the sputtered atom spends in the interaction region. It results in a simple angular dependence, which agrees with experimental observations (Sect. 3.2.2b). This is in addition to any angular anisotropy associated with the substrate crystallinity. In general, a surface-binding energy correction must be introduced to convert the measured escape velocity to the velocity at the crossing point. For low (< 10 eV) kinetic energy secondary ions, such correction can be substantial [3.57, 192].

Expression (3.28) also gives, for a small range of $\Delta\Phi$, an exponential dependence of Φ and $v_\perp^{-1}$ [3.57, 191]:

$$P^+ \simeq \exp\left[-\frac{I-\Phi}{\varepsilon_p}\right] \quad \text{and} \quad P^- \simeq \exp\left[-\frac{\Phi-A}{\varepsilon_n}\right], \qquad (3.29)$$

where ε_p, ε_n are proportional to $v\cos\theta$ evaluated at z_c. By assuming that the sputtered ion has to overcome a Morse potential on the surface, *Lang* [3.57] calculated the theoretical values of ε_n as a function of the measured normal component of the emission velocity for O^-, and found good agreement with the experimental data (Fig. 3.6). Hence the theory can reasonably explain the observed dependences of the ionization probabilities on work function, emission velocity, and emission angle for sputtered atoms from adsorbed layers on metallic surfaces (Sect. 3.2.2a,b). No analytical expression has been derived yet for cases where ε_a does not cross ε_F rapidly. This may happen to pure metals, where ε_a converges to a value close to ε_F at $z = 0$ [3.193]. Further investigation is needed.

The electron-tunneling transition is a one-electron process. Recent numerical calculations [3.194, 195] suggest that for favorable situations, the contribution from Auger transitions may be comparable to single-electron tunneling. This result warrants an experimental investigation.

The electron-tunneling model is very useful for explaining many general trends in secondary ion emission from metals. Up to this point, the model was developed for ideal situations. For example, the sputtered atom is assumed to interact with a perfect metal surface. In reality, during ejection there are structural changes and electronic disturbances that can modify the ionization probability. *Nourtier* et al. [3.196, 197] recently pointed out that this correction can become significant when the ionization probability calculated by (3.28) for the ideal case is below 10^{-6}. They considered sputtering from a pure metal (Cu) where the sputtered atom is pushed off by another atom. The interaction between the two atoms weakens the coupling to the metallic substrate, making the sputtered atom more likely to retain a nonequilibrium charge state.

The other concern is the electronic excitations generated in the collision cascade as the primary ion dissipates its energy. Recently, there were also discussions on the effect of the excitation generated by the tunneling electron [3.198]. *Sroubek* [3.199] modeled the substrate by a continuum of electronic states with a Fermi level ε_F, and postulated that the electrons in the cascade region are excited to a temperature T_s. To calculate T_s, he estimated the electronic energy density dissipated by the primary ions in the collision cascade for Cu, Si, and GaAs and equated it to the energy density of the excitations at temperature T_s. Within the present model, T_s was estimated to be on the order of a few 10^3 K. For Cu, which has large carrier concentrations, T_s is several times smaller than for Si and GaAs, suggesting that electronic excitations are most important for semiconductors. Also, T_s was found to be time dependent and nonuniform across the collision cascade as the latter evolves. In this model, T_s is also virtually independent of the primary ion energy E_0, because the volume of the collision cascade increases with E_0. This result is in contrast to the strong primary ion energy E_0 dependences found experimentally in semiconductors (Section 3.2.2d), which are frequently regarded as the signature for the effect of electronic excitations. This discrepancy still has to be resolved. In practice, T_s is often treated as a fitting parameter for the experimental data.

With the electrons at a finite temperature, a Fermi factor

$$f(\varepsilon, T_s) = \left[1 + \exp\left(\frac{\varepsilon - \varepsilon_F}{k_B T_s}\right)\right]^{-1} \tag{3.30}$$

has to be included in the integrand of P_2 in (3.24). Recently *Kasai* et al. [3.200] analyzed the available experimental data and found T_s to be about 900 K for the sputtering of Cs^+ from Al, and about 2000 K for the sputtering of O^- from vanadium.

In the high T_s limit, *Zavadil* [3.201] and *Geerlings* et al. [3.202] showed that the result of the tunneling model can be expressed by a rate equation:

$$\frac{dP}{dt} = -2\Delta(z(t))\left[P - P_0(z(t))\right], \tag{3.31}$$

where P_0 is the equilibrium population at distance z and can be approximated by $P_0 = \exp[(\varepsilon_F - \varepsilon_a(z))/k_B T_s]$. *Sroubek* [3.93, 94, 203, 204] has used this rate equation to investigate the effect of substrate excitations in secondary ion emission. Making the usual assumption (3.25) that Δ decreases exponentially with increasing distance z from the surface, and that the sputtered atom moves with uniform velocity v such that $z = v_\perp t$, the ionization probability in the limit of high T_s is given by

$$P^+ = \exp\left[-\frac{\varepsilon_F - \varepsilon_a(\chi_0)}{k_B T_s}\right], \tag{3.32}$$

where χ_0, the "freezing distance" [3.205], equals $(1/\gamma) \ln(2\Delta_0/\gamma v_\perp)$. The velocity dependence of P^+ depends on how the position of the atomic level $\varepsilon_a(z)$ varies with z as discussed above. *Sroubek* [3.203] assumed that $\varepsilon_a(z)$ varies linearly from $\varepsilon_a(0)$ to $\varepsilon_a(\infty)$ in a distance Γ^{-1}, then P^+ increases exponentially with the work function and P^+ has a power-law relation with $v_\perp$. The exponent is $[\varepsilon_a(0) - \varepsilon_a(\infty)]\Gamma/\gamma k_B T_s$. This exponent decreases linearly with increase in work function. Although Sroubek did not discuss the negative-ion formation, the derivation should be similar.

Another interesting subject is the local work-function effects. Work-function changes are usually induced by the adsorption of atoms or molecules that have electric dipole moments on the surface. The electric dipole field is long ranged. *Geerlings* et al. [3.206] illustrated this experimentally. They reported that during the scattering of Li^+ by Cs-covered W(110), the neutralization cross section of Cs is on the order of 600 bohr^2, which is equivalent to a circle of about 15 Å in diameter. Hence the potential at any point is the sum of the contributions of a large number of dipoles on the surface. According to the calculations of *Barlow* and *MacDonald* [3.207], the inhomogeneous electric field from these individual dipoles decays on the order of $r/2\pi$ where r is the distance between dipoles. The field inhomogeneity effect would be of concern only if the crossing distance z_c in

(3.28) becomes comparable with or smaller than the decay distance. For example, if the adsorbate coverage exceeds a tenth of a monolayer ($r \leq 10$ Å), the local-field inhomogeneity is not a first-order effect even when z_c is only 1.6 Å. For instances when either the adsorbate (dipole) coverage or the crossing distance is very small (e.g., when $I \gg \Phi$), the local work-function effect would be large.

3.3.2 Secondary Ion Formation by Local Interactions

In this section, we focus on the chemical effects in secondary ion emission. We briefly review the local thermal equilibrium model, the molecular dissociation model, the surface polarization model, and the bond-breaking model. Of these models, the bond-breaking model has received the most attention.

a) Local Thermal Equilibrium Model

This was the first model proposed to rationalize the apparent Boltzmann-like relation between P^+ and I (3.13) found experimentally. *Andersen* [3.119, 121, 124] postulated a local thermal equilibrium plasma of temperature T_p to exist at the sputtering site. Then the Boltzmann relation (3.13) follows naturally. Equation (3.13) has been used frequently for the semiquantitative analysis of SIMS data [3.120, 122, 208]. Although some theoretical analysis of this model has been made [3.209], there is still no experimental evidence for a plasma at the sputtering site. This model also does not provide a mechanistic explanation for the extreme sensitivity of the ionization process to the surface chemical state.

b) Dissociation of Sputtered Excited Molecules

Oxide molecule emission is frequently observed during the sputtering of oxides. *Thomas* [3.210] and *Gerhard* and *Plog* [3.211–213] proposed that the oxygen-enhancement effect originates from the dissociation of highly excited sputtered oxide molecules ($MO^* \rightarrow M^+ + O^-$) at a distance from the surface where the electronic influence of the surface, such as electron tunneling, is negligible. This idea is consistent with the constant M^+/O^- ratios observed during the oxidation of Sn and Si [3.214]. *Wittmaack* [3.113] pointed out that bond breaking on the surface (Sect. 3.2.3b) is also consistent with this proportionality between M^+ and O^- yields. A direct experimental verification of equal M^+ and O^- yields, still unavailable, would be a more definitive test of this dissociation model. This model also fails to explain the opposite responses (suppression/enhancement) of the Si^+ and O^- yields from oxidized silicon to the lowering of the work function [3.64]. More theoretical work is needed to quantify this model for comparison with the experimental data on ion yields and energy distributions.

c) Surface Polarization Model

Williams and *Evans* [3.100] proposed a local surface polarization model to rationalize the observation of the enhancement of metallic negative secondary ions by oxidation (Sect. 3.2.3b). When oxygen reacts with the metal surface under ion bombardment, the oxygen atoms can be on or beneath the layer of the surface atoms. The electric dipole moments of the metal-oxygen bonds hence orient both in and out of the surface. *Williams* and *Evans* postulated that when oxygen is adsorbed on the surface, it creates an electron-retention site that favors the emission of positive secondary ions. When oxygen is incorporated beneath the surface, it creates an electron-emissive site that favors the emission of negative secondary ions. This model was used to explain why during the Ar^+ sputtering of Al-Au alloys in the presence of oxygen [3.100], the oxidation of Al enhanced the emission of both Au^+ and Au^-. *Williams* and *Evans* proposed that Au^+ was formed at the electron-retention site, while Au^- was formed at the electron-emissive site. *Williams* [3.106] also speculated that these negative ions could originate from antisite defects where the metal atom resides in an anion site. He proposed that the metal atom at this site would have an electron affinity similar to the anion itself and thus may be sputtered as a negative ion. This idea still has to be verified experimentally.

d) Bond-Breaking Model

Slodzian [3.27, 105] first suggested the bond-breaking concept to explain the large secondary ion emission observed during the sputtering of ionic solids. Later *Williams* [3.23, 25] proposed that it may also be appropriate for compounds like oxides, where the bonds formed have only a partial ionic character. They both pointed out that the ionization of a sputtered metal atom, during the breaking of the bond with an electronegative atom on the surface, is qualitatively similar to the Landau-Zener curve-crossing picture [3.215, 216] for charge exchange processes in atomic collisions.

Our discussion will follow *Yu* and *Mann's* recent model developed with these concepts [3.107]. During the sputtering of M, a cation vacancy X is created on the surface. It is assumed that during the sputtering of M^+, the cation vacancy can trap the electron left behind, with an electron affinity A, for at least the sputtering time ($\simeq 10^{-13}$ s). Within this time scale, the excess electron does not necessarily occupy an equilibrium state in the solid. Figure 3.17 shows the diabatic potential-energy curves involved. Charge exchange can happen at the crossing of the diabatic covalent potential curve $M^0 + X^0$ and the diabatic ionic potential curve $M^+ + X^-$ at a distance R_c from the surface. According to *Landau* [3.215] and *Zener* [3.216], the transition probability is determined by the magnitude of the wave functions and the shape of the *diabatic* curves at the crossing point, which is a distance away from the equilibrium position, and not the *adiabatic* potential curves for the equilibrium state. Hence, as *Williams* pointed out [3.106], the transition probability is not sensitive to the details of

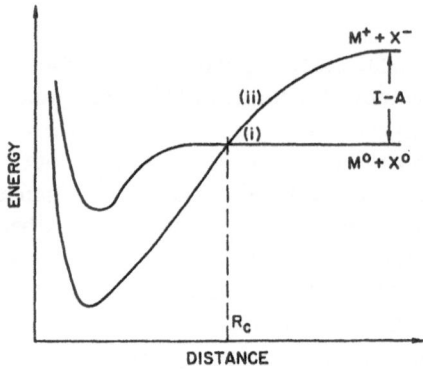

Fig. 3.17. Schematic energy diagram showing the diabatic covalent energy curve (*i*) $M^0 + X^0$ and the ionic potential energy curve (*ii*) $M^+ + X^-$. Crossing occurs at a distance R_c

the equilibrium bond. Still, the determination of these potential energy surfaces is formidable. The treatment is nontrivial even for an isolated Cu-O molecule [3.217]. The only attempt to calculate the ionization probability from the first principles is by *Passeggi* et al. [3.218]. This very recent result has not yet been experimentally verified.

However, it is possible, with certain simplifying assumptions, to determine the salient features of the bond-breaking model. Following *Yu* and *Mann's* approach [3.107], the covalent force is assumed to be very short ranged so that in the region of crossing, the covalent potential energy (Curve 1) is independent of R, the distance of the sputtered atom from the surface (Fig. 3.17). The ionic potential energy (Curve 2) is given by the Coulombic attraction between M^+ and the electron left at the cation vacancy. To calculate the ionization probability P^+, the Landau-Zener formula (in atomic units) is used:

$$P^+ \cong G \exp\left(-\frac{2\pi H_{12}^2}{v|a|} \right)_{R=R_c} , \tag{3.33}$$

where H_{12} is the transition-matrix element, v is the escape velocity, and $|a|$ is the difference in the first derivatives of the potential curves. All quantities are evaluated at the crossing point. Here G is the ratio of the degeneracies of M^+ and M^0 [3.219], and is very close in value to Z_+/Z_0 in (3.13) [3.125]. At infinity the ionic curve (Curve 2) lies above the covalent curve (Curve 1) by $I - A$. At the crossing point, the Coulombic potential exactly balances this energy difference. Hence the crossing distance R_c simply equals $(I - A)^{-1}$ in atomic units. The wave function for the atomic level of the sputtered atom is simulated by the decaying region of a hydrogen 1s wave function at the crossing point. The wave function for the electron trapped at the cation vacancy site is assumed to be the same as that of a negative ion in the presence of M^+ with an amplitude parameter C and an electron affinity A. Both C and A are quantities characteristic of the chemical bond. But without a sound understanding of the cation vacancy, these parameters have to be determined by fitting with experimental data.

In (3.33) P^+ is directly dependent on the velocity at the crossing point R_c, which is related through energy conservation to the emission energy E_1 by

$$v(R_c) = \left[\frac{2(E_1 + I - A)}{M_2} \right]^{1/2} . \tag{3.34}$$

As a result, P^+ has a finite value even at $E_1 = 0$, but approaches an exponential dependence on $E_1^{-1/2}(v^{-1})$ at high E_1. For the usual energy range of interest (few tens of eV), P^+ approximates a power-law dependence on E_1, in agreement with the observation of *Krauss* and *Gruen* [3.74]. Also, P^+ is related to the isotope mass M_2 through $v(R_c)$. The model predicts that P^+ will decrease linearly with the increase in the isotope mass, and the isotope effect will decrease gradually at higher emission energies (velocities). Both agree well with the data of *Lorin* et al. [3.130].

In general P^+ decreases rapidly with increasing I. Figure 3.18 shows the calculated P^+/G values for the 11 metallic elements (filled circles) in the fourth period of the periodic table, and with $C = 0.25$, $A = 1.463$ eV, and $E_1 = 10$ eV [3.107]. Here P^+ follows closely an exponential dependence on I, as would be predicted by (3.13). The parameter T_p derived from (3.13) equals 6120 K, in close agreement with reported values [3.126]. The apparent Boltzmann-like dependence of P^+ on I is thus reproduced. It is an approximate result for this small range of I, and T_p is a parameter and not a temperature of the system. No thermal equilibrium concept needs to be introduced. The value of T_p rises to 10 800 K for secondary ions with $E_1 = 50$ eV, which is qualitatively consistent with experimental observations [3.126]. In addition, the atomic mass M_2 is important. The lighter elements, B, Si, and Mg have larger P^+/G values, while the heavier elements Nb and Mo have smaller P^+/G values than the corresponding values for the fourth-period elements. This accords with reported data (Fig. 3.10) and is related to the velocity factor in (3.33).

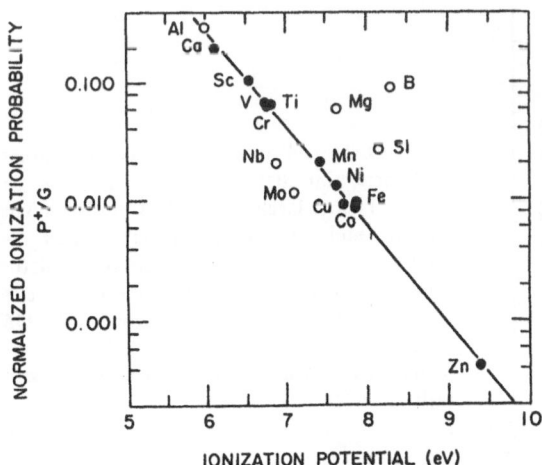

Fig. 3.18. Theoretical P^+/G values for the 11 fourth-period elements (●) and six metals in other periods (○) plotted as a function of the ionization potential. The best fit line for the fourth-period elements is characterized by a parameter T_p of 6120 K [3.107]

This model is qualitatively consistent with many experimental observations, and it offers physical insight into the physics of the ionization process. Aside from some details, the bond-breaking model is analogous to the tunneling model, with the valence band replaced by the level of the trapped electron in the cation vacancy. A more detailed comparison between the two models is given in [3.219]. Computation of ionization probabilities at a confidence level acceptable for use in SIMS analysis would, however, require a much better theoretical understanding of the relevant parameters, such as the nature of the cation vacancy, for a given experimental condition. The problem is complex and simple solutions are not expected.

3.3.3 Secondary Ion Formation by Energetic Collisions

The experimental evidence summarized in Sect. 3.2.4 suggests that the presence of core holes is the major reason for the ionization of sputtered atoms in energetic collisions. The most widely accepted mechanism for the production of such inner-shell excitations is the electron-promotion model developed originally by *Fano, Lichten* and coworkers [3.220–223] for atomic collisions.

During the close encounter between the incident ion (or atom when neutralized on the surface) and target atom, the electron shells interpenetrate. The electrons of the atomic orbitals (AO) can be promoted to the diabatic molecular orbitals (MO) of this diatomic system with higher principal quantum numbers. Then, as the atoms recede from each other, transitions can occur at the crossings of the MOs, leaving the promoted electrons in higher levels after a collision. To illustrate this mechanism, Fig. 3.19 shows schematically the correlation diagram of the electronic levels for the Ar → Si system. On the right-hand side are the

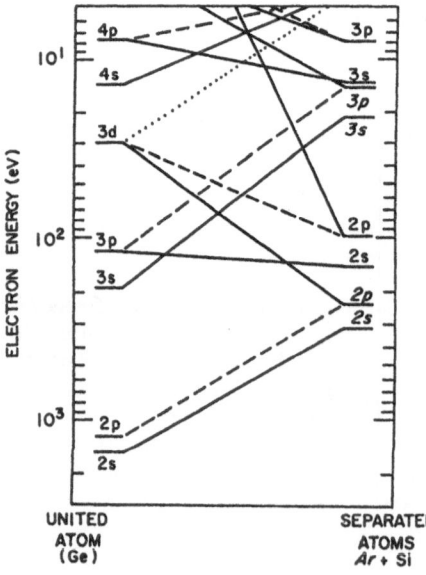

Fig. 3.19. United atom (UA)–separated atoms (SA) correlation diagram for the electrons of Ar and Si. The Ar levels are marked by italics. The diabatic molecular orbitals (MO) connect the levels of the SA with those of the UA (Ge) with the same value of $n - l$. For example, the Si 2p level is connected with the 3p and 4f (beyond scale) UA levels. Here MOs with $m = 0, 1, 2$ (σ, π, δ) are denoted by solid, dashed, and dotted lines respectively [3.147]

energy levels of the AOs of the separated atoms (SA) (interatomic distance $= \infty$). On the left-hand side are the energy levels of the "united atom" (UA). In this case the UA is equivalent to Ge. The two sets of energy levels are linked through diabatic MOs as represented schematically by the lines in between. The selection rule is that the MO must have the same $n - l$ values in both the UA and SA limits, where n is the principal quantum number and l is the angular momentum quantum number, so that the number of nodes in the "radial" wave function is conserved [3.222]. Electronic transition can occur when two MOs cross each other. When Ar and Si collide, the Si $2p$ electron can be promoted to the $3d\pi$ or the $4f\sigma$ MOs. If the distance of closest approach is sufficiently small, the $4f\sigma$ crosses a number of excited-state MOs as the Si atom leaves the Ar atom. It can make transitions at these crossings into other states through radial coupling (conserving angular momentum quantum number, $\Delta m_l = 0$), and rotational coupling ($\Delta m_l = \pm 1$) [3.222]. Selection rules are relaxed if there are correlated transitions of more than one electron. If the final state of the Si atom is an autoionizing state – e.g., $2p^5 3s^2 3p^3$ – the Si atom would be ionized. For a certain final state to occur, the collision has to be energetic enough to have the distance of closest approach less than the crossing distance, and the energy in the center-of-mass system higher than the excitation energy required [3.165]. Hence, the observed strong dependence of the ion yield on E_0 and the dependence on the primary species in direct recoil sputtering can easily be rationalized.

The energy required for electron promotion is particularly small when energy levels of collision partners match each other in energy [3.165]. This condition is automatically fulfilled in the symmetric collision of target/target atoms in collision cascades. A kinetic model that explains the emission of the multiply charged secondary ions by electron promotion was proposed by *Joyes* [3.224–226]. According to this model, the $2p$ core electrons of the target atom T (Mg, Al, or Si) can be promoted into the $4f\sigma$ MO during an energetic symmetric TT collision in the collision cascade. If the collisional energy is sufficient, the $4f\sigma$ orbital can be raised *above* the Fermi level of the target and the electron can escape into the unoccupied region of the valence band continuum, leaving a $2p$ core hole on one of the atoms when they recede from each other. The lifetimes of such core holes in the solid are on the order of 10^{-13} to 10^{-14} s [3.155], which is in the same time scale of a sputtering event. Hence it is possible for that atom to be sputtered with a core hole [3.167]. Then the T^{2+} produced by the subsequent Auger decay is accompanied by the emission of an L-shell Auger electron. Hence the strong correlation between the multiply charged secondary ion yields and the atomic-like Auger electron yields is an immediate consequence of the model. Since the number of TT neighbors is proportional to the square of the concentration in an alloy, the observed proportionality between the Al^{2+} yield and the square of the Al concentration in the Al-Cu alloys (Sect. 3.2.4b) can easily be understood. Also an additional $3p$ hole can be generated simultaneously with a $2p$ core hole during TT collisions by the same electron-promotion process, yielding a triply charged secondary ion. This is *consistent with the observation that the* Si^{3+} yield is correlated with the Si^{2+}

yield when E_0 is varied. As the secondary ion with the core hole leaves the surface, the charge state can still be changed by charge exchange between the sputtered atom and the valence band of the solid on the exit trajectory [3.147]. A doubly ionized sputtered atom can pick up a valence electron on its exit trajectory and contributes to the singly charged secondary ion yield, as was observed (Sect. 3.2.4b).

The electron-promotion model is applicable only to the production of positive secondary ions. *Rabalais* and *Chen* [3.147] have suggested that negative ions are not likely to survive such high-energy close encounters. The negative ionization probability should then be proportional to the product of the probability of being neutral after the violent collision, and that of picking up an electron through valence–electron transitions. However, the observation that D^- and Ni^- yields in direct-recoil sputtering increased with the recoil velocity (energy) (Sect. 3.2.4a, Fig. 3.12) disproves this speculation.

3.4 Emission of Excited Atomic Species

The presence of excited atomic species in the sputtered flux is most readily detected by the photons they emit during their radiative decay in front of the sample surface. Besides these excited states with short lifetimes, sputtered atoms were also found to be excited into the metastable multiplets of the ground state and other metastable states. The metastable excited states are those where de-excitation by electric dipole transition is forbidden. The sputtering of both excited neutral and excited ionic atomic species has been observed.

The sputtering of excited atoms shares many features with secondary ion emission [3.31, 35]. The most interesting phenomenon is the sensitivity of the excited atom yields to the surface chemical states. Similar to the secondary ion yields, after reacting with electronegative elements like oxygen, the excited atom yields from metallic and semiconductor surfaces are frequently orders of magnitude higher than those from clean surfaces [3.7, 31, 34, 227, 228]. Equally important is that the secondary ion yield can be more than one order of magnitude larger than the excited atom yields of the same element [3.59], although less energy is required to excite than to ionize the sputtered atom. Our present understanding of the excitation mechanisms is much less developed, compared with that of secondary ions.

3.4.1 Experimental Methods

The sputtering arrangement is very similar to that for secondary ion emission. Most reported experiments use high-fluence sputtering. Low-fluence sputtering data became available only recently, when laser techniques were used [3.229, 230]. The major difference from secondary ion emission experiments is the detection scheme, which can be subdivided into two categories.

a) Detection of Sputtered Excited Atoms with Short Lifetimes

Sputtered excited atoms with short lifetimes can readily be detected by observing the photons emitted during their decay in vacuum. Figure 3.20 shows the conventional approach. A quartz window on the vacuum chamber allows the emitted photons to be focused by a quartz lens into a grating spectrometer. Limited by the collection angle and the transmission loss of the spectrometer, the overall collection efficiency of the system is typically about 0.02% [3.231]. The photons can then be detected by a photomultiplier. Both photon counting and analog detection have been used. The spectral range covered is usually from 200 to 800 nm. Measurements of absolute photon yields have been reported by *Tsong* and *Yusuf* [3.232], who calibrated their detection system with a standard lamp. Relative photon yields were more frequently reported.

The secondary photon yield $N(\lambda_{jk})$ per incident primary ion for the transition $j \rightarrow k$ at wavelength λ_{jk} is given by

$$N(\lambda_{jk}) = \frac{I(\lambda_{jk})}{\eta(\lambda_{jk})} \; . \tag{3.35}$$

Here $I(\lambda_{jk})$ is the signal of the optical transition from level j to level k at wavelength λ_{jk}, and $\eta(\lambda_{jk})$ is the efficiency of detection. The measured photon yield is in part from the $j \rightarrow k$ transitions of the atoms sputtered in the j state, and in part from atoms sputtered in higher states, which then cascaded into the j state. Hence the photon yield $N(\lambda_{jk})$ is related to the yields Y_i^* of sputtered atoms at the ith excited state by the following relation:

$$N(\lambda_{jk}) = b_{jk}\left[Y_j^* + \sum_{i \neq j} b_{ij} Y_i^* \right] , \tag{3.36}$$

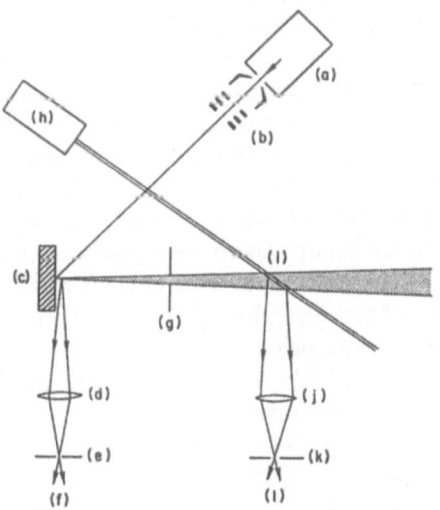

Fig. 3.20. Experimental arrangement for the detection of sputtered excited atoms: (*a*) primary ion source, (*b*) primary ion beam optics, (*c*) sample. For the detection of unstable excited atoms: (*d*) lens optics, (*e*) movable slit to define detection volume, (*f*) monochrometer. For the detection of metastable excited atoms by laser fluorescence: (*g*) slit to define sputtered beam, (*h*) tunable dye laser, (*i*) intersection volume between sputtered atom beam and laser beam, (*j*) lens optics, (*k*) slit to define detection volume, (*l*) monochrometer

where b_{ij} and b_{jk} are the branching ratios for the transitions $i \rightarrow j$ and $j \rightarrow k$, respectively. To obtain Y_j^* from $N(\lambda_{jk})$, a correction for the cascade contribution $\sum_{i \neq j} b_{ij} Y_i^*$, which involves the yields of many upper levels, has to be applied. Usually, the population of levels decreases rapidly with increasing excitation energy. A correction on the order of 20% has been quoted [3.233]. Implicit in these equations is the assumption that the photons emitted are unpolarized and the emission is isotropic. Experimental observations support this assumption [3.7, 234].

Two experimental approaches have been adopted to determine the energy (velocity) distributions of these excited atomic species. The first one, often called the light-versus-distance method, derives the information by measuring the spatial decay of the photon intensity in front of the target. The excited atoms usually have a very short lifetime of typically less than a microsecond. For an emission velocity of 10^5 cm/s, most of the excited atoms would decay in the distance of a few millimeters in front of the sample. If there is no cascade contribution, the intensity of photon emission at a distance z from the target $I(z)$ for a beam of emerging excited atoms is given by

$$I(z) = I_0 \exp\left(-\frac{z}{v_\perp \tau}\right), \qquad (3.37)$$

where I_0 is the intensity at the target surface, $v_\perp$ is the normal component of the velocity of the excited atoms, and τ is the lifetime of the excited state. Experimentally the spatial profile of the secondary photon emission can be observed by defining the observation region with narrow slits ($\simeq 150$ μm) or by an optical multichannel analyzer with about 5-μm resolution [3.34]. The signal is proportional to the photon yield integrated over a thin slab defined by the slit and the plane parallel to the sample surface.

For the actual data interpretation, we must model the sputtering process by assuming the angular and velocity distributions of the sputtered excited atoms. A number of functional forms for these distributions have been suggested by *Dzioba* et al. [3.235, 236]. This light-versus-distance technique was discussed in detail in several review articles [3.33, 34]. Its most serious limitation comes from the cascade contributions. Since cascading depends on the lifetimes of the upper levels, (3.37) no longer holds. Neglecting this contribution results in an overestimation of the emission velocity if (3.37) is used for the data analysis. This has led to a previously prevailing view that excited atoms can result only from very energetic collisions in the ion bombardment event. There have been several analyses of this complication [3.237, 238]. Useful schemes to correct for the cascade contributions are not yet available. Therefore, the light-versus-distance technique is now rarely used for velocity measurements.

An alternative experimental technique to obtain velocity information is based on measuring the Doppler broadening of the emission line. This technique was first used by *Snoek* et al. [3.7], later by *van der Weg* and *Bierman* [3.239], *White* et al. [3.240], *Hippler* et al. [3.241], and recently by *Betz* [3.36]. If the sputtered excited atom has a velocity component $v_\parallel$ toward the spectrometer which is viewing in a plane parallel to the sample surface, the frequency v of the emitted photon would be Doppler shifted by an amount equal to $v_0(1 + v_{\|}/c)$, where v_0 is the emission frequency when the atom is at rest. Hence the line profile is a measure of the flux component $dY^*/dv_{\|}$, which involves an integration of all the contributions over the whole range of the normal component $v_\perp$ of the velocity. The most detailed analysis reported so far [3.36] assumed the Thompson formula (3.3) for the velocity and angular distributions, and treated the surface binding energy U_0 as a parameter in the modeling. A grating spectrometer with a resolution of about 0.01 nm was used. The broadening is on the order of 0.1 nm. Cascade contributions simply add to the primary population as in (3.36). No correction scheme has been proposed yet. If the velocity distributions of the upper levels are basically the same as the level of interest, the correction should be small.

b) Detection of Metastable Excited Atoms

All the reported experimental work on metastable atoms has been done with the laser-induced fluorescence (LIF) technique and its extension, the Doppler-shift laser fluorescence (DSLF) technique. These techniques were originally developed for sputtered ground-state atoms. For metastable excited species with a lifetime on the order of milliseconds or longer, the radiative decay in the first centimeter in front of the sample is negligible and the technique is still useful.

A typical experimental setup (Fig. 3.20) has been described in detail by *Husinsky* and coworkers [3.242, 243]. A tunable laser beam is directed to intercept a beam of sputtered metastable excited atoms. The frequency of the laser beam is set to excite the sputtered atoms to an upper intermediate level, which subsequently decays and emits a fluorescence photon. In the two-level scheme (Fig. 3.21a), the final state is the same as the initial state (1). In the three-level scheme (Fig. 3.21b), the final state (3) is different from the initial state (1). With the proper knowledge of the interaction volume geometry and laser

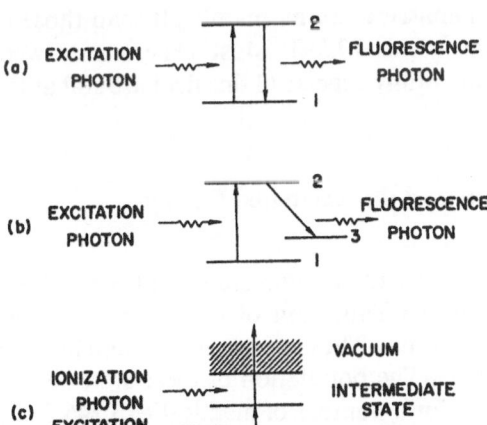

Fig. 3.21. Schematic energy diagrams for (a) the two-level scheme, (b) the three-level scheme in laser fluorescence detection, and (c) multiphoton resonant ionization for detecting metastable atoms

power, we can deduce the amount of sputtered atoms from the fluorescence signal. In addition, the excitation frequency is Doppler shifted because of the velocity component of the sputtered particle in the direction of the laser beam. Hence, by scanning the laser frequency, the velocity distribution of a selected kind of stable or metastable species can be obtained. Since the sampling volume and the direction of the laser beam are completely at the discretion of the experimentalist, modeling of the sputtering process is usually not required. Cascade contributions from upper levels will add to the population and cannot be distinguished. The correction to the velocity distribution is again not expected to be large for the same reason as in the Doppler linewidth broadening technique discussed above.

With the present laser technology, continuous-wave dye lasers are available with linewidth on the order of 1 MHz and a tunable range of about 30 GHz, depending on the dye used. Certain power stabilization during frequency scanning is usually available. By using a low-power CW dye laser with the above characteristics, measurement of the velocity distributions of the sputtered atom can be quite straightforward. Velocity resolution on the order of 10^3 cm/s can be achieved, but at the expense of sensitivity. High-power cw lasers and pulsed dye lasers can increase the sensitivity. In such situations, corrections for the power broadening of the linewidth, the depopulation of the levels (saturation effects), and the differentiation between flux and number density have to be considered. The complexity in the interpretation of DSLF data has been reviewed by *Wright* et al. [3.244] and *Bay* [3.245].

Multiphoton resonant ionization (MPRI) is another recent technique for detecting metastable atoms [3.230]. Selectivity is achieved by tuning the energy of one photon to match the transition from the initial state to an intermediate state (Fig. 3.21c). Then further absorption of photons excites the electron to the vacuum.

3.4.2 Experimental Results

Fundamental studies of excited atom emission are more difficult than those of secondary ions because of the lower yields [3.59]. Most experiments were performed with high-fluence sputtering. Many aspects of this field are still at the exploratory stage.

a) Dependences on the Quantum State and the Excitation Energy of the Sputtered Atom

The lowest lying excited states observed are the ground-state multiplets. These electronic states have the ground-state configuration of the sputtered atom, except for the spin and/or angular momentum. They are metastable and have an excitation energy on the order of 0.1 eV. The population of these states in the sputtered flux has been studied by LIF for a number of metals: Fe [3.246, 247], Ti [3.229, 248, 249], U [3.250], and Zr [3.244, 251, 252]. The populations of these levels are large and can reach up to 60% of the ground-state occupation. Empirically, the relative populations for an atomic species can be fitted to a Boltzmann distribution with the parameter T_p typically around 1000 K [3.245]:

$$\frac{n_i^*}{n_0} = \frac{g_i}{g_0} \exp\left(-\frac{E_i}{kT_p}\right) . \tag{3.38}$$

Here n_i^* and g_i are the sputtered-atom density and the multiplicity, respectively, of the ith excited state. The zeroth state is the ground state. Figure 3.22 shows a semilog plot of the normalized population versus excitation energy of ground-state multiplets for Fe. Note that for Fe [3.246], $T_p \simeq 980$ K for the a^5D multiplets, while $T_p \simeq 2000$ K for the a^5F multiplets. That two vastly different values of T_p were obtained in the same sputtering experiment excludes the existence of a thermal equilibrium at the collision cascade as an explanation for (3.38). Hence T_p is not a temperature. *Craig* et al. reported an exception to this large population of excited ground-state multiplets [3.230]. They found by the MPRI technique that there was no evidence for the population of the excited In ($^2P_{3/2}$) state during the sputtering of In by 5 keV Ar$^+$. The ground state is In ($^2P_{1/2}$). The authors proposed that the $5p$ electrons of In atoms interacted with the conduction electrons strongly, and the atoms in the $^2P_{3/2}$ state all de-excited to the $^2P_{1/2}$ state.

The sputtering of metastable excited atomic species Ba I (^1D, 1.4 eV) [3.253], Ca I ($^3P_2$, 1.8 eV) [3.254, 255], Fe I ($a\ ^5F_5$, 0.86 eV) [3.256], and Ti I ($a\ ^1D_2$, 0.9 eV) [3.249] from the pure elements by Ar$^+$ and Kr$^+$ has been observed by the laser-induced fluorescence technique. *Schweer* and *Bay* [3.256] estimated that about 1.2% of the sputtered neutral Fe atoms are in the metastable $a\ ^5F_5$ state.

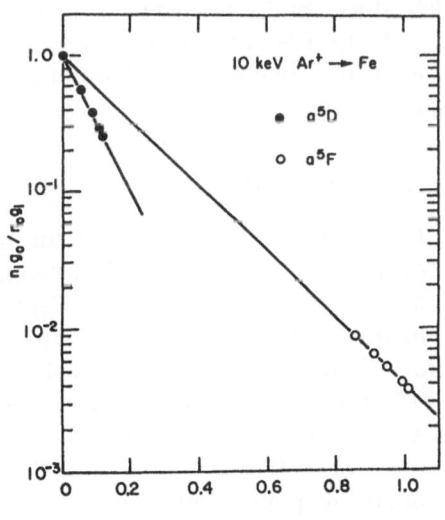

Fig. 3.22. Excited-state occupation divided by the multiplicity, normalized to the ground-state value, as a function of excitation energy for the ground-state multiplets of sputtered Fe [3.246]

Excited atoms in states with short lifetimes are usually detected by their radiative decay. No absolute excitation probability measurement has been reported for pure metallic targets. The excitation probabilities of the sputtered atoms from oxide glasses have been estimated to be on the order of 10^{-3} [3.232]. If we assume that the chemical-enhancement effect of the oxygen is 10^2, the excitation probabilities of atoms sputtered from pure metals would be on the order of 10^{-5}.

There have been extensive studies of the relative populations of the various short-lifetime excited levels of the same species, and occasionally of the states of different elements. Empirically, the excitation probability P_i^* follows roughly an exponential dependence on the excitation energy E_i in accordance with (3.38), where T_p is on the order of 3000 K. This has been demonstrated for the sputtering of Fe I and Ni I excited states from NBS steel [3.77, 257], Ti I excited states from Ti metal [3.258], and Fe I, Zn I from pure elements [3.259]. However, appreciable deviations from (3.38) have also been reported by *Andersen* et al. [3.260] for levels in Ga I, Tl I, and Mg II when they bombarded elemental targets with 80 keV Xe$^+$.

Jensen and *Veje* [3.234] and *Andersen* et al. [3.260] have tried to correlate the population of the different excitation levels with the principal quantum number n and the orbital angular momentum quantum number l. In general, the population varies inversely with n, which is expected, since the excitation energy increases monotonically with n. Power-law dependences on n have also been reported [3.261]. For the same n, the population of the d states is usually higher than those of the s and p states. This is probably related to the higher statistical weight $(2l + 1)$. However, it was observed for both Mg II and Zn II [3.260] that the population decreases steadily from the $d(l = 3)$ states to the $g(l = 5)$ states. No explanation has been proposed yet.

Veje [3.233, 262] also noticed that levels of high spin (s) multiplicity are preferentially populated beyond the statistical weight $(2s + 1)$ when compared with corresponding levels of low spin multiplicity. For example, the triplets in Cd I [3.233], Mg I, Ca I, and Al II [3.262] are generally populated substantially more than three times the corresponding singlets.

b) Relation to the Global Electronic Structure of the Surface

To determine whether resonant electron transfer (Sect. 3.3.1c) is important in sputtered-atom excitation, there were several attempts to correlate the sputtered excited atom yields with the positions of the excited levels relative to the Fermi levels of the metallic substrates. *Thomas* and *de Kluizenaar* [3.263] bombarded Al and Cu targets in an ultrahigh vacuum with 10 keV Kr$^+$ and observed the emission of the 309.2 nm Al I line and the 324.7 nm Cu I line. The intensity of the 309.2 nm line is proportional to the population of the $3d$ ^2D Al I level, which is 1.96 eV below the vacuum level, or 2.32 eV above the Al Fermi level. The intensity of the 324.7 nm line is proportional to the population of the $4p$ $^2P^0_{3/2}$ Cu I level, which lies 3.9 eV below the vacuum level, or 0.75 eV above the Cu

Fermi level. *Thomas* and *de Kluizenaar* then deposited Cs vapor via a nozzle onto the metal surface to lower the work function Φ during sputtering, so that Φ was 1.7 eV and 1.55 eV for the Al and Cu targets respectively. Hence the excitation levels were shifted below the Fermi levels, and the resonant de-excitation channel should be blocked (Fig. 3.23). They expected the excited-state yields to increase. However, a 20% decrease in the intensities was actually measured. They concluded that the populations of these excited levels are insensitive to the positions of the atomic levels relative to the valence band. Resonant electron transfer cannot be important.

Veje [3.233, 262, 264], however, reported data suggestive of resonant electron transfer. He noticed that for the group-II and III elements B, Be, Mg, Al, Zn, and Cd, the singly charged ions are excited much more efficiently than the neutral atoms, as shown in Table 3.2. Comparing the positions of the atomic (ionic) energy levels with the valence band of the metals, he found that the atomic levels that face the valence band usually have the larger populations:

1) In all the cases in Table 3.2, the positions of the photon-emitting atomic energy levels lie above the valence band of the metals, while the lowest lying excited levels of the ions, which also have the highest populations, are situated well below the top of the valence band [3.233].
2) Cd I has a triplet state $5p\ ^3P$, which faces the valence band of Cd, and a singlet state $5p\ ^1P$, which is above the valence band. The population of the triplet state was estimated to be 100 times larger [3.233].
3) The $2s\ ^2 3s$ level of neutral boron, which faces the boron band gap, was only weakly populated [3.233, 264].
4) The $3p$ level of Al^{2+} lies practically 6 eV below the valence band of Al, while the $4p$ level lies within the valence band. The population of the $3p$ level was at least a factor of three less than that of the $4p$ level [3.262].

The relation of the population of the excited states with the valence band of the solid is hence still unresolved. In all these experiments, no shifting of the energy levels by the interaction with the surface has been considered.

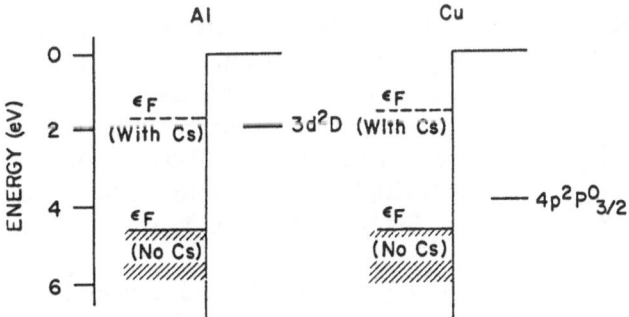

Fig. 3.23. Relative positions of the $3d^2 D$ Al I and $4p^2 P^0_{3/2}$ Cu I levels with respect to the Fermi levels ε_F of Al and Cu respectively

Table 3.2. Relative populations of excited levels of sputtered atoms from group-II targets under 50 keV Ar^+ bombardment [3.233]

Target	Level		Relative population
Be	Be I	$2s3s\ ^3S$	0.13
		$2s3d\ ^3D$	0.25
		$2s3d\ ^1D$	0.017
		$2s4d\ ^1D$	0.0010
		$2p3s\ ^3P$	0.0008
	Be II	$2p\ ^2P$	1.0
		$4d\ ^2D$	0.012
		$4f\ ^2F$	0.0020
Mg	Mg I	$3s3p\ ^1P$	0.061
	Mg II	$3p\ ^2P$	1.0
		$3d\ ^2D$	0.25
Zn	Zn I	$4s4p\ ^1P$	0.20
	Zn II	$4p\ ^2P$	1.0
Cd	Cd I	$5s5p\ ^1P$	0.14
	Cd II	$5p\ ^2P$	1.0
		$5d\ ^2D$	0.045

c) Chemical Effects

Often the yields of sputtered excited metallic atomic species have been modified by the reaction of the metal surfaces with electronegative elements or molecules. Oxygen adsorption has been studied most frequently. Nitrogen [3.227], ammonia, SF_6 [3.265], and carbon monoxide [3.266, 267] have also been observed to induce chemical effects.

The effect of oxidation on the ground-state atom yields was studied in detail for Ba [3.253], Ca [3.255], Cr [3.255], and Ti [3.249]. Not only is the total sputtering yield usually lowered by oxidation in these systems; also the ground-state atom yield is reduced by an additional order of magnitude. The yields of the low lying excited states of the ground-state multiplet are also correspondingly reduced by oxidation. The relative occupations of these multiplets show a small increase in favor of the higher excited levels. *Dullni* [3.249] reported that when the population of Ti ground-state multiplets was fitted to the Boltzmann expression (3.38), T_p increased from ~ 700 K for metallic Ti to 1400 K for heavily oxidized Ti surfaces.

The yields of metastable atoms not belonging to the ground-state multiplets have been observed to increase by about one order of magnitude for Ba [3.253] and Ca [3.254], and remain unchanged for the $a\ ^1D_2$ Ti I state of Ti [3.249] during oxidation. However, after correction for the reduction in the total sputter yield, the excitation probabilities were all enhanced by oxidation. Similar enhancement has also been observed for Ba II [3.253].

The short-lived excited atom yields from clean metal targets were frequently enhanced by reaction of the metal surfaces with electronegative elements or molecules. *Tsong* and *Yusuf* reported the most extensive study of the absolute sputtered excited atom yields from oxides [3.232]. They sputtered an NBS standard oxide glass SRM 611, which contains 61 elements at 50 ppm concentration, with 20 keV Ar^+, and measured the photon yields of the major atomic lines. Table 3.3 shows the excited atom yields after correction for the branching ratios, but not for the cascade contributions. A prominent feature is that the yields are usually quite small ($\sim 10^{-3}$) compared with the secondary ion yields from oxides, which can be as high as 10^{-1}.

Table 3.3. Excited state yields of sputtered atoms from SRM 611 NBS glass bombarded by 20 keV Ar^+ [3.232]

Element	Wavelength [nm]	Excitation Energy [eV]	Yield ($\times 10^3$)
Ag	328.1	3.77	10.1
Al	396.2	3.14	3.8
B	249.7	4.96	0.4
Ba	553.5	2.24	1.4
Be	234.8	5.27	0.2
Ca	422.7	2.93	3.0
Cd	228.8	5.41	0.2
Ce(II)	418.7	3.51	0.1
Co	345.3	4.01	0.4
Cr	425.4	2.91	1.7
Cs	455.5	2.72	4.6
Cu	324.7	3.81	3.9
Fe	358.1	4.31	0.8
Ge	303.9	4.95	1.0
In	451.1	3.02	4.5
K	766.5	1.61	2.7
Li	670.8	1.85	8.4
Mg	285.2	4.34	7.0
Mn	403.4	3.07	2.5
Mo	379.8	3.26	5.3
Na	589.0	2.10	7.4
Ni	341.5	3.65	0.7
Pb	405.8	4.37	0.1
Re	346.0	3.58	0.8
Si	288.2	5.07	1.3
Sr	460.7	2.69	4.2
Ta	331.1	4.43	0.09
Ti	365.3	3.44	0.2
Tl	351.9	4.48	4.7
Zn	213.8	5.79	0.01
Zr	360.1	3.59	0.1

Williams et al. [3.59] measured the absolute secondary ion yields and absolute sputtered excited atom yields from oxidized Si and Ni surfaces. Table 3.4 shows the result. The sputtered excited atom yields are clearly much smaller than the singly charged secondary ion yields. In fact, the secondary ion yields probably should be scaled upward ($Y_{Si}^+ \sim 0.4$) [3.58]. This is an important result because it indicates that for sputtering from compounds, ionization is not a simple extension of excitation to the continuum states (i.e., with $E_i > I$).

Since reaction with electronegative species (e.g., oxidation) changes appreciably the surface electronic structure, researchers have tried hard to establish a correlation between these changes and the chemical effects in excited atom emission, analogous to secondary ion emission. The most discussed idea is that (oxide) band-gap formation can block the resonant de-excitation channel [3.103, 188], although such changes as the work function [3.228] and the possibility of Auger de-excitation [3.103, 268] have also been considered. A lower de-excitation probability means a larger excited atom yield. The usual procedure is to compare the excited atom yields for levels that face the valence band with those that may lie above or below the valence band. Several authors reported some correlations: SiO [3.103], TiO [3.269], CuO [3.188], CaO [3.270], and Al_2O_3 [3.227]. These usually mean that levels facing the (filled) valence band or the band gap have comparatively larger yields. However, some researchers report lack of correlation: Al_2O_3 [3.271], oxidized Cr [3.272], and TiO [3.258].

The more satisfactory correlation is with the local chemical bonds, as in secondary ion emission (Sect. 3.2.3b). A convincing case was reported by *MacDonald* et al. [3.266, 267], who adsorbed submonolayers of CO on polycrystalline Ni surfaces to enhance the secondary ion and excited atom yields during 2 keV Ne$^+$ bombardment. They found that during sputtering, the Ni I (352.5 nm) photon signal, similar to the Ni$^+$ (direct-recoil) secondary ion signal, varies linearly with the CO coverage, as shown in Fig. 3.24. This is consistent

Table 3.4. Secondary ion and photon yields from Si and Ni targets saturated with oxygen during 17.5 keV and 20 keV Ar$^+$ bombardment respectively [3.59]

Species	Ionization potential or excitation energy [eV]	Ion or photon yield
Si$^+$	8.15	1×10^{-2}
Si^{2+}	16.34	1×10^{-6}
Si I (288.2 nm)	5.08	1.6×10^{-3}
Si I (251.6 nm)	4.95	2.2×10^{-4}
Si II (385.6 nm)	10.07	1.4×10^{-5}
Si II (634.7 nm)	10.07	1.0×10^{-5}
Ni$^+$	7.63	6×10^{-4}
Ni I (341.5 nm)	3.63	1.0×10^{-4}
Ni I (232.6 nm)	5.49	3.1×10^{-7}

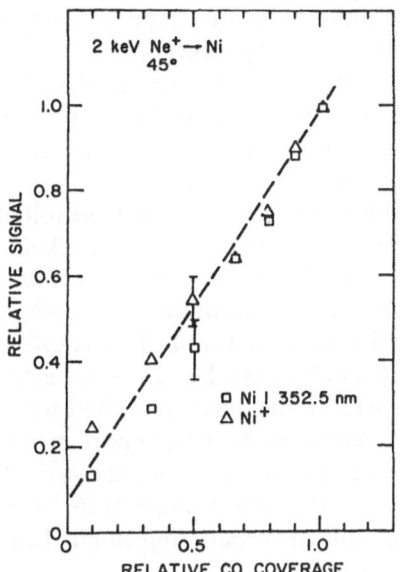

Fig. 3.24. Variation of the Ni^+ secondary ion and Ni I (352.5 nm) yields with CO coverage or Ni. Both signals and the CO coverages are normalized to the respective maximum values [3.267]

with the local bond-breaking picture (Sect. 3.2.3b). The signals at zero CO coverage are the clean Ni values. *Shimizu* et al. [3.273] also found that the Al I (396.1 nm) and Al^+ yields are proportional to each other during the sputtering of oxidized Al by 10 keV Ar^+.

The linear relation between the excited atom yield and the secondary ion yield holds where only one type of bonding site exists. When there are many types of chemical bonding sites, the chemical effects on the ion and excited atom yields are usually different. Such behaviors have been observed during the sputtering of Cr [3.272], Cr-Fe alloys [3.257], Cu, Ni, and Cu-Ni alloys [3.274] in the presence of oxygen. These observations can be rationalized in the bond-breaking picture. The sputtered excited atom yield for an excited level is a sum of the contributions from the various chemical states (i), as for the secondary ions (3.8):

$$Y_M^* = \sum_i f_i P_i^* Y_i \ . \tag{3.39}$$

Comparing the yields of excited atoms and secondary ions is interesting:

$$\frac{Y_M^*}{Y_M^+} = \frac{\sum\limits_i f_i P_i^* Y_i}{\sum\limits_i f_i P_i^+ Y_i} \ . \tag{3.40}$$

Since the ratio P_i^*/P_i^+ can change with the chemical state i, Y_M^*/Y_M^+ depends very much upon the distribution of f_i. Hence the excited atom and the secondary ion yields can have very different dependences on the concentration of the electronegative species – e.g., oxygen – when more than one type of surface compound is formed, as in dynamic mode sputtering experiments.

The magnitude of the chemical effect is also electronic-state specific [3.275]. For example, the Mg II yields have been observed to be reduced by oxidation [3.276]. *Braun* et al. [3.271] also reported that Al II- and Al III-excited atom yields are not affected by oxidation. No explanation has been offered.

Very little has been done to relate the yields of excited atoms to their initial valence states. *Pellin* et al. [3.229, 248] reported very little change in the relative excitation probabilities of the various ground-state multiplets, when the amount of oxygen on the Ti surface was decreased from three to one monolayer. *Gade* et al. [3.277] investigated the effect of the initial valence states on the excitation probabilities by sputtering Cu_2O, CuO, $CuCl$, $CuCl_2$, FeO, Fe_2O_3, and Fe_2O_4 with 80 keV Ar^+. They found that while the yields of the different excited states of Cu I, Cu II, Fe I, Fe II, and Fe III were enhanced over those from pure metals to different degrees, they were identical within a factor of two among the various compounds; i.e., there is no strong correlation with the valence state. This experiment, however, is not conclusive, since the Ar^+ sputtering might have changed the chemical state on the compound surface.

Data on sputtered excited electronegative atomic species are very rare. Usually no oxygen spectral lines were observed during the sputtering of oxides. *Veje* [3.278] attributed this to the negative ionlike state of oxygen in the oxide lattice. An exception has been reported by *Kelly* et al. [3.279], who observed O II emission during boron sputtering in the presence of oxygen. *Veje* proposed that the O II emission they observed is related to the covalent nature of the boron-oxygen bond [3.278].

d) Velocity Distributions

The velocity distributions of the sputtered atoms in the ground-state multiplets have been measured for Fe [3.247] and Zr [3.251] using the DSLF technique. No difference was found between those of the ground state and the excited counterparts. Velocity distributions of sputtered atoms in the metastable states outside the ground-state multiplets have been studied for Ba [3.253], Ca [3.254], Fe [3.256], and Ti [3.249]. In all these cases, the velocity distributions of the metastable atoms are broader and peak at higher velocity values than the velocity distributions of the respective ground-state sputtered atoms. Velocity data are frequently fitted to the Thompson distribution (3.3), with the value of the parameter U_0 indicating the typical emission energy. For all these metals studied, U_0 range between 10 and 20 eV. *Grischkowsky* et al. [3.253] compared the velocity distributions of sputtered Ba (^1D) and ground-state Ba (^1S) as in (3.1) and found that P^* for Ba (^1D) approaches an exponential dependence on v^{-1} at high velocities (Fig. 3.25),

$$P^* \propto \exp(-v_0/v) ,$$

(3.41)

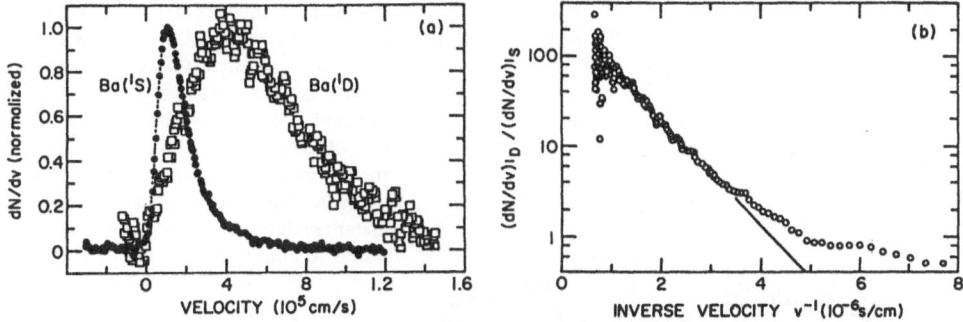

Fig. 3.25. (a) Velocity distributions of ground-state ^1S and excited state ^1D Ba atoms sputtered from Ba metal by 10 keV Ar$^+$. (b) The ratio of the two velocity distributions in (a) versus the reciprocal of the emission velocity [3.253]

where $v_0 \simeq 1.4 \times 10^6$ cm/s. Deviations are observed at velocities $< 2.5 \times 10^5$ cm/s, which corresponds to a 4.5 eV emission energy.

Since velocity measurements by the light-versus-distance technique are not reliable, we shall only summarize the velocity data obtained by measuring the Doppler-broadened line profiles of the photons emitted by the excited atoms (Sect. 3.4.1a). Early attempts using line-profile measurements without detailed modeling of the Doppler shift gave only rough estimates of the velocities. *Snoek* et al. [3.7] bombarded Cu with 60 keV Ar$^+$. For the 324.7 nm Cu I line, they estimated the velocities of the emitted Cu atoms to be a few times 10^7 cm/s, which corresponds to emission energies exceeding 3 keV. Measuring the Si I 288.2 nm line profile during the sputtering of Si by 5 keV Ar$^+$, *White* et al. [3.240] reported the Si* emission velocities to be on the order of 10^7 cm/s, or emission energies exceeding 1.6 keV. *Hippler* et al. [3.241] made measurements on Cu, Zn, and Al during 300 keV Ar$^+$ bombardment, and they also reported line profiles that would correspond to emission energies up to a few keV. More recent measurements, however, gave significantly lower typical energies. *Loxton* et al. [3.258, 280] studied the line profile of three Ti I emission lines. All three gave an average velocity around 2×10^4 cm/s, with a corresponding energy of about 100 eV. Recently, *Betz* [3.36] fitted his line-profile data to the Thompson velocity distribution (3.3) and obtained, for the bombardment of Ca by 15 keV Ar$^+$, a value for U_0 of 12 eV for Ca I, as compared with the 1.3 eV obtained for the ground-state Ca sputtered atoms. He also reported that U_0 was 4 eV for the Al I (396.2 nm) emission during the sputtering of Al (Fig. 3.26).

An alternative approach is to assume the Thompson distribution (3.3) (or a typical velocity) for the velocities of the sputtered atoms, and (3.41) for the velocity dependence of the excitation probability in the calculation of the Doppler-broadened line profile. The value of v_0 in (3.41) is then determined by fitting the theoretical curve with the experimental data; v_0 was determined in this way to be 2×10^6 cm/s for the Cu I 324.7 nm line [3.239, 281], and 1 to 3×10^7 cm/s for the Be I and Be II states [3.282]. *Hippler* et al. [3.241] reported

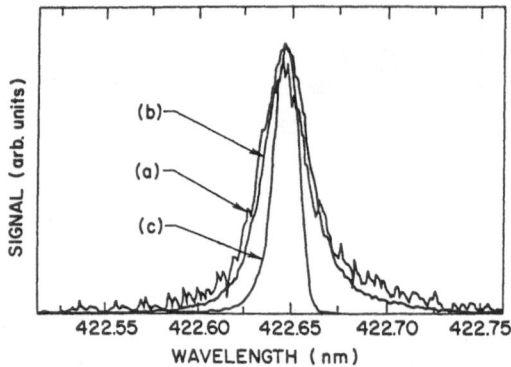

Fig. 3.26. Doppler-broadened line profiles for the excited Ca atoms (Ca I, 422.7 nm) sputtered from Ca metal by 15 keV Ar$^+$ near normal incidence for different states of the surface: (*a*) clean Ca, (*b*) oxygen-covered Ca. Curve (*c*) is the line profile for light from a hollow cathode lamp showing the spectrometer resolution. A 0.1 nm shift corresponds to about 7.1×10^6 cm/s [3.36]

that $v_0 = 5 \times 10^5$ cm/s for the 307.59 nm line and $v_0 = 5 \times 10^6$ cm/s for the 481.05 nm line, both of Zn.

Very careful data analysis is necessary to extract the correct information from line-profile measurements. The most recent data [3.36] show that the sputtered excited atoms have velocity distributions broader and mean velocities higher than those of the ground-state sputtered atom, but they are qualitatively similar to those of the secondary ions. The yield of excited atoms with high emission energy is significant only when direct-recoil sputtering becomes important [3.36].

While the discussion above is concerned with the sputtering from clean metals, the velocity distributions are affected by the chemical state of the target surface. With the sputtering of oxidized Ba [3.253], Ca [3.255], Cr [3.255], and Ti [3.249], the velocity distributions of ground-state neutral atoms were much broader than those for clean metal surfaces. When fitted to the Thompson distribution, the parameter U_0 increases from 2 eV for clean Ba to 22 eV for oxidized Ba. *Dullni* [3.249] reported that the velocity distributions of the excited Ti ground-state multiplets also have the same broadening. A significant portion of the detected ground-state atoms probably were actually cascade contributions from atoms originally sputtered in the higher unstable excited states.

The velocity distributions of atoms sputtered in metastable states outside the ground-state multiplet have been measured for Ba sputtered from BaF$_2$ [3.283] and oxidized Ba [3.253], and for Ca [3.254], Fe [3.256], and Ti [3.249] from the oxidized elements Fe, Ca, and Ti. In all cases, the distributions peak at velocities that correspond to energies around 10 eV, and are qualitatively similar to the velocity distributions of secondary ions. When fitted with the Thompson formula, the values of U_0 are typically between 10 to 20 eV. Historically these measurements established for the first time that sputtered excited atoms can be ejected through low-energy collisions in the usual collision cascade process, rather than only by energetic collisions.

Chemical effects on the velocity distribution of unstable excited atoms can be studied by comparing the line profiles of emitted photons obtained from clean and oxidized surfaces. For Ca I [3.36], there was essentially no change in the line profile (velocity distribution) between clean and oxidized Ca surfaces. For Cr I [3.36], Si I [3.240], and Ti I [3.258], the profiles for clean surfaces were *broader* than for the oxidized surfaces. For example, the "average energies" of Ti I atoms sputtered from Ti under the same condition as TiO in [3.258] were 40 and 85 eV for the 521.0 nm and 399.8 nm lines respectively. For Al I [3.36], the profile for the clean surfaces was *narrower* than for the oxidized surfaces. When fitted to the Thompson expression (3.3), the parameter U_0 equals 4 or 20 eV for a clean or oxidized Al surface, respectively.

Wright et al. [3.250] also found that the parameter v_0 in (3.41) changed with oxidation. They reported that while v_0 was above 1 to 3×10^7 cm/s for Be I and Be II sputtered from clean Be, it decreased to 5 to 7×10^6 cm/s upon O_2^+ bombardment for the Be I states, but remained unchanged for the Be II state studied. In their analysis, v in (3.41) was taken to be the maximum value attainable by collision with the primary ion.

e) Dependences on the Primary Ion and Impact Conditions

In the first approximation, changing the primary ion energy and/or the incident angle would affect the population of every excited level to the same extent as the sputtering yield. *Andersen* et al. [3.260] observed that the Au I levels (6p, 6d, 6s6p) from an Au target all showed the same relative change when the energy of the bombarding Xe$^+$ beam increased from 20 to 110 keV. This was also observed for Cu, Zn, and Mg targets, but not for Ga. The population of 6p Ga I and the 5p Ga II states actually decreased as the Xe$^+$ primary beam energy increased from 20 to 40 keV, while all other states showed increases in population.

Larsen and *Veje* [3.284] studied the atomic excitation during the sputtering of Be, B, Mg, and Au, as a function of the primary ion's angle of incidence. Qualitatively, the excitation intensities increase with increasing angle of incidence (with respect to the sample normal), as we would expect from the angular dependence of the total sputtering yield. However, the actual increase is a function of the excitation level. Figure 3.27 shows that the intensities for the Be $2p^2$ triplet and the 2p singlet P levels rise more rapidly with increasing angle of incidence than for the other Be levels. Since a larger angle of incidence favors the sputtering of atoms by energetic collisions, the data suggest that some levels are selectively excited through these collisions. *Larsen* and *Veje* proposed an electron-promotion process as an explanation [3.284].

Large dependences of the excited atom yields on the angle of incidence have been observed during Ar$^+$ and Ne$^+$ bombardment of Al [3.39], Cu [3.285], and Ni single crystals [3.286]. This is believed to be related to the channeling of the primary ions into the crystal. The magnitude of the effect depends on the specific excited state.

To study the effect of electronic excitations generated in the solid by the primary ion on the sputtering of excited atoms, *Harris* et al. [3.287] have studied the secondary photon emission from alkali chlorides during H$_n^+$ ($n = 1, 2, 3$) and

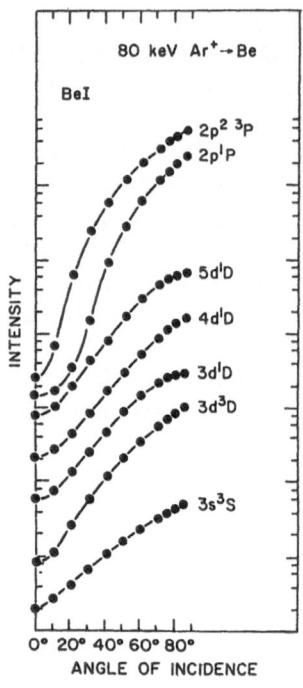

Fig. 3.27. Relative populations of the various Be I levels for Be atoms sputtered from Be metal as a function of the angle of incidence [3.284]

Ar^+ bombardment. The idea is that the electronic excitations generated by the ion bombardment in alkali halides have lifetimes comparable to or longer than the sputtering time, so that the sputtered atom can interact with an electronically excited environment along its exit trajectory. They observed no correlation between the observed fluorescence signal from the sputtered atoms and the electronic stopping power of the material or the intensity of the continuum radiation. Both are good indicators for the amount of electronic excitations produced in the target. The authors, however, recognized that the experiment could be complicated by the formation of alkali metal layers on the sample surface as the halogens were ejected preferentially through electronic sputtering.

3.4.3 Models for Sputtered-Atom Excitation

Our knowledge of the mechanisms for the excitation of sputtered atomic species is still very rudimentary. The local thermal equilibrium model for secondary ions was first extended to include the production of excited atoms [3.77, 257]. This model's validity is now very much in doubt (Sect. 3.3.2a). Here we shall review three proposed mechanisms for atomic excitations during sputtering: electronic excitation created by the primary ion in the collision cascade, electron tunneling, and bond breaking. None of these models has been rigorously tested. Usually only the variation of the excitation probability with the excitation energy was considered. The effects of angular momentum and spin quantum numbers are beyond our present understanding.

a) Excitation by Energetic Collisions

It was proposed by *Kelly* [3.288] that excited atoms can be sputtered through energetic collisions in the collision cascade. During the energetic collision, an amount of energy ΔE_e proportional to the collision energy E is assumed to be transferred to the electrons of the sputtered atom:

$$\Delta E_e = K_e E \; , \tag{3.42}$$

where K_e depends on the collision partners. Kelly used the expressions of *Firsov* [3.289] and *Kishinevski* [3.290] for electronic loss in atomic collisions to estimate K_e. In this model, all the atomic levels with (free-atom) excitation energy E_i less than ΔE_e are assumed to be populated with the probability proportional to the multiplicity g_i. The following expression for the absolute excited atom sputter yield was derived:

$$Y_i^* = 2g_i\, Y_T\, U_0\, K_e \frac{\displaystyle\sum_{k=i}^{\infty} G_k^{-1}(E_k^{-1} - E_{k+1}^{-1})}{\displaystyle\sum_{j}^{k} g_j} \; , \tag{3.43}$$

where Y_T is the total sputter yield, U_0 is the surface-binding energy, and $E_j \leq \Delta E_e$. In this model, no de-excitation through atom-solid interaction is considered. *Kelly* postulated that the chemical effect prevents any radiationless de-excitation of the sputtered excited atoms. Hence this model is applicable for chemical systems, and not for pure metals. He compared numerically the prediction of (3.43) with the experimental values reported by *Tsong* and *Yusuf* [3.232] for the sputtering from glasses (oxides) and found agreement to within a factor of 3 for 12 elements.

Despite this good agreement, one fundamental issue is unresolved. Since the values of K_e used in the calculations are about 10^{-2}, to excite levels with E_i on the order of a few eV, the collision energy has to exceed a threshold of a few hundred eV for excitation to occur. In fact, this theory predicts the sputtered excited atoms to have average kinetic energies on the order of 500 eV and higher. Experimentally, energy (velocity) distributions of sputtered excited atoms peak at energies about 10 eV (Sect. 3.4.2d), well below the expected collision thresholds from this theory.

Sputtered atoms can also be excited by energetic collisions in the collision cascade through the electron-promotion mechanism. *Larsen* and *Veje* [3.284] proposed that the $1s$ orbital of Be can be promoted to the $2p$ orbital during Be-Be collisions. According to the authors, excited atoms formed by energetic collisions should be very sensitive to the incident angle. These excited species are not expected to survive inside the solid and only those created at the surface, for example by direct recoil, can be ejected without de-excitation. This is consistent with the strong incident angle dependence of the yields of sputtered Be atoms with the $2p$ and $2p^2$ configurations (Fig. 3.27).

b) Atom-Surface Electron Transfer

According to several schemes the excitation levels of the sputtered atom can interact with the valence band of a metal. *Veje* [3.233, 262, 291] followed *Prival* [3.179] in assuming that the excited levels are populated by resonant electron transfer (tunneling) between the sputtered atom and the metallic substrate (Fig. 3.28a). Hence the model favors the population of levels that lie within the energy range of the valence band. *Veje* observed this correlation during the sputtering of group-II elements (Sect. 3.4.2b). Note that the broadening and shifting of the atomic levels near the surface have to be considered in a more accurate description. However, the insensitivity of the excited-state population to work-function changes seems to argue against this model (Sect. 3.4.2b). More experimental data are required.

An excited atom with its excitation level lying above the Fermi level can de-excite nonradiatively by electron tunneling into the unoccupied states in the metal (Fig. 3.28b). All such single-electron tunneling involves the change of the charge state by ± 1. The charge state can be preserved if a second electron can tunnel from the valence band into the ground state when the latter lies in the valence band, or can acquire an electron by Auger neutralization if the ground state lies below the valence band (Fig. 3.28c). The Auger electron may be ejected into the vacuum, depending on the energy balance. Again these processes, which

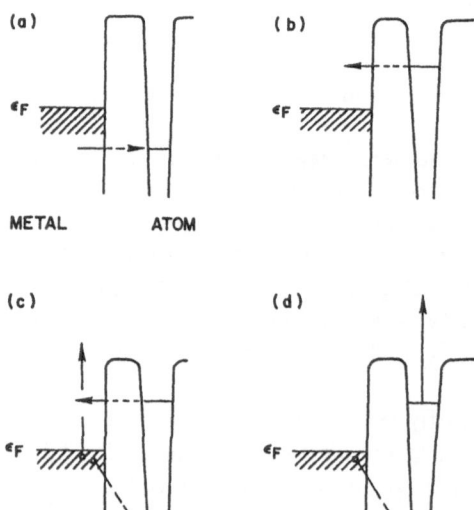

Fig. 3.28. Various possible electronic transitions: (a) excitation by electron tunneling, (b) de-excitation by electron tunneling, (c) de-excitation by electron tunneling followed by Auger neutralization, (d) Auger de-excitation

involve resonant electron transfer, should be sensitive to the position of the Fermi level relative to the atomic levels, and hence depend on the work function. An alternative that is less sensitive to the work function is to fill the ground level of the atom nonresonantly by an electron from the valence band, and then eject the electron in the excited level of the atom into the vacuum (Fig. 3.28d), so that the energy balance can be satisfied. All these processes have been observed in other experiments not related to sputtering [3.292, 293],

With the available experimental data, we cannot judge the relative importance of these numerous atom-surface electron exchange channels. The only quantum-mechanical treatment for sputtering was reported by *Sroubek* [3.294]. In his model, the sputtered atom, which moves away from the substrate with velocity v, has one ground state and one excited state of energy E_i. The Fermi level is assumed to be intermediate between these two electronic levels. The interaction potential V decreases exponentially with distance and has a decay length γ^{-1}. Hence the sputtered atom is subjected to a time-dependent perturbation $V_0 \exp(-\gamma v t)$, causing a substrate electron to be transferred up to the excitation level. With a substrate with a continuum of levels of constant density of state, which is more appropriate for sputtering from solid surfaces, the following expression for the excitation probability was found [3.294]:

$$P^* = \frac{\pi E_i}{2\gamma \hbar v} \exp\left(-\frac{\pi E_i}{2\gamma \hbar v}\right) . \tag{3.44}$$

Usually $\gamma \hbar v \ll E_i$, so P^* is often small. The equation predicts an exponential dependence of P^* on E_i and v^{-1} in agreement with (3.38, 41).

c) Bond-Breaking Model

Like secondary ion emission, the bond-breaking concept was introduced specifically to explain the chemical effects on the yields of excited atoms not belonging to the ground-state multiplet. As the atom leaves the surface, electronic transition occurs at the crossing of the appropriate potential curves. Three schemes have been proposed.

Thomas [3.210] considered the collision between the metal atom and the (lighter) electronegative atom on the chemically reacted sample surface. Momentum conservation dictates that both atoms would be ejected simultaneously as a quasi molecule. He proposed that if sufficient energy were available in the collision, curve-crossing processes in the dissociating quasi molecule could lead to the population of the excited molecular states. This could then yield excited or ionized sputtered atoms when the energy is high enough to break the molecular bond.

Williams et al. [3.59] extended the above concept and proposed that during the collision of the two atoms of the quasi molecule, excited atoms can be produced at the crossing of the steep repulsive side of the diabatic covalent excited curves with the ionic curve. They further postulated a Coulombic interaction between the solid surface and the departing quasi molecule to depress the ionization potential, so that the excited-state potential curves lie above the ionic curve, and no further curve crossing occurs. Under this scheme, only high-energy collisional events can reach the crossing point. The model in its present form cannot explain the low emission energies (Sect. 3.4.2d) of the excited atoms sputtered from oxides.

Blaise [3.295] proposed a simple extension of the bond-breaking concept in secondary ion formation (Sect. 3.3.2d). The diabatic covalent excited-state curve $M^{0*} + X^0$ lies above the covalent ground-state curve $M^0 + X^0$ by the excitation energy E_i (Fig. 3.29). It crosses the attractive side of the ionic curve at $R_{cx} = (I - A - E_i)^{-1}$ in atomic units; M^+ acts as the precursor for M^{0*}, and two curve crossings have to be considered. The excitation probability P^* is the product of the intermediate ionization probability P_I^+ at R_c and the intermediate neutralization probability P_I^* at the second crossing point R_{cx}. The additional collisional energy required is only E_i. It is hence applicable to atoms ejected by the usual cascade sputtering.

Recently *Yu* [3.296] adopted the bond-breaking model (Sect. 3.3.2d) and used the Landau-Zener curve-crossing formula (3.33) to calculate P_I^*:

$$P_I^* = g * \frac{1 - p}{(1 - p)g_0 + pg_+} \qquad \text{where}$$

$$p = \exp\left(\frac{- 2\pi H_{23}^2}{v|a|}\right)_{R = R_{cx}} . \tag{3.45}$$

Here g_* is the degeneracy of the excited state. Again assuming simple s wave functions, H_{23}^2 is found to decay exponentially with increasing R_{cx}. Also P_I^* decreases rapidly with E_i, since R_{cx} increases correspondingly. For typical

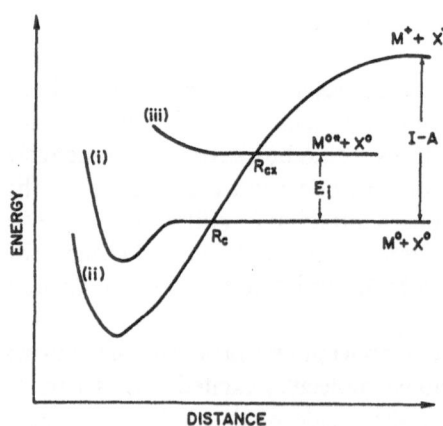

Fig. 3.29. Diabatic potential energy curves (i) ground state $M^0 + X^0$, (ii) ground state $M^+ + X^-$, (iii) excited state $M^{0*} + X^0$

values of E_i of several eV, P_i^* is usually small. This model hence predicts that the excited atom yield should be much smaller than the secondary ion yield. As an example, for Ca atoms sputtered with $U_0 = 5$ eV and a Thompson-like energy distribution, and with cation-vacancy parameters $C = 0.24$, $A = 1.46$ eV (Sect. 3.3.2d), the energy-averaged P^* for Ca I (2.93 eV) is 1.9×10^{-3}, while P^+ for Ca is 3.4×10^{-1}, according to this model.

Although this model cannot be used for the accurate quantitative calculation of excitation probabilities, it shows that excited atoms can be ejected in usual cascade sputtering. Also qualitatively explained is the observation [3.59] that for the sputtered excited atoms from chemical compounds, the excitation probability is much smaller than the ionization probability. While the model is promising for the excitation process, it also has certain limitations. It cannot describe any excitations with $E_i > I - A$, since there would not be any curve crossing, as was pointed out by *Blaise* [3.295]. Here, other mechanisms have to be introduced.

3.5 Concluding Remarks

Our understanding of the charge states of sputtered atoms is far from being complete, but it is in reasonable order. Semiquantitative models of ionization have been worked out for the various boundary conditions of the sputtering event. The dependences of the ionization probability on relevant parameters – such as the ionization potential, electron affinity, and the emission velocity – can be understood mostly within the framework of reasonable physical models. Progress toward more quantitative understanding of the ionization process is still slow. What is missing is the experimental determination of the electronic state at the sputtering site during the time of atom ejection, and information on the atomic trajectory of the sputtered atom.

Our understanding of excited atom sputtering is very rudimentary. Many fundamental properties – for example, the velocity distributions and the work-function dependences – are still not firmly established. Most available experimental data is from high-fluence sputtering experiments under insufficiently characterized surface conditions. More experiments on well-defined, radiation-damage-free surfaces will be helpful. Also, more experiments are needed to reveal the relation between the excitation process and the electronic state of the target. Progress in excitation theories is hindered by inadequate experimental information. This field is still wide open, with exciting opportunities.

Acknowledgments. I am indebted to Drs. R. Behrisch, K. Wittmaack, and W. Reuter for their valuable comments on this review. My thanks also go to Mrs. M. O'Brien for patiently preparing many versions of this manuscript.

References

3.1   R. Behrisch (ed.): *Sputtering by Particle Bombardment I*, Topics Appl. Phys., Vol. 47 (Springer, Berlin, Heidelberg 1981)

3.2   R. Behrisch (ed.): *Sputtering by Particle Bombardment II*, Topics Appl. Phys., Vol. 52 (Springer, Berlin, Heidelberg 1983)

3.3   R.E. Honig: J. Appl. Phys. **29**, 549 (1958)

3.4   C.W. White, D.L. Simms, N.H. Tolk: Science **177**, 481 (1972)

3.5   R. Courant: *Differential and Integral Calculus II* (Wiley, New York 1936)

3.6   G. Slodzian, J.F. Hennequin: C. R. Acad. Sc. **263**, 1246 (1966)

3.7   C. Snoek, W.F. van der Weg, P.K. Rol: Physica **30**, 341 (1964)

3.8   A. Benninghoven, J. Okano, R. Shimizu, H.W. Werner (eds.): *Secondary Ion Mass Spectrometry SIMS II*, Springer Series in Chemical Physics, Vol. 9 (Springer, Berlin, Heidelberg 1979)

3.9   A. Benninghoven, J. Giber, J. Laszlo, M. Riedel, H.W. Werner (eds.): *Secondary Ion Mass Spectrometry SIMS III*, Springer Series in Chemical Physics, Vol. 19 (Springer, Berlin, Heidelberg 1982)

3.10   A. Benninghoven, J. Okano, R. Shimizu, H.W. Werner (eds.): *Secondary Ion Mass Spectrometry SIMS IV*, Springer Series in Chemical Physics, Vol. 36 (Springer, Berlin, Heidelberg 1984)

3.11   A. Benninghoven, R.J. Colton, D.S. Simons, H.W. Werner (eds.): *Secondary Ion Mass Spectrometry SIMS V*, Springer Series in Chemical Physics, Vol. 44 (Springer, Berlin, Heidelberg 1986)

3.12   A. Benninghoven, A.M. Huber, H.W. Werner (eds.): *Secondary Ion Mass Spectrometry SIMS VI* (Wiley, New York 1988)

3.13   N.H. Tolk, J.C. Tully, W. Heiland, C.W. White (eds.): *Inelastic Ion-Surface Collisions* (Academic, New York 1977)

3.14   R. Kelly (ed.): Surf. Sci. **90** (1979)

3.15   E. Taglauer, W. Heiland (eds.): *Inelastic Ion-Surface Collisions*, Springer Series in Chemical Physics, Vol. 47 (Springer, Berlin, Heidelberg 1980)

3.16   P. Sigmund (ed.): Phys. Scripta **T6** (1983)

3.17   P. Williams (ed.): Nucl. Instrum. and Meth. **B14** (1986)

3.18   D.M. Gruen, A.R. Krauss, M.J. Pellin, C.E. Young (eds.): Nucl. Instrum. and Meth. **B27** (1987)

3.19   A. Benninghoven, F.G. Rudenauer, H.W. Werner: Secondary Ion Mass Spectrometry (Wiley, New York 1987)

3.20   H.W. Werner: Vacuum **24**, 493 (1975)

3.21   K. Wittmaack: In *Inelastic Ion-Surface Collisions*, ed. by H.H. Tolk, J.C. Tully, W. Heiland, C.W. White (Academic, New York 1977) p. 153

3.22   G. Blaise, A. Nourtier: Surf. Sci. **90**, 495 (1979)

3.23   P. Williams: Surf. Sci. **90**, 588 (1979)

3.24   V.I. Veksler: Radiat. Eff. **51**, 129 (1980)

3.25   P. Williams: Appl. Surf. Sci. **13**, 241 (1982)

3.26   N. Winograd: Prog. Solid State Chem. **13**, 285 (1982)

3.27   G. Slodzian: Phys. Scripta **T6**, 54 (1983)

3.28   M.L. Yu, N.D. Lang: Nucl. Instrum. Meth. **B14**, 403 (1986)

3.29   M.L. Yu: Comments At. Mol. Phys. **20**, 79 (1987)

3.30   Z. Sroubek: In *Secondary Ion Mass Spectrometry SIMS VI*, ed. by A. Benninghoven, A.M. Huber, H.W. Werner (Wiley, New York 1988) p. 17

3.31   C.W. White, E.W. Thomas, W.F. van der Weg, N.H. Tolk: In *Inelastic Ion-Surface Collisions*, ed. by N.H. Tolk, J.C. Tully, W. Heiland, C.W. White (Academic, New York 1977) p. 201

3.32   G.E. Thomas: Surf. Sci. **90**, 381 (1979)

3.33   R.J. MacDonald, C.M. Loxton, P.J. Martin: In *Inelastic Ion-Surface Collisions, Springer Series in Chemical Physics*, Vol. 47, ed. by E. Taglauer, W. Heiland (Springer, Berlin, Heidelberg 1980) p. 224

3.34 I.S.T. Tsong: In *Inelastic Ion-Surface Collisions, Springer Series in Chemical Physics*, Vol. 47, ed. by E. Taglauer, W. Heiland (Springer, Berlin, Heidelberg 1980) p. 258

3.35 D.M. Gruen, M.J. Pellin, C.E. Young, M.H. Mendelsohn, A.B. DeWald: Phys. Scripta T6, 42 (1983)

3.36 G. Betz: Nucl. Instrum. Meth. B27, 104 (1987)

3.37 M. Bernheim, G. Slodzian: Surf. Sci. 40, 169 (1973)

3.38 E. Zwangobani, R.J. MacDonald: Radiat Eff. 20, 81 (1973)

3.39 P.J. Martin, R.J. MacDonald: Radiat. Eff. 32, 177 (1977)

3.40 S.P. Holland, B.J. Garrison, N. Winograd: Phys. Rev. Lett. 43, 220 (1979)

3.41 R.A. Gibbs, S.P. Holland, K.E. Foley, B.J. Garrison, N. Winograd: J. Chem. Phys. 76, 684 (1982)

3.42 A. Benninghoven: Surf. Sci. 35, 427 (1973)

3.43 A. Benninghoven: Surf. Sci. 53, 596 (1975)

3.44 K. Wittmaack, J. Maul, F. Schulz: Int. J. Mass Spectrom. Ion Phys. 11, 23 (1973)

3.45 A.R. Bayly, R.J. MacDonald: J. Phys. E10, 79 (1977)

3.46 P.H. Dawson, P.A. Redhead: Rev. Sci. Instrum. 48, 159 (1977)

3.47 A.R. Krauss, D.M. Gruen: Appl. Phys. 14, 89 (1977)

3.48 K. Komori, J. Okano: Int. J. Mass Spectrom. Ion Phys. 27, 379 (1978)

3.49 M.A. Frisch, W. Reuter, K. Wittmaack: Rev. Sci. Instrum. 51, 695 (1980)

3.50 R.J. Colton, J.E. Campana, T.M. Barlak, J.J. DeCorpo, J.R. Wyatt: Rev. Sci. Instrum. 51, 1685 (1980)

3.51 R.A. Gibbs, N. Winograd: Rev. Sci. Instrum. 52, 63 (1981)

3.52 K. Wittmaack: Vacuum 89, 65 (1982)

3.53 M.W. Siegel, M.J. Vasile: Rev. Sci. Instrum. 52, 1603 (1981)

3.54 Y.L. Wang, R. Levi-Setti, J. Chabala: In *Secondary Ion Mass Spectrometry SIMS VI*, ed. by A. Benninghoven, A.M. Huber, H.W. Werner (Wiley, New York 1988) p. 53

3.55 R.A. Gibbs, S.P. Holland, K.E. Foley, B.J. Garrison, N. Winograd: Phys. Rev. B24, 6178 (1981)

3.56 M.J. Vasile: Phys. Rev. B29, 3785 (1984)

3.57 N.D. Lang: Phys. Rev. B27, 2019 (1983)

3.58 P. Sander, U. Kaiser, R. Jede, D. Lipinsky, O. Ganchow, A. Benninghoven: J. Vac. Sci. Technol. A3, 1946 (1985)

3.59 P. Williams, I.S.T. Tsong, S. Tsuji: Nucl. Instrum. Meth. 170, 591 (1980)

3.60 K. Wittmaack: Appl. Surf. Sci. 9, 315 (1981)

3.61 M.L. Yu: Nucl. Instrum. Meth. B15, 151 (1986)

3.62 C.A. Anderson: Int. J. Mass Spectrom. Ion Phys. 3, 413 (1970)

3.63 M.L. Yu, N.D. Lang: Phys. Rev. Lett. 50, 127 (1983)

3.64 M.L. Yu: Phys. Scripta T6, 67 (1983)

3.65 M.L. Yu: Phys. Rev. B29, 2311 (1984)

3.66 M.L. Yu: Phys. Rev. Lett. 40, 574 (1978)

3.67 M.L. Yu: Phys. Rev. B24, 1147 (1981)

3.68 M.L. Yu: Phys. Rev. B24, 5625 (1981)

3.69 M.L. Yu: Phys. Rev. B26, 4731 (1982)

3.70 M. Bernheim, F. LeBourse: Nucl. Instrum. Meth. B27, 94 (1987)

3.71 M. Bernheim, G. Slodzian: J. Microsc. Spectrosc. Electron. 6, 141 (1981)

3.72 T.R. Lundquist: J. Vac. Sci. Technol. 15, 684 (1978)

3.73 R.G. Hart, C.B. Cooper: Surf. Sci. 94, 105 (1980)

3.74 A.R. Krauss, D.M. Gruen: Surf. Sci. 92, 14 (1980)

3.75 R.F. Garrett, R.J. MacDonald, D.J. O'Conner: Nucl. Instrum. Meth. 218, 333 (1983)

3.76 R.J. MacDonald: Surf. Sci. 43, 653 (1974)

3.77 R.J. MacDonald, R.F. Garnett: Surf. Sci. 78, 371 (1978)

3.78 G.A.v.d. Schootbrugge, A.G.J. de Witt, J.M. Fluit: Nucl. Instrum. Meth. 132, 321 (1976)

3.79 A.R. Bayly, R.J. MacDonald: Radiat. Eff. 34, 169 (1978)

3.80 A. Wucher, H. Oechsner: In *Secondary Ion Mass Spectrometry SIMS VI*, ed. by A. Benninghoven, A.M. Huber, H.W. Werner (Wiley, New York 1988) p. 143

3.81   M.J. Vasile: Surf. Sci. **115**, L141 (1982)
3.82   H.D. Hagstrum: Phys. Rev. **96**, 336 (1954)
3.83   M.L. Yu: Phys. Rev. Lett. **47**, 1325 (1981)
3.84   J.J. Cuomo, R.J. Gambino, J.M.E. Harper, J.D. Kuptsis: J. Vac. Sci. Technol. **15**, 281 (1978)
3.85   M. Riedel, T. Nenadovic, B. Perovic: Acta Chim. Acad. Sci. Hung. **97**, 187 (1978)
3.86   H. Rodriguez-Murciaudin, H.E. Beske: Rept. No. 1292 KFA, Jülich (1976)
3.87   G. Blaise, G. Slodzian: J. de Phys. **31**, 93 (1970)
3.88   G. Blaise, G. Slodzian: Rev. Physique Appl. **8**, 105 (1973)
3.89   G. Blaise, G. Slodzian: Rev. Physique Appl. **8**, 247 (1973)
3.90   G. Blaise, G. Slodzian: J. de Phys. **35**, 243 (1974)
3.91   M.L. Yu, W. Reuter: J. Vac. Sci. Technol. **18**, 570 (1981)
3.92   W.F. Croydon, M.G. Dowsett, R.M. King, E.H.C. Parker: J. Vac. Sci. Technol. B**3**, 604 (1985)
3.93   Z. Sroubek: Appl. Phys. Lett. **42**, 514 (1983)
3.94   Z. Sroubek: Nucl. Instrum. Meth. **218**, 336 (1983)
3.95   W. Reuter: In *Secondary Ion Mass Spectrometry SIMS IV, Springer Series in Chemical Physics*, Vol. 36, ed. by A Benninghoven, J. Okano, R. Shimizu, H.W. Werner (Springer, Berlin, Heidelberg 1984) p. 54
3.96   W. Reuter: Anal. Chem. **59**, 2081 (1987)
3.97   W. Reuter, J. Clabes: Anal. Chem. **60**, 1404 (1988)
3.98   K. Mann, M.L. Yu: In *Secondary Ion Mass Spectrometry SIMS V, Springer Series in Chemical Physics*, Vol. 44, ed. by A. Benninghoven, R.J. Colton, D.S. Simons, H.W. Werner (Springer, Berlin, Heidelberg 1986) p. 26
3.99   K. Mann, M.L.Yu: Phys Rev. B**35**, 6043 (1987)
3.100 P. Williams, C.A. Evans, Jr.: Surf. Sci. **78**, 324 (1978)
3.101 K. Wittmaack: to be published
3.102 M.L. Yu: J. Vac. Sci. Technol. A**1**, 500 (1983)
3.103 P.J. Martin, A.R. Bayly, R.J. MacDonald, N.H. Tolk, G.J. Clark, J.C. Kelly: Surf. Sci. **60**, 349 (1976)
3.104 M.L. Yu, J. Clabes, D.J. Vitkavage: J. Vac. Sci. Technol. A**3**, 1316 (1985)
3.105 G. Slodzian: Surf. Sci. **48**, 161 (1975)
3.106 P. Williams: Int. J. Mass Spectrom. Ion Phys. **53**, 101 (1983)
3.107 M.L. Yu, K. Mann: Phys. Rev. Lett. **57**, 1476 (1986)
3.108 P. Sigmund: In *Secondary Ion Mass Spectrometry SIMS IV, Springer Series in Chemical Physics*, Vol. 36, ed. by A. Benninghoven, J. Okano, R. Shimizu, H.W. Werner (Springer, Berlin, Heidelberg 1984) p. 2
3.109 A.R. Krauss, D.M. Gruen: Surf. Sci. **90**, 564 (1979)
3.110 G. Blaise, M. Bernheim: Surf. Sci. **47**, 324 (1975)
3.111 V.R. Deline, W. Katz, C.A. Evans, Jr., P. Williams: Appl. Phys. Lett. **33**, 832 (1978)
3.112 H. Oechsner, Z. Sroubek: Surf. Sci. **127**, 10 (1983)
3.113 K. Wittmaack: Surf. Sci. **112**, 168 (1981)
3.114 H. Gnaser: Nucl. Instrum. Meth. **218**, 312 (1983)
3.115 A.E. Morgan, H.A.M. de Grefte, N. Warmoltz, H.W. Werner, H.J. Tolle: Appl. Surf. Sci. **7**, 372 (1981)
3.116 W. Reuter: Nucl. Instrum. Meth. B**15**, 173 (1986)
3.117 M.L. Yu: Surf. Sci. **71**, 121 (1978)
3.118 M.L.Yu: J. Vac. Sci. Technol. **15**, 668 (1978)
3.119 C.A. Andersen: Int. J. Mass Spectrom. Ion Phys. **2**, 61 (1969)
3.120 C.A. Evans, Jr.: Anal. Chem. **44**, 67A (1972)
3.121 C.A. Andersen, J.R. Hinthorne: Anal. Chem. **45**, 1421 (1973)
3.122 Z. Jurela: Int. J. Mass Spectrom. Ion Phys. **12**, 33 (1973)
3.123 R. Shimizu, T. Ishitani, Y. Ueshima: Jap. J. Appl. Phys. **13**, 249 (1974)
3.124 C.A Andersen: In *Secondary Ion Mass Spectrometry*, ed. by K.F.J. Heinrich, D.E. Newbury (NBS SP-427, USGPO, Washington, D.C., 1975) p. 79
3.125 D.S. Simons, J.E. Baker, C.A. Evans, Jr.: Anal. Chem. **48**, 1341 (1976)

3.126 A.E. Morgan, H.W. Werner: Anal. Chem. **48**, 699 (1976)
3.127 A.E. Morgan, H.W. Werner: Anal. Chem. **49**, 927 (1977)
3.128 V.R. Deline, C.A. Evans, Jr., P. Williams: Appl. Phys. Lett. **33**, 578 (1978)
3.129 H. Gnaser, I.D. Hutcheon: Phys. Rev. **B35**, 877 (1987)
3.130 J.C. Lorin, A. Havetti, G. Slodzian: In *Secondary Ion Mass Spectrometry SIMS III, Springer Series in Chemical Physics*, Vol. 19, ed. by A. Benninghoven, J. Giber, J. Laszlo, M. Riedel, H.W. Werner (Springer, Berlin, Heidelberg 1982) p. 140
3.131 H. Gnaser, I.D. Hutcheon: In *Secondary Ion Mass Spectrometry SIMS VI*, ed. by A. Benninghoven, A.M. Huber, H.W. Werner (Wiley, New York 1988) p. 29
3.132 N. Shimizu, S.R. Hart: J. Appl. Phys. **53**, 1303 (1982)
3.133 S.A. Schwarz: J. Vac. Sci. Technol. **A5**, 308 (1987)
3.134 M.L. Yu: Nucl. Instrum. Meth. **149**, 559 (1978)
3.135 K. Wittmaack: Surf. Sci. **53**, 626 (1975)
3.136 B.A. Tsipinyuk, V.I. Veksler: Vacuum **29**, 155 (1979)
3.137 W. Eckstein: Nucl. Instrum. Meth. **B27**, 78 (1987)
3.138 H.J. Barth, E. Mühling, W. Eckstein: Appl. Surf. Sci. **22/23**, 136 (1985)
3.139 W. Eckstein, H.J. Barth, E. Mühling: Nucl. Instrum. Meth. **B14**, 507 (1986)
3.140 H.J. Barth, E. Mühling, W. Eckstein: Surf. Sci. **166**, 458 (1986)
3.141 E. Mühling, W. Eckstein, H. Verbeek: Surf. Sci. **177**, 565 (1986)
3.142 J.A. Schultz, S. Contarini, Y.S. Jo, J.W. Rabalais: Surf. Sci. **154**, 315 (1985)
3.143 J.A. Schultz, Y.S. Jo, J.W. Rabalais: Nucl. Instrum. Meth. **B10/11**, 713 (1985)
3.144 J.A. Schultz, Y.S. Jo, J.W. Rabalais: Solid State Comm. **55**, 957 (1985)
3.145 J.A. Schultz, C.R. Blakely, M.H. Mintz, J.W. Rabalais: Nucl. Instrum. Meth. **B14**, 500 (1986)
3.146 J.A. Schultz, Y.S. Jo, S. Tachi, J.W. Rabalais: Nucl. Instrum. Meth. **B15**, 134 (1986)
3.147 J.W. Rabalais, J.N. Chen: J. Chem. Phys. **85**, 3615 (1986)
3.148 J.F. Hennequin: J. de Phys. **29**, 655 (1968)
3.149 J.F. Hennequin: J. de Phys. **29**, 957 (1968)
3.150 p.D. Brochard, G. Slodzian: J. de Phys. **32**, 185 (1971)
3.151 J. Maul, K. Wittmaack: Surf. Sci. **47**, 358 (1975)
3.152 J.F. Hennequin, P. Viaris de Lesegno: Surf. Sci. **42**, 50 (1974)
3.153 J.F. Hennequin: J. de Phys. **29**, 1053 (1968)
3.154 W.A. Metz, K.O. Legg, E.W. Thomas: J. Appl. Phys. **51**, 2888 (1980)
3.155 T.D. Andreadis, J. Fine, J.A. Matthew: Nucl. Instrum. Meth. **209/210**, 495 (1983)
3.156 R.A. Baragiola, E.V. Alonso, H.J.L. Raite: Phys. Rev. **A25**, 1969 (1982)
3.157 J.F. Hennequin, R.L. Inglebert, P. Viaris de Lesegno: In *Secondary Ion Mass Spectrometry SIMS V, Springer Series in Chemical Physics*, Vol. 44, ed. by A. Benninghoven, R.J. Colton, D.S. Simons, H.W. Werner (Springer, Berlin, Heidelberg 1986) p. 60
3.158 J.F. Hennequin, J.L. Bernard: In *Secondary Ion Mass Spectrometry SIMS VI*, ed. A. Benninghoven, A.M. Huber, H.W. Werner (Wiley, New York 1988) p. 25
3.159 J.F. Hennequin: Surf. Sci. **203**, 245 (1988)
3.160 J.F. Hennequin, R.L. Inglebert, P. Viaris de Lesegno: Surf. Sci. **140**, 197 (1984)
3.161 K. Wittmaack: Surf. Sci. **90**, 557 (1979)
3.162 P. Joyes, J.F. Hennequin: J. de Phys. **29**, 483 (1968)
3.163 K. Wittmaack: Nucl. Instrum. Meth. **170**, 565 (1980)
3.164 K. Wittmaack: Surf. Sci. **85**, 69 (1979)
3.165 J.J. Vrakking, A. Kroes: Surf. Sci. **84**, 153 (1979)
3.166 K. Wittmaack: Nucl. Instrum. Meth. **B2**, 674 (1984)
3.167 V.I. Veksler: Sov. Phys. Solid State **24**, 997 (1982)
3.168 K. Wittmaack: Phys. Rev. Lett. **43**, 872 (1979)
3.169 P. Williams: Phys. Rev. **B23**, 6187 (1981)
3.170 M.L. Knotek, P.J. Feibelman: Phys. Rev. Lett. **40**, 969 (1978)
3.171 T.E. Madey, J.T. Yates, Jr.: J. Vac. Sci. Technol. **8**, 525 (1971)

3.172 J.A. Schultz, P.T. Murray, R. Kumar, H.K. Hu, J.W. Rabalais: In *Desorption Induced by Electronic Transitions DIET I, Springer Series in Chemical Physics*, Vol. 24, ed. by N.H. Tolk, M.M. Traum, J.C. Tully, T. Madey (Springer, Berlin, Heidelberg 1983) p. 191
3.173 P.G. Blauner, R.A. Weller: Phys. Rev. **B35**, 1485 (1987)
3.174 P.G. Blauner, R.A. Weller: Phys. Rev. **B35**, 1492 (1987)
3.175 J.M. Schroeer, T.N. Rhodin, R.C. Bradley: Surf. Sci. **34**, 571 (1973)
3.176 Z. Sroubek: Surf. Sci. **44**, 47 (1974)
3.177 M. Cini: Surf. Sci. **54**, 71 (1976)
3.178 J. Antal: Phys. Lett. **55A**, 493 (1976)
3.179 H.G. Prival: Surf. Sci. **76**, 443 (1978)
3.180 Z. Sroubek, K. Zdansky, J. Zavadil: Phys. Rev. Lett. **45**, 580 (1980)
3.181 B.J. Garrison, A.C. Diebold, J.H. Liu, Z. Sroubek: Surf. Sci. **124**, 461 (1983)
3.182 J.A. Olson, B.J. Garrison: Nucl. Instrum. Meth. **B14**, 414 (1986)
3.183 R.L. Erickson, D.P. Smith: Phys. Rev. Lett. **34**, 297 (1975)
3.184 E.C. Goldberg, J. Ferron, M.C.G. Passeggi: Phys. Rev. **B30**, 2448 (1984)
3.185 J.A. Olson, B.J. Garrison: J. Chem. Phys. **81**, 1355 (1984)
3.186 J.A. Olson, B.J. Garrison: J. Chem. Phys. **83**, 1392 (1985)
3.187 N.D. Lang, J.K. Norskov: Phys. Scripta T**6**, 15 (1983)
3.188 W.F. van der Weg, P.K. Rol: Nucl. Instrum. Meth. **38**, 274 (1965)
3.189 A. Blandin, A. Nourtier, D.W. Hone: J. Phys. **37**, 396 (1976)
3.190 R. Brako, D.M. Newns: Surf. Sci. **108**, 253 (1981)
3.191 J.K. Norskov, B.I. Lundqvist: Phys. Rev. **B19**, 5661 (1979)
3.192 B.J. Garrison: Surf. Sci. **167**, L225 (1986)
3.193 D.M. Newns, K. Makoski, R. Brako, J.N.M. van Wunik: Phys. Scripta T**6**, 5 (1983)
3.194 R. Hentschke, K.J. Snowdon, P. Hertel, W. Heiland: Surf. Sci. **173**, 565 (1986)
3.195 K.J. Snowdon, R. Hentschke, A. Närmann, W. Heiland: Surf. Sci. **173**, 581 (1986)
3.196 A. Nourtier, J. Quazza, J.P. Jardin: In *Secondary Ion Mass Spectrometry SIMS VI*, ed. by A. Benninghoven, A.M. Huber, H.W. Werner (Wiley, New York 1988) p. 147
3.197 A. Nourtier, J.P. Jardin, J. Quazza: Phys. Rev. **B37**, 10628 (1988)
3.198 Z. Sroubek, G. Falcone: Surf. Sci. **166**, L136 (1986)
3.199 Z. Sroubek: Appl. Phys. Lett. **45**, 850 (1984)
3.200 H. Kasai, H. Nakanishi, A. Okiji: J. Phys. Soc. Japan **55**, 3210 (1986)
3.201 J. Zavadil: Surf. Sci. **143**, L383 (1984)
3.202 J.J.C. Geerlings, J. Los, J.P. Gauyacq, N.M. Temme: Surf. Sci. **172**, 257 (1986)
3.203 Z. Sroubek: Phys. Rev. **B25**, 6046 (1982)
3.204 Z. Sroubek: Nucl. Instrum. Meth. **194**, 533 (1982)
3.205 B. Rasser, J.N.M. Van Wunnik, J. Los: Surf. Sci. **118**, 697 (1982)
3.206 J.J.C. Geerlings, L.F. Tz. Kwakman, J. Los: Surf. Sci. **184**, 305 (1987)
3.207 C.A. Barlow, Jr., J.R. MacDonald: J. Chem. Phys. **43**, 2575 (1965)
3.208 Z. Jurela: Int. J. Mass Spectrom. Ion Phys. **37**, 67 (1981)
3.209 J.N. Coles: Surf. Sci. **79**, 549 (1979)
3.210 G.E. Thomas: Radiat. Eff. **31**, 185 (1977)
3.211 W. Gerhard, C. Plog: Z. Phys. B**54**, 59 (1983)
3.212 C. Plog, W. Gerhard: Z. Phys. B**54**, 71 (1983)
3.213 W. Gerhard, C. Plog: Surf. Sci. **152/153**, 127 (1985)
3.214 C. Plog, G. Roth, W. Gerhard, W. Kerfin: In *Secondary Ion Mass Spectrometry SIMS V, Springer Series in Chemical Physics*, Vol. 44, ed. by A. Benninghoven, R.J. Colton, D.S. Simons, H.W. Werner (Springer, Berlin, Heidelberg 1986) p. 29
3.215 L. Landau, Z. Phys. Sov. **2**, 46 (1932)
3.216 C. Zener, Proc. Roy. Soc. London A**137**, 696 (1932)
3.217 C. Coudray, G. Slodzian: In *Secondary Ion Mass Spectrometry SIMS VI*, ed. by A. Benninghoven, A.M. Huber, H.W. Werner (Wiley, New York 1988) p. 45
3.218 M.C.G. Passeggi, E.C. Goldberg, J. Ferron: Phys. Rev. **B35**, 8330 (1987)
3.219 M.L. Yu: Nucl. Instrum. Meth. **B18**, 542 (1987)

3.220 U. Fano, W. Lichten: Phys. Rev. Lett. **14**, 627 (1965)
3.221 W. Lichten: Phys. Rev. **164**, 131 (1967)
3.222 M. Barat, W. Lichten: Phys. Rev. **A6**, 211 (1972)
3.223 W. Lichten: J. Phys. Chem. **84**, 2102 (1980)
3.224 P. Joyes: J. de Phys. **30**, 243 (1969)
3.225 P. Joyes: J. de Phys. **30**, 365 (1969)
3.226 P. Joyes: Radiat. Eff. **19**, 235 (1973)
3.227 R. Kelly, C.B. Kerkdijk: Surf. Sci. **46**, 537 (1974)
3.228 J.P. Meriaux, R. Goutte, C. Guillaud: Appl. Phys. **7**, 313 (1975)
3.229 M.J. Pellin, C.E. Young, M.H. Mendelsohn, D.M. Gruen, R.B. Wright, A.B. DeWald: J. Nucl. Mat. **111/112**, 738 (1982)
3.230 B.I. Craig, J.P. Baxter, J. Singh, G.A. Schick, P.H. Kobrin, B.J. Garrison, N. Winograd: Phys. Rev. Lett. **57**, 135 (1986)
3.231 I.S.T. Tsong, A.C. McLaren: Spectrochimica Acta **30B**, 343 (1975)
3.232 I.S.T. Tsong, N.A. Yusuf: Appl. Phys. Lett. **33**, 999 (1978)
3.233 E. Veje: Surf. Sci. **110**, 533 (1981)
3.234 K. Jensen, E. Veje: Z. Physik **269**, 293 (1974)
3.235 S. Dzioba, O. Auciello, R. Kelly: Radiat. Eff. **45**, 235 (1980)
3.236 S. Dzioba, R. Kelly: Surf. Sci. **100**, 119 (1980)
3.237 M. Szymonski, A. Paradzisz, L. Gabla: In *Inelastic Ion-Surface Collisions, Springer Series in Chemical Physics*, Vol. 17, ed. by E. Taglauer, W. Heiland (Springer, Berlin, Heidelberg 1980) p. 322
3.238 R.F. Garrett, R.J. MacDonald, D.J. O'Connor: Surf. Sci. **131**, L399 (1983)
3.239 W.F. van der Weg, D.J. Bierman: Physica **44**, 206 (1969)
3.240 C.W. White, D.L. Simms, N.H. Tolk, D.V. McCaughan: Surf. Sci. **49**, 657 (1975)
3.241 R. Hippler, W. Kruger, A. Scharmana, K.H. Schartner: Nucl. Instrum. Meth. **132**, 439 (1976)
3.242 W. Husinsky, R. Bruckmuller, P. Blum, F. Viehbock, D. Hammer, E. Benes: J. Appl. Phys. **48**, 4734 (1977)
3.243 D. Hammer, E. Benes, P. Blum, W. Husinsky: Rev. Sci. Instrum. **47**, 1178 (1976)
3.244 R.B. Wright, M.J. Pellin, D.M. Gruen: Surf. Sci. **110**, 151 (1981)
3.245 H.L. Bay: Nucl. Instrum. Meth. **B18**, 430 (1987)
3.246 B. Schweer, H.L. Bay: In *Proc. 4th Int. Conf. on Solid Surf. Sci. and 3rd Europ. Conf. on Surf. Sci.*, ed. by D.A. Degras, M. Costa (Societe Francaise du Vide, Paris 1980) p. 1349
3.247 C.E. Young, W.F. Calaway, M.J. Pellin, D.M Gruen: J. Vac. Sci. Technol **A2**, 693 (1984)
3.248 M.J. Pellin, D.M. Gruen, C.E. Young, M.D. Wiggins: Nucl. Instrum. Meth. **218**, 771 (1983)
3.249 E. Dullni: Appl. Phys. **A38**, 131 (1985)
3.250 R.B. Wright, MJ. Pellin, D.M. Gruen, C.E. Young: Nucl. Instrum. Meth. **170**, 295 (1980)
3.251 M.J. Pellin, R.B. Wright, D.M. Gruen: J. Chem. Phys. **74**, 6448 (1981)
3.252 R.B. Wright, C.E. Young, M.J. Pellin, D.M. Gruen: J. Vac. Sci. Technol. **20**, 510 (1982)
3.253 D. Grischkowsky, M.L. Yu, A.C. Balant: Surf. Sci. **127**, 315 (1983)
3.254 W. Husinsky, G. Betz, I. Girgis: Phys. Rev. Lett. **50**, 1689 (1983)
3.255 W. Husinsky, G. Betz, I. Girgis: J. Vac. Sci. Technol. **A2**, 698 (1984)
3.256 B. Schweer, H.L. Bay: Appl. Phys. **A29**, 53 (1982)
3.257 R.J. MacDonald, P.J. Martin: Surf. Sci. **66**, 423 (1977)
3.258 C.M. Loxton, R.J. MacDonald, P.J. Martin: Surf. Sci. **93**, 84 (1980)
3.259 P.J. Martin, R.J. MacDonald: Surf. Sci. **62**, 551 (1977)
3.260 N. Andersen, B. Andresen, E. Veje: Radiat. Effect **60**, 119 (1982)
3.261 K. Kierkegaard, S. Ludvigsen, B. Patterson, E. Veje: Nucl. Instrum. Meth. **B13**, 388 (1986)
3.262 E. Veje: Phys. Rev. **B28**, 5029 (1983)
3.263 G.E. Thomas, E.E. de Kluizenaar: Nucl. Instrum. Meth. **132**, 449 (1976)
3.264 E. Veje: Phys. Rev. **B28**, 88 (1983)
3.265 W. Husinsky, P. Wurz, B. Strehl, G. Betz: Nucl. Instrum. Meth. **B18**, 452 (1987)
3.266 R.J. MacDonald, W. Heiland, E. Taglauer: Appl. Phys. Lett. **33**, 576 (1978)
3.267 R.J. MacDonald, E. Taglauer, W. Heiland: Appl. Surf. Sci. **5**, 197 (1980)

3.268 W. Heiland, J. Kraus, S. Leung, N.H. Tolk: Surf. Sci. **67,** 437 (1977)

3.269 V.V. Gritsyna, T.S. Kijan, A.G. Koval, Ya.M. Fogel: Radiat. Eff. **14,** 77 (1972)

3.270 T.S. Kiyan, V.V. Gritsyna, Ya.M. Fogel: Nucl. Instrum. Meth. **132,** 435 (1976)

3.271 M. Braun, B. Emmoth, R. Buchta: Proc. 7th Int. Vac. Congr. & 3rd Int. Conf. on Solid Surface (Vienna, 1976) p. 501

3.272 R.J. MacDonald, P.J. Martin: Surf. Sci. **67,** 237 (1977)

3.273 R. Shimizu, T. Okutani, T. Ishitani, H. Tamura: Surf. Sci. **69,** 349 (1977)

3.274 C.M. Loxton, I.S.T. Tsong, H.W. Pickering: Nucl. Instrum. Meth. **218,** 340 (1983)

3.275 E. Veje: Z. Phys. B. **70,** 55 (1988)

3.276 C.B. Kerkdijk, R. Kelly: Radiat. Eff. **38,** 73 (1978)

3.277 D. Gade, K.B. Larsen, K.T. Palle, E. Veje: Nucl. Instrum. Meth. **B18,** 570 (1987)

3.278 E. Veje: Surf. Sci. **109,** L545 (1981)

3.279 R. Kelly, S. Dzioba, N.H. Tolk, J.C. Tully: Surf. Sci. **102,** 486 (1981)

3.280 C.M. Loxton, R.J. MacDonald, E. Taglauer: Surf. Sci. **102,** L76 (1981)

3.281 C.W. White, N.H. Tolk: Phys. Rev. Lett. **26,** 486 (1971)

3.282 R.B. Wright, D.M. Gruen: J. Chem. Phys. **73,** 664 (1980)

3.283 M.L. Yu, D. Grischkowsky, A.C. Balant: Phys. Rev. Lett. **48,** 427 (1982)

3.284 P. Larsen, E. Veje: Phys. Rev. **B28,** 5011 (1983)

3.285 W.F. van der Weg, N.H. Tolk, C.W. White, J. Kraus: Nucl. Instrum. Meth. **132,** 405 (1976)

3.286 I.S.T. Tsong, N. Tolk, T.M. Buck, J.S. Kraus, T.R. Pian, R. Kelly: Nucl. Instrum. Meth. **194,** 655 (1982)

3.287 P. Harris, J.O. Madsen, E. Veje: Nucl. Instrum. Meth. **B18,** 566 (1987)

3.288 R. Kelly: Phys. Rev. **B25,** 700 (1982)

3.289 O.B. Firsov: Zh. Eksp. Teor. Fiz. **36,** 1517 (1959) [Sov. Phys.-JETP **36,** 1076 (1959)]

3.290 L.M. Kishinevski, Izv. Akad. Nauk SSR, Ser. Fiz. **26,** 1410 (1962) [Bull. Acad. Sci. USSR, Phys. Ser. 26, 1433 (1962)]

3.291 E. Veje: Nucl. Instrum. Meth. **194,** 593 (1982)

3.292 H.D. Hagstrum: Phys. Rev. Lett. **43,** 1050 (1979)

3.293 H. Conard, G. Ertl, J. Kuppers, W. Sesselman, H. Haberland: Surf. Sci. **100,** 461 (1980)

3.294 Z. Sroubek: Phys. Scripta **T6,** 24 (1983)

3.295 G. Blaise: Surf. Sci. **60,** 65 (1976)

3.296 M.L. Yu: In *Secondary Ion Mass Spectrometry SIMS VI,* ed. by A. Benninghoven, A.M. Huber, H.W. Werner (Wiley, New York 1988) p. 41

4. Surface and Depth Analysis Based on Sputtering

Klaus Wittmaack

With 47 Figures

Sputtering is an almost universal technique for controlled erosion of solid materials. The merits of sputtering for analytical applications became fully recognized some 20 years ago. In this chapter we discuss a variety of methods for compositional analysis of solids. All use sputtering for controlled etching of the sample. Two types of approaches can be distinguished: The first group consists of methods based on sputtered-particle analysis and includes secondary ion mass spectrometry (SIMS) and sputtered neutral mass spectrometry (SNMS). The second group comprises methods by which the near-surface composition of the sample is probed intermittently between deliberately chosen periods of sputter erosion, for example, by Auger electron spectroscopy (AES) or ion scattering spectrometry (ISS).

The implications of using ion bombardment for sputter etching are discussed in detail. The advantages and limitations of the various analytical techniques are described rather briefly. SIMS is the only technique that works with all elements of the periodic table. Because an inherent background signal is absent and high secondary ion yields can be produced, the sensitivity achieved with SIMS (and with laser-based SNMS) is one to five orders of magnitude higher than with other techniques. However, the elemental sensitivities observed in SIMS depend strongly on the sample composition. This "matrix effect" is much less severe in SNMS, AES, or ISS.

With all sputter-based analytical techniques, depth calibration can be a problem because the erosion rate depends on the sample composition and the bombardment parameters, aspects that deserve particular attention with multi-layer samples. Ion-bombardment-induced broadening effects constitute another important issue in sputter depth profiling. Various physical and chemical processes such as collisional mixing, radiation-enhanced diffusion, segregation, and element-differential ("preferential") sputtering may contribute to the observed relocation of atoms in the sample. Moreover, the depth resolution obtained in sputter profiling of polycrystalline materials degrades rapidly because of growing surface roughness.

Last but not least we discuss how to identify and minimize various types of background signals, either related to the sputtering process ("memory effects") or associated with nonideal vacuum conditions. Some ideas concerning future trends in sputter-based analysis are outlined at the end of the chapter.

Topics in Applied Physics, Vol. 64

Sputtering by Particle Bombardment III Eds.: R. Behrisch · K. Wittmaack

© Springer-Verlag Berlin Heidelberg 1991

4.1 Historical Background

Sputtering was discovered more than 130 years ago [4.1–3]. Its use for analytical applications can be traced [4.4] to the beginning of this century, when *J.J. Thomson* reported emission of positively charged secondary rays from a metal plate under the impact of primary *Kanalstrahlen* [4.5]. Some other observations of secondary ion emission were published in the 1930s [4.3, 4]. The first "real" SIMS study was described by *Sloane* and *Press* [4.6] in 1938. World War II interrupted work in this field for about ten years [4.7], and activities remained at a rather low level for another 20 years [4.8]. Toward the end of this period, however, two outstanding technical achievements were reported: the ion microscope by *Castaing* and *Slodzian* [4.9] and the raster scanning ion microprobe by *Liebl* [4.10]. The development of SIMS instrumentation has been reviewed repeatedly [4.11–13].

In the early 1960s *Lutz* and *Sizmann* demonstrated the use of sputtering for microsectioning of single crystals [4.14]. They showed that range profiles of implanted radioactive ions can be determined quite precisely by measuring the residual activity of the sample after repeated removal of thin layers by low-energy ion bombardment. At about the same time, ultrahigh vacuum equipment became commercially available. These developments formed the necessary technical basis for a rapid development of SIMS and many other current surface-sensitive techniques. The successful application of quadrupole mass filters gave SIMS additional momentum [4.13, 15–19]. An important result of basic research was the finding that the degree of ionization of positively or negatively charged sputtered particles can be greatly enhanced by oxygen [4.8, 20–22] or cesium [4.15, 23, 24], respectively, at the bombarded surface.

In the early 1970s *Benninghoven* and coworkers [4.17, 25–28] reported on the use of SIMS for analyzing adsorbed layers as well as for investigating chemical reactions at surfaces. Various other groups began to use SIMS for analyzing thin-film structures [4.29, 30] and for measuring range profiles of implanted ions in solids, notably of boron in silicon [4.31–35]. The early depth profiling studies already revealed some of the problems one may encounter when using sputtering for microsectioning solid samples, for example, beam-induced profile tailing [4.36] and broadening [4.30], or surface roughening of polycrystalline films [4.37].

At the same time the potential of AES in combination with sputtering was demonstrated [4.38–40]. This technique soon became very popular for depth profiling of thin-film structures [4.41, 42], the reason being twofold. First, the cylindrical mirror analyzer (CMA) was shown to be a very useful device for the spectroscopy of electrons emitted from solid surfaces [4.43]. Due to its high transmission, good energy resolution, and ease of operation, this device soon became a standard tool in many laboratories [4.41, 44]. Second, compared with SIMS, the elemental sensitivity in AES depends little on the chemical environment of the component of interest; i.e., matrix effects are of minor importance

[4.45, 46]. Therefore, multielement compositional analysis by AES is much simpler than by SIMS. But because of its much higher sensitivity, SIMS was able to maintain a strong position in surface and in-depth analysis. Moreover, remarkable progress was made during the past decade in SIMS instrumentation [4.12, 47–55] and quantification [4.56–60] and in understanding the mechanisms involved in secondary ion production [4.61–66] (Chap. 3). As to analysis, the demonstration of a dynamic range of 10^6 in depth profiling [4.52] and the accomplishment of a lateral resolution of better than 100 nm [4.53–55] are two highlights of recent SIMS development.

In the early and mid 1970s two groups began to investigate the applicability of plasma discharges for ionizing sputtered neutral atoms and molecules. *Coburn, Kay*, and *Taglauer* [4.67, 68] used a conventional rf glow discharge operated at rather high pressures ($\sim 10^{-1}$ mbar), whereas *Oechsner*'s group [4.69, 70] employed a magnetically confined electrodeless rf plasma, which can be sustained at considerably lower pressures ($\sim 10^{-3}$ mbar). The two analytical techniques were termed glow discharge mass spectrometry (GDMS) and secondary (or sputtered) neutral mass spectrometry (SNMS), respectively. Even though some interesting depth profiling data evolved from GDMS studies [4.71], work in this field has been continued only on a very low level, probably because of background and contamination problems associated with the high pressure in the glow discharge. Currently SNMS is used for large-area depth profiling of high-concentration constituents [4.72].

Another avenue of sputter-based analysis that began in the early 1970s is optical spectroscopy, either in the ion beam mode (bombardment-induced light emission, BLE) [4.73] or in the glow discharge mode. In glow discharge optical spectroscopy (GDOS) the plasma serves for both sputter erosion of the sample and excitation of the ejected particles [4.74, 75]. Soon it became evident that experiments involving BLE suffer from the same kind of complication as SIMS (e.g., large oxygen-induced yield enhancement [4.76–79]). Nevertheless, research was stimulated by the idea that BLE might outpace SIMS in sensitivity [4.80]. Detailed measurements, however, showed that under conditions of oxygen saturation ion yields may exceed photon yields by up to one order of magnitude [4.80, 81]. As a result of this finding, which is now fairly well understood (Chap. 3), investigations using BLE for analysis have vanished almost completely. By contrast, GDOS is still used in different laboratories for routine analysis [4.82–84].

The early and mid 1970s also saw a lot of activity in the field of (low-energy) ion scattering spectrometry (ISS). This technique was originally developed as a surface analytical tool [4.85] but later was used quite successfully with sputtering [4.86–89]. Some very interesting work has also been reported recently in a closely related field, direct-recoil sputtering [4.90, 91].

Around 1975 important progress was made in organic mass spectrometry. *Macfarlane* and coworkers [4.92] and *Benninghoven* et al. [4.93] were able to show that the impact of high-energy fission fragments [4.92] or low-energy ions [4.93] can cause ejection of large ionized biomolecules from a solid backing without fragmentation. These observations initiated remarkably intense research, which is reviewed in detail in Chap. 5.

During the past few years there has been rapid development in the field of electron and laser beam (post) ionization of sputtered particles (electron and laser beam SNMS). The idea of using electron beams (e-beams) dates back to some preliminary efforts in the late 1950s [4.8] and the early 1960s [4.94]. Recent studies [4.95, 96] revealed two important advantages of e-beam SNMS: matrix effects are almost absent, and compared with SIMS, the ion yields depend only very little on the ionization potential of the element under study. Because the electron beam approach achieves low ionization efficiency ($\sim 10^{-4}$) [4.96], this technique will be really useful only for basic studies on secondary ion and neutral emission, as well as for special analytical problems such as depth profiling of inert gas distributions in solids [4.95]. A much more promising method for sputtered-particle ionization is to use a laser instead of an electron gun. During the past few years several groups have demonstrated the potential [4.97], the applicability [4.98–100], and the extreme sensitivity [4.101, 102] of resonant multiphoton ionization and nonresonant multiphoton ionization [4.103, 104]. *Young* et al. have reviewed the current status of this exciting field [4.105].

A very recent development in depth profile analysis is the combination of Rutherford backscattering spectrometry (RBS) with controlled in situ sputtering [4.106]. An exciting aspect of the novel approach is that the good near-surface depth resolution achievable in RBS can also be obtained at larger depths if a layer of the desired thickness is removed from the sample. Despite the use of sputtering, one can convert the measured spectra to depth profiles in which mixing effects do not show up. This is accomplished by taking into account only those sections of the RBS spectra that relate to the region beyond the shallow zone of mixing underneath the eroded surface (Sect. 4.5.2).

4.2 Basic Considerations

4.2.1 Objective of an Analysis and the Ideal Case

Usually the purpose of a sputter-based analysis is to determine the concentration $c_i(z)$ of one or more elements at the very surface and/or as a function of the distance z normal to the original surface. The depth resolution in such a measurement should be on the order of a few monolayers while the area of interest may range from 1 μm^2 or less to 1 mm^2 or more. Whenever the sample composition is nonuniform in the plane of the surface (xy plane) and below, three-dimensional characterization is usually desirable. If the nonuniformity is due merely to local contamination, it will suffice to discriminate between contaminated and noncontaminated areas. Finally, determining topographic features of the sample can be of interest.

One way of assessing the quality of a sputter-based compositional analysis is to start with a description of an ideal experiment. In such an experiment the impact of an ion beam would cause controlled removal of material, the top atomic layer of the sample first, then the second layer, and so forth. The atomic arrangement in layers not yet removed must not be affected by ion impact. If sputtered particles provide the information about the sample's composition, a measurement of the sputtered flux of each element or isotope versus the bombardment fluence would exactly map the concentration-versus-depth for the species of interest. For this ideal situation to be true the recorded intensity $I_i(z)$ must be directly proportional to the elemental concentration, i.e.,

$$I_i(z) = B_i(z)c_i(z) \ , \tag{4.1}$$

where $B_i(z)$ is a conversion factor reflecting the properties of the instrument used as well as the sensitivity for the element or isotope under study. Equation (4.1) should also hold if a separate analytical technique is used to probe the composition of the material not yet sputtered. Then the ultimate requirement would be a depth resolution of one atomic layer. With the assumed layer-by-layer removal characteristics, it would suffice if the information depth of the analytical technique were restricted to the top monolayer at the instantaneous surface.

4.2.2 Aspects of Sputtering at Low, Intermediate, and High Bombardment Fluences

Performing the ideal experiment sketched above requires an erosion technique with which atoms can be removed very gently from a sample. Unfortunately, knockon sputtering is by no means a gentle process. Briefly, sputtering can be described as follows [4.107, 108].

If a primary ion with an energy on the order of 0.1–10 keV enters a solid it will lose its energy in collisions with atoms and electrons of the target. Eventually the projectile comes to rest at some distance from the surface. Since the cross sections for elastic energy transfer are very large for the particle velocities of interest here [4.107, 109], an incident ion will set many target atoms in motion, either directly or via collisions of primary knockon atoms with other target atoms. Such an event is commonly termed a "collision cascade" (Fig. 4.1). If the energy transferred to a target atom exceeds a certain limit, the struck atom may be displaced permanently from its original site, leaving a vacancy behind [4.110, 111]. Sputtering will occur if a struck atom residing near the surface is set in motion in the direction of the vacuum half-space with an energy sufficient to overcome the surface-binding energy.

Note for the analytical application considered here that the (mean) number of sputtering events per incident ion – i.e., the sputtering yield – is generally small compared with the mean number of relocation events in the cascade volume. This means that in terms of energy consumption sputtering is very inefficient [4.112, 113]. At best one can try to confine the cascade to a shallow region near

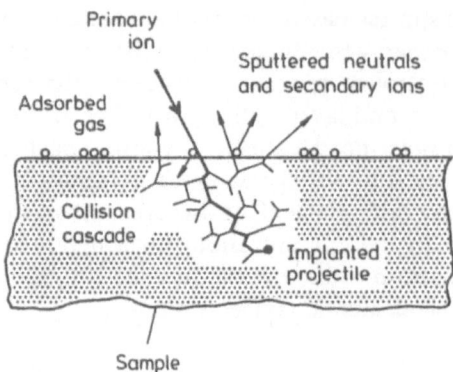

Primary
ion

Sputtered neutrals
and secondary ions

Adsorbed
gas

Collision
cascade

Implanted
projectile

Sample

Fig. 4.1. Generation of a collision cascade and the sputtering process

the surface, so that the energy consumed in ejection (surface-binding energy plus kinetic energy of sputtered particles) constitutes a sizable fraction of the deposited energy (up to about 10%) [4.113]. Typically, however, this fraction amounts to less than 1%, whereas the remaining fraction is consumed in generating damage and heat within the sample.

Now we should distinguish between several cases of sputter-based analysis in terms of the applied bombardment fluence. In the low-fluence case, the number n_0 of primary ions (or neutrals) hitting the sample is kept so small that only a minor fraction of the exposed area A is permanently altered by the bombardment. If the impact of one individual ion results in compositional and structural changes over a mean area $\tilde{\sigma}$, the condition for low-fluence bombardment can be written in the simple form (Chap. 5, Fig. 5.2)

$$\tilde{\sigma}\Phi_0 \ll 1 , \tag{4.2}$$

where $\Phi_0 = n_0/A$ is the primary ion fluence. If (4.2) holds, the "quality" of a sputter-based analysis will be affected only little, if any, by the damage produced in the bulk of the sample. In that case the key quality parameter is the distribution of the depth of origin of sputtered particles, i.e., the normalized depth-differential sputtering yield $y_i(z)$ of an element or isotope i in a multicomponent sample,

$$y_i(z) = \frac{c_i^{-1}\partial Y_i}{\partial z} , \tag{4.3}$$

where Y_i denotes the partial sputtering yield of i atoms. Using (4.3) we can define a mean depth of origin of sputtered i atoms,

$$L_{s,i} = \frac{\int_0^\infty z y_i(z)\,dz}{\int_0^\infty y_i(z)\,dz} . \tag{4.4}$$

If L_s is shown to be on the order of one monolayer, we would in fact be able to perform an ideal experiment, at least as far as microsectioning is concerned. Quantification of the recorded signals might remain a delicate problem.

In the high-fluence case (i.e., for $\tilde{\sigma}\Phi_0 \gg 1$) the sample is sputter etched to a depth that frequently exceeds the maximum range of the primary ion. Due to the damage generated concurrently with sputtering, an atom originally located at some depth from the original surface will have been relocated many times before the instantaneous surface has advanced to the point where this atom is eventually ejected. If this "cascade mixing" is treated as a random-walk problem [4.114, 115], the bombardment-induced broadening Δz_R^∞ becomes [4.113]

$$\Delta z_R(z \to \infty) = \Delta z_R^\infty = 0.37 R_d \left(\frac{E_n}{Y E_{dp}} \right)^{1/2} . \tag{4.5}$$

where R_d is the mean distance over which a permanently displaced atom is relocated, E_{dp} is the displacement energy, E_n is the fraction of the primary ion energy which is deposited into nuclear motion, and Y is the sputtering yield. Inserting reasonable numbers in (4.5) we find that at energies around or below 1 keV, Δz_R may range from 2 to 4 nm [4.113, 115]. More pronounced broadening is expected at higher energies.

Note that estimates of beam-induced mixing based on (4.5) constitute a lower limit because only collisional relocation is taken into account. Due to the high density of vacancies and interstitials generated in the cascade volume, diffusional transport of material generally is important in the redistribution.

In the intermediate-fluence range (i.e., for $\tilde{\sigma}\Phi_0 \simeq 1$), one would expect a gradual transition from escape-depth-governed erosion to fully established damage. Considering again only the effect of cascade mixing, the broadening produced after erosion to depth z is [4.113]

$$\Delta z_R(z) = \Delta z_R^\infty \left(\frac{E_n(z)}{E_n} \right)^{1/2} \quad \text{with} \tag{4.6}$$

$$E_n(z) = \int_0^z F_n(z') dz' , \tag{4.7}$$

where $F_n(z) = dE_n/dz$ is the nuclear energy deposition per unit depth. Clearly, (4.6) makes sense only beyond a characteristic sputtered depth z^* defined by $\Delta z_R(z^*) = L_s$. Otherwise, Δz_R could become vanishingly small at small depths of erosion.

Another concern at moderate fluences is that the points of impact of primary ions on the sample surface are distributed statistically. This means that as material is removed from the top layer of the sample there is an increasing probability that the next incident ion will hit an eroded spot rather than the remainder of the nonsputtered top layer. Due to this effect, sputter removal of the topmost layer will proceed exponentially [4.116] rather than according to a step function.

It has sometimes been assumed that this statistical aspect of sample erosion by ion impact will also determine the depth resolution achievable at a depth corresponding to the removal of tens or hundreds of monolayers [4.117]. Such an approach completely ignores the counteracting effect of a variety of features and processes (e.g., site-dependent sputtering probability [4.107] or surface and bulk diffusion [4.62, 118]), which will prevent the evolution of the pronounced microroughening predicted by the statistical model. In fact, the characteristic monolayer thicknesses derived by rigorously applying the model to experimental data [4.118] turned out to be completely uncorrelated with, and sometimes smaller than, the respective lattice parameters. This is not surprising at all since, for the sputtered layer thicknesses considered here, the observed depth resolution is determined by beam-induced relocation phenomena and element-differential (preferential) sputtering. Attempts to save the statistical model by introducing additional fitting parameters [4.119, 120] will not bring a better understanding of the processes that determine the depth resolution in a particular experiment.

Sputter erosion can result in rather pronounced roughening of the surface, specifically with polycrystalline metal samples [4.121–124]. This undesirable modification of the surface topography is most often due to the channeling effect, which causes the sputtering yield of individual grains to depend on their orientation with respect to the primary ion beam [4.125]. The extent of roughening in polycrystalline metals can be reduced by simultaneous bombardment with more than one ion beam [4.126–128] or by sample rotation during sputtering [4.128, 129]. Sometimes roughening can be partly or fully suppressed by bombardment conditions that enforce amorphization of the polycrystalline sample (e.g., argon ion sputtering at elevated oxygen pressure [4.37], or bombardment with oxygen [4.130, 131] or nitrogen ion beams [4.131–133]).

For completeness we mention the effect of blistering or flaking, which also gives rise to severe surface roughening at high fluences [4.134]. This phenomenon is observed with primary ions of permanent gases, specifically inert gases, which are characterized by a very low solid solubility. The mechanisms involved have been studied in great detail for light-ion bombardment [4.134]. With heavy inert-gas ions the effect may be observed beyond a critical energy ($\sim$ 100 keV for Ar^+ on Si [4.135]). At energies below about 10 keV, beam-induced outdiffusion through the nearby surface [4.136, 137] usually keeps the concentration of (heavy) inert gases like Ar below the level necessary for blistering. Bubbles, however, may still be formed at 20 keV, as shown by Rutherford backscattering spectrometry (Ar bombardment of PtSi [4.138]). Even at rather low energies there is evidence for bubble formation, as indicated by time-of-flight mass spectrometry of Ar released from Si during sputtering at 3 keV [4.137], as well as by scanning tunneling microscopy of the surface topography of Si after 0.7 keV Ar bombardment [4.139].

4.2.3 Depth of Origin of Sputtered Particles

During the past decade many aspects of sputtering have been explored by computer simulations; recently the merits and limitations of such studies have been discussed [4.140–142]. Of prime interest here is a detailed knowledge about the depth-differential sputtering yield $y_i(z)$. Results of simulations using different codes have been reported in the literature [4.140, 143–149]. Figure 4.2 shows examples. Most sputtered particles (more than 80%) originate within a depth of about 0.5 nm from the surface. This holds true even for impact energies as large as 90 keV (Fig. 4.2a). Raising the primary ion energy enlarges the maximum depth of ejection [4.144, 147]. At 90 keV (Fig. 4.2a [4.145]) or 100 keV [4.147] the relative contribution to the sputtered flux from depths exceeding 1 nm is still on the order of 1%. Very important for the prospects of sputter-based analyses are the results of Fig. 4.2b, according to which the lighter element in a two-component target can escape from a significantly larger depth than the heavier one [4.148]. For comparable bombardment conditions, the results obtained using different codes are almost identical [4.143, 147].

Analytical estimates for the distribution of the depth of origin [4.150] suggest an exponential behavior with a characteristic depth $z_{1/e} = 0.42 \times 10^{16} \, \text{cm}^{-2}/N$, where N is the number density of target atoms. For $N = 5 \times 10^{22}$ atoms/cm^3 we have $z_{1/e} = 0.8$ nm. Accordingly, the depth from which 80% of the sputtered atoms originate would correspond to 1.3 nm, which is a factor of almost three larger than the corresponding number derived from numerical simulations (Fig. 4.2). *Sigmund* has recently discussed this discrepancy and the implications of a refined escape depth for a transport theory of sputtering [4.151]. In a refined theory the energy spectrum of sputtered particles and the sputtering yield would have to be insensitive to the form of the low-energy cross section, in contrast to the original assumptions [4.107, 108].

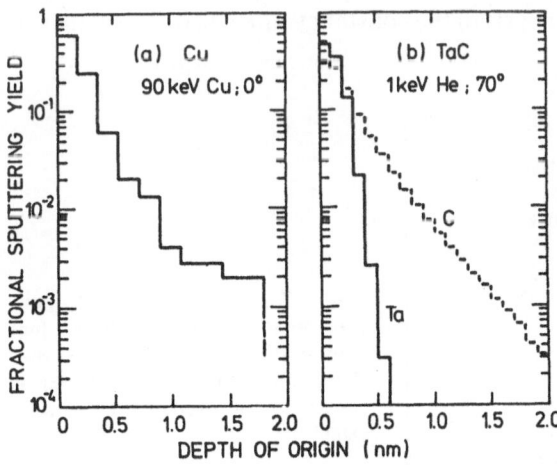

Fig. 4.2. Distributions of the depth of origin of sputtered particles: (a) elemental target [4.145], (b) compound target [4.148]

Kelly, Falcone and *Oliva* have advocated the idea that most of the sputtered particles are ejected from the topmost monolayer [4.152–154]. The conclusions were based on theoretical arguments as well as on an analysis of experimental data, which, however, are quite difficult to interpret. *Falcone* put forward the novel idea that the escape depth derived from transport theory should be reinterpreted as a collisional mean free path [4.155].

Unfortunately only very few experimental data are available in the literature from which the depth of origin of sputtered particles can be clearly assessed. *Dumke* et al. [4.156] performed interesting experiments on the gallium-indium eutectic alloy, both in liquid and in solid form. The results suggest that under 15 keV Ar^+ bombardment 85% of the sputtered atoms originate in the surface monolayer of the liquid alloy. The quantitative aspects of that work are debatable: (i) Calibration of the surface composition by ISS and AES was carried out in separate systems. (ii) The foils used for collecting sputtered material contained rather large holes so that the uncertainty in evaluating the peak and the integral (!) of the angular distributions of sputtered particles was quite large. In fact the surprisingly large effect of the Ar^+ energy on the form of the Ga distributions might be due partly to an incorrect assignment of the peak intensity.

In an interesting set of experiments *Niehus, Prigge*, and *Bauer* [4.157, 158] used a combination of SIMS and AES to determine the depth of information in SIMS. AES was employed to control the growth of Cu, Ag, Pd, and Y films on W(110), which are known to grow layer by layer. For all systems investigated the (low-fluence) W^+ SIMS signal was found to decrease rapidly with increasing coverage, as shown in Fig. 4.3. Upon completion of the first monolayer (ML) of Cu, Ag, and Pd, the W^+ signal amounted to between 3% and 12% of the signal for the clean backing. Since the signals of Cu^+, Ag^+, and Pd^+ were directly proportional to the coverage (up to 1 ML), the ionization probability P^+ for the overlayer species was constant. Despite a small depression of the work function Φ of W(110), induced by deposition of Cu, Ag, and Pd (e.g., $\Delta\Phi_{max} = -0.5$ eV for Pd), we can conclude on the basis of the constancy of P^+ (overlayer) that the

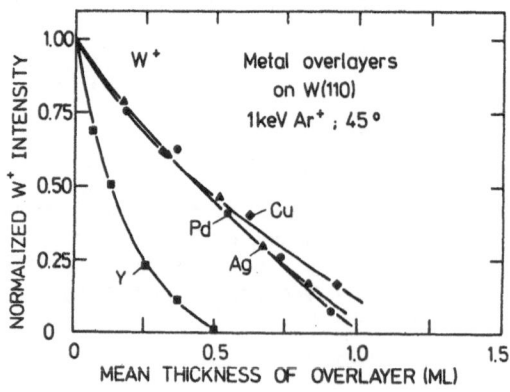

Fig. 4.3. Intensity of W^+ secondary ions sputtered from W(110) as a function of the thickness of different overlayers [4.157, 158]. The data for Pd relate to an overlayer annealed at 800 K

ionization probability of W^+ (also) remained unaffected during growth of the overlayers (Chap. 3). For these systems the coverage dependence of the W^+ intensity in Fig. 4.3 thus suggests that about 90% of the sputtered particles originate from the top monolayer. This interpretation is supported by the experimental observation that the overlayer signal usually saturated at a coverage between 1 ML (Ag^+[4.157]) and 2 ML (e.g., Pd^+ [4.158]). The relatively large increase in the secondary ion yield observed in the latter case at a coverage between 1 and 2 ML can be attributed to a change in the bond strength (Pd-Pd versus Pd-W).

In contrast to copper, silver, and palladium, deposition of yttrium caused the W^+ signal to disappear already at a coverage of 0.5 ML (Fig. 4.3). This rapid decay of the substrate signal correlated with a pronounced lowering of the work function Φ of W(110), $\Delta\Phi_{max} = -2.2$ eV at 0.5 ML [4.158]. In fact, $P^+(Y)$ decreased by a factor of about 70 in passing from the lowest coverage (~ 0.01 ML) to 0.5 ML. These complications suggest that the data for W^+ emission from yttrium-covered tungsten should be left out of consideration here.

Results similar to those of *Bauer* and coworkers [4.157, 158] have been obtained recently by *Burnett* et al. [4.159], who used nonresonant laser ionization to measure partial sputtering yields of Cu and Ru during 3.6 keV Ar^+ bombardment of Cu/Ru(0001) for Cu coverages up to 1.4 ML. The results suggest that roughly two thirds of the sputtered atoms originated in the first atomic layer.

In summary all the cited data [4.156–159] provide rather safe evidence that sputtering can successfully remove material almost exclusively from the topmost layer of a sample. In that sense sputtering can be considered an almost ideal technique. Note, however, that this favorable situation holds only for low-fluence bombardment (Sect. 4.2.2).

4.2.4 Effect of Element-Differential Ejection and Beam-Induced Relocation on the Sputtered-Particle Flux

Without specifying the experimental procedure used to measure the flux of sputtered particles of a certain element or isotope i, we can introduce a formalism that describes, in general terms, the intensity $I_i(z(t))$ recorded as a function of the time t of bombardment. Here $I_i(z(t))$ might also be termed an uncalibrated sputter depth profile. Assume we want to analyze a sample in which the concentration (atomic fraction) c_i of species i is uniform in planes parallel to the (flat) surface, but varies in some way as a function of depth z, $c_i(x, y, z) = c_i(z)$. Then [4.113, 160]

$$I_i(z) = \eta_i P_i(z) A_0 J Y_i(z) , \qquad (4.8)$$

where η_i is an instrumental factor (transmission and detection efficiency) that reflects the integration over the solid angle and the energy-band pass, as defined in (3.1). Here P_i is a factor that specifies the (depth-dependent) fraction of sputtered i atoms present in a state necessary for analysis (for SIMS and SNMS analysis, P_i would be the ionization probability P_i^+ or P_i^-), A_0 is the sample area that contributes to the measured signal (field of view of the spectrometer or "gated" area), J is the mean primary ion flux (ions/cm^{2}s) across A_0, and $Y_i(z(t))$ is the time-dependent partial sputtering yield, or tracer yield

$$Y_i(z) = \int_0^\infty y_i(z')\tilde{c}_i(z + z')dz' \ . \tag{4.9}$$

In (4.9) $\tilde{c}_i(z)$ represents the concentration distribution altered as a result of bombarding the sample during sputter erosion to depth z. If the original concentration distribution were not affected during sputter erosion, (4.9) could be written in a simplified form, i.e.,

$$Y_i(z) = Y(z)\bar{c}_i(z) \ , \tag{4.10}$$

where $Y(z)$ is the total sputtering and $\bar{c}_i(z)$ the impurity concentration averaged over the depth of origin of sputtered particles. Neglecting the difference between $\bar{c}_i(z)$ and $c_i(z)$ we come back to the ideal case of (4.1). The factor $B_i(z)$ in (4.1) is then found by comparison with (4.8, 10),

$$B_i(z) = \eta_i P_i(z) A_0 J Y(z) \ . \tag{4.11}$$

Unfortunately, the effect of ion bombardment is usually so severe that the undisturbed concentration distribution $c_i(z)$ cannot be derived from (4.8) through (4.9) or (4.10). We do not have the means to predict $Y_i(z)$ with sufficient accuracy, largely because we do not know the relative importance of the various processes that may contribute to material transport in an ion-bombarded sample.

a) Collisional (Ballistic) Mixing

Numerical estimates of tracer yields, based on a transport theory for calculating collisional mixing, have been presented by *Littmark* and *Hofer* [4.161]. Figure 4.4 shows examples for step-function-type tracer distributions with two different thicknesses. The data were calculated for 5 keV Ar atoms normally incident on a silicon sample containing matrix (Si) and tracer (Si*) atoms of identical mass. Comparing the calculated flux of sputtered tracer atoms (i.e., the tracer profile) with the original distribution reveals a pronounced bombardment-induced broadening effect. The rather qualitative notation "broadening" can be quantified: define, with reference to Fig. 4.4, characteristic parameters of the tracer profiles, for example, the depth interval $\Delta z_i^\uparrow$ sputtered in passing from the 10% level of the normalized tracer signal $I/\hat{I}(\hat{I} = \text{peak intensity})$ to the leading edge of the original distribution, the peak shift $\Delta \hat{z}_i$ observed with thin tracers, and the slope λ_i characterizing the exponential tail of the tracer profile in a given matrix.

Generally, the extent of broadening, as specified by $\Delta z_i^\uparrow$, $\Delta \hat{z}_i$, and λ_i, will depend on the dopant-matrix combination studied, as well as on experimental

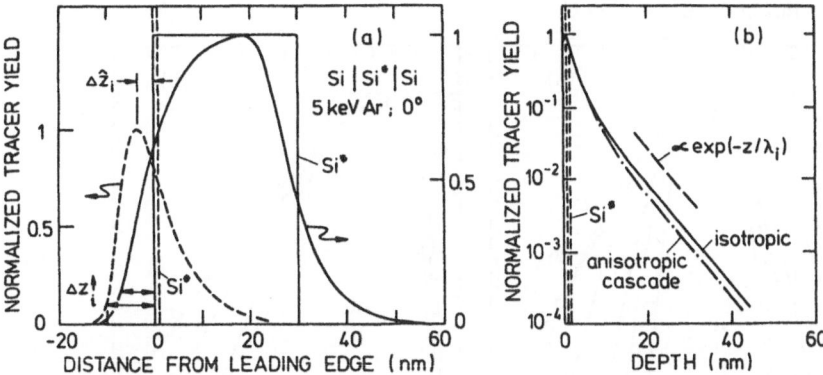

Fig. 4.4. Calculated flux of tracer atoms sputtered from markers of different thickness; (a) linear scale, (b) logarithmic scale [4.161]

parameters (energy, mass, and impact angle of the primary ions; target temperature; residual gas pressure; etc.). The broadening parameter $\Delta z_i^{\dagger}$ should be closely related to the width of the damage distribution generated by the incident ion [4.161]. In fact, calculations of the kind shown in Fig. 4.4 suggest that $\Delta z_i^{\dagger}$ decreases with increasing primary ion mass [4.162].

b) Shape of a Sputter Profile and Peak Shift

At first sight it might be somewhat surprising that the peak of the calculated sputter profile in Fig. 4.4a does not appear at the original location of the narrow tracer. However, the occurrence of a peak shift, as well as the evolution of an asymmetric profile, can be understood using a very simple argument suggested by *Tsong* and *Sankey* [4.163]. The model is based on the idea that cascade mixing can be treated as a one-dimensional diffusion problem [4.164]. Additional assumptions are that mixing and sputter erosion can be treated separately and that the diffusion coefficient D is independent of depth.

Consider a very thin tracer (or marker) initially located at a depth z_i^0 from the surface. At time $t = 0$ we switch on bombardment, which causes diffusional broadening of the marker (sputter erosion will be added subsequently). Solution of Fick's second law yields the well-known result [4.165] for the time dependence of the (internal) marker distribution,

$$c_i(\xi, t) \propto t^{-1/2} \exp\left[-\frac{(\xi - \xi_i^0)^2}{t}\right],$$

(4.12)

where $\xi = z/2D^{1/2}$ is a reduced depth. Figure 4.5a shows the evolution of the internal profiles according to (4.12). From these data we now derive a sputter profile by artificially removing material from the sample at a speed $\dot{z}$. The profile is found from Fig. 4.5a by determining c_i at those points in depth and time

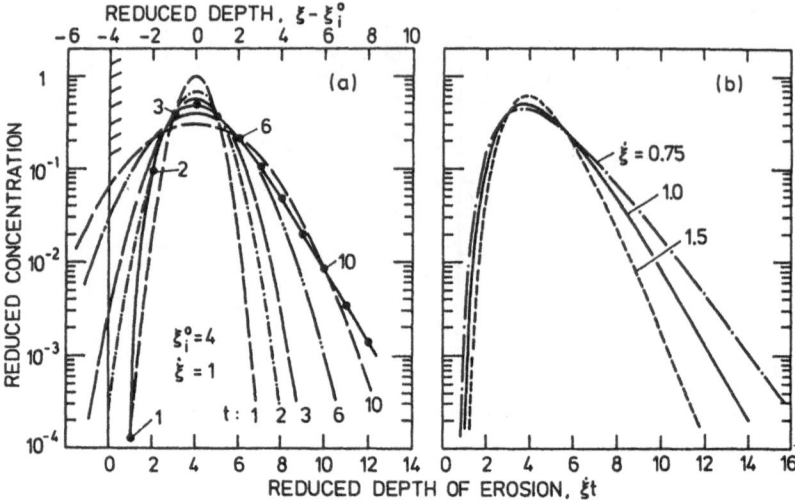

Fig. 4.5. (a) Diffusional broadening of a thin marker layer initially located at the reduced depth $\zeta_0 = 4$ (dashed and dash-dotted curves). The full curve represents the hypothetical sputter profile measured at a reduced erosion rate $\dot{\zeta} = 1$. The full circles denote the point of intersection between the receding surface and the internal diffusion profile at time t. **(b)** Hypothetical sputter profiles for different values of the reduced erosion rate $\dot{\zeta}$

where $\zeta = \dot{\zeta}t$ (full circles and full curve). Evidently, the hypothetical sputter profile reveals the main features of Fig. 4.4a, i.e., the asymmetry and the peak shift. Moreover, the exponential tail seen in Fig. 4.4b is also reproduced approximately; compare (4.12) and Fig. 4.5b. As one would expect intuitively, the hypothetical sputter profiles become narrower and steeper on either side as the erosion rate increases, i.e., as the time for diffusion decreases.

Two shortcomings of the above model deserve attention. It is implied that the amount of tracer material subject to diffusion remains constant during sputter profiling and that the extent of diffusional broadening is unlimited. If these aspects are taken into account, the peak shift may still be explained qualitatively. Once the instantaneous surface has receded to the point that the cascade intersects the marker, atoms from the marker will be transported to the surface where they get sputtered. This loss of i atoms will go on as the surface advances to depth z_i^0. At the same time, intermixing of marker and host atoms continues until a maximum spreading of i atoms is achieved at $z = z_i^0$. As a result of the simultaneous action of the two processes, the instantaneous surface concentration will pass through a maximum at a depth $z < z_i^0$. Due to the small depth of origin of sputtered particles, this maximum gives rise to a corresponding peak in the sputtered flux of i atoms.

The arguments outlined above concerning the shape of the sputter profile of a thin layer of tracer atoms apply in modified form to a thick layer as well. The peak in the thin-layer case corresponds to the maximum slope at the leading edge of a thick-layer sputter profile. As one can tell from comparing the profiles in Fig. 4.4a, the peak and the maximum slope are shifted by the same amount. This is not surprising since a thin layer may be considered a differential probe for the broadening to be expected with a thick tracer layer.

c) Exponential Tail and Decay Length

The diffusion approach (Fig. 4.5) can be applied more precisely in computer simulations. Results reported by *King* and *Tsong* [4.164] are depicted in Fig. 4.6a. Supporting previous assumptions [4.166, 167], the calculations show that in the "final" state of mixing observed beyond z_i^0 (i.e., beyond the original depth of location of a thin tracer or beyond the maximum depth of doping of a thick tracer), the internal distribution of i atoms attains a shape that does not change as bombardment proceeds and is independent of the original distribution of i atoms (i.e., it is the same for tracer layers of different thickness). The evolution of such a final state of mixing has been confirmed by Rutherford backscattering analysis of sputter-etched samples [4.168]. The only effect of continued sputter erosion is a further reduction of the areal density N_i^A (atoms/cm^2) of i atoms in the sample. At this state of erosion we may thus factorize $\tilde{c}_i$ in the form

$$\tilde{c}_i(z + z') = N^{-1} N_i^A(z) p_i(z') , \qquad (4.13)$$

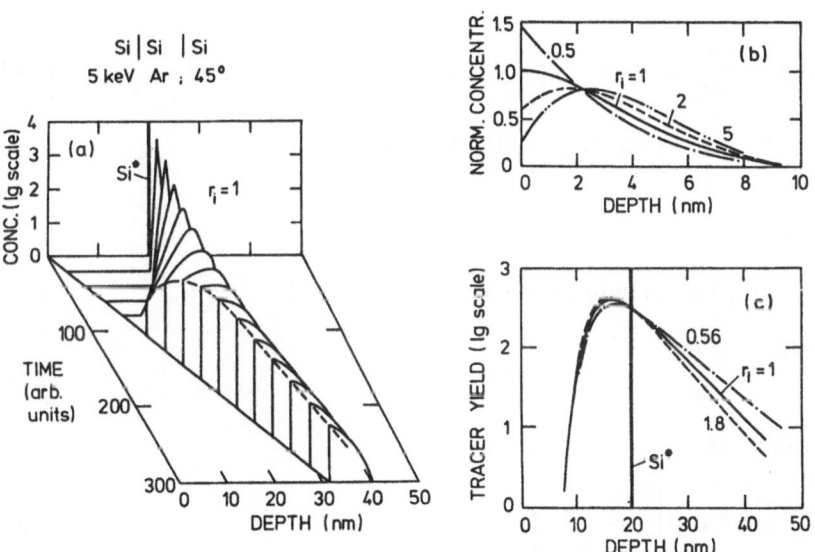

Fig. 4.6. (a) Calculated evolution of the internal tracer profiles during a sputtering experiment. The sample consisted of an Si matrix with a thin Si* tracer located at a depth of 20 nm. The data are based on a diffusion model of beam-induced mixing (relative component sputtering yield $r_i = 1$). The dashed curve resembles the external profile expected in a measurement involving the detection of sputtered atoms [4.164]. (b) Stationary internal tracer profiles ("mixing profiles") and (c) tracer sputter profiles for different values of r_i

where N (atoms/cm^3) is the number density of matrix atoms and $p_i(z')$ the stationary internal distribution of i atoms ("mixing profile"), which is normalized in the form

$$\int_0^\infty p_i(z')dz' = 1 \ . \tag{4.14}$$

We can also assume that in this final state of mixing the impurity concentration is very small ("dilute limit", i.e., $\tilde{c}_i \ll 1$), so that the matrix sputtering yield Y is constant. Inserting (4.13) in (4.9) we get

$$Y_i(z) = N^{-1} N_i^A(z) \int_0^\infty y_i(z')p_i(z')dz' \ . \tag{4.15}$$

Since $y_i(z')$ and $p_i(z')$ are independent of z the integral in (4.15) is an element-specific constant with the dimension of an inverse length. Normalizing $Y_i(z)$ to Y we may define a decay length λ_i in the form

$$\lambda_i = \frac{Y}{\displaystyle\int_0^\infty y_i(z')p_i(z')dz'} \ . \tag{4.16}$$

The partial sputtering yield $Y_i(z)$ is also defined as the incremental loss $-\,dN_i^A$ per fluence interval $d\Phi$,

$$Y_i(z) = -\frac{dN_i^A}{d\Phi} \ . \tag{4.17}$$

Similarly, the total sputtering yield Y can be written

$$Y = \frac{N\,dz}{d\Phi} \ . \tag{4.18}$$

Inserting (4.15, 16, 18) in (4.17) we get

$$-\frac{dN_i^A}{dz} = \frac{N_i^A(z)}{\lambda_i} \ . \tag{4.19}$$

Integration yields

$$N_i^A(z) = N_{i,r}^A \exp\left[-\frac{z - z_r}{\lambda_i}\right] , \tag{4.20}$$

where $N_{i,r}^A = N_i^A(z = z_r)$ is the remaining areal density of i atoms at some reference depth z_r. From (4.20) it follows that the signal measured beyond the maximum depth of doping decreases exponentially with a characteristic decay length λ_i. Such behavior has in fact been observed experimentally [4.33, 113, 169, 170] and in numerical simulations [4.161] (Fig. 4.4b).

According to (4.16) the decay length λ_i depends on the element-specific sputtering behavior of i atoms in the matrix of m atoms, as well as on the shape of the mixing profile. Since the mean depth of origin of sputtered atoms is so small (Sect. 4.2.3), we may, to a first-order approximation, define a mean value of p_i near the surface, $p_i(z < L_s) \sim p_i^s$. Equation (4.16) can then be written

$$\lambda_i \simeq \frac{Y}{p_i^s Y_i^c} = \frac{1}{r_i p_i^s} , \qquad (4.21)$$

where Y_i^c is the component sputtering yield, i.e., the partial sputtering yield of i atoms normalized to unit concentration ($c_i = 1$),

$$Y_i^c = \int_0^\infty y_i(z) dz \qquad \text{and} \qquad (4.22)$$

$$r_i = \frac{Y_i^c}{Y} . \qquad (4.23)$$

The ratio r_i may be termed a relative component sputtering yield. The commonly used notation "preferential" sputtering factor (or ratio) is quite misleading because it does not cover the regime $r_i < 1$, where we are dealing with restrained rather than preferred sputtering of i atoms relative to the matrix atoms.

As one might expect, (4.21) tells that λ_i is inversely proportional to the product of r_i and p_i^s. Even if r_i were known, this formal relation would be of little help for predicting λ_i because according to numerical simulations [4.164], p_i^s depends on r_i. The calculated mixing profiles (Fig. 4.6b) show that with increasing r_i the region near the instantaneous surface becomes progressively depleted of i atoms; i.e., p_i^s decreases. As a result, even pronounced variations of r_i have only a moderate effect on λ_i (Fig. 4.6c).

These findings reveal a very important aspect of the employed method of sputter depth profiling [4.167]: *in principle, it is impossible to derive any detailed information about the internal distribution of i atoms from only the sputter profile.* To understand the observed amount of broadening and to determine, say, the mixing profile $p_i(z)$, additional experiments are necessary. Sometimes sputter depth profiling under *modified conditions* can yield useful information (e.g., in experiments performed at a lower sample temperature [4.171, 172] or by changing the chemical character of the primary ion [4.173, 174]). But a reliable evaluation of the internal distribution of intermixed atoms is possible only using Rutherford backscattering spectrometry in a *high-resolution* geometry [4.175, 176]. Then, however, we are limited to heavy dopants in light-atom matrices.

Finally we consider how the bombardment parameters affect a sputter depth profile. We already saw that the width $\Delta z_i^\dagger$ is expected to scale with the extension of the collision cascade in the direction normal to the surface. Accordingly, $\Delta z_i^\dagger$ should decrease with decreasing energy and more glancing beam incidence. The expected effect of the impact energy has in fact been observed

[4.177]. Also, at a given energy, there is hardly any difference in the $\Delta z_i^\uparrow$ values observed for different impurity elements [4.177, 178]. This result agrees with Fig. 4.6c, suggesting that $\Delta z_i^\uparrow$ is practically independent of r_i.

As to the bombardment parameters' effect on the decay length λ_i, we note that according to presently available data, r_i depends only little on the impact energy [4.179]. Consequently, any change of λ_i observed as a result of ion energy [4.177, 178] or angle of incidence variation [4.180] should be due primarily to a change in p_i^s (4.21). Because of normalization (4.14), p_i^s will increase as the collision cascade shrinks toward the surface, for example, by reducing the ion energy or by going to a more glancing angle of beam incidence. Again, however, we can argue only in qualitative terms because the bombardment parameters' effect on the complex interrelation between r_i and p_i^s is not known in any detail.

d) Sputtering of Alloys and Compounds: The Altered-Layer Model

The definition of a decay length (4.16) implies that an internal mixing profile of well-defined shape is established once the instantaneous surface has receded beyond the maximum doping depth. Essentially the same idea has been used to explain the early observation [4.181] that ion bombardment forms an altered layer at the surface of a solid containing more than one element. The effect has been attributed to element-differential (preferential) sputtering of the constituents [4.182]. *Winters* and *Coburn* [4.183] suggested a simple phenomenological model based on the assumption that the compositional modifications generated during bombardment can be represented as the product of a time-dependent amplitude function and a time-independent spatial function. This corresponds to the factorization of $c_i(z + z')$ in (4.13) where $N_i^A(z(t))$ constitutes the time-dependent amplitude function and $p_i(z')$ the time-independent spatial function. In analogy to (4.20) the altered-layer model for a two-component system predicts an exponential time (i.e., depth) dependence of the change in concentration of either component [4.183].

If the depth of origin of sputtered atoms is assumed to be one monolayer, the change in surface concentration δc_1^s of component 1 can be written in the form [4.183]

$$\delta c_1^s(\Phi) = \delta c_1^s(\infty)\exp\left(-\frac{\Phi}{N\zeta_a}\right) , \qquad (4.24)$$

where $\delta c_1^s(\Phi) = c_1^s(\Phi) - c_1^s(\infty)$, $\delta c_1^s(\infty) = c_1 - c_1^s(\infty)$, and $\Phi = Jt$. Moreover,

$$c_1^s(\infty) = \frac{c_1 Y_2^c}{c_1 Y_2^c + c_2 Y_1^c} \qquad \text{and} \qquad (4.25)$$

$$\zeta_a = \frac{z_a^*}{c_1 Y_2^c + c_2 Y_1^c} . \qquad (4.26)$$

Here c_1 and c_2 are the bulk concentrations of the two components, and $c_1 + c_2 = 1$ in an initially homogeneous sample, i.e., $c_1^s(\Phi = 0) = c_1$. (In a refined model the surface concentrations c_1^s and c_2^s may be replaced by numbers averaged over the distributions of the depth of origin of each component [4.184, 185]). The quantity z_a^* in (4.24) represents a "mean" depth of the altered layer [4.185] defined by

$$z_a^* = \frac{1}{\delta c_1^s(\infty)} \int\limits_0^\infty \delta c_1(z', \infty)dz' \ . \tag{4.27}$$

Equation (4.27) corresponds to the normalization (4.14). Similarly ζ_a in (4.24) is the equivalent of the decay length λ_i in (4.20). The main difference between the two approaches is that, in contrast to the decay-length concept, the altered-layer model allows for a time-dependent variation of the erosion rate brought about by the presence of two components with different component sputtering yields Y_1^c and Y_2^c. The analogy of the two approaches becomes fully evident, however, if we consider the altered-layer model in the dilute limit, i.e., $c_1 \ll c_2 \simeq 1$ and $Y \simeq Y_2^c$. Then the exponent in (4.24) reduces to $zY_1^c/z_a^* Y$. Comparison with (4.20, 21) shows that z_a^* has the same meaning as the inverse of p_i^s.

Note that the model itself does not predict the magnitude of z_a^*, ζ_a, Y_1^c, or Y_2^c. Measurements of $\delta c_{1,2}^s(\Phi)$ provide some information [4.183–185]. To determine the forces responsible for material transport in the altered layer it is necessary, however, to measure the concentration distribution within the layer, i.e., not just at the surface [4.171, 172].

A final remark concerns the stationary fluxes of sputtered atoms of the two components which are directly proportional to the respective stationary partial sputtering yields, $Y_1(\infty)$ and $Y_2(\infty)$. By definition,

$$Y_1(\infty) = c_1^s(\infty)Y_1^c \ . \tag{4.28}$$

From (4.25, 28) and analogous equations for component 2 we get

$$\frac{Y_1(\infty)}{Y_2(\infty)} = \frac{c_1}{c_2} \ . \tag{4.29}$$

This condition for a stationary state of sputtering is thus contained in the model, as necessary.

4.2.5 Depth of Information and the Resulting Broadening Effect in Analyses Involving Electron-Spectroscopy Techniques

The aspects discussed in the preceding section are important for all analytical techniques involving sputter erosion of the sample. An additional complication comes into play if, rather than inspecting the composition of the sputtered-particle flux, we use an independent technique to determine the sample composition near the instantaneous surface at deliberately chosen stages of sputter

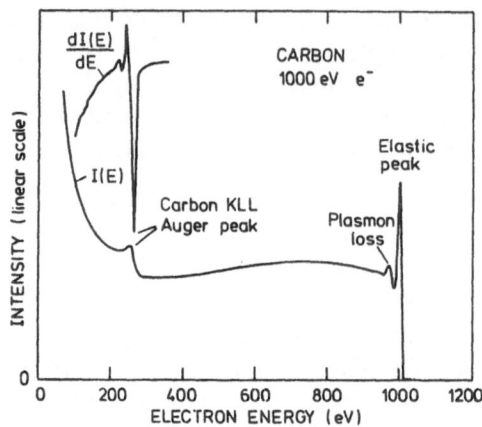

Fig. 4.7. Energy spectrum of secondary electrons and backscattered primary electrons observed during bombardment of a carbon sample with 1000 eV electrons [4.186]

erosion. Then the depth resolution or the depth of information of the method used will add to the broadening produced by ion bombardment.

In Auger electron spectroscopy (AES) and X-ray-induced photoelectron spectroscopy (XPS) the sample is exposed to a beam of electrons or monochromatic X-rays. The incident radiation causes the ejection of electrons with energies extending from zero to a maximum determined by the primary electron or photon energy. The energy spectrum shows element-specific narrow peaks superimposed on a broad "continuum". The location and intensity of the characteristic peaks can be used to identify elements in a sample and to determine their concentration [4.41, 42, 44–46, 186]. Figure 4.7 shows an example [4.186]. Note that in the direct spectrum $I(E)$ the characteristic carbon AES peak is not very high compared with the secondary electron background. To ease signal identification and quantification, the spectra are commonly recorded in derivative form, dI/dE.

a) Escape Depth of Electrons

In AES and XPS the surface sensitivity is brought about by the relatively short mean free path for inelastic electron scattering [4.46, 187, 188]. Experimentally, the surface sensitivity can be demonstrated by measuring the AES or XPS intensity due to substrate or overlayer atoms as a function of the thickness of the overlayer deposited on a substrate. Usually the substrate signal I^e is found to decrease exponentially with increasing overlayer thickness z [4.189]. Accordingly, the AES intensity I_i^e recorded for a sample containing an impurity i with a depth-dependent concentration $c_i(z)$ can be written [4.46]

$$I_i^e(\theta_{\text{out}}) = I_0 \sigma_i^e(E_0)\{1 + R_e\}\eta \int_0^\infty c_i(z)\exp\left(-\frac{z}{L_0 \cos \theta_{\text{out}}}\right)dz \ . \tag{4.30}$$

Here I_0 is the primary electron current; $\sigma_i^e(E_0)$ is the cross section for ionization of the respective inner shell of the impurity atom by electrons of energy E_0; R_e is an effective (composition-dependent) backscattering yield reflecting the contribution of the backward flux of injected electrons to the ionization yield; η is the instrument transmission and detection efficiency; and θ_{out} is the angle of electron emission to the surface normal. The parameter L_0 in (4.30) is the electron-escape depth as determined from overlayer experiments. The expression "attenuation length", also used for L_0, defines the average distance that an electron travels between subsequent inelastic collisions, assuming that elastic scattering is negligible [4.188].

Equation (4.30) relates to the situation where the solid angle of detection is so small that the emission angle of electrons reaching the detector can be represented sufficiently well by the orientation of the spectrometer entrance aperture with respect to the point of Auger electron emission. If a large aperture is used, (4.30) must be integrated over the appropriate solid angle.

For a homogeneously doped sample, integration of (4.30) yields

$$I^e_{i,h}(\theta_{out}) = I_0 \sigma^e_i(E_0)\{1 + R_e\}\eta c_i L_0 \cos\theta_{out} \; . \tag{4.31}$$

Since σ^e_i, R_e, and η are usually not known with sufficient accuracy, quantification of the AES signal for an unknown sample is commonly achieved by reference to a standard of known concentration c_i. A knowledge of L_0 is needed to evaluate the contribution of subsurface layers to the measured signal. Complications arise from the fact that L_0 depends not only on the energy of the Auger electron but also on the elements making up the analyzed samples. For energies E in the range $300 < E < 4000$ eV the following empirical formula has been proposed [4.190]

$$L_0 = k_e E^m \; . \tag{4.32}$$

Depending on the sample material the parameters k_e and m in (4.32) fall into range $0.01 < k < 0.12$ nm and $0.35 < m < 0.78$ [4.191]. A previously suggested "universal" formula like (4.32), but with $k = 0.41\, a^{1.5}_{ML}$ nm (a_{ML} = monolayer thickness in nm) and $m = 0.5$ [4.187] does not describe experimental data with sufficient accuracy [4.191].

In AES and XPS analysis one commonly uses characteristic lines appearing at kinetic energies between about 200 and 2000 eV, in which case L_0 ranges from about 1 to 10 nm. Escape depths exceeding a few nanometers can introduce a sizable broadening effect, in addition to ion-beam-induced profile broadening. The exponential signal attenuation (4.30) not only smears out the tracer distribution but also shifts it toward the surface [4.192, 193]. Figure 4.8 illustrates the effect for three different idealized distributions, ignoring any broadening that might occur during sputter etching of the sample. Essentially, the calculated profile shifts because in AES and XPS the mean depth of information normal to the sample surface is not vanishingly small, but equal to the "effective" escape depth L. This, for the simple case of (4.30), can be defined as

$$L = L_0 \cos\theta_{out} \; . \tag{4.33}$$

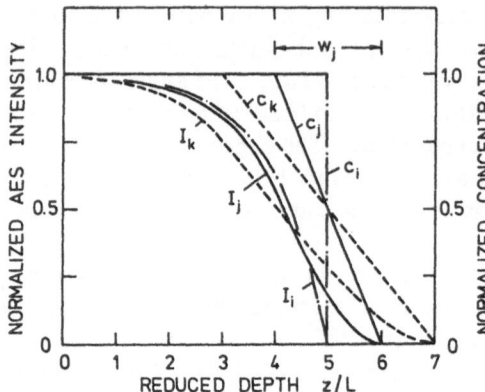

Fig. 4.8. Comparison of the composition $c_{i,j,k}$ of three different tracer layers with the calculated AES sputter profiles $I_{i,j,k}$ (superscript "e" omitted for clarity). The tracer layers have a mean thickness of $5L$ but different composition gradients at the crossing to the substrate (interface widths $w_{i,j,k}/L = 0$, 2, and 4, respectively). Sample erosion is assumed to proceed without damage. Moreover, any effect of the substrate on the tracer AES intensity is ignored

The nonvanishing escape depth results in nonconservation of mass at the early stage of depth profiling. For sufficiently thick layers, the mass deficit is directly proportional to the areal density of tracer atoms contained within the depth $z < L$. This apparent loss of material, or shift in depth scale, has to be accounted for in corrected profiles. (These problems do not arise with techniques involving analysis and detection of the sputtered-particle flux. Then conservation of mass is not violated, but the constancy of P_i may not be easy to accomplish (4.8)).

Note that L is a quantity reflecting the fundamental electron-solid interaction as well as the experimental geometry used. Therefore, numbers for L measured in different laboratories can be compared only if obtained with the same experimental arrangements.

b) Feasibility of Deconvolution

As in (4.8, 9) we now have to account for the fact that in sputter depth profiling the original concentration distribution $c_i(z)$ will be altered to become $\tilde{c}_i(z)$. Due to the exponential signal attenuation, the AES or XPS intensity $I_i^e(z)$ measured during sample erosion will be further broadened. Using (4.30, 31, 33) we have

$$I_{i,n}^e(z) = L^{-1} \int_0^\infty \tilde{c}_i(z + z') \exp\left(-\frac{z'}{L}\right) dz' \ , \tag{4.34}$$

where $I_{i,n}^e(z)$ is normalized to the signal for $c_i = 1$; i.e.,

$$I_{i,n}^e(z) = \frac{I_i^e(z)}{I_{i,h}^e(c_i = 1)} \ . \tag{4.35}$$

Concentration-dependent variations of the electron backscattering coefficient that might occur in AES are neglected in (4.34). Substituting $s = z + z'$ and differentiating with respect to z, one immediately finds the following relation

$$\tilde{c}_i(z) = I^e_{i,n}(z) - \frac{L dI^e_{i,n}(z)}{dz} . \tag{4.36}$$

Iwasaki and *Nakamura* first suggested this deconvolution formula to derive the altered distribution $\tilde{c}_i(z)$ from XPS sputter depth profiling measurements [4.194]. A similar formula has also been given for stepwise analysis [4.195], rather than for the quasi-continuous case implied in deriving (4.36). If sample erosion could be accomplished in an ideal manner, (4.36) would exactly recover $\tilde{c}_i(z)$; i.e., even the nonconservation of mass would be removed.

Equation (4.36) has been applied by different groups [4.195–198] to derive corrected sputter depth profiles from AES measurements. Unfortunately, however, the approach is not applicable to sample erosion by sputtering. The mathematical treatment leading to (4.36) implies the existence of a distribution $\tilde{c}_i(z)$, which remains unaltered during erosion. This is certainly not true in sputter profiling, because the concentration $\tilde{c}_i(z = z_d)$ at a fixed depth z_d is altered continuously as long as z_d is located within the cascade volume. Since the evolution of the altered distribution $\tilde{c}_i(z, t)$ cannot be predicted, applying an appropriate correction is impossible. In other words, no method allows us to determine the concentration at the receding surface from measurements based on a technique sensitive to the shape of the unknown internal profile.

Even though *Kirschner* and *Etzkorn* [4.193, 198] were concerned about these problems, they used (4.36) to derive "corrected" sputter profiles from measurements involving AES. While the error introduced by depth-dependent modifications of the internal profiles is generally hard to quantify, a deconvolution is definitely inadequate in the exponential tails of a sputter profile. In that case the measured intensity is found by inserting (4.13) in (4.34),

$$I^e_{i,n}(z) = \frac{N^A_i(z)}{b_i N L} , \tag{4.37}$$

where b_i is a factor that depends on the effective attenuation length and the shape of the mixing profile,

$$b_i^{-1} = \int_0^\infty p_i(z') \exp\left(-\frac{z'}{L}\right) . \tag{4.38}$$

Note that the same exponential falloff will be observed, irrespective of whether we measure the sputtered-particle flux or residual concentration (4.20, 37).

A rigorous application of (4.36) to (4.20, 37) yields the result

$$\tilde{c}_i(z) = \left(1 + \frac{L}{\lambda_i}\right) I^e_{i,n}(z) . \tag{4.39}$$

Fortunately, the deconvolution procedure does not alter the form of the exponential falloff. However, the "corrected" intensity derived in [4.193, 198] will exceed the measured intensity by a factor $(1 + L/\lambda_i)$.

c) In Situ Determination of the Mean Escape Depth

Despite the problems outlined above, a knowledge of L is always desirable in sputter depth profiling experiments. *Kirschner* and *Etzkorn* [4.199] have suggested an in situ method for determining the effective escape depth averaged over all angles of acceptance of the cylindrical mirror analyzer. The procedure was investigated using Ge/Si sandwich layers sputtered under bombardment with 0.5 keV Ne (Fig. 4.9a). The Si "backing" became observable in the AES spectrum at a distance of about 10 nm from the interface. An exponential increase of the signals I_{Si}^e and $I_{Ge}^e(\infty) - I_{Ge}^e$ was observed during sputter removal of the first 7 nm of Ge ($I_{Ge}^e(\infty)$ is the Ge signal measured for the as-prepared, almost infinitely thick sample). Deviations from the exponential behavior were observed near the Ge-Si interface. These were attributed to beam-induced intermixing of Ge and Si [4.193, 198, 200]. Results similar to those of Fig. 4.9a have been obtained for sputter profiling of Ge on Ta_2O_3 [4.198].

The data in Fig. 4.9a reveal a fairly nice exponential behavior at a large distance from the interface. Nevertheless, caution is needed when using the results to evaluate the position of the interface between two layers by way of extrapolating the exponential part to the 100% concentration level, i.e., to the level $I_{Si}^e = 1$ or $I_{Ge}^e(\infty) - I_{Ge}^e = 1$ [4.199]. Figure 4.9b shows AES intensities

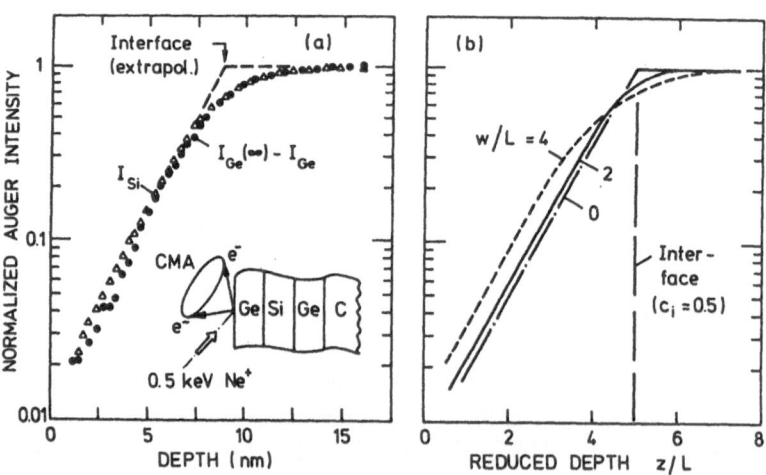

Fig. 4.9. (a) Variation of the normalized AES signals due to Ge and Si observed during sputter etching of a Ge/Si/Ge layer structure deposited on polished graphite [4.199]. (b) Calculated substrate-specific AES signal (escape depth L) as a function of the reduced depth of erosion z/L. The sample structure is the same as in Fig. 4.8

calculated for very simple – i.e., linearly varying – concentration distributions as in Fig. 4.8 (interface width w). Except for a sharp interface, $w/L = 0$, it is evident that the interface position derived by extrapolation does not coincide with the true position $z(c_i = 0.5)/L = 5$. Since sputter depth profiling will always produce a broadened interface, the extrapolation technique should be used only if the broadening is small compared with L. The situation in sputter depth profiling is particularly difficult because the extent of broadening increases during erosion (4.6). Whereas at large distances from the interface the AES intensity should vary according to the case $w/L = 0$ (Fig. 4.9b), curves for gradually increasing values of w/L will apply as we approach the original interface.

4.2.6 Mass and Depth Resolution in Ion Scattering Spectrometry

In ion scattering spectrometry the sample is bombarded with a beam of ions of mass M_1 and energy E_0. Information about the sample composition is obtained by measuring the energy spectrum of particles scattered into a small solid angle $\Delta\Omega$ around a well-defined scattering angle θ. If the scattering center, an atom of mass M_i, is located right at the surface, the energy E_i^r of the scattered particle after a single elastic collision is [4.85]

$$E_i^r = K_i E_0 , \qquad \text{where} \tag{4.40}$$

$$K_i = \frac{[\cos\theta + (\mu^2 - \sin^2\theta)^{1/2}]^2}{(1 + \mu)^2} , \tag{4.41}$$

with $\mu = M_i/M_1 > 1$. Here K is the kinematic scattering factor which, for $\theta = 90°$ and $180°$, takes the simple form

$$K_i = \left(\frac{\mu - 1}{\mu + 1}\right)^n , \tag{4.42}$$

where $n = 1$ and 2 for $\theta = 90°$ and $180°$, respectively.

Although it is evident from (4.40–42) that identification of M_i is very simple if M_1 and θ are known, a severe limitation of all ion scattering techniques results from the fact that the mass resolution $\partial K/\partial\mu$ becomes very poor for $\mu > 10$,

$$\left.\frac{\partial K}{\partial\mu}\right|_{\theta = 90°} = \frac{2}{(\mu + 1)^2} . \tag{4.43}$$

For $\theta = 180°$ the right-hand side of (4.43) has to be multiplied by a factor $2(\mu - 1)/(\mu + 1)$. The best mass resolution is obtained for μ between 1 and 2 (for $\theta = 90°$ and $180°$, respectively).

Another limitation is that a final scattering angle θ may come about not only from single but also from dual, plural, or multiple scattering. Unless single scattering dominates, identification of M_2 becomes impossible. Moreover, a primary ion may enter the sample and travel some distance before suffering a large-angle scattering event. Then the energy loss ΔE experienced along the ingoing and outgoing trajectory must be taken into account. To be useful analytically, ion scattering spectrometry has to be performed in such a way that, as far as the elastic energy loss is concerned, particles observed at an angle θ have experienced only one binary collision.

a) Low-Energy Ion Scattering Spectrometry

In ion scattering spectrometry (ISS) at low impact energies (~ 300–2000 eV) [4.85, 201–209] one wants to analyze the outermost surface layer only. This is accomplished through specific aspects of neutralization in ion-solid collisions [4.204–208] and the large scattering cross section $d\sigma_i/d\Omega$ [4.204, 205]. The intensity I_i^{r} of ions reflected in single binary collisions from a (mono-) layer of i atoms with an areal density N_i^{A} may be written [4.204, 205]

$$I_i^{\mathrm{r}} = \beta_i N_i^{\mathrm{A}} I_0 P_i \frac{d\sigma_i}{d\Omega} \Delta\Omega \; , \qquad\qquad (4.44)$$

where β_i is the detector efficiency, I_0 (ions/s) the primary ion current, $d\sigma_i/d\Omega$ the differential cross section for scattering into the solid angle $\Delta\Omega$ of the detection system, and P_i the probability that an incident ion retains its charge state (i.e., escapes without being neutralized). The important point is that for inert-gas ions (but not for hydrogen [4.208] and alkali ions [4.207]) the escape probability P_i in a single binary collision is typically two to three orders of magnitude higher than in multiple scattering events [4.207]. This means that, specifically for He^+, the intensity observed in an ISS spectrum is almost exclusively due to scattering from atoms that form the "top" layer.

The thickness of this top layer becomes smaller, the stronger the interaction potential between the incident ion and the scattering center. A quantitative description is commonly given in terms of the so-called shadow cone [4.85, 204, 205], i.e., a region behind the scattering center into which no ion can penetrate, because of the repulsive potential acting between the projectile and the target atom. The change in the substrate intensity I_i^{r} due to deposition of an adsorbate layer of areal density N_a^{A} may be taken into account by replacing N_i^{A} in (4.44) by $(N_i^{\mathrm{A}} - \alpha_a N_a^{\mathrm{A}})$, where α_a is a shadowing coefficient.

If atoms deposited on a substrate have a high atomic number, the scattering cross section $d\sigma_a/d\Omega$ and the shadowing coefficient α_a become rather large, so complete shadowing of the substrate can be achieved at monolayer coverage. Figure 4.10 shows this situation: the ISS signal due to an Ni substrate decreases rapidly as the coverage with vapor-deposited Pb atoms increases [4.210]. Upon completion of the first layer of Pb, the Ni signal has vanished almost completely. In other words, the depth resolution in this ISS experiment was one monolayer.

For adsorbates of low atomic number the smaller interaction potential results in relatively small values of $d\sigma_a/d\Omega$ and α_a. Then effective shadowing of the substrate can be enlarged using glancing angles of beam incidence θ_{in}. (The

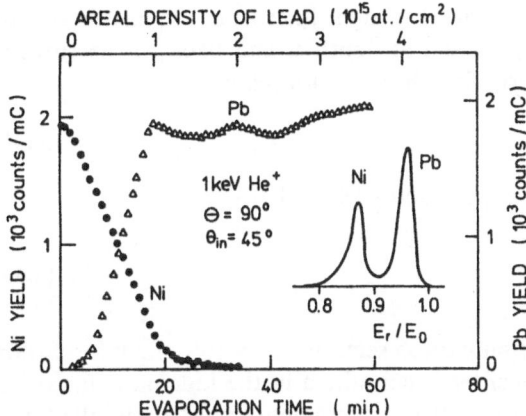

Fig. 4.10. ISS signals measured during vapor deposition of lead on a nickel substrate. Calibration of the areal density of Pb (top scale) is based on RBS analysis of a sample for which evaporation was interrupted at an areal density of 1×10^{15} atoms/ cm^2 [4.210]

other choice of large takeoff angles θ_{out} is not recommended for positive ions, because $P_i^+ \propto \exp(-v_0/v_n)$, where v_n is the velocity of the escaping particle normal to the surface [4.208]). The effective shadowing coefficient will, to first order, increase as $\alpha_a(\theta_{in}) = \alpha_a(0°)/\cos\theta_{in}$ (θ_{in} and θ_{out} are measured to the surface normal). Accordingly, a high surface sensitivity can be achieved even with light-atom adsorbates, provided θ_{in} is large [4.201].

b) High-Energy (Rutherford) Backscattering Spectrometry

In contrast to ISS, backscattering spectrometry at high impact energies is a technique used for compositional analysis from the surface up to a certain depth [4.211]. At primary ion energies above a few hundred keV the quantitative aspects of scattering can be described rather well by the Rutherford scattering cross section [4.212, 213]. Therefore, this analytical technique is usually referred to as Rutherford backscattering spectrometry (RBS) [4.211]. Sometimes a distinction is made between medium-energy ($\sim$ 50–300 keV) and high-energy ($>$ 300 keV) ion scattering, MEIS and HEIS. For quantitative analysis the high-energy regime is usually preferred.

In RBS the scattered intensity can be written in analogy to (4.44). If a surface-barrier detector is used, all scattered particles hitting the active area will be detected irrespective of their charge state. Accordingly β_i and P_i in (4.44) equal unity, so the scattered intensity $I_i^r(z)$ due to a layer of i atoms of areal density $N_i^A(z)$ located at depth z is

$$I_i^r(z) = I_0 N_i^A(z) \frac{d\sigma_i}{d\Omega} \Delta\Omega \ . \tag{4.45}$$

Here $\Delta\Omega$ is defined by the active area of the detector and the spacing between the sample and the detector. Information about the composition as a function of depth is derived from the inelastic (electronic) energy loss suffered by the probing particle in passing along (almost) straight lines from the instantaneous surface to the scattering center i, located at depth z, and back to the surface. Including a depth-dependent energy-loss term, (4.40) reads

$$E_i^r(z) = K_i E_0 - [\varepsilon]_i N z \ . \tag{4.46}$$

where $[\varepsilon]_i$ is the stopping cross-section factor [4.211]

$$[\varepsilon]_i = \frac{K_i \varepsilon_{in}}{\cos \theta_{in}} + \frac{\varepsilon_{out}}{\cos \theta_{out}} \ . \tag{4.47}$$

Here ε denotes the electronic stopping cross section. As in ISS, the information about the mass of the scattering center is contained in the kinematic factor K_i (4.41). To get a high depth resolution, two requirements must be fulfilled: $[\varepsilon]_i$ must be as large as possible and the energy resolution δE of the spectrometer must be as good as possible. According to (4.47) $[\varepsilon]_i$ increases with increasing ε, θ_{in}, and θ_{out}. Inspection of data compilations [4.214] shows that for a probing beam of He the stopping cross section passes through a maximum at energies between 400 keV and 1 MeV, where ε_{max} ranges from about 40 to 120 eV/10^{15} at/cm^2 for light- and heavy-atom targets like carbon and platinum, respectively. At beam energies of this order and with a "standard" scattering geometry (near normal beam incidence and $|\theta_{in} - \theta_{out}| = 30°$), $[\varepsilon]_i$ will amount to some 200 eV per monolayer ($\sim 10^{15}$ at/cm^2) for a scattering center like copper ($K_i \sim 0.8$). Using electrostatic energy analyzers the FOM-group has been able to achieve an energy resolution $\delta E/E$ of 0.3%, i.e., $\delta E = 300$ eV for $E = 100$ keV [4.215, 216]. This corresponds to a depth resolution δz between 0.3 and 0.4 nm. Resolution values close to these numbers have been reported by *Feuerstein* et al. [4.217] using 250 keV He$^+$ ions at oblique beam incidence ($1/\cos \theta_{in} \sim 9$, $\delta E/E = 0.7\%$).

The quoted data for the energy resolution relate to surface scattering. With increasing penetration depth, the resolution degrades rapidly because of energy-loss straggling and multiple scattering [4.211, 217, 218]. Depending on the target material, a resolution of 2 nm can be obtained only up to depths between 5 and 10 nm [4.216, 217]. Therefore, to push the zone of good depth resolution to larger depth, it makes sense to combine RBS analysis with controlled sputtering of the sample.

4.3 Experimental Procedure: General Aspects

The objective of a sputter-based analysis is a three-dimensional mapping of the elemental concentrations in the sample under study. To accomplish this goal the method has to satisfy at least three essential requirements: (i) the number of primary ions or atoms striking the sample per unit area for sputtering must be known as a function of position (x, y) and analysis time t; (ii) the position or area from which the recorded signal originates must be well under control; (iii) at each stage of erosion the measured signal heights should be convertible into an elemental concentration.

Whereas the first two issues are merely technical, the third issue relates to the basic physical phenomena responsible for the observed signal. In the following sections we will discuss these aspects in detail.

4.3.1 Controlled Bombardment

The basic equation relating the depth of erosion z at a point (x, y) on the sample to the bombardment and sample parameters J, Y, and N reads

$$z(x, y, t) = \frac{Y(x, y, z)J(x, y)t}{N(x, y, z)} \ . \tag{4.48}$$

Equation (4.48) looks fairly complex, mostly because we have allowed for a sample with a three-dimensional nonuniformity, $N = N(x, y, z)$, which will result in a corresponding spatial variation of the total sputtering yield, $Y = Y(x, y, z)$. The extent by which Y will vary within the analyzed volume depends on the bombardment parameters (energy, mass, and impact angle of the primary ions) as well as on the composition and structure (crystalline or amorphous) of the sample. Even with some means to control the magnitude of the Y variation (for example, by suppressing effects due to the local crystalline structure of the sample [4.37, 130–134]), problems will arise from a three-dimensionally nonuniform sample structure.

With these potential problems the complexity of (4.48) must be reduced so that the bombardment fluence is uniform across the area of interest. The only reliable way to accomplish this is to use controlled raster scanning of a focused, constant-current beam. "Controlled scanning" characterizes a beam positioning program in which the beam stays at each point (or pixel) of the scan field the same amount of time while passing through one frame. If the scan width is sufficiently large compared with the size of the beam, the time-averaged primary ion flux $J(x, y)$ will be very uniform within the central part of the scan field, as shown in Fig. 4.11a [4.219]. Unless imprecise performance of the electronic scan unit sets a less favorable limit [4.220], the relative deviation $\Delta J/J$ can easily be kept below 10^{-3} [4.219].

Application of the raster scanning techniques to laterally uniform samples yields the desired result, i.e., a sputtered crater with a flat bottom (Fig. 4.11b [4.219]). Within the area across which the average current density J is sufficiently constant, (4.48) can thus be simplified to read

$$z(x, y, t) = \frac{Y(x, y, z)}{N(x, y, z)} Jt \ . \tag{4.49}$$

According to (4.49) a calibration of the sputtered depth requires a knowledge of the ratio Y/N at each point within the eroded volume.

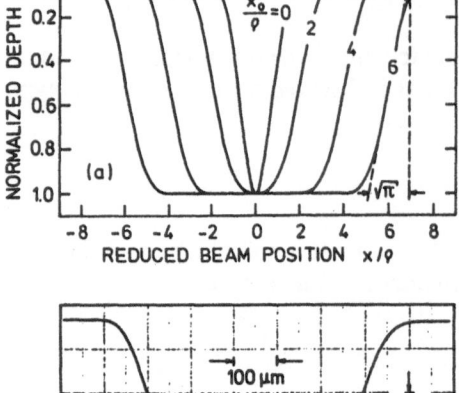

Fig. 4.11. (a) Calculated crater profile produced by raster scanning bombardment using a focused beam with a Gaussian current-density distribution. Parameter is the scan width x_0 in units of the beam radius ϱ. The profile for $x_0/\varrho = 0$ represents the contour of the static beam. (b) Cross section of a sputtered crater in a garnet material produced by raster scanning bombardment with 12 keV Ar^+ ions (beam radius 60 µm). The shape of the crater was determined using a surface microprofilometer [4.219]

4.3.2 Calibration of the Sputtered Depth

Ideally, one would like to control sputter erosion in situ. A number of methods do allow a direct measurement of the sputtered depth in favorable cases. An optical method developed by *Kempf* [4.221] uses the phase difference between two orthogonally polarized laser beams sensing the sputtered and the unsputtered areas. For opaque samples the sputtered depth z is linearly related to the phase difference ψ,

$$z = \frac{\lambda_0 \psi}{4\pi} , \qquad (4.50)$$

where λ_0 is the wavelength of the laser. For transparent films the situation is less simple because of interference effects. If the optical constants of the material(s) are known, the state of erosion can be determined by comparing measured and calculated values of the phase difference and the reflectance [4.221, 222]. However, a change in depth by at least $\lambda/4n_i$ is needed for an accurate determination of the erosion rate (n_i is the real index of refraction).

Up to now this sophisticated optical technique has been applied mostly to basic studies on the sputter yield variation at the initial stage of bombardment. These transients have been known for some time [4.36, 179, 223–228] and are still of concern in attempts to derive properly corrected sputter depth profiles [4.229–232]. Interpretation of the optical data [4.221, 222] is difficult, because the movement of the bombarded surface is measured, rather than the loss of atoms. If the generation of damage and/or the incorporation of primary beam particles causes the sample to swell [4.233, 234], the actual amount of sputtered material is hard to evaluate accurately. Nevertheless, *Kempf* et al. [4.235] have been able to show that, on the basis of reasonable assumptions, laser interferometry data can be used to quantify sputter erosion, even during the initial period of oxygen buildup in silicon.

The laser technique can be used only on optically flat samples. In routine analyses is often impossible to comply with this requirement. Also, the complexity and cost of the laser technique will hardly allow it to become a standard add-on tool for sputter depth profiling instruments.

Kirschner and *Etzkorn* have described an alternative in situ method for measuring the status of erosion in thin-film sputtering [4.193, 198, 200, 225]. They added an X-ray detector to a conventional AES sputter depth profiling device. Using 10 keV electrons for excitation, characteristic X-rays from most of the elements are produced with sufficient intensity. The relation between the X-ray intensity and the film thickness can be calibrated. For thicknesses small compared with the electron range, an (almost) linear relationship is observed [4.224, 236, 237]. Since Auger electrons and X-rays are produced simultaneously, two types of signals with vastly different depths of information are available at the same time, the former for measuring the near-surface concentration of the elements of interest and the latter for measuring the film thickness at each stage of sputtering.

In much the same way as with electrons one can use protons for X-ray excitation and thickness calibration [4.238, 239]. However, the proton energy must exceed 100 keV (preferably > 1 MeV) to get a reasonable signal or to maximize sensitivity [4.240]. Protons and helium ions with energies of that order have also been employed in many RBS studies on sputter erosion of sandwich layers or ion-implanted samples [4.166, 179, 241–246]. Only very recently, however, has the analytical potential of RBS in combination with sputtering been fully realized [4.106].

The advantage of electron and ion techniques for controlling the state of sputter etching is that by using small-spot beams we can determine $z(x, y, t)$ at any point on the eroded sample. On the other hand, the methods suffer from a severe limitation: they are applicable only for layered samples. Moreover, the high cost of ion accelerators will hardly justify implementation of such auxiliary tools on sputter depth profiling instruments.

Thus the two most common methods for depth calibration are based on the application of a uniform primary ion flux (4.49). If Y/N is constant to a certain depth z_m, we can use multiple-beam interferometry or the stylus technique (i.e., a surface profilometer) to measure the crater depth z_m, or a fraction thereof [4.247]. This is illustrated in Fig. 4.11b. Then

$$z(t) = \frac{z_m}{t_m} t \; , \tag{4.51}$$

where t_m is the time elapsed during sputtering to depth z_m (with $J = $ const).

For multilayer samples the procedure becomes more elaborate because the erosion rate dz/dt has to be determined for each compositionally different region in the sample. This can be done as sketched in Fig. 4.12, i.e., by producing

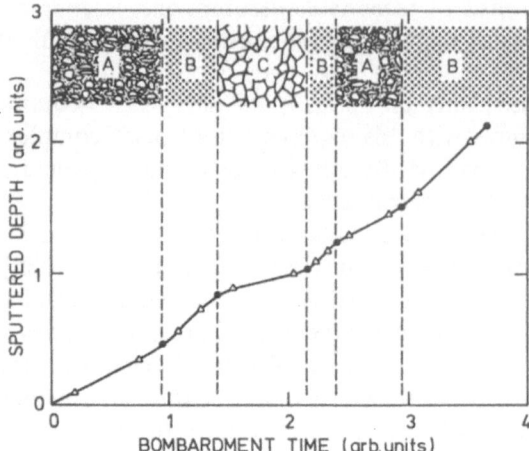

Fig. 4.12. The progress of sputter erosion in a sample composed of several layers of different composition. The filled circles and open triangles represent data points required to achieve a calibration of the sputtered depth

a number of sputtered craters so that bombardment is always stopped at the interface between layers of different composition (filled circles). Then we need a signal to identify the time of breakthrough between adjacent layers. This may not be a simple procedure, because broadening effects and sputtering yield changes will occur at each interface. A safer but even more elaborate way to determine erosion rates for individual layers is to measure $z(t)$ for a sufficiently large number of craters, so that each region of constant erosion rate becomes well characterized (at least two data points per layer, open triangles in Fig. 4.12). Samples with layers of the same composition at different depths require a calibration procedure only once.

Knowing the composition of all layers in a sample, one can also calibrate the erosion rate separately with suitable standards. In essence this corresponds to determining the sputtering yield Y. To be quantitative the beam current must be measured correctly, i.e., with a Faraday cup. Sputtering yields derived from crater volume measurements have been reported repeatedly [4.175, 226, 249–253]. Studies of this kind have shown that the total sputtering yield (or sputtering rate) for a homologous series of compounds (e.g., $Ga_xAl_{1-x}As$) varies monotonically with changing composition [4.252–254]. Accordingly, the sputtering yield may often be determined with sufficient accuracy by way of interpolation.

Very recently *Voigtmann* and *Moldenhauer* [4.255] suggested a new technique for determining the depth dependence of the ratio Y/N which, according to (4.49), determines the erosion rate dz/dt for a known mean current density J (keeping the energy, mass, and impact angle of the primary ions fixed). The method involves the preparation of only one calibration crater (rather than many, as in the case shown in Fig. 4.12). The idea is to program the motion of the focused ion beam so that a wedge-shaped crater is formed. This can be accomplished by a linear increase in the mean current density along one scan direction, e.g., the x direction, so that $J(x, y)$ takes the form

$$J(x, y) = J'_x x \ ,\tag{4.52}$$

where $J'_x = dJ/dx$. If the sample is uniform parallel to the x, y plane but nonuniform in the z direction, the crater contour, measured through the center and along the x direction after sputtering time t, will be of the form

$$z(x) = \frac{Y(z)}{N(z)} J'_x x t \ .\tag{4.53}$$

Layers of constant sputtering yield or, more correctly, of constant ratio Y/N, will thus show up in the crater contour as regions of constant gradient dz/dx,

$$\frac{Y(z)}{N(z)} = \frac{1}{J'_x t} \frac{dz}{dx}\bigg|_t \ .\tag{4.54}$$

Noted that this approach is applicable only for sufficiently small gradients dz/dx. Otherwise the angular dependence of the sputtering yield would make the analysis difficult, if not impossible. A real limitation of the technique is brought about by the nonnegligible size of the primary beam. For a reasonably accurate evaluation of the ratio Y/N, the product of the layer thickness and the respective inverse slope $dx/dz|_t$ must exceed at least three beam diameters. Unless a rather large x scan is used, the calibration factor Y/N can be evaluated only for a few layers; otherwise the slopes dz/dx are not sufficiently well defined. For the same reason a very large x scan will be required to determine Y/N for a thin layer located at a large depth.

4.3.3 Localization of the Analyzed Area: Imaging and Gating

Apart from being able to perform an experiment under well-controlled conditions of sputter erosion, it is essential to know precisely the area from which the signal originates. The two methods for localizing the analyzed area are direct emission microscopy and microprobe analysis [4.11].

In direct emission microscopy the sample can be bombarded with a broad beam, but raster scanning is often employed to ensure uniform illumination. The desired spatial resolution is achieved by using a dedicated optical system to transport secondary ions so that a mass-resolved, magnified image S' of the object S_0 is formed in the detection plane (Fig. 4.13a). This concept has been verified in secondary ion microscopes [4.9, 51] where the lateral resolution is determined by the width of the energy-band pass and the strength of the extraction field. For reasonable transmission, the lateral resolution is limited to a few micrometers [4.51, 256]. An advantage of direct imaging systems is that the quality of the primary beam has little importance. The only requirement is that the imaged area should be illuminated uniformly. With a broad beam the whole area under study is imaged simultaneously. This reduces the recording

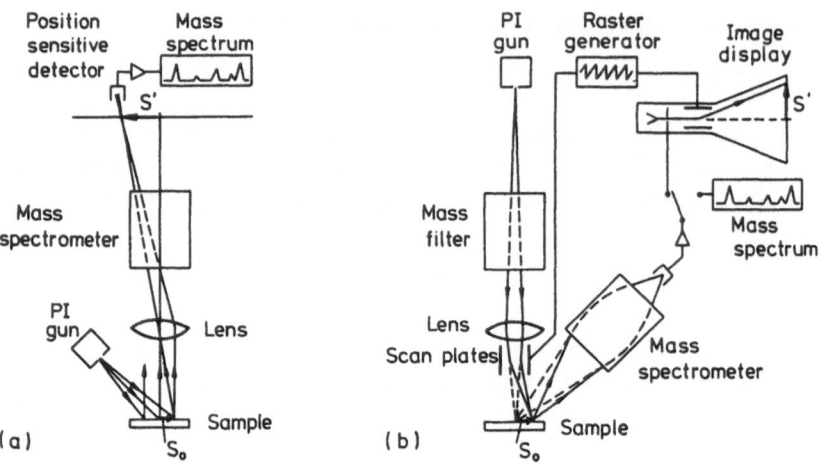

Fig. 4.13. Two methods of microanalysis: (a) direct imaging combined with mass analysis and (b) microprobe technique [4.11]. PI = primary ion, S_0 = object, S' = image

time to a minimum set by the concentration and ionization probability of the element or isotope of interest.

Using an ion microscope for depth profiling, signals from regions outside the area of interest can be discriminated against by inserting a mechanical aperture of the desired size at an appropriate position in the secondary ion column (e.g., in the image plane). However, because the high-transmission mode achieves limited discrimination, depth profiling measurements aiming at a high dynamic range are usually performed by raster scanning bombardment and imaging in combination with electronic gating [4.257].

In microprobe analysis the lateral resolution is determined essentially by the size of the primary beam (ions, electrons, or photons). Localization is accomplished by moving the focused beam in a controlled manner across the sample and by synchronizing the channel of detection with the position of the beam on the target (Fig. 4.13b). In the first SIMS instruments of this kind [4.10, 258] scanning was accomplished in a line-by-line fashion. Images were obtained by driving the xy deflection of a cathode ray tube synchronized with the beam raster, and by using the detected signal to modulate the oscilloscope intensity.

For signal generation the scan speed can be considered a free parameter set according to the requirements of the actual analysis. Limitations in scan speed must be taken into account, however, if the time τ_d elapsed between primary particle impact and detection of the ejected secondary particle becomes comparable to the line scan period τ_s. Due to time-of-flight effects, the recorded image point will be displaced from the true point of emission by the amount δ_s, which can be written in the form [4.259]

$$\delta_s = \frac{\tau_d w_s}{\tau_s} = \tau_d v_s \; , \tag{4.55}$$

where w_s is the width of the image field parallel to the lines of scanning and $v_s = w_s/\tau_s$ is the scan speed along a line.

Unwanted time-of-flight effects can occur in quadrupole-based scanning ion microprobes because mass filtering is performed at low secondary ion energies (typically 10 eV [4.13]). For medium-mass secondary ions ($M \simeq 50$ u) and traveling distances of about 30 cm, τ_d will amount to some 50 μs [4.259]. To keep the relative image displacement δ_s/w_s below 1%, τ_s should exceed 5 ms.

High-quality depth profiling in the raster scanning mode is accomplished with a so-called electronic gate (or electronic aperture), an approach first mentioned by *Whatley* et al. [4.260]. An electronic aperture is a logical unit that allows data acquisition only while the beam hits a predetermined area within the scan field [4.219, 261, 262]. The first devices of this type only discriminated against unwanted signals originating from the rim of the raster-scanned area [4.219], but the rapidly increasing storage capacity of computer memories now allows profiles to be recorded on a much larger number of areas within the same scan field. This "checkerboard" technique of data acquisition and evaluation was introduced by *von Criegern* et al. [4.263], who used 8×8 cells. More recently *Frenzel* et al. [4.264] described an instrument featuring 16×16 acquisition cells. The great advantage of the checkerboard technique is that any unwanted signal appearing in the flat bottom part of the crater (e.g., due to contamination [4.180] or the formation of etch pits or hillocks [4.265]) may not only be identified after the measurement, but can also be removed from the set of useful data. Such contamination blanking [4.263] allows the analyst to make ultimate use of the stored data (maximum dynamic range and ultimate statistical significance).

In some types of analyses, alternatives to the commonly used line-scan beam rastering procedure can be quite beneficial. A spiral scan, for example, allows mass and energy spectra to be recorded at high scan rates [4.264] without suffering from time-of-flight effects in the manner described by (4.55). In fact, using a spiral scan with one central gate, one may formally replace τ_s in (4.55) by the (much longer) residence time of the beam in the central data acquisition area.

The procedures described above for beam rastering and electronic gating can be used whenever the signal is generated by (or with) the beam used for sputter erosion. This is the case not only in SIMS but quite often also in ISS [4.89, 266, 267]. In AES the probing electron beam is usually small compared with the size of the eroded area [4.268, 269]. With a static electron beam we merely have to ensure that the same area is probed during erosion. Electron-beam-induced desorption or local heating effects [4.270–273] often make a raster-scanned electron beam desirable. Scanning AES may also be used to map the lateral distribution of elements present at the instantaneous surface [2.269]. For XPS analysis with controlled sputter erosion, only static X-ray beams covering a fairly large area (several mm^2) have been used [4.268, 274].

4.4 Instrumentation: Concepts and Performance

4.4.1 Secondary Ion Mass Spectrometry

Compared with all other sputter-based techniques, SIMS has seen by far the most technical developments. The history of SIMS instrumentation has been reviewed by *Liebl* [4.10, 11], *Honig* [4.275], and, more recently, in the SIMS compendium by *Benninghoven* et al. [4.276]. Current equipment can be classified according to the method of mass analysis. The three most common mass separating devices are the magnetic sector field, the electric quadrupole filter and time-of-flight systems. Instruments belonging to the third category are described in detail in Chap. 5. Here we discuss only relevant aspects of SIMS devices from the first two categories.

Figure 4.14a shows a modified version [4.277] of the Cameca IMS-3F secondary ion microscope [4.49, 278], a magnetic sector field instrument built according to *Slodzian*'s concept [4.51]. The instrument was designed for high secondary ion transmission at moderate mass resolution ($M/\Delta M \simeq 400$). At the expense of transmission it can operate in the high-resolution mode ($M/\Delta M$ up to 10000). Like many modern SIMS systems the modified version of the

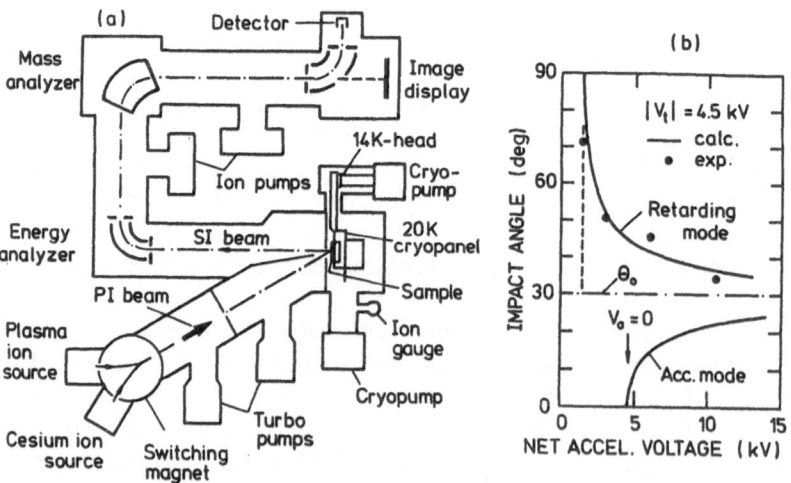

Fig. 4.14. (a) Magnetic sector field secondary ion microscope (Cameca IMS-3F) in the modified form featuring a 20K cryopanel near the sample [4.277]. (b) Correlation between the impact angle θ_{in} and the final acceleration voltage $|V_0 - V_t|$ of the primary ions (V_0 is the ion source terminal voltage, and V_t the target bias)

IMS-3F features two primary ion guns [4.279], one to generate ions of permanent gases, specifically oxygen, and the other to produce cesium ions. Beams of O_2^+ (or O^-) and Cs^+ are commonly used to enhance the degree of ionization of positive or negative secondary ions, respectively (Chap. 3) [4.22–24, 56–62, 280–285]. In the modified IMS-3F, a switching magnet provides primary ion mass analysis and, at the same time, directs either one of the two beams onto the common axis of the ion optical column.

The sophisticated ion optical approach [4.51] requires many active and passive elements, which only a skilled, experienced operator can handle successfully. Although designed as a high-performance ion microscope, the IMS-3F is more commonly used for SIMS depth profiling. With particular emphasis on semiconductor analysis, many interesting applications have been reported [4.256, 278, 286–291]. To overcome the background problems arising from the moderate vacuum achievable in the sample chamber (low to mid 10^{-7} Pa), *Homma* and *Ishii* [4.277, 289] have added a 20K cryopanel near the sample (Fig. 4.14a). Thereby the detection limits for H, C, and O could be improved considerably (Sect. 4.8.3).

The concept of the IMS-3F [4.51] implies that the target is maintained at a relatively high bias potential with respect to ground, $|V_t| = 4.5$ keV [4.278]. This feature can cause significant cross-contamination or memory effects (Sect. 4.8.2) [4.292]. Also of concern is the fact that because of the high target bias the actual angle of ion impact θ_{in} will deviate from the tilt angle θ_0 defined by the axis of the primary ion column and the surface normal ($\theta_0 = 30°$). The relation between θ_{in}, θ_0, V_t and V_0, the acceleration voltages of the primary ions, reads [4.293]

$$\sin \theta_{in} = \left(\frac{V_0}{V_0 - V_t} \right)^{1/2} \sin \theta_0 \ . \tag{4.56}$$

Accordingly, θ_{in} and the impact energy $q(V_0 - V_t)$ cannot be chosen independently (q is the charge of the primary ion) [4.286, 294]. Figure 4.14b shows the variation of θ_{in} with $|V_0 - V_t|$. A few experimental data points reported for the retarding mode, i.e., for $V_t/V_0 > 0$, are also included [4.286]. Evidently in both the retarding and the accelerating modes the accessible impact energy has a theoretical lower limit (the accelerating mode has to be used, for example, when analyzing negative secondary ions sputtered by Cs^+ bombardment). Take into account the following: the generally observed energy dependence of the maximum beam current [4.286], mostly due to space-charge effects [4.295]; the loss of beam quality at low impact energies and large impact angles [4.286]; and the beam displacement due to retardation or acceleration [4.286, 293]. Then the practical lower limits of $|V_0 - V_t|$ are ~ 3 kV and ~ 8 kV in the retarding and the accelerating modes, respectively. An impact energy as low as 1.5 keV [4.286] is achievable only if the incident beam is deflected before retardation so that $\theta_0 < 30°$. Without deflection, ions retarded to 1.5 keV should ideally propagate parallel to the sample surface, i.e., $\theta_{in} = 90°$ (Fig. 4.14b).

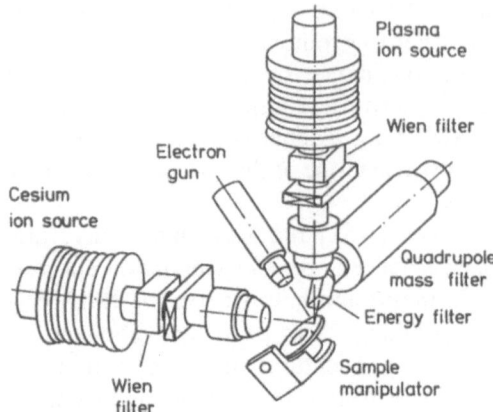

Fig. 4.15. Three-dimensional view of the quadrupole-based dual-beam raster scanning ion microprobe Atomika 6500 [4.296]

As an example of quadrupole-based SIMS instruments Fig. 4.15 shows a dual-beam raster scanning ion microprobe, the Atomika 6500 [4.264, 296]. Other recently developed quadrupole-based SIMS instruments have been described [4.297–299] and reviewed for performance [4.13]. The Atomika 6500 has two independent primary ion columns supplying mass-filtered beams of either Cs^+ or of ions of permanent gases, e.g., O_2^+ or Ar^+. This new concept allows simultaneous impact of the two ion beams on the target, thereby opening a new avenue of research in SIMS. Many advantageous prospects can be envisaged, e.g., the use of the second beam for improved control and removal of crater edge effects or for studies on synergistic effects.

The Atomika 6500 also allows us to choose the primary ion energy and the angle of beam incidence according to the requirements of the experiment. This flexibility results from the fact that in instruments of this kind (e.g., the DIDA ion microprobe [4.13]) the target bias is on the order of only 10 V. Moreover, the secondary ion extraction field is rather weak (< 5 V/mm) [4.13, 300]. As a result, any displacement of the beam due to deliberate changes of the target bias (secondary ion energy offset [4.301] for suppression of molecular ions) or the polarity of the extraction field (switching from positive to negative secondary ion detection or vice versa) can be compensated for easily by an appropriate static offset superimposed on the scan voltage.

The low target potentials and weak extraction fields common in quadrupole-based SIMS systems are also advantageous if low-energy electron beams are used for charge compensation in SIMS analysis of insulators [4.219, 302–310]. In such systems all electron impact parameters – i.e., angle of incidence, energy, current density, raster size and position relative to the point of primary ion impact – can be chosen deliberately. Even with a very dedicated design [4.311], this free choice of all relevant electron beam parameters is not possible in magnetic sector field instruments [4.312, 313].

Electric quadrupole mass filters are in principle capable of providing a mass resolution $M/\Delta M = 1500$ or more [4.314]. Due to the relatively small size and the limited power and frequency of commercial rf voltage supplies, quadrupole-based SIMS instruments can usually be operated only in the range $M/\Delta M < 500$. To achieve $M/\Delta M$ values exceeding 1000, the energy filter preceding the quadrupole must be able to transmit low-energy secondary ions within a narrow energy band [4.13, 315]. Much as with magnetic sector field instruments [4.51], this can be accomplished only at the expense of transmission.

A more detailed discussion of the performance of current SIMS instrumentation is beyond the scope of this chapter. Within the limits set by the concept of the respective instrument, a rather high degree of sophistication has often been achieved. All SIMS machines, however, have the same shortcoming: it has not been possible to establish a method of bombardment that will make the ionization degree of a sputtered atom reasonably independent of the matrix composition.

4.4.2 Mass Analysis of Postionized Sputtered Neutrals

The severe, yet unpredictable matrix effect in SIMS (Chap. 3) has encouraged many groups to design instruments for sputtered-flux mass analysis in which sputtering and ionization are decoupled from each other. In principle this can be accomplished by generating an ionizing medium that intersects the trajectories of the sputtered neutrals. Having passed through this region, the postionized secondaries can be analyzed in essentially the same type of mass spectrometer as in SIMS. Accordingly, all techniques involving mass spectrometric analysis of postionized sputtered neutrals may be included under the generic term "sputtered (or secondary) neutral mass spectrometry" (SNMS). Taking into account the specific method of ionization, the following subdivision appears appropriate: electron-beam SNMS, electron-gas SNMS, Penning and charge transfer ionization, and laser ionization.

a) Electron-Beam SNMS

Before tunable, powerful lasers became available, postionization by electron impact was the method of choice [4.8, 94–96]. Basically, the technique involves an electron beam (e-beam) set up to intersect the flux of sputtered particles (Fig. 4.16a). To maximize sensitivity, the electron energy should range between 50 to 150 eV, where the ionization cross sections are known to pass through a maximum [4.316]. Calculations suggest that an ionization efficiency on the order of 10^{-3} can be obtained for elements with an ionization potential below 10 eV [4.317]. This is significantly lower than the degree of ionization achieved in SIMS under conditions of optimum surface chemistry (Chap. 3). However, compared with SIMS [4.56–59], the relative sensitivity factors observed in e-beam SNMS exhibit only a rather small variation between different elements (factor of ~ 10 between C and Al [4.96]).

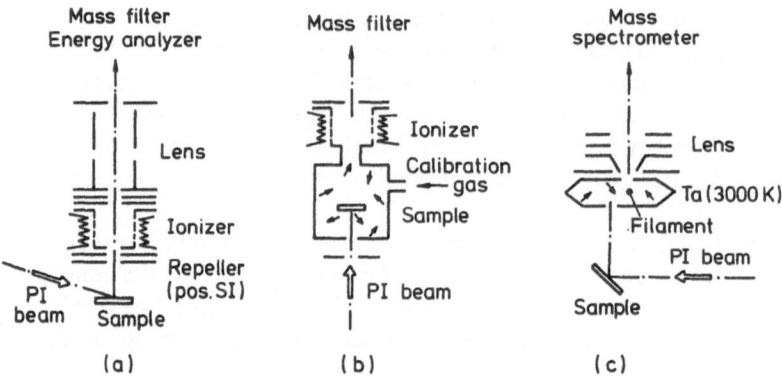

Fig. 4.16. Different techniques used for electron beam postionization of sputtered neutrals: (a) direct-impact mode (standard e-beam SNMS) [4.95], (b) gas ion probe [4.318], and (c) high-temperature thermalization and dissociation with e-beam ionization [4.322] (PI = primary ion, SI = secondary ion)

An important problem in e-beam SNMS is the removal of secondary ions and the suppression of signals due to ionized residual gases [4.95, 96, 317]. One can also envisage that sample contamination due to repelled secondary ion can be quite severe in such instruments. Further investigations are required for a more detailed assessment of the technique's analytical capabilities.

b) Electron-Beam Ionization of Thermalized Gases

A factor limiting the ionization efficiency in e-beam SNMS is the relatively high velocity of sputtered atoms [4.317]. With this problem in mind *Kiko* et al. [4.318] set up a so-called gas ion probe that essentially constitutes an e-beam SNMS system, but with ionization after thermalization of the sputtered inert gases. As shown in Fig. 4.16b the sputtered gases can reach the ionization volume only after multiple collisions with the walls of the target box, whereby thermalization is accomplished. In situ calibration is obtained by bleeding a known amount of gas into the target box.

The gas ion probe has been used to determine ranges of ^{4}He and ^{22}Ne implanted in terrestrial ilmenites ($FeTiO_3$). Moreover, concentration profiles of ^{4}He as well as of noncondensing gases could be measured in lunar minerals from the Apollo 17 landing site. Note, however, that the gas ion probe has only a very limited dynamic range because, as pointed out by *Gnaser* et al. [4.95], the long time constant of the technique ($\sim$ 1 ms [4.318]) prevents the application of the raster scanning and electronic gating scheme [4.219]. Note also that the assumption of energetic rare-gas emission underlying the design of the gas ion probe is not justified. In fact, e-beam SNMS [4.95, 319] and, more specifically, time-of-flight measurements [4.137] have shown that inert gases released by ion bombardment have thermal energies (unless they originate from bubbles, in which case the energy is on the order of 0.1 eV [4.137]).

c) Electron-Impact-Assisted Ionization in a High-Temperature Cell

Extending the idea of thermalization to analyzing solids, *Blaise* and *Castaing* [4.320, 321] have suggested the use of a high-temperature cell for ionization of sputtered atoms, as well as for dissociation and ionization of sputtered molecules. The experimental procedure is sketched in Fig. 4.16c. The sputtered material enters the high-temperature cell through a small hole. The exit aperture is displaced from the axis defined by the center of the bombarded target area and the entrance aperture, so that injected particles can escape from the cell in the direction of the mass spectrometer only after many collisions (~ 100). Accordingly, complete thermalization is accomplished [4.322]. Moreover, the high temperature of the Ta cell (up to 3000 K) causes molecules, including those with a high binding energy like UO and UO_2, to dissociate almost completely [4.321]. Even at 3000 K, however, thermal ionization is inefficient for elements with an ionization potential above 8 V. Therefore, the original approach [4.319, 321] has been extended to include electron impact ionization within the high-temperature cell (in principle, other external methods of ionization could also be used) [4.322]. Application of this technique to a variety of alloys showed that elements like Mg, Al, Cu, and Zn can be analyzed with relative sensitivity factors independent of the host matrix [4.322, 323]. This truly quantitative procedure allows in situ calibration of the sputtered depth in terms of the amount of material eroded per unit area [4.322].

Unfortunately, application of the technique is limited to certain areas because it suffers from a few disadvantages [4.322]: (i) the total detection efficiency or fractional ion yield is low (10^{-6}), mostly because of the small solid angle of acceptance of the cell. (ii) The background intensity is rather high, so that the dynamic range in depth profiling does not exceed 10^2. (iii) The time constants involved prevent lateral (micro-) analysis. (iv) Using high-temperature cells of Ta or W, elements like B, C, and O cannot be analyzed because they are efficiently dissolved in the cell walls.

d) Electron-Gas SNMS

Rather than using a biased, heated filament to generate the required electron flux, ionization of sputtered neutrals can also be accomplished in a plasma containing electrons of sufficiently high energy. This idea of electron-gas SNMS was introduced by *Oechsner* and *Gerhard* [4.69]. The ionizing medium is contained in a small box filled with Ar at a pressure of about 10^{-3} mbar (~ 0.1 Pa). Taking advantage of electron cyclotron wave resonance, an inductively excited high-frequency plasma is generated in a weak dc magnetic field. The electron temperature T_e corresponds to an energy of about 10 eV ("hot" electron gas [4.72]), depending upon the excitation conditions. The energy of the ions generated in the plasma is only slightly above thermal. Due to the relatively low operating pressure, the mean free path for collisions between sputtered neutrals and atoms or ions in the plasma is significantly larger than the length of travel through the ionizing medium (typically 5 cm [4.72]).

Two versions of electron-gas SNMS have been described repeatedly, the direct-bombardment mode (Fig. 4.17a) [4.69, 70, 72, 324–327] and the external-bombardment mode (Fig. 4.17b) [4.72, 327–331]. The direct-bombardment mode has the advantage that sputter erosion can be performed at ion impact energies down to about 50 eV, while maintaining current densities on the order of 1 mA/cm². However, only normal ion incidence is possible under these conditions. Recently the potential advantage of using low impact energies was found to be offset by pronounced differences between the angular distributions of atoms sputtered from a variety of alloys. This effect is much more severe at 250 eV than at 2 keV [4.332].

The direct bombardment also implies that the advantage of the electronic gating scheme cannot be utilized. Accordingly, the dynamic range achieved in sputter depth profiling hardly exceeds one order of magnitude [4.72, 325–327, 333]. Even this rather limited performance is possible only if sputter erosion extends over a relatively large area (typically 5 mm in diameter). Note also that for a given position of the sample with respect to the aperture shielding the plasma, the current density distribution of the ions hitting the sample depends critically on the applied potential [4.325]. The desired uniform current density is achieved only at a certain critical acceleration voltage (Fig. 4.18). These limitations could be avoided in the external-bombardment mode. The latter mode has been used for basic studies on oxidation and ionization phenomena [4.328–331], but apparently not for sputter depth profiling.

The ionization efficiency in electron-gas SNMS is estimated to be on the order of 10^{-2} [4.72], about one order of magnitude better than in e-beam SNMS [4.319]. Because of the small acceptance angle of laboratory [4.324–331] and commercial instruments [4.333], the fractional ion yield is only on the order

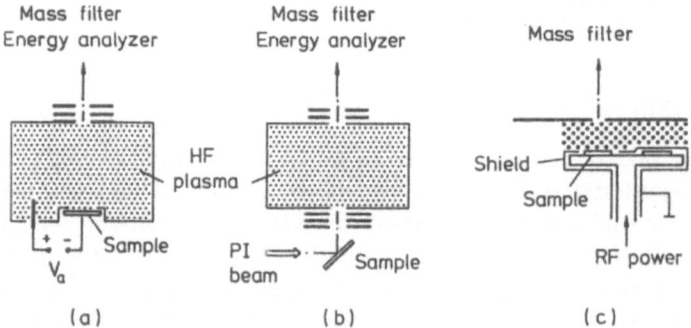

Fig. 4.17. Postionization in a low-pressure, hot-electron gas: (a) direct-bombardment mode, (b) external-bombardment mode (PI = primary ion) [4.72]. (c) Postionization in a high-pressure glow discharge [4.68]

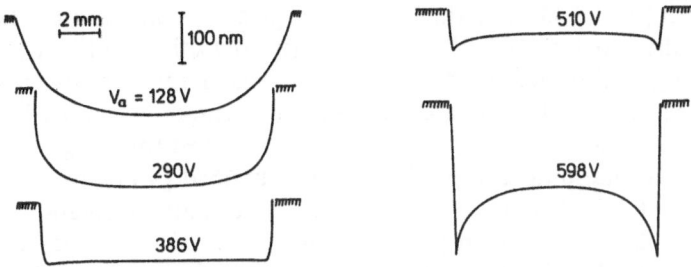

Fig. 4.18. Profiles of sputter craters produced in the direct-bombardment mode of electron gas SNMS. Parameter is the ion acceleration voltage applied using a fixed bombardment geometry [4.322]

of 10^{-9} [4.333, 334]. While this low yield could be tolerated in quantitative survey analysis of impurities at concentration levels above 1 ppm, it is rather disappointing to see that, according to recently reported studies [4.333, 335], the original idea of matrix-independent relative sensitivity factors [4.72] could not be verified up to now. Reasonably accurate quantification still requires standards.

e) Glow Discharge Mass Spectrometry

In essentially the same manner as in the direct-bombardment mode of electron-gas SNMS [4.69, 70, 72, 324–327], one can also use a glow discharge for sputter erosion of the sample, as well as for ionization of the ejected neutrals. The characteristic features of glow discharges are described in Chap. 7. The use of an rf glow discharge for mass spectrometric analysis of solid samples was first reported by *Coburn* et al. [4.67, 68, 71]. Figure 4.17c illustrates the basic configuration of the technique for a planar diode geometry. The operating conditions in glow discharge mass spectrometry (GDMS) as well as the mechanisms responsible for ionization of sputtered neutrals are distinctly different from electron-gas SNMS.

In GDMS the sample (together with the sample holder) constitutes an active element in the discharge. To generate a stable plasma the operating pressure has to be on the order of 0.1 mbar (~ 10 Pa) for gases like Ar or Ne [4.336, 337]. Under these conditions the mean free path of sputtered neutrals is much smaller than the distance between the sample and the exit aperture. The other important aspect is that the plasma contains a sizable fraction of metastable gas atoms (Ar^m or Ne^m) with excitation energies between 11.5 and 16.7 eV. Accordingly, Penning ionization of sputtered M atoms, i.e., the mechanism $Ar^m + M \rightarrow M^+ + Ar + e$, is usually the dominating ionization process [4.336].

The following advantages of GDMS are noteworthy: (i) all elements of the periodic table can be analyzed, except for He and Ne. The detection sensitivities relative to Fe range from about 2 for Zr to 0.2 for C, P, and S [4.338]; (ii) matrix effects are very small [4.68, 338]; (iii) the impact energy of the ions striking the sample is only on the order of 100 eV [4.71, 337]; (iv) detection limits in the parts-per-billion range are attainable [4.337, 338]; (v) erosion rates up to about 100 nm/min can be achieved at moderate rf power; (vi) memory effects are less severe than expected [4.68]; (vii) using small samples attached to a large-area sample holder, high-resolution depth profiles can be obtained [4.71].

The limitations of GDMS are as follows: (i) there is no spatial resolution in the plane of the sample surface; (ii) in the high-depth-resolution geometry the entire sample is consumed and the sample holder is etched during the analysis [4.68]; (iii) the high gas load requires differential pumping [4.337]; (iv) background gas species and sputtered molecules (nondissociating during ionization) sometimes complicate the analysis [4.68].

f) Ionization via Charge Transfer Processes

A somewhat unusual method of sputtered-neutral ionization involves the transfer of a primary ion's charge to a particle just being ejected from the sample as a result of a preceding impact. Charge transfer processes have been studied in great detail in dedicated gas-phase collision experiments [4.339]. The cross section for charge transfer depends on the chemical identity and the excitation state of the colliding species, as well as on their relative velocity. Cross sections can be as large as 10^{-15} cm^2.

In principle, charge transfer processes between primary ions and sputtered neutrals can take place in any SIMS instrument without an additional ionizing beam or medium [4.340, 341]. The ionization probability P^+, however, increases as the square of the primary ion current density J [4.342, 343]. An increase in secondary ion yield due to this effect becomes observable at current densities in excess of about 1 mA/cm^2. The relation $P^+ \propto J^2$ can easily be understood by realizing that, for charge transfer to occur, an incident ion as well as a sputtered particle must be found within the volume element dV' above the surface. The probability for either species being present in dV' during the time interval $t \ldots t + dt$ is proportional to J. Thus the probability for charge transfer is proportional to J^2.

Apart from their usual current-density dependence, charge transfer processes may be identified in SIMS experiments by inspecting the energy spectrum of secondary ions [4.340–345]. If an acceleration voltage V_a is applied between the sample and the entrance electrode of the mass spectrometer (spacing D_s), ions generated at a distance d above the sample surface will suffer from a deficit in acceleration voltage δV_a, which, compared with the secondary ions originating at the very surface, amounts to

$$\delta V_a = \frac{d}{D_s} V_a .$$

(4.57)

This deficit in potential energy causes a corresponding shift in the kinetic energy of the secondary ions propagating through the energy and mass analyzer. In instruments with high extraction fields V_a/D_s (like the Cameca IMS-3F, in which $V_a/D_s \sim 1$ kV/mm [4.278]) the energy deficit may frequently exceed 10 eV [4.344, 345]. Low-velocity sputtered neutrals are postionized preferentially because of their long residence time in dV' [4.343], so energy deficits exceeding $\sim$ 1eV will cause a detectable shift into the "negative" branch of the apparent energy spectrum [4.344, 345]. A separation of "true" secondary ions from postionized sputtered species is thus easily accomplished by appropriate tuning of the energy-band pass.

An interesting feature of postionization via charge transfer is the rather high probability for generation of doubly charged ions [4.343–345]. The physics behind this process are not yet understood. Except for the special case of light elements like Mg, Al, and Si, doubly charged species are usually not found in the secondary ion flux (Chap. 3). Therefore doubly charged ions, produced by postionization, can provide quantitative analysis of a sample's major constituents, as demonstrated by *Williams* and *Streit* [4.345]. More detailed investigations into the basic mechanisms of ionization, as well as into the use of charge transfer for analytical applications, are desirable.

g) Laser-Based SNMS

Undoubtedly the most promising technique for sputtered-neutral ionization involves laser irradiation. Since the ionization potential of most elements is larger than the photon energy of presently available lasers, sputtered-neutral ionization usually requires interaction with two or more photons. Two different methods of multiphoton ionization (MPI) have been explored: resonance MPI [4.98–102] and nonresonance MPI [4.103, 104].

In resonance MPI the flux of sputtered neutrals is intersected by a pulsed, relatively broad laser beam (diameter between 1 and 10 mm), tuned to a resonant frequency of the element of interest (Fig. 4.19). To maximize the ion yield the laser beam propagates in close proximity to the sample surface. A second laser of

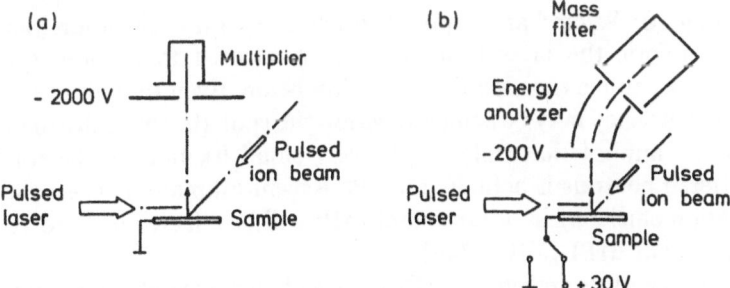

Fig. 4.19. Arrangements for sputter-based analysis using resonant multiphoton ionization: (a) simple time-of-flight system and (b) instrument with energy and mass analysis of secondary ions (sample at ground) and ionized neutrals (sample biased) [4.99]

sufficiently high frequency raises the energy of the resonantly excited sputtered species to a value exceeding the ionization potential (two-color scheme, Fig. 3.21c). Within a few 10^{-13} s after absorption of the ionizing photon, the excited atom or molecule becomes an ion. In certain cases the same laser beam may serve for resonant excitation as well as for ionization (one-color scheme) [4.99].

Postionized neutrals can be separated from secondary ions generated at the sample surface by applying an electric field between the sample and the entrance aperture of the mass spectrometer [4.98, 105]. As with charge transfer ionization discussed above, the separation in space is translated into an easily measurable separation in energy [4.57].

The ionization efficiency depends on several factors; the most important one is the geometrical overlap of the laser beam with the sputtered-particle cloud generated by a short primary ion pulse. The fraction χ of neutrals in the photoionization volume is usually estimated [4.105, 346] on the basis of the standard energy distribution of particles ejected from a collision cascade [4.107, 108, 347]. Depending on the assumptions entering into the calculations, χ-values range from 6% [4.346] to 42% [4.105] for a medium-mass species like Fe. Because of the variation of ion velocity with mass (at a fixed energy), χ is expected to scale with particle mass roughly as $M^{1/2}$ [4.346].

Other factors of concern are the Doppler broadening, spontaneous decay losses, and the timing of the ion and the laser beam pulses. Mean ionization efficiencies per ground-state atom, estimated under reasonable assumptions, amount to about 0.4χ [4.346]. With this high efficiency, detection sensitivities below 1 ppb could be demonstrated [4.102]. However, matrix effects may come into play in resonant MPI because the fraction of neutrals ejected in the ground state has been found to be rather sensitive to the surface chemistry. Oxidation of metals causes the population of ground-state atoms to decrease by up to two orders of magnitude or more (Chap. 3) [4.348]. This matrix effect in resonant MPI, which may be considered an inverse to the strong oxygen-induced yield enhancement in SIMS, will aggravate quantification in the analysis of samples with varying oxygen content.

Nonresonant MPI is accomplished with pulsed high-power ultraviolet laser beams. Because of the low cross section for multiphoton ionization, power densities exceeding 10^9 W/cm^2 are required to achieve saturation in ion yield [4.349, 350]. Therefore, the laser beam has to be focused to a small spot (diameter < 0.2 mm) above the area hit by the ion beam. Even though saturation in ionization efficiency may be achieved within the spot, the small size of the interaction volume implies that only a relatively small fraction of the total number of sputtered neutrals is actually ionized. Recent estimates suggest that the mean ionization efficiency in nonresonant MPI will be a factor of 50 to 100 lower than in resonant MPI [4.105, 346].

The big advantage of nonresonant MPI is that all elements of the periodic table can be ionized with a comparatively small degree of selectivity. Even inert gases like Xe have been ionized successfully [4.350]. The form of the "calibration" curve (i.e., the dependence of the ion yield on the laser power density) is not the same for elements of significantly different ionization potential and electronic structure [4.349, 350]. Even so, quantification is possible without reference standards, through a one-time calibration of the relative ionization efficiencies for a fixed laser condition. The uncertainty involved will depend on the matrix dependence of the sputtered-particle energy spectrum.

The pulsed ion and laser beams make MPI-SNMS a slow depth profiling technique. A significant gain in erosion rate may be obtained by pulsing the raster-scanned ion beam at a higher rate than the laser beam. A depth profiling experiment should be performed so that the laser is pulsed only while the ion beam resides within a small central fraction of the sputtered area [4.334, 346]. For removal of edge effects this approach is equivalent to the electronic gating scheme in SIMS or ISS. Even with the improved technique, however, depth profiling based on MPI will be at least one order of magnitude slower than SIMS.

Much effort is currently being devoted to further exploring and improving the MPI-SNMS technique. At this stage it is hard to predict the ultimate performance in the different fields of sputter-based analysis (surface and bulk analysis, depth profiling). The future of this technique depends largely on the ability of manufacturers of scientific equipment to offer MPI-SNMS systems that allow routine analysis with the same degree of sophistication as current SIMS or AES machines.

4.4.3 Optical Analysis of Sputtered Particles

As an alternative to SIMS and GDMS, many groups have studied whether the optical radiation emitted from sputtered particles [4.351] can be used for analysis. Sputter-based optical analysis does have several interesting aspects that justify a thorough exploration of the technique. First, measurements using optical spectroscopy are relatively easy to perform, specifically in comparison with mass spectrometry. Second, the number of emission lines is relatively small [4.73], so that identification of a certain element is generally straightforward. Third, a large fraction of the energy distribution of sputtered particles can be used with tolerable Doppler broadening of the emission line. Finally, analysis of insulators is less difficult than in SIMS, because the kinetic energy of excited sputtered neutrals is not affected by a charging of the sample.

a) Bombardment Induced Light Emission

Figure 4.20 shows the experimental arrangement used for analyzing optical radiation emitted in front of an ion-bombarded sample [4.73, 352, 353]. The vacuum chamber housing the sample has a quartz window through which photons can escape without significant absorption. A quartz lens serves to focus the emitted light into an appropriate monochromator. The spectral distribution of the radiation is detected with a cooled photomultiplier. The measured intensity is recorded using single-photon counting. The large spacing between

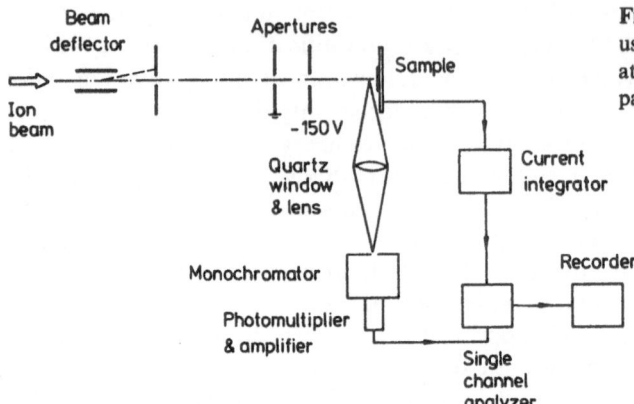

Fig. 4.20. Experimental setup used for analyzing optical radiation emitted from sputtered particles [4.352]

the sample and the quartz lens makes it easy to implement other surface analytical techniques like SIMS or AES in the same chamber [4.354].

About ten years ago different groups used bombardment induced light emission (BLE) with some success for sputter depth profiling [4.78, 352, 353]. Since large-area static beams were used [4.355], the dynamic range, observed at the trailing edge of a sandwich layer, did not exceed two orders of magnitude. In other basic studies BLE was found to suffer from the same kind of chemical effects as SIMS. The analogy was particularly evident in studies on the oxygen-induced yield enhancement (Chap. 3) [4.8, 351, 353, 354]. Rather disappointing was the finding that, in contrast to original assumptions, the yield of photons was generally lower than the yield of ions [4.80, 81]. Largely because of this finding hardly any research has been done in BLE since about 1980. I think that the analytical potential of BLE should be further explored, one argument being that the experimental equipment used in the early studies was fairly primitive. Significant improvements in data quality can be envisaged. In fact only one apparatus has been described in the literature in which ion bombardment was performed with a raster-scanned beam so that BLE imaging could be demonstrated successfully [4.356, 357].

b) Glow Discharge Optical Spectroscopy

As discussed in Sect. 4.4.2e, a glow discharge is very useful for producing sputter ablation of a solid sample, as well as for ionizing the ejected particles. The discharge also effectively excites neutrals and ions (Chap. 7). In glow discharge optical spectroscopy (GDOS) the light emitted from the excited species is used for quantitative analysis.

Two different discharge geometries have been explored for GDOS, either a "conventional" geometry with a planar anode and a planar cathode (sample) facing each other [4.75, 358, 359] as in Fig. 4.17c, or the Grimm discharge lamp [4.74, 82–84, 360] shown in Fig. 4.21. In the planar geometry the emitted light is

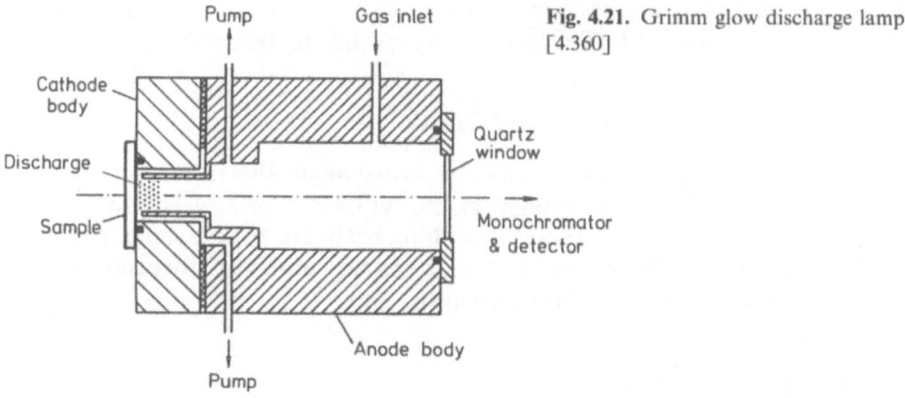

Fig. 4.21. Grimm glow discharge lamp [4.360]

viewed parallel to the sample surface. Because of the large anode-sample spacing (~4 cm) a stable discharge can be sustained at a pressure of about 0.1 mbar. [4.358]. The maximum emission intensity is observed in the cathode glow region, a few millimeters away from the surface [4.75].

In the design based on *Grimm*'s work [4.74], the discharge volume is rather small (diameter of the anode between 4 mm [4.83] and ~10 mm [4.74], anode-sample spacing 0.2 mm). Accordingly, pressures on the order of 10 mbar are required for a stable discharge, which is commonly operated in the abnormal glow region (Chap. 7). The geometry of the device (Fig. 4.21) causes the discharge to be established between the inner wall of the anode and the part of the sample facing the cylindrical anode cavity. Rather uniform erosion of the sample can be obtained using this approach [4.83]. The emitted light is viewed along the discharge axis.

For both types of geometries quantitative elemental analysis was found to be possible with no detectable evidence for a matrix effect [4.83, 358]. The emission intensity for metals and oxides is practically the same (e.g., Fe light emission from Fe and FeO [4.83]). The planar geometry has been used with some success for sputter depth profiling [4.75, 358, 359, 361], the dynamic range being limited to about two orders of magnitude. As in GDMS the need for large samples and the lack of lateral resolution constitute severe limitations. Moreover, upon discharge initiation an appreciable time (up to minutes) elapses before stationary sputtering conditions are attained [4.358, 359]. The transient behavior aggravates a calibration of the origin of the depth scale, the effect being more important for shallow structures. Taken together, these limitations are apparently responsible for the fact that during the last decade hardly any work has been reported using the planar GDOS geometry.

The Grimm discharge, on the other hand, is still being used successfully to analyze various types of samples. One advantage of the technique is the remarkable erosion rate of about 0.1 μm/s, which can be obtained under standard operating conditions [4.83, 84]. Accordingly, depth profiling of a sample's major constituents over a depth of several 10 μm can be achieved within a few minutes.

Since sample size restrictions are not too severe (diameter > 4 mm), we can envisage a wide field of future applications of this analytical technique.

4.4.4 Analysis of the Sputter-Etched Sample

Here we discuss sputter-based analysis instruments that measure the near-surface composition of the sample at appropriate stages of sputter erosion. Usually, the sputtering beam and the probing beam are not the same. Therefore, the experimental conditions for sputtering can be optimized independent of the conditions used for compositional analysis.

a) Electron Spectroscopy

The main components of an instrument for sputter depth profiling in combination with electron spectroscopy are an ion gun, an excitation source, an electron energy analyzer, and a versatile sample stage (Fig. 4.22). Preferably, the ion gun should provide a mass-analyzed beam of inert-gas ions with energies in the range 0.5–3 keV. To achieve a time-averaged uniform current density over a sufficiently large sample area, raster scanning of the beam is desirable. Depending on whether photoelectrons or Auger electrons are used for analysis, excitation is performed with a monochromatic X-ray beam [4.362, 363] or a focused 1–10 keV electron beam [4.40–44]. Various energy analyzers have been employed in electron spectroscopy [4.362]. In commercial instruments the axial cylindrical mirror analyzer (CMA) and the concentric hemispherical analyzer (CHA) are preferred. Double-pass CMA systems are used for improved definition of the excitation volume [4.364].

Since the sample has to be mounted close to the entrance aperture of the CMA, this analyzer is commonly equipped with an internal electron gun mounted in an on-axis position [4.44, 364]. A system with a hybrid electron-ion gun has been described recently (Fig. 4.23) [4.365]. The gun can be switched within 0.1 s from ion to electron beam operation and vice versa. Despite the narrow space available within the CMA, beam focusing and rastering could also be implemented.

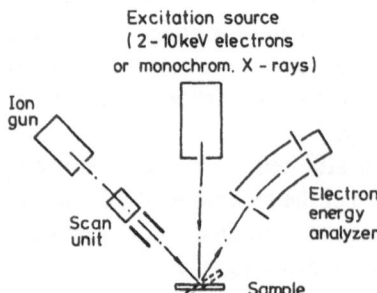

Fig. 4.22. Basic configuration of an instrument for sputter depth profiling in combination with electron spectroscopy

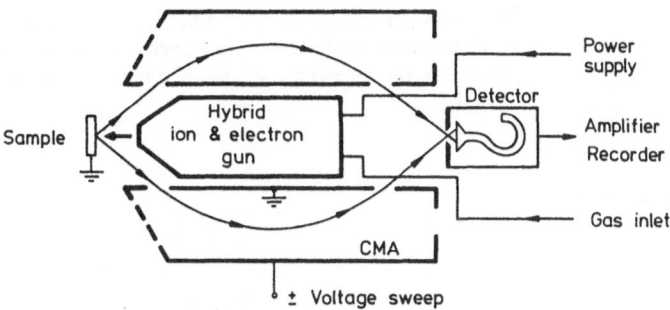

Fig. 4.23. Spectrometer based on a cylindrical mirror analyzer (CMA) with an axially integrated hybrid electron-ion gun [4.365]

b) Low-Energy Ion Scattering Spectrometry

The basic arrangement for depth profiling in combination with low-energy ion scattering spectrometry (ISS) is essentially the same as in Fig. 4.22, except that the excitation source is an ion gun and the potentials on the energy analyzer are reversed (since positively charged ions are commonly analyzed). For simplicity the probing and the sputtering beam are often the same [4.86–89, 171, 266, 267, 366] or are mixed together [4.267]. In most systems electrostatic sector fields have been used for energy analysis. Occasionally, cylindrical mirror analyzers have served the purpose [4.203, 365]; the setup in Fig. 4.23 is an example. Note, however, that ISS experiments based on use of a CMA can be performed at one scattering angle only (138°) [4.365].

One characteristic of ISS is the poor mass resolution (Sect. 4.2.6). This limitation can be overcome with a spectrometer that provides both energy and mass analysis, i.e., a setup similar to or identical with common SIMS instruments. Ion scattering spectrometry experiments involving mass analysis of the scattered particles were first described by *Grundner* et al. [4.367] and, in more detail, by *Bernheim* and *Slodzian* [4.368]. Combined SIMS and mass-resolved ISS depth profiling studies have also been reported [4.369]. Very recently I used the quadrupole-based DIDA ion microprobe [4.13] for mass-resolved ISS studies at scattering energies below ~ 100 eV, i.e., at mass ratios $M_2/M_1 < 1.5$ (impact energy ≤ 3 keV, scattering angle $\sim 150°$) [4.370]. Since the effect of oxygen [4.371] (or chlorine [4.366]) on the ISS yield of metals is considered negligible, combined SIMS and mass-resolved ISS studies allow an in situ calibration of the SIMS signal at the oxide-metal interface.

c) High-Energy Backscattering Spectrometry

The analytical potential of high-energy (Rutherford) backscattering spectrometry in combination with sputtering ("sputter/RBS") has been realized only very recently [4.106]. In principle, a sputter/RBS instrument can be put together

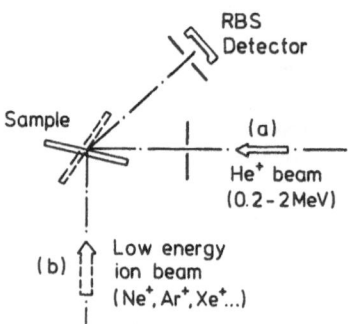

Fig. 4.24. Procedure used in RBS analysis combined with low energy sputtering: (**a**) glancing-incidence RBS geometry, (**b**) large-angle sputtering geometry [4.106]

simply by adding a low-energy ion gun of adequate performance to an existing RBS system. The analysis is performed by recording RBS spectra at appropriately chosen intervals of sputter erosion. Figure 4.24 illustrates an arrangement in which the sample can be tilted about an axis normal to the two final beam axes. This geometry [4.106] has the advantage that the angle of incidence of either beam can be chosen deliberately. To avoid edge effects, careful alignment of the two beams with respect to each other is mandatory (size of the probing beam smaller than the sputtered area).

If we want high depth resolution in the RBS analysis, glancing angles of beam incidence or of signal takeoff (or both) are required. Using the first approach we have to take into account that the size of the beam normal to the tilt axis increases as $1/\cos \theta_{in}$. To circumvent the need for nonquadratic, large-area sputtering, the probing beam should be defined by a slit with an aspect ratio $1/\cos \theta_{in}$.

4.5 Miscellaneous Examples of Analytical Applications

4.5.1 Results Obtained by SIMS, Laser-based SNMS, and ISS

In this section the analytical capabilities of some methods of sputter-based analysis are illustrated. Space permits discussion of only very few representative examples.

Figure 4.25 shows the results of a SIMS analysis of deuterium diffusion in hydrogenated amorphous silicon [4.372]. This study demonstrates the ability of SIMS to analyze hydrogen as well as to discriminate between the two stable isotopes, ^1H and ^2D. The sample consisted of three layers of amorphous silicon (a-Si:H) deposited on a stainless steel substrate. The center layer was made using SiD_4 with a purity of 98%, whereas the other two layers were made from SiH_4. One portion of the sample was annealed for 20 min at 330 °C. A comparison of the depth profiles for the as-prepared and the annealed portions shows significant diffusion of the two isotopes. From such data the diffusion constants

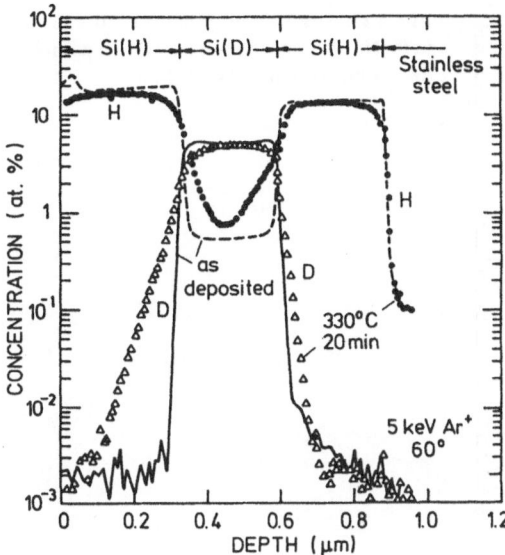

Fig. 4.25

Fig. 4.25. SIMS depth profiles of H and D in a three-layer sample of hydrogenated silicon. Solid and dashed curves: as-prepared sample; full circles and open triangles: after annealing [4.372]

Fig. 4.26. SIMS depth profiles of a Ni tracer layer embedded in a crystalline Cu matrix, measured before and after thermal treatment: (a) raw data, (b) data plotted in a form suitable for diffusion constant evaluation [4.374]. Here $\hat{z}$ denotes the peak position of the Ni profile

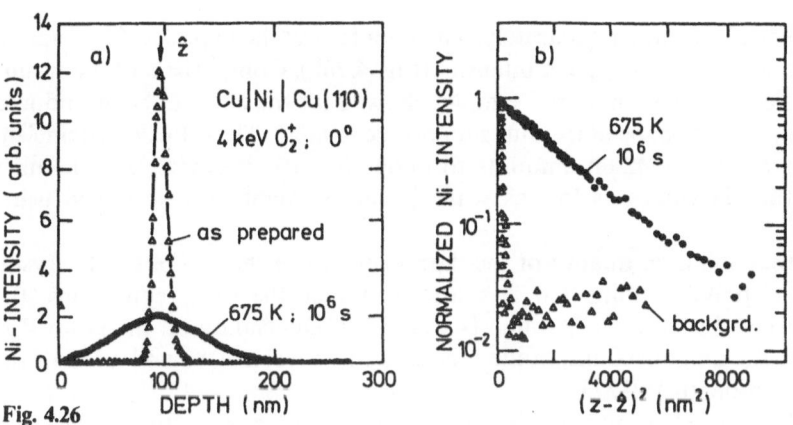

Fig. 4.26

for H and D in a-Si:H as a function of hydrogen concentration as well as activation energies could be obtained [4.372]. Implantation profiles of H and D in Si have also been reported [4.373].

Figure 4.26 shows another example of the use of SIMS to determine very small diffusion constants in single crystalline metals [4.374]. The sample was prepared by sputter depositing trace amounts of Ni onto Cu(110). Then the tracer was covered by an epitaxial layer of pure Cu (~ 100 nm) grown by vapor deposition. Such specimens were depth profiled by SIMS before and after thermal treatment. The results (Fig. 4.26) illustrate that the diffusional spreading of the Ni tracer atoms, occurring in the substrate within a depth of only 100 nm,

214   *Klaus Wittmaack*

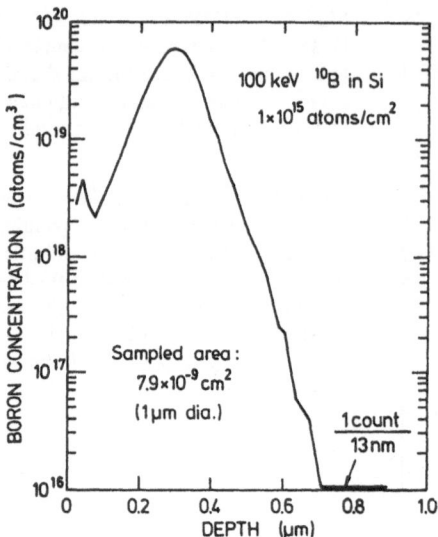

Fig. 4.27. SIMS depth profile of ^{10}B implanted in Si. Note the small sampled area (experimental details not specified) [4.375]

can be determined rather accurately. The data follow the expected Gaussian fit down to about 3% of the peak intensity (Fig. 4.26b). Comparison of the results for the diffused and as-prepared samples shows that the effect of beam-induced broadening on the shape of the diffusion profile is small. The diffusion coefficient D derived from Fig. 4.26b amounts to about 5×10^{-18} cm^2/s. This number suggests that D values as low as $\sim 10^{-19}$ cm^2/s might be determined using SIMS.

Note that the high quality of the data arises from the use of single-crystal material. If polycrystalline Cu is sputtered under the same conditions, the eroded surface becomes very rough [4.122, 374]. Depending on the grain size, the maximum roughness amplitude amounts to between 20% and 40% of the mean eroded depth [4.122].

The ability of SIMS to provide depth analysis from areas with a diameter of the order of 1 μm is illustrated in Fig. 4.27 [4.375]. In this measurement the fractional ion yield ηP^+ of B$^+$ amounted to about 1%; i.e., out of 100 boron atoms sputtered, one B$^+$ secondary ion was actually detected. Taking into account an estimated degree of ionization P^+ of 10%, this result implies that the overall transmission η of the instrument was about 10%. Under these conditions a volume of 1×10^{-14} cm^3 has to be sputtered to record, on the average, one count for an impurity present at a concentration level of 1×10^{16} atoms/cm^3.

Even if the fractional ion yield in SIMS analysis could be improved further, the destructive nature of the technique sets a lower limit to the detection sensitivity in microanalysis (because each atom may at best give rise to a single pulse at the detector). According to [4.8, 10, 18] the signal $I_i \Delta t$ accumulated while sputtering a layer of thickness Δz is

$$I_i \Delta t = \eta_i P_i A_0 \Delta z N c_i . \tag{4.58}$$

For $I_i \Delta t = 1$ count the required impurity concentration can be derived from the following relation:

$$N c_i > (\eta_i P_i A_0 \Delta z)^{-1} . \tag{4.59}$$

On samples containing features with small A_0, as in very large scale integration (VLSI) technology, an acceptable detection limit $(N c_i)_{min}$ will often be achievable only by integration over several features of the same type. Background problems, however, can raise $(N c_i)_{min}$ significantly above the level set by (4.59) [4.376].

As a final example of SIMS analysis, Fig. 4.28 shows high-resolution raster scanning ion images obtained with a 40 keV Ga$^+$ primary ion beam [4.377, 378]: an Li$^+$ image of an Al-Li alloy microsphere and an Al$^+$ image recorded to determine the size of the probing beam. Analysis of the latter image shows that the falloff at the edge of the Al$^+$ feature occurs within 36 nm (90 % falloff). A beam diameter of about 70 nm is deduced from this experiment. The ultimate resolution is determined by chromatic aberration [4.378]. To obtain probe sizes below 100 nm the beam-defining aperture must be reduced to about 25 μm [4.378]. The beam current in a 100 nm spot is then limited to about 100 pA [4.379].

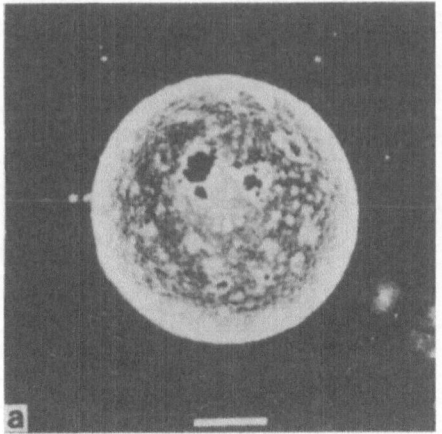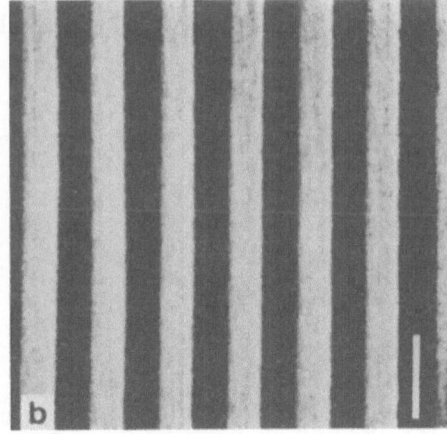

Fig. 4.28. SIMS raster scanning imaging obtained with a 50 pA, 40 keV Ga$^+$ primary ion beam. (a) Li$^+$ image of an Al-Li alloy microsphere deposited on a graphite substrate (4×10^6 counts recorded in 32 s; scale bar 5 μm). (b) Al$^+$ image of an Al test pattern on Si (0.4 μm wide bars; 1.5×10^6 counts recorded in 128 s; scale bar 1 μm) [4.377]. (Reproduced with permission of the author and Scanning Microscopy International)

Clearly, instruments with a lateral resolution of 100 nm or less can be used in several fields. However, a quantitative interpretation of the recorded images is very difficult because of chemical, crystallographic, and topographic effects [4.380–383].

Figure 4.29 shows the analytical potential of resonant multiphoton ionization for sputter depth profiling [4.102]. The experiment relates to the problem of ^{56}Fe analysis in Si. Due to isobaric interference with ^{28}Si$_2^+$ ($M = 55.95386$ u) only a rather poor detection limit for ^{56}Fe$^+$ ($M = 55.93494$ u) can be obtained using SIMS at high mass resolution. The selectivity achieved with resonant MPI allows ^{56}Fe to be measured at concentration levels below 10^{15} atoms/cm^3. From measurements performed in addition to those of Fig. 4.29, *Pellin* et al. [4.102] deduce a detection limit of 1×10^{14} atoms/cm^3, i.e., 2 ppb.

Note that the near-surface part of the ^{56}Fe profile can be determined in great detail, (inset of Fig. 4.29). Such quantitative measurements are practically impossible with SIMS. To evaluate the Fe concentration through the native surface oxide by resonant MPI, one has to make sure that the fraction of ground-state atoms is not affected by the presence of oxygen. A correct depth calibration is also difficult in this region.

Figure 4.30 depicts results obtained by nonresonant MPI. The section of the mass spectrum shown in Fig. 4.30a demonstrates that the sensitivity for Ga and As, observed during steady-state sputtering of GaAs, is essentially the same [4.104]. This finding deviates favorably from the common observation in SIMS: as a consequence of differences in ionization potential (IP), the sensitivity for Ga (IP $= 6.0$ eV) is more than three orders of magnitude higher than for As (IP $= 9.8$ eV) [4.95, 384, 385].

Figure 4.30b shows a comparison of depth profiles for a sample of GaAs covered with a 100 nm layer of Au [4.386]. The nonresonant MPI measurements were performed on an unannealed and an annealed sample. Diffusion of

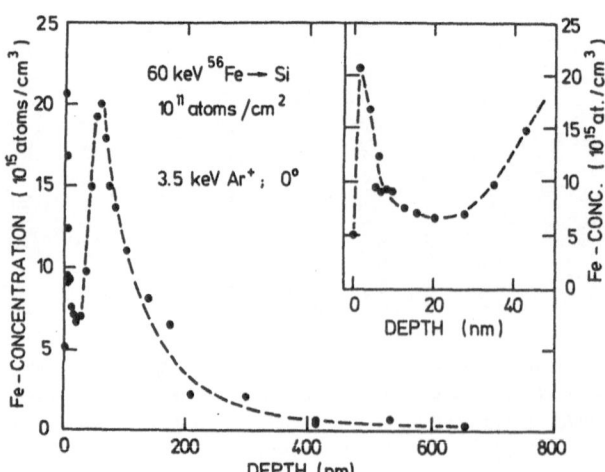

Fig. 4.29. Depth profile of ^{56}Fe implanted in Si, measured by sputtering in combination with resonant multiphoton ionization. The inset shows the near-surface part of the profile on an expanded depth scale [4.102]

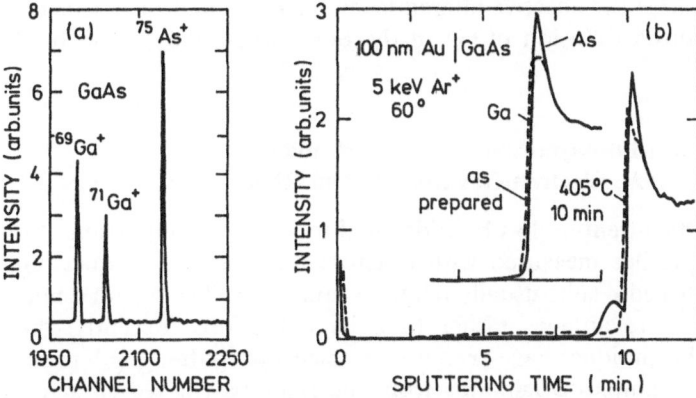

Fig. 4.30. Detection of sputtered Ga and As atoms by nonresonant multiphoton ionization: (a) section of the mass spectrum of GaAs observed under stationary bombardment conditions [4.104]; (b) depth profiles of Ga and As measured for a UHV-cleaved GaAs substrate covered with an in situ deposited Au overlayer. The two sets of profiles relate to depth profiling of the sample before and after annealing [4.386]

As as well as of Ga into the Au overlayer is evident in the annealed sample. Somewhat surprisingly, the Ga and As signals exhibit a pronounced peak in passing from the overlayer to the substrate. This effect is not understood; local charging of the sample has been suggested [4.386].

Figure 4.31 addresses a completely different issue. *Hoff* and *Lam* [4.387] have used ISS with sputtering to investigate the surface and subsurface compositions of a Ni-10 at % Ge alloy at temperatures between 25° and 700 °C. When the alloy was heated, Ge atoms were found to segregate to the surface (Gibbsian adsorption). The steady-state subsurface compositional profiles, established after prolonged bombardment at a fixed temperature, were measured after rapid cooling to room temperature. From the profiles and the observed changes with

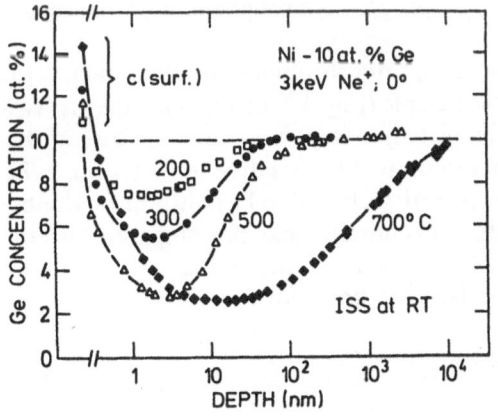

Fig. 4.31. Steady-state Ge profiles measured by ISS at room temperature after prolonged high-temperature bombardment of an Ni-Ge alloy. The numbers denote the temperatures during pre-analysis bombardment [4.387]

temperature (Fig. 4.31) the effective altered-layer thickness and the coefficients for radiation-enhanced diffusion of Ge in the bombarded alloy could be deduced.

4.5.2 Comparison of Multilayer Sputter Depth Profiles Measured by AES, Electron-Gas SNMS, and RBS

In the past very little attention has been devoted to a direct comparison of the quality of depth profiles measured with different analytical techniques. One reason is that published results usually relate to samples well suited for demonstrating a method's advantages rather than its limitations. Recently, three different analytical techniques have been used to measure sputter depth profiles of Ta-Si multilayer samples originating from the same batch. We discuss the results in some detail here because they allow an assessment of the relative merits of the different techniques.

The samples were prepared by alternating sputter deposition of Ta and Si layers on an Si substrate. The thicknesses of the Ta and Si layers were chosen so that a silicide with the composition $TaSi_2$ could be produced by annealing [4.388]. The sample considered here comprised ten double layers of Ta and Si, the nominal thickness of one double layer being 20 nm.

Figure 4.32a shows depth profiles of Si measured by AES in combination with glancing-incidence, low-energy Ar^+ sputtering [4.389]. To reveal the most essential features of the profiles, the data for the central five layers have been omitted. The following paragraphs discuss some noteworthy aspects.

The individual Si and Ta layers are well resolved. (Since the oxygen contamination observed in AES sputter profiling did not exceed 1 at % [4.389], the Ta profiles are simply the complement of the Si profiles). In the top layer, however, the apparent Si concentration is as high as in the substrate. Moreover, the peaks are seen to be slightly asymmetric. These features are probably due to bombardment-induced relocation of target atoms, element-differential (preferential) sputtering, and, to some extent, to the nonnegligible escape depth of the Auger electrons ($L_0 < 1$ nm).

The depth resolution is essentially independent of the sputtered depth. This finding suggests that there is hardly any progressive roughening of the sample during sputter erosion.

The characteristic time Δt required to sputter through one individual layer is almost constant throughout the whole sample (Fig. 4.32b). Only for the last two Si layers (i.e., the first two Si layers deposited) there is a sizeable increase in Δt_{Si} compared with the other layers (increase up to 17% relative to layers no. 2–8).

Within experimental accuracy the Δt ratio is found to be constant for double layers no. 2–9, $\Delta t_{Si}/\Delta t_{Ta} = 1.70 \pm 0.05$. This number may be compared with the ratio expected on the basis of [4.49]. Neglecting transients in the total sputtering yield, which might occur at each interface, we have, for $J = $ const,

$$\frac{\Delta t_{Si}}{\Delta t_{Ta}} = \frac{N_{Si}}{N_{Ta}} \frac{\Delta z_{Si}}{\Delta z_{Ta}} \frac{Y_{Ta}}{Y_{Si}} .$$

$$(4.60)$$

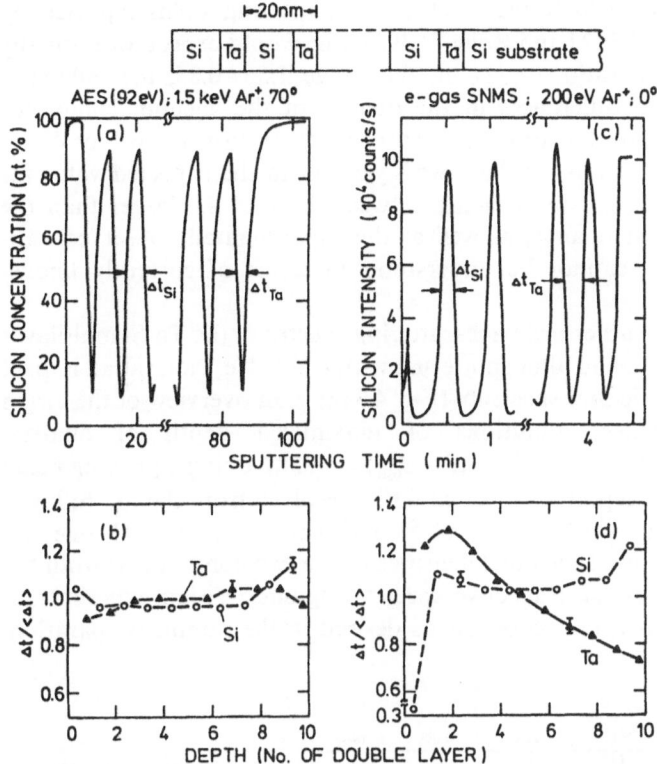

Fig. 4.32. Sections of the sputter depth profiles of a Ta-Si multilayer measured by (a) AES [4.389] and (c) electron-gas SNMS [4.391]. The sample structure containing a total of ten Ta-Si double layers is specified at the top. (b, d) Normalized time required to sputter through one individual layer in the depth profiling measurements (a, c)

Inserting $N_{Si}\Delta z_{Si}/N_{Ta}\Delta z_{Ta} = 2.0 \pm 0.1$, the expected ratio is $\Delta t_{Si}/\Delta t_{Ta} = 2 Y_{Ta}/Y_{Si}$. Available sputtering yield data [4.390] for normally incident 1 keV Ar suggest that Y_{Ta}/Y_{Si} is close to unity. Even though this yield ratio may vary somewhat with increasing impact angle [4.390], the agreement between the observed and the expected Δt ratio is satisfactory.

The profiles obtained by electron-gas SNMS (Fig. 4.32c) [4.391], deviate significantly from the AES data. Particularly surprising is the fact that only a small peak is observed in passing through the top Si layer. Moreover, the time required for sputter removal of the top Si layer is very short compared with the mean value $\langle \Delta t \rangle_{Si}$ (Fig. 4.32d). These two features suggest that electron-gas SNMS suffers from rather pronounced initial transients in the ionization efficiency, as well as in the erosion rate. Such effects are known from GDOS measurements [4.358, 359]. The presence of transients in SNMS apparently was ignored in the previous work by *Oechsner* [4.327, 391].

Another rather irritating aspect of the data in Fig. 4.32c, d is that the ratio $\Delta t_{Si}/\Delta t_{Ta}$ increases continuously from 0.42 for double layer no. 2 to 0.68 for double layer no. 9, i.e., a 62% increase! Using sputtering yields reported by *Laegreid* and *Wehner* [4.392] and *Zalm* [4.393], one would expect for normally incident 200 eV Ar$^+$ a ratio Y_{Ta}/Y_{Si} in the range 1.56−0.85, the difference probably being due to differences in the quality of the vacuum conditions [4.394], which determine the oxygen surface concentration in the ion-bombarded sample [4.395]. Even for the lowest yield ratio the expected value for $\Delta t_{Si}/\Delta t_{Ta}$ would be 1.7, i.e., by a factor between 2.5 and 4.0 larger than the SNMS results. This discrepancy, as well as the experimentally observed variation of the Δt ratio, cannot be understood in terms of currently known sputtering artifacts.

Figure 4.33 shows Rutherford backscattering spectra of the Ta-Si multilayer sample, measured before and after sputtering with 4 keV Ne$^+$ ions at an impact angle of 60° [4.106]. The first spectrum (Fig. 4.33a) is an overview of the virgin sample, recorded under conditions of maximized depth of analysis ($\theta_{in} = 28°$, $\theta_{out} = -17°$, $\theta = 180° - (\theta_{in} - \theta_{out}) = 135°$). Using a probing beam of 700 keV He^{2+}, the deepest Ta layer could be resolved from the top Si layer. Knowing the stopping cross sections of Si and Ta for He [4.214], one can determine the mean ratio of the areal densities of the two components from the average signal height in the respective spectral regions. Within experimental accuracy the ratio turned out to be 2.0, as desired in the sample preparation.

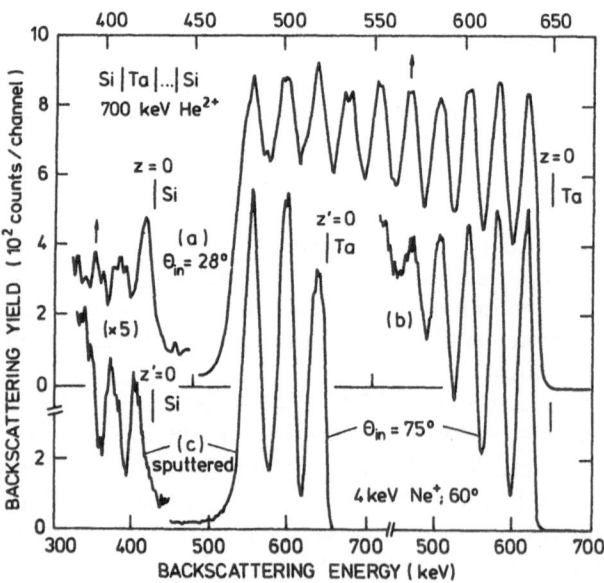

Fig. 4.33. Rutherford backscattering spectra of the same Ta-Si multilayer sample as in Fig. 4.32: (**a, b**) virgin sample analyzed at two different angles of incidence of the probing He beam. The vertical bars denote the He energy for scattering off Si and Ta atoms, respectively, at the surface, i.e., $z = 0$. (**c**) Spectrum taken after sputter removal of more than seven Ta-Si double layers. The position of the sputter-etched surface is denoted by $z' = 0$ [4.106]

Moreover, the total thickness of the ten double layers was (185 ± 10) nm, very close to the desired value of 200 nm.

In the first spectrum (Fig. 4.33a) the depth resolution was largely determined by the energy resolution of the detector ($\delta E = 7$ keV). In the region close to the surface a significantly improved depth resolution could be obtained by performing the RBS analysis at a glancing angle of beam incidence, $\theta_{in} = 75°(\theta_{out} = 30°)$ (Fig. 4.33b). However, a 2.3-fold increase in energy loss of the probing beam per unit sample thickness made the effect of energy straggling much more important at glancing beam incidence than at $\theta_{in} = 28°$. Therefore, the improvement in depth resolution was limited to the top three Ta layers. To explore the complete sample structure with good depth resolution, stepwise removal of material was necessary.

The bottom spectrum (Fig. 4.33c) shows RBS data observed after sputter erosion to the point where, counting from the top, removal of the eighth Ta layer had just begun. The observed depth resolution is the same as for the virgin sample, except for the surface region where intermixing of Ta and Si is evident from the tail on the Si feature. The last two double Ta-Si layers are 10%–15% thicker than the eight other layers. (Note that for a Si-to-Ta ratio of 2 and for He energies between 500 and 800 keV the mean stopping cross section is constant to within a few percent [4.214]. Therefore, the depth scale is linear in the Ta region of Fig. 4.33; $\varepsilon_{Ta}/\varepsilon_{Si} \simeq 1.7$.)

A very important aspect of the sputter/RBS technique is that we can explore the sample composition with good to excellent depth resolution, while circumventing the problems caused by sputter-induced surface compositional changes. Whereas these changes may cause severe profile distortions whenever we use sputtering with a surface-sensitive analytical technique, RBS measurements provide means to look beyond the altered layer.

Summarizing this comparison we can say that there is very good agreement between the results obtained by AES and RBS in combination with sputtering. The significant transients seen in the electron-gas SNMS profiles cannot be explained. However, compared with AES and RBS, electron-gas SNMS allows very fast sputter depth profiling (see abscissa of Figs. 4.32a, c). With this advantage, experiments aiming at an understanding and control of the observed SNMS artifacts would be highly desirable.

4.6 Beam-Induced Profile Broadening

4.6.1 Effect of Bombardment Parameters

To determine a sample's composition as a function of depth with the least distortion from the sputter-etching technique, we must know the influence of the bombardment parameters on depth profile quality. In the development of sputter-based analytical techniques, many investigations were performed to clarify this issue. About six years ago I published a comprehensive review of beam-induced broadening effects in sputter depth profiling [4.113], which was updated two years later [4.178]. Therefore, only the essential aspects of the topic are summarized here.

a) Primary Ion Energy

The effect of the primary ion energy on the depth resolution is well documented in the literature. With apparently only one exception [4.396], all published experimental data show that, for a given primary ion species and a fixed impact angle, profile broadening increases with increasing beam energy.

The influence of the beam energy (or any other bombardment parameter) on the depth resolution can be evaluated quantitatively only if other disturbing effects are reduced. Most important is minimizing the surface roughness of the virgin and the bombarded samples. With polycrystalline metals this implies the use of fine-grained films with a thickness of less than about 30 nm [4.192].

Figure 4.34 shows AES sputter profiles of a thin Ni overlayer deposited on a high-purity Fe foil [4.192]. The two sets of profiles denoted by full and open symbols reflect AES measurements at either "low" energies ($E < 100\,\text{eV}$) or "high" energies ($E > 500\,\text{eV}$). The effective escape depths L were 0.4 and $0.8 - 1.0$ nm, respectively. In quantitative agreement with Fig. 4.7, the profiles recorded with the large AES probing depth are shifted toward the surface

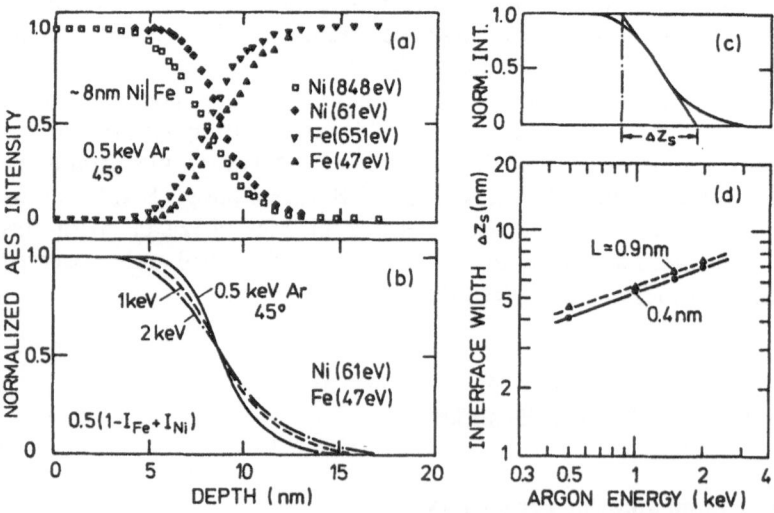

Fig. 4.34. (a) AES sputter depth profiles of Ni and Fe measured during low-energy sputtering of a Ni-covered Fe foil. The full and open symbols show data recorded using AES lines at "low" and "high" energies, respectively. (b) Averaged profiles as a function of the Ar energy. (c) Definition of the interface width Δz_s. (d) Interface width as a function of the Ar energy. Parameter is the "effective" escape depth of the Auger electrons (4.33) [4.192]

relative to the profiles pertaining to a small probing depth. In either case, the profiles have an asymmetric form.

The effect of the ion energy E_0 on the shape and width of the profiles is illustrated in Fig. 4.34b, d. The data in Fig. 4.34d relate to a definition of the interface width Δz which is based on an evaluation of the slope dI/dz at the 50% level of the normalized intensity I,

$$\Delta z_s = \left(\frac{dI}{dz}\right)^{-1}\Bigg|_{I=0.5}, \tag{4.61}$$

as illustrated in Fig. 4.34c. This width can be shown to be roughly equal to the width $\Delta z_{0.1}^{0.9}$, i.e., the depth interval sputtered in passing from the normalized intensity $I' = 0.9$ to $I'' = 0.1$ (or vice versa) [4.118, 192]. Although evaluations of the interface width based on a measurement of the depth (or time) interval $|z(I') - z(I'')|$ with $I' + I'' = 1$ suggest a symmetric profile shape, (4.61) can be applied to all profile shapes without restriction.

The data in Fig. 4.34d can be described by the relation

$$\Delta z_s = k_s E_0^m, \tag{4.62}$$

where k_s is a factor specific to the layer/substrate combination being analyzed, as well as to other parameters (sample temperature, interface location, primary ion mass, and impact angle). From Fig. 4.34d we find $m \simeq 0.4$. The observed width Δz_s constitutes a convolution of the intrinsic interface width Δz^0, the beam-induced broadening and the exponential AES sensitivity function (4.30). Since Δz^0 may have been as large as a few nanometers, the "true" beam-induced broadening would increase more rapidly with E_0 than in Fig. 4.34d; i.e., $m > 0.4$.

Exponents in the range $0.4 < m < 0.5$ have been reported for (or can be deduced from the published results on) a variety of layer/substrate combinations, e.g., SiO_2/Si [4.397], Si/Al_2O_3 [4.48, 63], $In_xGa_{1-x}As/GaAs$ [4.398], Ta_2O_5/Ta and Nb_2O_5/Nb [4.399], Cr/Ni [4.127, 400–402], and $GaAs/Al_{0.3}Ga_{0.7}As$ [4.403]. Exponents in this range have also been derived from theoretical models on the basis of the assumption of purely collisional mixing [4.115, 162]. This agreement, however, cannot be considered conclusive in the sense that other mechanisms of interface broadening are not operative during sputter etching. In fact, for polycrystalline samples we have to argue in terms of an energy dependence of surface roughening, rather than in terms of mixing.

A problem encountered in a quantitative comparison between theoretical models and experiments on overlayer or multilayer samples is that we are dealing with rather high concentrations of "impurity" atoms in the respective "matrix" (substrate atoms transported into the overlayer and vice versa). There-fore, pronounced concentration-dependent effects must be expected. Moreover, a calibration of the true depth scale in such an experiment is rather difficult. Only very few authors have at least tried to determine the relative erosion rates for the overlayer and substrate materials [4.127, 192, 401]. Usually they assume that the erosion rate does not change in passing from the overlayer to the

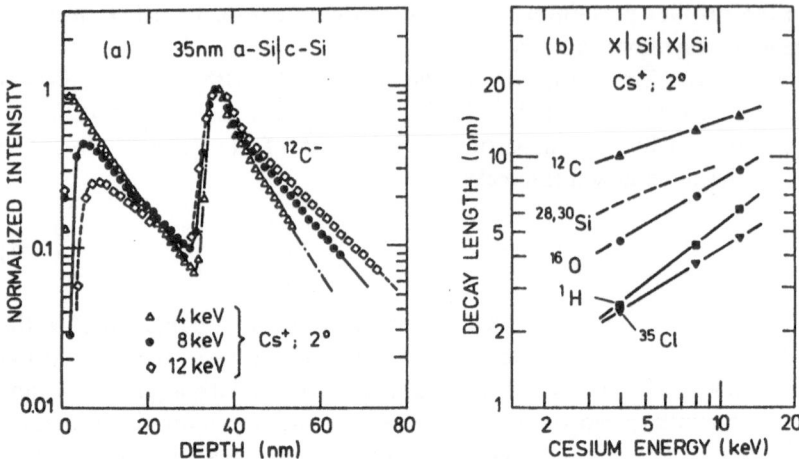

Fig. 4.35. (a) SIMS depth profiles of adsorbed carbon present on either side of an overlayer of amorphous Si (a-Si) deposited on a crystalline Si backing (c-Si). Parameter is the energy of the Cs^+ primary ion beam. The interfacial C^- peaks were assumed to be located at exactly the same depth [4.177, 407]. (b) Decay length for various impurities in Si as a function of the primary ion impact energy [4.170, 177]. Results derived from depth profiling studies involving isotopically pure multilayers of ^{28}Si and ^{30}Si are shown for comparison [4.408]

substrate, so that interface width can be deduced from the respective sputtering time interval [4.113, 268].

With these difficulties in mind we should look for alternative prototype samples. Thin buried layers may be considered very useful test structures [4.30, 177, 404–407]. Figure 4.35a shows depth profiles of C measured on a sample composed of a crystalline Si backing and an amorphous, vapor-deposited Si overlayer [4.177, 407]. The observed C signal is attributed to carbon containing impurities adsorbed at the surface before and after Si deposition. Several features are evident from Fig. 4.35a: (i) the profiles due to interfacial carbon are quite asymmetric, exhibiting a rather steep rise at the leading edge and a very slow exponential decay at the trailing edge; (ii) the profile steepness on either side of the interface decreases with increasing energy; (iii) the decay length characterizing the falloff of the C signal near the surface is the same as for the interfacial peak.

According to Fig. 4.35a the depth resolution achievable for thin impurity layers is largely determined by the steepness of the exponential falloff. Often the depth resolution may thus be assessed simply by measuring the decay length [4.169]. Experimental data for several surface impurities on (or in) Si [4.170, 177, 407] are plotted in Fig. 4.35b as a function of the Cs^+ primary ion energy. Also shown are results for the "ideal" case of a Si tracer embedded in a Si matrix. The latter experiment was performed using multilayer films of isotopically purified ^{28}Si and ^{30}Si grown by low-energy ion beam deposition [4.408]. Similar to the interface width in Fig. 4.34, the decay length λ is seen to increase with increasing impact energy in general accordance with (4.62); i.e., $\lambda \propto E_0^m$. However, not only the absolute magnitude of λ but also the power m exhibits a pronounced element-specific character (m ranges from 0.33 for C to 0.80 for H). Qualitatively speaking, the results suggest that C impurities are more difficult to remove from Si than the matrix atoms themselves, whereas the opposite appears to hold true, for e.g., Cl impurities. More detailed conclusions concerning the relative extent of beam-induced mixing and element-differential sputtering are hardly possible if nothing but decay-length data are available (Sect. 4.2.4c).

b) Primary Ion Mass

Relatively few experiments have been performed to investigate the effect of primary ion mass on profile broadening. The lack of data is partly due to the fact that SIMS studies are mostly devoted to low-concentration dopant distributions. With these the use of primary ions like O_2^+ or Cs^+ is mandatory because of the required high secondary ion yield. Although inert-gas primary ions are thus rarely employed in SIMS, some groups have used these species in AES work to study the dependence of the apparent interface width on the primary ion mass.

Unfortunately, a consistent picture of the mass effect is not evident from previous investigations. In some cases a reduction of the interface width was observed with increasing mass [4.127, 200, 397], but the opposite trend has also been found [4.403]. Figure 4.36 compiles representative data. The specified interface width $\Delta z_{2\sigma}$ relates to a frequently used interface definition [4.398, 409] based on the assumption that the profile can be described by an error function (Gaussian integral) with a standard deviation σ:

$$\Delta z_{2\sigma} = |z(I' = 0.84) - z(I'' = 0.16)| . \tag{4.63}$$

The large mass effect evident for the Ni/Cr interface width might be due to

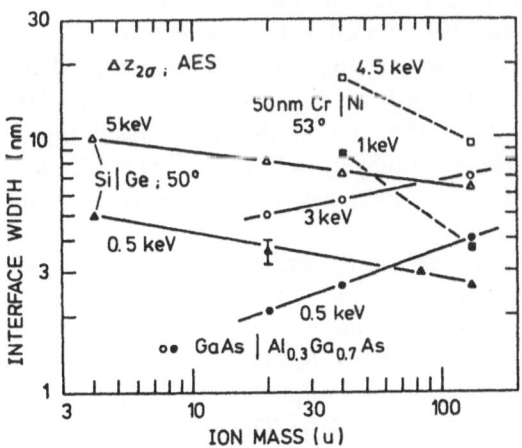

Fig. 4.36. Interface width measured by AES for a variety of overlayer-substrate combinations as a function of the mass of the inert-gas primary ions. In each case data are shown for two different bombardment energies. Cr/Ni [4.127], Si/Ge [4.200], GaAs/$Al_{0.3}Ga_{0.7}As$ [4.403]

mass-dependent differences in the amount microroughening. The decrease in the Si/Ge interface width observed with increasing primary ion mass is somewhat less pronounced than expected on the basis of models involving cascade mixing [4.113, 162, 200]. No presently available model explains the positive mass effect reported for the interface width between GaAs and $Al_{0.3}Ga_{0.7}As$ [4.403].

c) Beam-Induced Chemistry

The depth resolution observed in sputter profiling can be affected significantly by compositional changes of the sample introduced by beam-induced incorporation of chemically active elements. Effects of this kind are frequently encountered in SIMS analyses because secondary ion yields are known to be extremely sensitive to the surface chemistry (Chap. 3). To maximize sensitivity it is common practice, therefore, to use special bombardment conditions to enforce the desired chemical modifications at the sputtered surface. Analyses involving the detection of positive secondary ions are usually carried out under oxygen bombardment or during simultaneous exposure of the sample to an ion beam and a flow of gaseous oxygen. Conversely, negative secondary ion yields can be maximized using cesium primary ions or a cesium jet in combination with inert-gas ion bombardment.

The effect of beam-induced oxidation of the analyzed sample on the depth resolution is rather complicated. In the analysis of polycrystalline metals, enforced oxidation often prevents roughening of the sputtered surface [4.37, 130, 131, 410] and thus greatly improves the achievable depth resolution [4.37]. Similar beneficial effects have also been observed occasionally using N_2^+ bombardment [4.132, 133]. Very recently it has been shown that where O_2^+ or N_2^+ primary ions do not provide the desired depth resolution, somewhat more exotic primary ion species like CF_3^+ may produce sputter craters with smooth surfaces [4.411], as well as enhance secondary ion yields [4.412].

Implanted primary ions may also improve the depth resolution by absorbing a certain fraction of the energy deposited in the sample. Ideally, material transport occurring in the subsystem of implanted atoms will not contribute to the mixing of sample constituents [4.226]. This advantageous aspect is counterbalanced to some extent by the reduction of the matrix sputtering yield that results from the incorporation of oxygen (Fig. 4.37) [4.250, 395, 413–416]. Nevertheless, recent experiments have shown that the depth resolution in sputter profiling of isotopically purified multilayers of ^{28}Si and ^{30}Si is a factor of two to three better using O_2^+ rather than inert gas or Cs^+ ions (at the same energy per atom; $\theta_{in} = 2°$) [4.408].

In contrast to the beneficial effects of suppressed surface roughening and partial absorption of the deposited energy, beam-induced oxidation can also give rise to the rather adverse phenomenon of impurity segregation. For a variety of metal impurities in Si, notably of noble metals, SIMS analyses performed under conditions of complete oxidation of the sample result in depth profiles with extremely long tails [4.174–177, 411, 417–419]. This profile tailing could

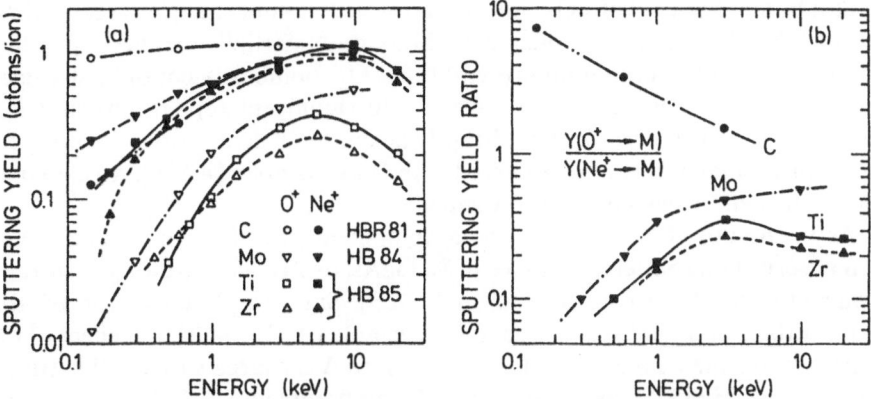

Fig. 4.37. (a) Comparison of sputtering yields of different materials bombarded with oxygen or neon ions, HBR 81 [4.413], HB 84 [4.414], HB 85 [4.415]. **(b)** Ratio of the sputtering yields measured under oxygen and neon bombardment. Yield ratios much below unity, as for the metals, indicate significant oxygen incorporation or oxide formation. By contrast, yield ratios above unity, as for carbon, can be attributed to the formation of a volatile compound, presumably CO [4.414, 416]

be shown to be due to the pileup of the respective impurity at the interface between the continuously replenished SiO_2 and the underlying Si substrate [4.173, 174, 176]. Figure 4.38 shows an RBS study of this process. As a result of segregation, the partial sputtering yield of the impurity becomes very small [4.176, 418, 419] and the decay length very large (4.21).

A quantitative understanding of the impurity transport out of and through the oxide is not available. Note, however, that impurity segregation does not

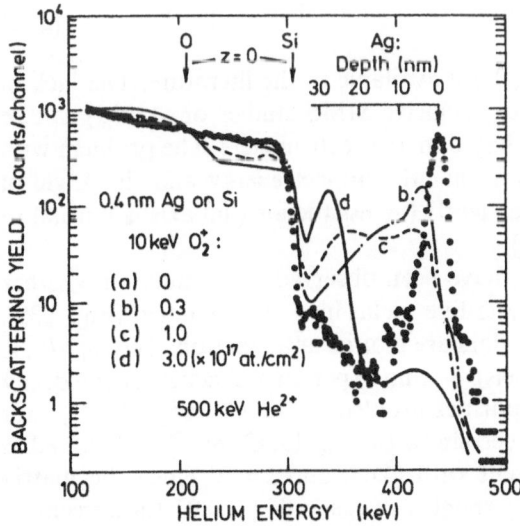

Fig. 4.38. RBS spectra of a Si sample covered with a thin layer of Ag, (*a*) as-prepared, (*b–d*) at various stages of oxygen bombardment. On completion of beam-induced oxidation, i.e., at an oxygen fluence of about 2×10^{17} atoms/cm^2, ~40% of the original Ag deposit has piled up at the SiO_2/Si interface. At this point the partial sputtering yield of Ag is more than three orders of magnitude lower than before the formation of a fully oxidized layer [4.176]

occur if bombardment with O_2^+ ions is performed at impact angles greater than 30° [4.173, 411, 418], in which case Si is only partially oxidized [4.420]. Unfortunately it has been found recently that O_2^+ bombardment of Si at angles in the range $30° < \theta_{in} < 45°$ can give rise to significant topography and ion yield changes, which occur rather abruptly at sputtered depth between about 2.5 and 4 μm. Similar effects were observed with GaAs samples [4.421]. An explanation for this effect has not been presented.

As to the use of Cs^+ beams, the development of cesium-rich features has been observed on sputtered samples of Si, GaAs, or InP after prolonged storage in vacuum [4.422] or exposure to air [4.423]. Since ripple formation on the actual surface of Si and GaAs was not detectable [4.421–423], the effect of Cs outgrowth on the depth resolution is not clear. Very large surface distortions were produced in GaSb as a result of Cs^+ bombardment [4.424]. These results suggest that the effect of beam-induced chemistry has to be discussed separately for each sample material, taking into account the sample treatment before ion bombardment [4.422].

d) Angle of Beam Incidence

Apart from the energy, the mass, and the chemical nature of the primary ions, the angle of beam incidence constitutes yet another important parameter determining the depth resolution in sputter-based analyses. If beam-induced broadening is mainly due to collisional relocation of target atoms (cascade mixing), simple arguments based on a random-walk approach [4.114, 115] suggest that the interface broadening will decrease with increasing sputtering yield as $\Delta z \propto Y^{-1/2}$ (4.5). Since sputtering yields vary with impact angle θ_{in} roughly as $\cos^{-s}\theta_{in}$, where $1 < s < 2$ for $\theta_{in} < 80°$ [4.108, 390], we would expect a θ_{in}-dependence of beam-induced broadening of the form

$$\Delta z(\theta_{in}) = \Delta z(0°)\cos^{s/2}\theta_{in} \ . \tag{4.64}$$

Some experimental data on $\Delta z(\theta_{in})$ are available in the literature. The lack of data is partly due to the fact that conclusive SIMS studies on $\Delta z(\theta_{in})$ can be performed only using quadrupole-based mass spectrometers. The problem with magnetic sector field instruments is that the impact energy and the angle of incidence cannot be chosen independently (unless the ion gun axis is normal to the sample surface, $\theta_0 = 0°$ (4.56)).

Rather controversial θ_{in} effects have been observed, depending on whether semiconductor samples or polycrystalline metal films were studied (Fig. 4.39). To ease comparison with (4.64) the data are plotted as a function of $\cos\theta_{in}$ (θ_{in} is specified at the top of the two panels). Within experimental accuracy the decay lengths for a variety of surface impurities profiled in Si and GaAs [4.180] vary linearly with $\cos\theta_{in}$ (Fig. 4.39a). In certain cases – e.g., for C and O in Si as well as O in GaAs – the data are consistent with (4.64) and $s = 2$. Since the matrix sputtering yields were found to vary roughly as $\cos^{-2}\theta_{in}$ [4.180], the agreement

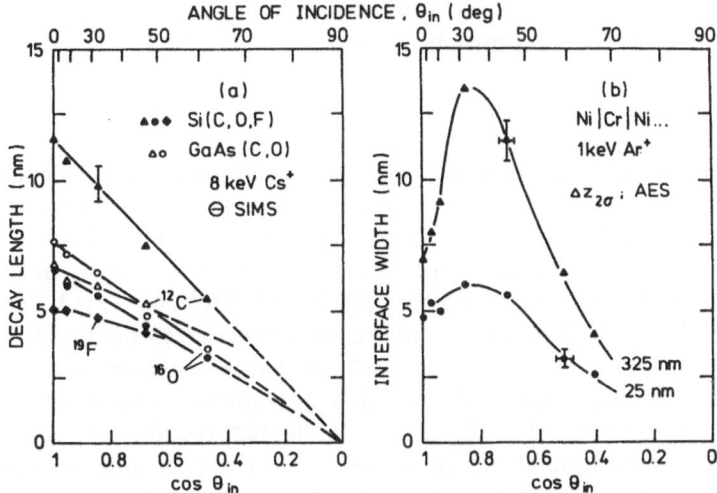

Fig. 4.39. (a) Decay length for various surface impurities in Si and GaAs versus the angle of incidence of the Cs^+ primary ion beam [4.180]. (b) Interface width for an Ni-Cr multilayer sample versus the angle of incidence of the Ar^+ primary ion beam. The numbers denote the interface location [4.401, 402]

with (4.64) is perfect. This agreement, however, does not necessarily imply that the model leading to (4.64) provides a correct description of the mixing process. Note that in general the form of the θ_{in}-dependence is different for different impurity elements in the same matrix (Fig. 4.39a).

Figure 4.39b shows a significantly different θ_{in}-dependence. Interface widths $\Delta z_{2\sigma}$ were measured at two different depths of a Cr/Ni multilayer sample (1 keV Ar^+) [4.401, 402]. Rather than decreasing monotonically with increasing impact angle, $\Delta z_{2\sigma}$ passes through a maximum around 35°. Similar behavior has been observed at larger beam energies (2 and 4 keV Ar^+, maximum around 30°) [4.400]. Transmission electron micrographs indicate that the observed interface broadening is closely related to the surface roughness established during sputter erosion [4.401]. Apparently, the variation of the surface roughness with ion energy, ion mass, angle of incidence, and possibly also the current density [4.400] constitutes a complex phenomenon that deserves further investigation. Practically speaking, we should keep in mind that, to optimize the depth resolution for polycrystalline metals, bombardment moderately off-normal ($10° < \theta_{in} < 50°$) is not be advisable. With the Cr/Ni system we should work either at normal beam incidence or at rather large angles, $\theta_{in} > 50°$.

4.6.2 Effect of Sample Parameters

Even for the same bombardment conditions, the depth resolution may depend significantly on the sample's intrinsic characteristics as well as on its temperature during analysis. Whereas little can be done about the "quality" of a given sample, the temperature during analysis can often be set to an optimum value without affecting the sample composition.

a) Sample Crystallinity

One of the most important factors determining the quality of a sputter depth profile is the crystalline structure of the sample. The situation is very favorable with amorphous materials like oxides, provided they are not rendered (poly-) crystalline as a result of ion bombardment. In fact, the best results in terms of depth resolution have been reported for SiO_2/Si [4.197, 424] and Ta_2O_5/Ta layers [4.399, 425, 426]. The high quality of the profiles is due to the uniform erosion of the oxides [4.427]. A favorable behavior has also been found for semiconductors like Si, Ge, and GaAs, which are rendered amorphous as a result of ion bombardment [4.428, 429]. The fluence required to cause complete amorphization within the range of the primary ions corresponds to a sputtered depth on the order of only one monolayer [4.430, 431].

Uniform sputter erosion can also be observed with monocrystalline samples, as illustrated above in Fig. 4.26 [4.374]. Ion bombardment of single crystals sometimes can result in an improved surface flatness [4.14, 122]. By contrast, sputter erosion of polycrystalline metals causes (micro-)roughening of an initially smooth surface, the effect being more pronounced the larger the depth (Fig. 4.39b) [4.121–124, 127, 129, 132, 133, 274, 374, 400–402, 427]. The amount of roughening has been shown to be strongly material dependent, and some correlation appears to exist with the sputtering yield [4.124]. As one might expect intuitively, the depth resolution degrades if the sample to be analyzed exhibits an intrinsic roughness [4.402, 432]. Roughening of polycrystalline metals can be reduced significantly by sample rotation during oblique ion bombardment [4.128, 401].

A compilation of selected experimental data for the depth dependence of the interface width is shown in Fig. 4.40. The material dependence discussed above is evident. Note that the evolution of the surface roughness may be quite

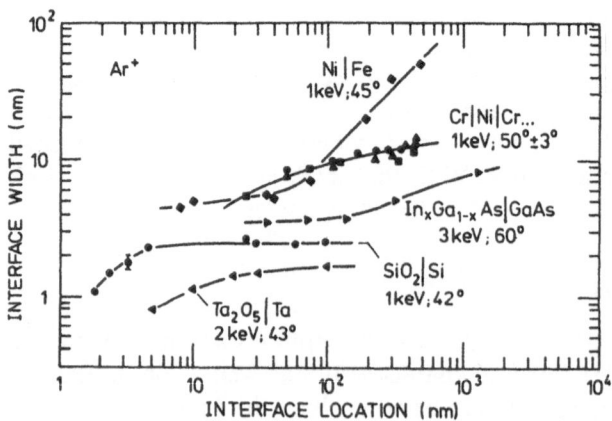

Fig. 4.40. Variation of the interface width with increasing depth of interface location: Ta_2O_5/Ta [4.425], SiO_2/Si [4.197, 424], $In_xGa_{1-x}As/GaAs$ [4.398], Cr/Ni multilayer [4.127, 400–402], Ni/Fe [4.192]

different for different polycrystalline materials. Whereas Δz for an Ni layer on Fe increases roughly linearly with depth for $z > 80$ nm [4.192], only a very gradual increase is observed for the Cr/Ni multilayer system [4.127, 400–402]. The physical origin of this difference is not known. An aspect deserving clarification is how, for the same two elements, the depth dependence of the interface width observed using a single overlayer compares with results obtained with multilayer samples.

b) Sample Temperature

Considering the possible effect of the sample temperature on the depth resolution, we must distinguish between materials that retain a smooth surface and those that become increasingly rough during sputter erosion. In the former case the depth resolution is determined essentially by bombardment-induced redistribution (mixing) of atoms in the sample. The amount of redistribution that actually occurs in addition to collisional relocation depends on the thermal mobility of the created defects, as well as on the chemical driving forces [4.433].

Occasionally an improvement in depth resolution has been reported when the specimen is cooled to low temperatures, e.g., for Ag/Cu multilayers profiled at ~ 100 K [4.434]. A rather pronounced improvement in surface flatness of Ar^+ sputtered InP was observed on changing from a water-cooled to a liquid-nitrogen-cooled sample holder [4.435]. Cooling also caused the erosion rate to increase by up to a factor of 2.9 (500 eV Ar^+ bombardment at normal incidence). Sputtering of InP at room temperature is known to result in severe surface roughening [4.423, 436–438], probably due to preferential loss of the volatile P [4.437] and simultaneous formation of In islands. As a result of cooling to the temperature of liquid nitrogen, this loss of P is apparently suppressed, and sputtering of InP rather than of metallic In determines the erosion rate.

Depth profiling studies involving a thin tracer layer of In embedded in single crystalline Cu [4.439] or an As implant in Si [4.440] did not reveal a decrease of the impurity decay length upon cooling. In both cases, however, the decay length increases upon heating the sample above room temperature (Fig. 4.41). The effect is very pronounced for In in Cu. The rapid increase of λ_{In} above ~ 340 K can be attributed to the onset of diffusion of vacancies and interstitials, which are being generated in the sample by ion bombardment. This radiation-enhanced diffusion substantially increases the mean transport length of In atoms beyond the value set by pure collisional relocation.

The situation is more complex for As in Si, because bombardment with O_2^+ at near-normal incidence causes a layer of SiO_2 to be formed at the sputtered surface [4.420]. Since As piles up at the interface between thermally grown SiO_2 and Si [4.441], this segregation effect could also play a role during sputter depth profiling. Metal impurities such as Cu [4.173], Ag [4.176], and Pd [4.174] have in fact been shown to pile up at the interface between the Si substrate and the oxide generated during sputter depth profiling at room

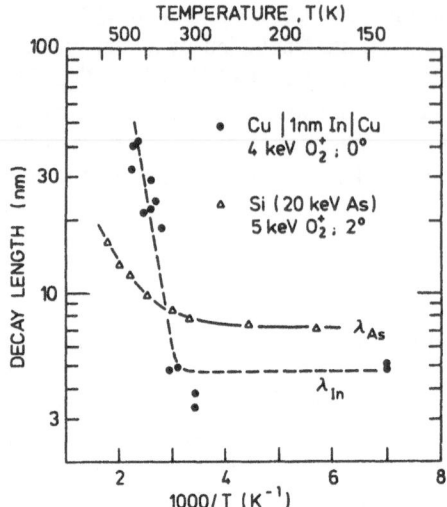

Fig. 4.41. Decay length for In in Cu [4.439] and As in oxidized Si [4.440] as a function of the (inverse) sample temperature during SIMS analysis

temperature. At 300 °C Cr shows a similar effect [4.174]. For As no pileup could be identified on the ("internal") mixing profiles measured under Cs^+ bombardment, not even at 300 °C [4.174]. The only effect of increasing the sample temperature was a more uniform distribution of As in the continuously replenished oxide. This change in the As mixing profile apparently gives rise to the moderate temperature effect seen in Fig. 4.41. (Note, with reference to (4.16), that λ increases if the mixing profile $p_i(z')$ extends to a larger depth.) The absence of an As pileup peak has been attributed to insufficient mobility of As in the oxide [4.174].

The effect of temperature on the broadening of metal/silicon interfaces sputtered under Ar^+ bombardment has been investigated between 80 and 775 K [4.442, 443]. Radiation-enhanced mixing was observed for Mo [4.442] and W [4.443] at temperatures above 675 K. No effect of this kind was found for Mg, Ti, and Ge. Thermally activated diffusion was apparently responsible for the observed high-temperature broadening of the Al and Ag profiles.

The examples discussed so far relate to the case where an increase in sample temperature had the effect of broadening the characteristic features of a sputter depth profile or, at best, leaving the feature unchanged. However, in certain cases elevated temperatures can have a beneficial effect. With polycrystalline samples, for example, the surface roughness may be lowered by surface diffusion. For Ag on Cu the decay length measured at room temperature was reduced by a factor between 2 and 2.6, when working at 423 and 498 K, respectively [4.444].

c) Thermodynamic Properties

The observation of a segregation effect in oxygen-bombarded silicon (preceding section) raises the question of the extent to which the thermodynamic properties of the sample constituents are relevant to the depth resolution in sputter profiling. The issue has been addressed only very recently. *Cheng* et al. [4.445] measured the interface width for a variety of metal substrates (Ti, Ni, Zr, Mo), each covered with a 15 nm overlayer of Pt (sputtering with 1 keV Ar^+ at 45°). The profiles, plotted as a concentration versus sputtering time, were significantly (factor 2 to 3) broader for Ti and Zr than for Ni and Mo substrates. The authors noted a correlation between the interface width $\Delta t_{2\sigma}$ and the mixing rates observed for the same bilayers under bombardment with 600 keV Xe^{2+}. In either case the mixing was attributed mostly to diffusion in thermal spikes. A recently developed model for this mixing process [4.446] led to the conclusion that the heat of mixing of binary alloys plays an important role in profile broadening.

Experiments of this kind are highly desirable to elucidate the relevance of thermodynamic properties in sputter depth profiling, but some caution is needed concerning the quantitative aspects of the measured profiles. In fact, the arguments were based on a determination of $\Delta t_{2\sigma}$ rather than of $\Delta z_{2\sigma}$. Available data on low-energy sputtering suggest that in the experiments of *Cheng* et al. [4.445] the erosion rates for Ni (and possibly also for Mo) may have been a factor of 2 higher than for Ti and Zr [4.392, 393, 413–415]. Corrections for these differences would cause the $\Delta z_{2\sigma}$ values to be much closer in magnitude than the numbers for $\Delta t_{2\sigma}$. Clearly, an improved data basis is necessary.

4.7 Aspects of Quantification and Insulator Problems

4.7.1 Dependence of Signal Heights on the Sample Composition

In this section we briefly discuss the effect of the sample composition on the elemental signal heights measured in surface analysis. The variation of elemental sensitivity factors with sample composition is usually referred to as the matrix effect.

a) Electron Spectroscopy

In electron spectroscopy the chemical environment around the atom to be analyzed affects the measured signal in two ways. First, the number, shape, and position of the characteristic AES and XPS lines depend strongly on the chemical bonding. In AES the peak shape also depends on the density of states in the valence band [4.41, 44, 46]. Rather severe quantification problems arise in AES if the conventional method of using derivative spectra is used to analyze samples of vastly different composition, e.g., oxides on metals. It has also been pointed out recently that significant errors in the derivative Auger amplitudes may result from an analyzer's insufficient energy resolution [4.447]. To minimize such quantitation problems, use the direct spectrum. Adequate background subtraction will then remain the problem of concern [4.46].

The second factor affecting quantitation in electron spectroscopy is the electron escape depth L_0 (4.31). Variations of L_0 with chemical composition have been discussed by *Seah* and *Dench* [4.187]. In view of the exponential dependence of the measured intensity on L_0 (4.21), a rather precise knowledge of L_0 is required to arrive at an accurate signal calibration, in particular if the analyzed samples contain regions of different chemical composition.

b) Ion Scattering Spectrometry

In ion scattering spectrometry the effect of the matrix composition on the signal height is described by the factor P_i in (4.44), which denotes the probability that an incident ion retains its original charge state on leaving the sample. It has been found that P_i scales with the normal component v_n of the velocity of the scattered particle. Experimental data can often be described with sufficient accuracy [4.208] by the relation

$$P_i \propto \exp\left(-\frac{A'}{a'v_n}\right), \tag{4.65}$$

which was originally developed to describe Auger neutralization [4.448]. A' and a' are a rate constant and a characteristic inverse distance, respectively. The ratio A'/a' depends on the ion-target combination; i.e., the matrix effect in ISS is described by (4.65).

Apart from the fact that (4.65) constitutes only a rather rough description of P_i, the problem encountered experimentally is that P_i also depends on the cleanliness of the surface [4.208]. In view of these difficulties, conversion of signal heights to concentrations is commonly accomplished using empirical calibration factors based on the use of standards [4.371].

c) Mass Spectrometry and Optical Spectroscopy

The effect of sample chemistry on the intensity observed in SIMS, resonant mutiphoton ionization (resonant MPI), and bombardment induced light emission (BLE) is discussed in detail in Chap. 3. The matrix effects that may be encountered in these fields of analysis are much more severe than in electron spectroscopy or ion scattering spectrometry. Particularly well known is the large change in elemental intensities produced by the presence of oxygen in metals or semiconductors. Whereas in SIMS and BLE the intensity may be enhanced by several orders of magnitude [4.20, 21, 26, 57, 60–66, 78–81, 250, 284], the inverse effect occurs in resonant MPI involving ground-state atoms [4.348]. With all three techniques quantitative analysis of samples containing varying concentrations of oxygen is not possible now. The best one might be able to do is to generate the same state of surface oxidation throughout the total depth of analysis. Such attempts are based on the use of oxygen primary ion beams and/or bombardment during exposure of the sample to an added flow of gaseous oxygen. The success of this procedure, however, depends on the sticking coefficient of oxygen and/or the ability of the matrix to retain incorporated oxygen under ion bombardment.

4.7.2 Matrix-Dependent Secondary Ion Yield Enhancement

Repeatedly it has been pointed out that in SIMS analyses a high degree of ionization is achieved only if a favorable chemical state can be established at the bombarded surface. While O_2^+ and Cs^+ beams are commonly used for this purpose, it has been noted for quite some time that the enhancement actually obtained depends strongly on the properties of the studied matrix.

Figure 4.42 shows the pronounced effect of Cs^+ bombardment on the yield of a variety of secondary ions emitted from a silicon target [4.285, 449] (corresponding data for oxygen bombardment are shown in Fig. 3.9). Initially the Cs^+ intensity increases monotonically with increasing time of bombardment (Fig. 4.42a). At a critical fluence, however, corresponding to ~ 700 s of bombardment, the Cs^+ intensity passes through a well-defined peak, and finally decreases to a level a factor of 25 below the peak. The Si^+ intensity shows an initial rapid falloff due to the removal of the native surface oxide, followed by a plateau. Somewhat before the Cs^+ peak the Si^+ intensity starts to decrease, attaining a stationary level a factor of 10 below the plateau.

The peak and subsequent falloff of the Cs^+ intensity suggests that the amount of Cs building up at the bombarded surface is large enough to reduce the work function by 2 eV or more (compare Fig. 3.2). This lowering of the work function enhances the yield of negative secondary ions (Fig. 4.42b). The most

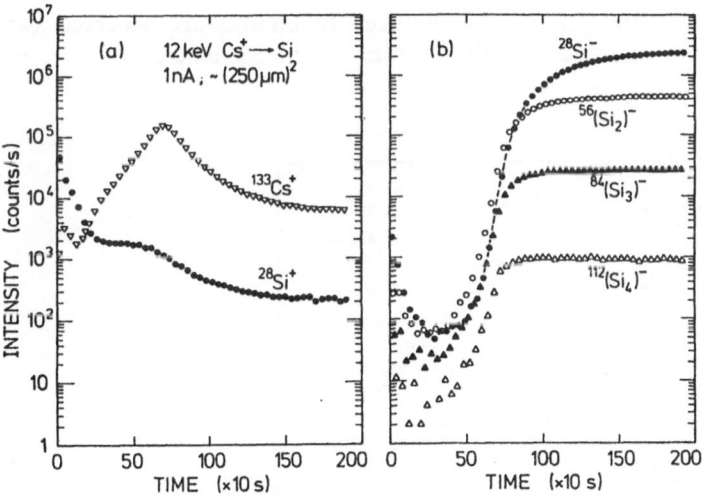

Fig. 4.42. Evolution of the intensity of (**a**) positive and (**b**) negative secondary ions during the buildup of Cs in Si. The intensities were recorded at the peak of the secondary ion energy distributions [4.285, 449]; 10 s of bombardment correspond roughly to a fluence of 1×10^{14} ions/cm^2

rapid rise of the Si⁻ intensity occurs at a bombardment fluence corresponding to the position of the Cs⁺ peak in Fig. 4.42a. The Si⁻ yield enhancement obtained under stationary bombardment conditions amounts to a factor of 4×10^4. Even larger enhancement factors can be achieved at lower impact energies. (Note that Cs implantation has only a very small effect on the sputtering yield of Si, as indicated by a comparison of data obtained under Xe⁺ and Cs⁺ bombardment [4.180].)

Very little is known about the Cs distribution in the first few layers of the bombarded sample. Rutherford backscattering spectrometry measurement at moderate depth resolution (~ 5 nm) revealed some interesting aspects [4.450], but the desired depth information can be obtained only under high-resolution conditions (< 2 nm). Semiquantitative information about the composition of the top monolayer(s) can be derived from the results obtained for the negative molecular ions Si_2^-, Si_3^-, and Si_4^-. As shown in Fig. 4.42b, a stationary intensity level for these species is observed at lower bombardment fluences, the larger the number of atoms in the molecule. This suggests that the Cs concentration becomes large enough to affect the formation probability of Si_n^- molecules. Even larger effects have been observed for Si_n^+ molecules emitted from oxygen-bombarded Si [4.451].

Silicon is a favorable target material because one can obtain exceptionally high secondary ion yield enhancement by bombardment with O_2^+ or Cs⁺. Much lower enhancement effects are often observed with other materials, not only for the matrix ions [4.281], but also for a given impurity sputtered from different matrices [4.56, 60, 249, 252, 282, 452]. The latter effect can be demonstrated nicely using standards prepared by ion implantation. Figure 4.43 shows ³¹P profiles measured under identical SIMS conditions for samples containing the same amount of phosphorus. Pronounced differences are observed with respect to the height and width of the profiles. In the extreme case – Si versus

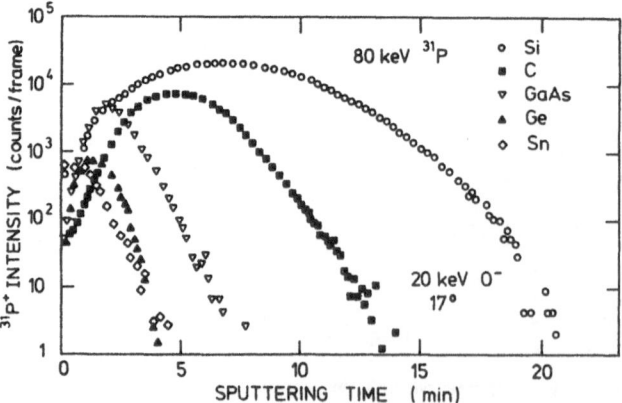

Fig. 4.43. SIMS depth profiles of P implanted in different matrices [4.56]. The implantation fluence (5×10^{15} atoms/cm² [4.282]) and the SIMS analysis conditions were the same for all profiles

Sn – the relative secondary ion yields, derived by integration over the complete P profiles, differ by more than two orders of magnitude [4.282].

Attempts to rationalize the results of Fig. 4.43 have been made on the basis of the assumption that the stationary surface concentration of oxygen is exclusively determined by the matrix sputtering yield [4.56, 281, 282]. Moreover, these studies assumed that the yield enhancement is the same for the same oxygen concentration in different matrices. Detailed experiments [4.249, 252] as well as an analysis of literature data [4.60] have shown, however, that in practice these assumptions are not justified. While the matrix sputtering yield is apparently one out of several parameters that determine the stationary surface concentration of oxygen [4.452], quantitative predictions concerning the magnitude of the matrix effects in an unknown sample are not possible solely on the basis of a knowledge of the sputtering yield of the matrix under study.

As to the calibration of secondary ion yields observed under Cs^+ bombardment, measurements involving the use of negative atomic species showed a matrix effect at least as large as under O_2^+ bombardment [4.282]. The magnitude of the Be matrix effect in $Al_x Ga_{1-x} As$ is much reduced if $AsBe^-$ ions are monitored [4.453]. Moroever, studies on a variety of impurities (M) in Si showed that ion yield variations are much less pronounced if MSi^- rather than M^- ions are recorded [4.454]. Only a relatively small matrix effect – if any – has been reported using CsM^+ ions [4.455]. The yields of these positive dimer ions are, however, smaller than the respective negative matrix ion yields [4.290]. Rather interesting is the observation that Cs^+ may form dimers with sputtered inert-gas atoms. Detection limits down to 4×10^{17} atoms/cm^3 were obtained, for example, using $CsAr^+$ secondary ions to profile Ar in Ge [4.456].

The lack of a reliable model for the matrix effect forces the use of standards in the calibration of secondary ion intensities of an unknown sample. The various current approaches have been discussed in some detail [4.457]. We can distinguish between bulk standards of uniform composition and standards produced by ion implantation. The latter class of reference samples is commonly prepared by "external" fabrication using an ion implanter. More recently, some groups have used in situ implantation in the ion microprobe to produce "internal" standards in the sample [4.284, 458–462]. With commercially available SIMS instruments application of this useful procedure is restricted to elements of which ions can be generated easily in plasma sources. To circumvent this limitation, a dedicated ion source has to be attached to the system [4.462]. Note that the accuracy of the in situ implantation approach depends on correctly measuring the ion beam current as well as the bombarded area.

4.7.3 Charge Buildup on Insulators

Ion impact on poorly conducting materials will generally result in charge buildup on the bombarded sample. The magnitude of the effect depends on such parameters as the resistivity of the sample, and the charge state, energy, and impact angle of the incident particle. In a simplified model the total amount of charge building up on the sample may be assessed by considering the balance of three different currents: the beam current, the ion-induced electron current, and tertiary currents involving charged particles released from electrodes adjacent to the sample [4.463]. With high-resistivity materials ($> 10^{14}$ Ωcm) the charging may become comparable to the terminal voltage of the ion gun, so that controlled bombardment of the sample is no longer possible. In SIMS analyses even moderate charging will usually cause such undesirable effects as deterioration of beam focusing, shift of secondary ion energy spectra, strong reduction or even complete loss of secondary ion signals, and migration of mobile ions in the sample.

The detrimental effects brought about by sample charging and the various methods used to minimize charge buildup have been discussed by *Werner* and *Warmoltz* [4.464]. The most generally applicable technique for charge compensation (or better, for charge stabilization) is based on the application of simultaneous electron bombardment [4.302–313]. Successful charge stabilization has been reported for a wide range of electron energies, from below 1 eV [4.312] to several keV [4.308]. The technique works well even for negative secondary ion detection [4.311]. Poor detection limits may, however, be obtained in the analysis of species that desorb from the sample as a result of electron impact, e.g., fluorine [4.310]. The problems encountered in SIMS analysis of insulators may be greatly alleviated – though not fully removed – by bombardment with neutral primary beams [4.465].

Even though auxiliary electron beams have been successful in SIMS analysis of poorly conducting materials, the status of this approach is hardly satisfactory. More often than not, depth profiles of the desired quality are obtained only after optimizing the experimental parameters (impact angle, energy, current, and current density of the electron beam) in a trial-and-error process. This uncertainty about the optimum operating conditions arises from our poor understanding of the compensation mechanism. For example, there have been no experiments clarifying the dependence of the surface potential on the electron emission coefficient. Enhancing conductivity with electron beams may be possible [4.305, 308]. Further studies are highly desirable.

4.8 Factors Determining Detection Limits

4.8.1 Overview

The signal I measured in sputter-based analysis is generally the sum of several contributions. Superimposed on the "true" intensity I_{true} reflecting the elemental concentration in the sample, there will be an "apparent" intensity I_{app} originating from various sources not related to the (original) sample composition:

$$I = I_{true} + I_{app} .$$

(4.66)

The apparent intensity may be split into the following contributions:

$$I_{app} = I_{dk} + I_{if} + I_{ns} + I_{st} + I_{cc} + I_{ad} \; . \tag{4.67}$$

These represent, respectively, the dark current of the detection system, the intensity due to mass or energy interference, the shot noise associated with the technique's inherent background, the signal generated by stray primary particles, the effect of cross contamination, and the contribution of residual gas adsorption. Clearly, the instrument should be designed so that I_{true} is maximized and all contributions to I_{app} are minimized.

Methods for reducing the last two terms in (4.67) will be discussed separately (Sects. 4.8.2, 3). Appropriate electronic design usually can reduce the dark current to less than 0.1 count/s, i.e., to a negligible level. The presence and magnitude of the interference term depends on the element under study, the matrix composition, and the characteristic features of the technique. In AES analysis of TiN_x, for example, interference of the N_{KLL} and the Ti_{LMM} lines constitutes a real challenge [4.466, 467]. Recently different approaches have been described to solve the problem and to accomplish adequate sample characterization. A well-known case in SIMS is the mass interference between $^{75}As^+$ ($M = 74.9217$ u) and $^{29}Si\,^{30}Si\,^{16}O^+$ ($M = 74.9452$ u) encountered in the analysis of As in SiO_2 or in oxygen-bombarded Si. Separation of the two mass lines requires a resolution $M/\Delta M > 5000$ not accessible to commercial quadrupole-based instruments. Good detection limits may nevertheless be achieved by energy discrimination [4.301], i.e., by making use of the fact that, compared with atomic secondary ions, the energy distributions of molecular ions decrease rather rapidly beyond the low-energy peak.

The noise term in (4.67) will be of concern whenever the true signal is superimposed on an inherent background, as is the case in AES and XPS. In the shot noise limit the AES signal-to-noise ratio may be written [4.44].

$$\frac{I_{true}}{I_{ns}} = \frac{H_e}{\omega_e} (g I_0 t)^{1/2} \; , \tag{4.68}$$

where H_e is the Auger yield for the transition of interest, ω_e is the inherent width of the Auger line, g is an instrumental factor comprising, e.g., the energy-band pass and the instrument transmission, I_0 is the primary electron current, and t is the signal integration time. The maximum signal-to-noise ratio derived from (4.68) amounts to about 100 ppm. Detection limits of this order can be achieved in favorable cases, provided the signal integration time is long enough [4.468].

In RBS an inherent background is encountered if the energy loss due to scattering from the (impurity) element of interest (mass M_i) exceeds the loss due to scattering from matrix species (M_m) [4.211]. For $M_i > M_m$ the matrix background is of concern whenever the M_i species are located at a depth $z > z_{i,m}$, where $z_{i,m}$ can be derived from (4.46),

$$z_{i,m} = \frac{(K_i - K_m) E_0}{N[\varepsilon]_i} \; . \tag{4.69}$$

In such a case, signal overlap can be removed by combining RBS analysis with controlled sputter etching of the sample up to the point where the impurity is found at a depth z' from the eroded surface, so that $z' < z_{i,m}$ [4.106].

In ISS the inherent background is due to secondary ions released from the sample with the scattered ions. This background can be removed by mass-resolved ISS [4.205, 367–370].

All techniques involving mass spectrometric analysis of sputtered particles are characterized by the absence of an inherent background; i.e., $I_{ns} = 0$. With these techniques any sizable contribution of I_{st} to I_{app} becomes of concern. The most important case encountered so far is the generation of secondary ions by the neutral component of a primary ion beam [4.52]. Since the area hit by the beam of energetic neutral particles is usually much larger than the area set for data acquisition ("gated area"), the signal recorded after sputtering through the depth of doping will significantly exceed I_{true}. The best way to solve this problem is to remove the neutral component from the primary beam by electrostatic filtering [4.52]. Secondary ions generated outside the area of interest may also be suppressed to some extent by ion optical means [4.13, 47].

4.8.2 Cross Contamination (Memory Effect)

Up to now we have been concerned with the fate of the sputtered material only to the extent to which it can be used for determining the composition of the bombarded sample. In general, however, only a small fraction of the sputtered material is transported into the analyzer, whereas the remainder is deposited on electrodes facing the sample. During a sputter-based analysis these electrodes are exposed to a flow of backscattered primary particles, as well as of sputtered neutrals and ions. Since not only the backscattered primaries but also the sputtered particles may have energies exceeding the threshold for sputtering (typically 20–50 eV [4.390]), we must expect that part of the previously deposited material will be released from the adjacent electrodes, with a small fraction coming to rest on the sample area under study. This cross contamination of the sample gives rise to a signal I_{cc}, which cannot be separated from I_{true}.

The first detailed study of the cross-contamination effect was performed by *Deline* [4.292] using a Cameca IMS-3F secondary ion microscope. His approach was first to contaminate the instrument deliberately by (pre)sputtering material from a certain matrix, e.g., GaAs. Subsequently depth profiling studies were performed on another kind of sample (Si) containing the contaminant (As) as a dopant (Fig. 4.44a). The presputtering of GaAs generated a very high cross-contamination signal which, however, decreased during Si sputtering.

The initial level of contamination, derived from these and other profiles by extrapolating to the beginning of a depth profiling run, are compiled in Fig. 4.44b. Data relating to Si analysis in GaAs after contamination of the IMS-3F with Si are shown for comparison. Also included are results obtained by *Clegg* using the quadrupole-based A-DIDA ion microprobe (Ga analysis in Si after GaAs sputtering) [4.469]. In all three cases the initial level of cross contamina-

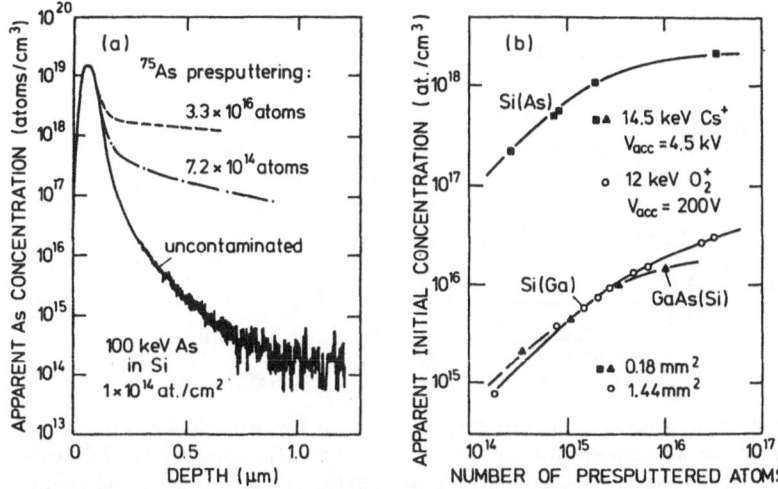

Fig. 4.44. (a) SIMS depth profiles of As implanted in Si, measured at different levels of As contamination of the ion microprobe. The numbers denote the amount of As sputtered from a GaAs sample before As depth profiling in Si [4.292]. (b) Apparent levels of cross contamination in different SIMS instruments as a function of the number of atoms sputtered before analysis of As in Si, Si in GaAs [4.292], and Ga in Si [4.469]

tion c_{cc} increases linearly with increasing number of presputtered atoms. In the IMS-3F, c_{cc} appears to saturate after removal of $\sim 3 \times 10^{16}$ atoms, whereas a stationary level of c_{cc} is not yet evident at this point in the A-DIDA.

A rather remarkable finding is that in the same instrument c_{cc} may differ by more than two orders of magnitude, depending on the species being analyzed. *Deline* [4.292] has related c_{cc} to the probability of negative-ion formation, which is expected to scale exponentially with the electron affinity (Chap. 3). With such elements as Si featuring a high electron affinity (1.39 eV compared with 0.8 eV for As) the flux of accelerated secondary ions striking the extraction electrode will be high, thus making c_{cc} (As) large. On the other hand, deposition of resputtered Si$^-$ on the negatively biased GaAs sample will be suppressed, so that c_{cc} (Si) is comparatively low.

I have investigated the cross-contamination process in detail, both experimentally and theoretically [4.470]. The calculations relate to a simplified geometry sketched in Fig. 4.45a. The apparent concentration derived from the model reads

$$c_{cross} = \gamma_{st} \, p' \, p'' \left(\frac{A}{R^2} \right) \frac{\theta_s \varepsilon_{cc}}{Y} \, , \qquad (4.70)$$

where γ_{st} is the sticking coefficient of backsputtered atoms striking the sample, p' is a numerical factor ($p < 0.3$), p'' is a geometrical (filling) factor, A is the total bombarded area, R is the electrode-sample separation (radius of the hemisphere

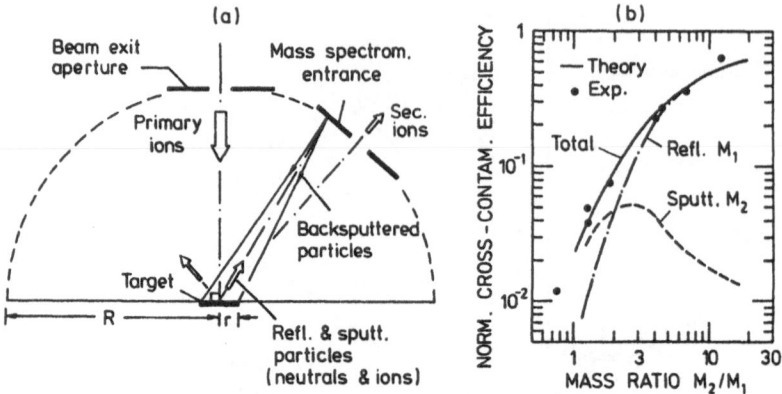

Fig. 4.45. (a) Schematic illustration of cross contamination. (b) Comparison of experimental and theoretical data for the normalized cross-contamination efficiency. The dot-dashed and the dashed curve represent the calculated contributions due to reflected primary particles and sputtered particles, respectively [4.470]

in Fig. 4.45a), θ_s is the fractional coverage of the electrode facing the sample, ε_{cc} is the total cross-contamination efficiency, and Y is the sputtering yield of the analyzed sample. The efficiency ε_{cc} contains the two contributions due to backscattered primaries and sputtered particles. Knowing the energy distributions of these two types of species as well as the energy dependence of the sputtering yield for the contaminated electrode, we can evaluate ε_{cc}.

Figure 4.45b compares experimental and theoretical cross-contamination efficiencies [4.470]. The measured data represent results of SIMS experiments performed using 8 keV O_2^+ primary ions ($\theta_{in} = 2°$) and a low extraction voltage ($V_{acc} = 120$ V). The calculated curves relate 3 keV Ne bombardment at normal incidence. The agreement between experiment and theory is surprisingly good.

The most important aspect of (4.70) is that c_{cc} is directly proportional to the bombarded sample area A. This proportionality has been verified experimentally [4.469, 470]. (Note that because of the larger bombarded area the cross-contamination effect for Ga analysis in Si shown in Fig. 4.44b is exaggerated by a factor of 8 compared with the other data.) For a given instrument with a certain level of contamination, the only way to reduce c_{cc} is to keep A as small as possible.

4.8.3 Adsorption and Incorporation of Residual Gases

It has been known for some time that imperfect vacuum conditions will degrade the detection limit for impurities also present in the residual gas. In AES analysis, for example, electron impact in the presence of carbon containing gaseous species results in the buildup of elemental carbon [4.451, 471, 472]. In SIMS only rather poor detection limits (e.g., 4×10^{19} atoms/cm³ for H and O analysis and 5×10^{18} atoms/cm³ for C analysis in Si) could be obtained at times when many instruments were still operated at base pressures between 10^{-8} and 10^{-7} mbar [4.473]. Since then much effort has been devoted to reducing the partial pressure of species like H_2O to as low as possible, either by using bakeable ultrahigh vacuum (UHV) equipment [4.474, 475] or by placing low-temperature cryoshields near the sample [4.277, 289, 476].

The apparent concentration c_{ad} generated by residual gas adsorption can be estimated using the idea that, in a stationary sputtering state, there will be a balance between the rates of gas uptake and removal so that

$$c_{ad} YJ = \gamma_{st} J_{rg} . \tag{4.71}$$

Here Y is the matrix sputtering yield, J the primary ion flux, γ_{st} the sticking coefficient, and J_{rg} the flux of residual gas species hitting the analyzed area (with adsorption of molecules we must take into account the equivalent flux of atoms). Equation (4.71) has the reasonable consequence that c_{ad} is lower, the larger the flux of sputtered atoms YJ released from the analyzed area.

The inverse proportionality between c_{ad} and the sputtering rate has been demonstrated in several experiments [4.277, 289, 475, 476]. Figure 4.46 illustrates the improvement in dynamic range obtained by changing from a moderate to a high (average) current density [4.289]. The inverse proportionality between c_{ad} (represented by the stationary intensity observed after sputtering through the implantation profile) and J is not fully obeyed because of crater edge effects encountered at the higher current density (J was varied by reducing the scan width to the point where the beam size became the limiting factor [4.277].)

When adsorption of a gaseous species is assisted by the incident beam c_{ad} cannot be assessed on the basis of (4.71) [4.395]. This process can formally be taken care of by replacing γ_{st} in (4.71) by $\gamma_{st} + \beta' J$, where β' defines the rate of

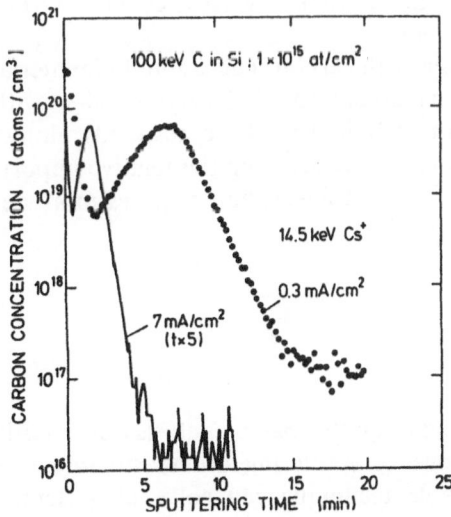

Fig. 4.46. SIMS depth profiles of C in Si for two different mean current densities of the primary ion beam [4.289]

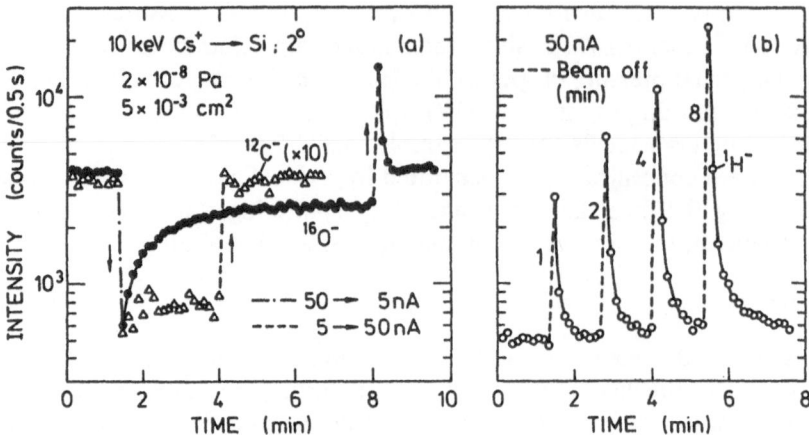

Fig. 4.47. (a) Background signal variation for C and O observed on sudden changes of the Cs$^+$ ion current. (b) Effect of interrupting ion bombardment on the H background signal. The numbers denote the duration of the beam-off interval [4.475]

beam-induced incorporation [4.475]. Another complication may arise from gaseous species liberated from electrodes near the sample, some of these species coming to rest on the analyzed area. This will result in a contribution to c_{ad} equivalent to a cross-contamination term (4.70).

Figure 4.47a illustrates the differences encountered with different residual gas species. The experiment was performed to elucidate the process of carbon and oxygen incorporation in Si bombarded with Cs$^+$ ions [4.475]. The O$^-$ signal variation observed on sudden changes of the Cs$^+$ current suggests that beam-induced oxygen uptake contributes little to the stationary concentration c_{ad}. This supposition is supported by beam-off experiments like those shown in Fig. 4.47b for H$^-$ [4.75]. In fact, the amount of hydrogen accumulated in the absence of the ion beam corresponds to about 90% of what is estimated from the stationary H$^-$ signal (70% for oxygen).

A rather different behavior is observed with carbon. The amount of material adsorbed in the absence of the ion beam is rather small ($\sim$ 20%) compared with the stationary signal. Consequently, the C$^-$ in Fig. 4.47a responds almost directly to beam current variations. A decision concerning the relative importance of cross contamination and beam-induced adsorption requires more detailed experiments.

4.9 Summary and Prospects

The data reviewed here show that the field of sputter-based analysis is now fairly mature. The advantages and limitations of sputtering are reasonably well understood. Since no presently available technique can solve all potential problems, the selection of the most appropriate technique will depend very much on the desired information, as well as on the basic characteristics of the analyzed sample.

At this point SIMS is clearly the method of choice for dealing with impurity concentrations below about 100 ppm. In the near future laser-based SNMS will probably become an alternative to SIMS, particularly to reduce the large SIMS-specific differences in elemental sensitivities. Limitations of laser-based SNMS still need to be worked out, e.g., the magnitude of memory effects.

In the high-concentration region, the technique currently used most frequently with sputtering is AES. We can expect increasing use of XPS here, along with steadily improving lateral resolution of the analyzers. Whenever extreme surface sensitivity is required, ISS will remain a very useful technique. The advantages of an in situ combination of SIMS and ISS do not seem to have been explored sufficiently. The technique of sputter/RBS has just been established. It will be interesting to apply this method to problems that cannot be solved with surface-sensitive tools.

As to the merits of techniques using a plasma for sputtering and excitation (GDMS and GDOS), the relatively small variation of elemental sensitivities and the high speed of analysis are certainly very attractive features whenever moderate detection limits and a relatively poor depth resolution can be tolerated. The future of electron-gas SNMS is not clear yet, because many of its basic features are awaiting exploration.

Acknowledgements. Dr. R. Behrisch's review of the manuscript is appreciated. The almost endless jobs of typing and retyping the manuscript and drawing the figures were carried out by Mrs. M. Englberger and Mrs. E. Schneider-Kracke, respectively. Their skill and patience is gratefully acknowledged.

References

4.1   W.R. Grove: Philos. Mag. **5,** 203 (1853)
4.2   R. Behrisch (ed.): *Sputtering by Particle Bombardment I*, Topics Appl. Phys., Vol. 47 (Springer, Berlin, Heidelberg 1981)
4.3   R. Behrisch (ed.): *Sputtering by Particle Bombardment II*, Topics Appl. Phys., Vol. 52 (Springer, Berlin, Heidelberg 1983)
4.4   P. Williams, Surface Sci. **90,** 588 (1979)
4.5   J.J. Thomson: Phil. Mag. **20,** 752 (1910)
4.6   R.H. Sloane, R. Press: Proc. R. Soc. (London) A **168,** 284 (1938)
4.7   R.F.K. Herzog, F.P. Viehböck: Phys. Rev. **76,** 855L (1949)
4.8   R.E. Honig: J. Appl. Phys. **29,** 549 (1958)
4.9   R. Castaing, G. Slodzian: J. Microscopie **1,** 395 (1962)
4.10   H. Liebl: J. Appl. Phys. **38,** 5277 (1967)
4.11   H. Liebl: Anal. Chem. **46,** 22A (1974)
4.12   H. Liebl: J. Phys. E: Sci. Instrum. **8,** 797 (1975)
4.13   K. Wittmaack: Vacuum **32,** 65 (1982)
4.14   H. Lutz, R. Sizmann: Phys. Lett. **5,** 113 (1963)

4.15 V.E. Krohn: J. Appl. Phys. **33**, 3523 (1962)
4.16 R.J. MacDonald, E. Dennis, E. Zwangobani: In *Atomic Collision Phenomena in Solids*, ed. by D.W. Palmer, M.W. Thompson, P.D. Townsend (North Holland, Amsterdam 1970) p. 307
4.17 A. Benninghoven, E. Loebach: Rev. Sci. Instrum. **42**, 49 (1971)
4.18 K. Wittmaack, J. Maul, F. Schulz: Int. J. Mass Spectrom. Ion Phys. **11**, 23 (1973)
4.19 R. Schubert, J.C. Tracy: Rev. Sci. Instrum. **44**, 487 (1973)
4.20 G. Slodzian, J.-F. Hennequin: C.R. Acad. Sci. (Paris) **263B**, 1246 (1966)
4.21 A. Benninghoven: Z. Naturforschg. **22a**, 841 (1967)
4.22 C.A. Andersen: Int. J. Mass Spectrom. Ion Phys. **2**, 61 (1969)
4.23 G. Hortig, P. Mokler, M. Müller: Z. Physik **210**, 312 (1968)
4.24 C.A. Andersen: Int. J. Mass Spectrom. Ion Phys. **3**, 413 (1970)
4.25 A. Benninghoven: Z. Physik **230**, 403 (1970)
4.26 A. Benninghoven: Surface Sci. **28**, 541 (1971); **53**, 596 (1975)
4.27 A. Benninghoven, S. Storp: Appl. Phys. Lett. **22**, 170 (1973)
4.28 A. Benninghoven, E. Loebach, C. Plog, N. Treitz: Surface Sci. **39**, 397 (1973)
4.29 C.A. Evans, Jr., J.P. Pemler: Anal. Chem. **42**, 1060 (1970)
4.30 J.A. McHugh: Radiat. Eff. **21**, 209 (1974)
4.31 V.M. Pistryak, A.K. Gnap, V.F. Kozlov, R.I. Garber, A.I. Fedorenko, Ya. M. Fogel: Sov. Phys. Sol. State **12**, 1005 (1970)
4.32 R.P. Gittins, D.V. Morgan, G. Dearnaley: J. Phys. D: Appl. Phys. **5**, 1654 (1972)
4.33 J. Maul, F. Schulz, K. Wittmaack: Phys. Lett. **41A**, 177 (1972)
4.34 W.K. Hofker, H.W. Werner, D.P. Oosthoek, H.A.M. de Grefte: Radiat. Eff. **17**, 83 (1973)
4.35 G. Schwarz, M. Trapp, R. Schimko, G. Butzke, K. Rogge: Phys. Stat. Sol. (a) **17**, 653 (1973)
4.36 F. Schulz, K. Wittmaack, J. Maul: Radiat. Eff. **18**, 211 (1973)
4.37 M. Bernheim, G. Slozdian: Int. J. Mass Spectrom. Ion. Phys. **12**, 93 (1973)
4.38 M.L. Tarng, G.K. Wehner: J. Appl. Phys. **42**, 2449 (1971)
4.39 T. Smith: Surface Sci. **27**, 45 (1971)
4.40 P.W. Palmberg: J. Vac. Sci. Technol. **9**, 160 (1972)
4.41 C.C. Chang: In *Characterization of Solid Surfaces*, ed. by R.F. Kane, G.B. Larrabee (Plenum, New York 1974) p. 509
4.42 J.M. Morabito, R.K. Lewis: In *Methods of Surface Analysis*, ed. by A.W. Czanderna (Elsevier, Amsterdam 1975) p. 279
4.43 P.W. Palmberg, G.K. Bohn, J.C. Tracy: Appl. Phys. Lett. **15**, 254 (1969)
4.44 A. Joshi, L.E. Davis, P.W. Palmberg: in *Methods of Surface Analysis*, ed. by A.W. Czanderna (Elsevier, Amsterdam 1975) p. 159
4.45 L.E. Davis, N.C. MacDonald, P.W. Palmberg, G.E. Riach, R.E. Weber: *Handbook of Auger Electron Spectroscopy*, 2nd ed. (Physical Electronics Division, Perkin Elmer Corp., Eden Prairie, Minnesota 1976)
4.46 M.P. Seah: In *Practical Surface Analysis by Auger and X-ray Photoelectron Spectroscopy*, ed. by D. Briggs, M.P. Seah (Wiley, Chichester 1983) p. 181
4.47 C.W. Magee, W.L. Harrington, R.E. Honig: Rev. Sci. Instrum. **49**, 477 (1978)
4.48 K. Wittmaack: In *X-ray Optics and Microanalysis*, ed. by D.R. Beaman, R.E. Ogilvie, D.B. Wittry (Pendell, Midland 1980) p. 311
4.49 J.M. Rouberol, M. Lepareur, B. Autier, J.M. Gourgout: ibid p. 322
4.50 R. Levi-Setti: In *Applied Charged Particle Optics*, Part A, ed. by A. Septier (Academic, New York 1980) p. 261
4.51 G. Slodzian: ibid, Part B, p. 1
4.52 K. Wittmaack, J.B. Clegg: Appl. Phys. Lett. **37**, 285 (1980)
4.53 M. Komuro, T. Kanayana, H. Hiroshima, H. Tanone: Appl. Phys. Lett. **42**, 908 (1983)
4.54 R. Levi-Setti, P.H. LaMarche, K. Lam, T.H. Shields, Y.-L. Wang: Nucl. Instrum. Methods **218**, 368 (1983)
4.55 A.R. Bayly, A.R. Waugh, K. Anderson: Nucl. Instrum. Methods **218**, 375 (1983)
4.56 P. Williams: IEEE Trans. Nucl. Sci. **NS-26**, 1807 (1979)

4.57   K. Wittmaack: Nucl. Instrum. Methods **168**, 343 (1980)
4.58   H.W. Werner: Surface Interf. Anal. **2**, 56 (1980)
4.59   A.E. Morgan: Surface Interf. Anal. **2**, 123 (1980)
4.60   K. Wittmaack: Appl. Surf. Sci. **9**, 315 (1981)
4.61   K. Wittmaack: In *Inelastic Ion-Surface Collisions*, ed. by N.H. Tolk, J.C. Tully, W. Heiland, C.W. White (Academic, New York, 1977) p. 153
4.62   G. Blaise: In *Materials Characterization Using Ion Beams*, ed. by J.P. Thomas, A. Chachard (Plenum, New York 1978) p. 143
4.63   K. Wittmaack: Surface Sci. **89**, 668 (1979)
4.64   P. Williams: Ann. Rev. Mater. Sci. **15**, 517 (1985)
4.65   Z. Sroubek: Nucl. Instrum. Methods **194**, 533 (1982)
4.66   M.L. Yu: Nucl. Instrum. Methods **B15**, 151 (1986)
4.67   J.W. Coburn, E. Kay: Appl. Phys. Lett. **19**, 350 (1971)
4.68   J.W. Coburn, E. Taglauer, E. Kay: J. Appl. Phys. **45**, 1779 (1974)
4.69   H. Oechsner, W. Gerhard: Phys. Lett. **40A**, 211 (1972); Surface Sci. **44**, 480 (1974)
4.70   H. Oechsner, E. Stumpe: Appl. Phys. **14**, 43 (1977)
4.71   J.W. Coburn, E.W. Eckstein, E. Kay: J. Appl. Phys. **46**, 2828 (1975)
4.72   H. Oechsner: In *Thin Film and Depth Profile Analysis*, ed. by H. Oechsner (Springer, Berlin, Heidelberg 1984) p. 63
4.73   C.W. White, D.L. Simms, N.H. Tolk: Science **177**, 481 (1972)
4.74   W. Grimm: Spectrochim. Acta **23B**, 443 (1968)
4.75   J.E. Greene, J.M. Whelan: J. Appl. Phys. **44**, 2509 (1973)
4.76   G.E. Thomas, E.E. de Kluizenaar: Int. J. Mass Spectrom. Ion Phys. **15**, 165 (1974)
4.77   R. Kelly, C.B. Kerkdijk: Surface Sci. **46**, 537 (1974)
4.78   C.W. White, D.L. Simms, N.H. Tolk, D.V. McCaughan: Surface Sci. **49**, 657 (1975)
4.79   R. Shimizu, T. Okutani, T. Ishitani, H. Tamura: Surface Sci. **69**, 349 (1977)
4.80   P. Williams, I.S.T. Tsong, S. Tsuji: Nucl. Instrum. Methods **170**, 591 (1980)
4.81   I.S.T. Tsong: In *Inelastic Particle-Solid Collisions*, ed. by E. Taglauer, W. Heiland, Springer Ser. Chem. Phys. Vol. 17 (Springer, Berlin, Heidelberg 1981) p. 258
4.82   R. Berneron, J.C. Carbonnier: Surface Interf. Anal. **3**, 134 (1981)
4.83   J. Pons-Corbeau: Surface Interf. Anal. **7**, 169 (1985)
4.84   M.G. Barker, I.E. Schreinlechner: Surface Interf. Anal. **9**, 371 (1986)
4.85   D.P. Smith: J. Appl. Phys. **38**, 340 (1967); Surface Sci. **25**, 171 (1971)
4.86   R.F. Goff, D.P. Smith: J. Vac. Sci. Technol. **7**, 72 (1970)
4.87   R.E. Honig, W.L. Harrington: Thin Sol. Films **19**, 43 (1973)
4.88   D.W. Hoffmann, R. Nimmagadda. J. Vac. Sci. Technol. **11**, 657 (1974)
4.89   T.W. Rusch, J.T. McKinney, J.A. Leys: J. Vac. Sci. Technol. **12**, 400 (1975)
4.90   P.J. Schneider, W. Eckstein, H. Verbeek: Nucl. Instrum. Methods **218**, 713 (1983)
4.91   Y.-S. Jo, J.A. Schultz, S. Tachi, S. Contarini, J.W. Rabalais: J. Appl. Phys. **60**, 2564 (1986)
4.92   D.F. Torgerson, B.P. Skowronski, R.D. Macfarlane: Biochem. Biophys. Res. Comm. **60**, 616 (1974)
4.93   A. Benninghoven, D. Jaspers, W. Sichtermann: Appl. Phys. **11**, 35 (1976)
4.94   A. Benninghoven, F. Kirchner: Z. Naturforschg. **189**, 1008 (1963)
4.95   H. Gnaser, J. Fleischhauer, W.O. Hofer: Appl. Phys. **A37**, 211 (1985)
4.96   D. Lipinsky, R. Jede, O. Ganschow, A. Benninghoven: J. Vac. Sci. Technol. **A3**, 2007 (1985)
4.97   G.S. Hurst, M.G. Payne, S.D. Kramer, J.P. Young: Rev. Mod. Phys. **51**, 767 (1979)
4.98   N. Winograd, J.P. Baxter, F.M. Kimock: Chem. Phys. Lett. **88**, 581 (1982)
4.99   F.M. Kimock, J.P. Baxter, N. Winograd: Nucl. Instrum. Methods **218**, 287 (1983)
4.100 J.E. Parks, H.W. Schmitt, G.S. Hurst, W.H. Fairbanks, Jr.: Thin Solid Films **8**, 69 (1983)
4.101 M.J. Pellin, C.E. Young, W.F. Calaway, D.M. Gruen: Surface. Sci. **144**, 619 (1984)
4.102 M.J. Pellin, C.E. Young, W.F. Calaway, J. Burnett, B. Jorgensen, E.L. Schweitzer, D.M. Gruen: Nucl. Instrum. Methods: **B18**, 446 (1987)
4.103 C.H. Becker, K.T. Gillen: Anal. Chem. **56**, 1671 (1984)

4.104 C.H. Becker, K.T. Gillen: J. Vac. Sci. Technol. **A3**, 1347 (1985)
4.105 C.E. Young, M.J. Pellin, W.F. Calaway, B. Jorgensen, E.J. Schweitzer, D.M. Gruen: Nucl. Instrum. Methods **B27**, 119 (1987)
4.106 K. Wittmaack, N. Menzel: Appl. Phys. Lett. **53**, 1708 (1988)
4.107 P. Sigmund: Phys. Rev. **184**, 383 (1969); **187**, 768 (1969)
4.108 P. Sigmund: In *Sputtering by Particle Bombardment I*, Topics Appl. Phys., Vol. 47, ed. by R. Behrisch (Springer, Berlin, Heidelberg 1981) p. 9
4.109 J. Lindhard, V. Nielsen, M. Scharff: K. Dan. Vidensk. Selsk. Mat. Fys. Medd. **36**, No. 10 (1968)
4.110 G.K. Kinchin, R.S. Pease: Rep. Progr. Phys. **18**, 1 (1955)
4.111 P. Sigmund: Appl. Phys. Lett. **14**, 114 (1969)
4.112 G. Carter, N.J. Nobes, D.G. Armour: Vacuum **32**, 509 (1982)
4.113 K. Wittmaack: Vaccum **34**, 119 (1984)
4.114 P.K. Haff, Z.E. Switkowski: J. Appl. Phys. **48**, 3383 (1977)
4.115 H.H. Andersen: Appl. Phys. **18**, 131 (1979)
4.116 A. Benninghoven: Z. Phys. **230**, 403 (1970)
4.117 S. Hofmann: Appl. Phys. **9**, 59 (1976); **13**, 205 (1977)
4.118 K. Wittmaack, F. Schulz: Thin Solid Films **52**, 259 (1978)
4.119 J. Erlewein, S. Hofmann: Thin Solid Films **69**, L39 (1980)
4.120 M.P. Seah, J.M. Sanz, S. Hofmann: Thin Solid Films **81**, 239 (1981)
4.121 B. Navinsek: Progr. Surface Sci. **7**, 49 (1976)
4.122 V. Naundorf, M.-P. Macht: Nucl. Instrum. Methods **168**, 405 (1980)
4.123 G. Carter, B. Navinsek, J.L. Whitton: In *Sputtering by Particle Bombardment II*, ed. by R. Behrisch (Springer, Berlin, Heidelberg 1983) p. 231
4.124 M.P. Seah, M.E. Jones: Thin Solid Films **115**, 203 (1984)
4.125 H.E. Roosendaal: In *Sputtering by Particle Bombardment I*, Topics Appl. Phys., Vol. 47, ed. by R. Behrisch (Springer, Berlin, Heidelberg 1981) p. 219
4.126 D.E. Sykes, D.D. Hall, R.E. Thurstans, J.M. Walls: Appl. Surface Sci. **5**, 103 (1980)
4.127 D.F. Mitchell, G.I. Sproule: Surface Sci. **177**, 238 (1986)
4.128 D. Coulman, A. Turner: Mat. Res. Soc. Symp. Proc. **69**, 135 (1986)
4.129 A. Zalar: Thin Solid Films **124**, 223 (1985)
4.130 K. Tsunoyama, Y. Ohashi, T. Suzuki, K. Tsuruoka: Japan. J. Appl. Phys. **13**, 1683 (1974)
4.131 K. Tsunoyama, T. Suzuki, Y. Ohashi: Japan, J. Appl. Phys. **15**, 349 (1976)
4.132 W.O. Hofer, H. Liebl: Appl. Phys. **8**, 359 (1975)
4.133 R.J. Blattner, S. Nadel, C.A. Evans, Jr., A.J. Brandmeier, Jr., C.W. Magee: Surface Interf. Anal. **1**, 32 (1979)
4.134 B.M.U. Scherzer: In *Sputtering by Particle Bombardment II*, Topics Appl. Phys, Vol. 52, ed. by R. Behrisch (Springer, Berlin, Heidelberg 1983) p. 271
4.135 K. Wittmaack, W. Wach: Appl. Phys. Lett. **32**, 532 (1978)
4.136 N. Menzel, K. Wittmaack: Nucl. Instrum. Methods **B7/8**, 366 (1985)
4.137 G.N.A. van Veen, F.H.M. Sanders, J. Dieleman, A. van Veen, D.J. Oostra, A.E. de Vries: Phys. Rev. Lett. **57**, 739 (1986)
4.138 Z.L. Liau, T.T. Sheng: Appl. Phys. Lett. **32**, 716 (1978)
4.139 R.M. Feenstra, G.S. Oehrlein: Appl. Phys. Lett. **47**, 97 (1985)
4.140 D.E. Harrison, Jr.: Radiat. Eff. **70**, 85 (1983)
4.141 H.H. Andersen: Nucl. Instrum. Methods **B18**, 321 (1987)
4.142 J.P. Biersack: Nucl. Instrum. Methods **B27**, 21 (1987)
4.143 T. Ishitani, R. Shimizu: Phys. Lett. **46A**, 487 (1974)
4.144 M.T. Robinson: J. Appl. Phys. **54**, 2650 (1983)
4.145 M. Rosen, G.A. Mueller, W.A. Fraser: Nucl. Instrum. Methods **209/210**, 63 (1983)
4.146 M. Rosen, R.H. *Bassel*: Nucl. Instrum. Methods **B2**, 592 (1984)
4.147 J.P. Biersack, W. Eckstein: Appl. Phys. **A34**, 73 (1984)
4.148 W. Eckstein, W. Möller: Nucl. Instrum. Methods **B7/8**, 727 (1985)
4.149 J.P. Biersack: Fusion Technol. **6**, 475 (1984)
4.150 G. Falcone, P. Sigmund: Appl. Phys. **25**, 307 (1981)

4.151 P. Sigmund: Nucl. Instrum. Methods B27, 1 (1987)
4.152 R. Kelly, A. Oliva: Nucl. Instrum. Methods B13, 283 (1986)
4.153 G. Falcone, R. Kelly, A. Oliva: Nucl. Instrum. Methods B18, 399 (1987)
4.154 A. Oliva, R. Kelly, G. Falcone: Nucl. Instrum. Methods B19/20, 101 (1987)
4.155 G. Falcone: Surf. Sci. 187, 212 (1987)
4.156 M.F. Dumke, T.A. Tombrello, R.A. Weller, R.M. Hously, E.H. Cirlin: Surface Sci. 124, 407
       (1983); see also T.A. Tombrello: Nucl. Instrum. Methods B27, 221 (1987)
4.157 H. Niehus, E.G. Bauer: Electron. Fisc. Apl. 17, 53 (1974)
4.158 S. Prigge, E. Bauer: In Advances in Mass Spectrometry, Vol. 8A, ed. by A. Quale (Heyden,
       London 1980) p. 543
4.159 J.W. Burnett, J.P. Biersack, D.M. Gruen, B. Jorgensen, A.R. Krauss, M.J. Pellin,
       E.L. Schweitzer, Y.T. Yates, Jr., C.E. Young: J. Vac. Sci. Technol. A6, 2064 (1988)
4.160 W. Wach, K. Wittmaack: Phys. Rev. B27, 3528 (1983)
4.161 U. Littmark, W.O. Hofer: Nucl. Instrum. Methods 168, 329 (1980)
4.162 H.W. Etzkorn, U. Littmark, J. Kirschner: In Symposium on Sputtering, ed. by P. Varga,
       G. Betz, F.P. Viehböck (Institut für Allgemeine Physik, TU Wien 1980) p. 542
4.163 I.S.T. Tsong, O.F. Sankey: private communication (1986)
4.164 B.V. King, I.S.T. Tsong: J. Vac. Sci. Technol. A2, 1443 (1984)
4.165 P.G. Shewmon: Diffusion in Solids (McGraw-Hill, New York 1963)
4.166 Z.L. Liau, B.Y. Tsaur, J.W. Mayer: J. Vac. Sci. Technol. 16, 121 (1979)
4.167 K. Wittmaack: J. Appl. Phys. 53, 4817 (1982)
4.168 K. Wittmaack, N. Menzel: Europ. Conf. on Applications of Surface and Interface Analysis,
       Fellbach (1987), Book of Abstracts p. 112
4.169 P. Williams: Appl. Phys. Lett. 36, 758 (1980)
4.170 K. Wittmaack: Nucl. Instrum. Methods 209/210, 191 (1983)
4.171 D.G. Swartzfager, S.B. Ziemecki, M.J. Kelly: J. Vac. Sci. Technol. 19, 185 (1981)
4.172 L.E. Rehn, N.Q. Lam, H. Wiedersich: Nucl. Instrum. Methods B7/8, 764 (1985)
4.173 P. Boudewijn, H.W.P. Akerboom, M.N.C. Kempeners: Spectrochim. Acta B39, 1567 (1984)
4.174 S.D. Littlewood, J.A. Kilner, J.P. Gold: In Secondary Ion Mass Spectrometry VI, ed. by
       A. Benninghoven, A.M. Huber, H.W. Werner, (Wiley, Chichester, Ner York 1988) p. 737
4.175 W. Vandervorst, F.R. Shepherd, M.L. Swanson, H.H. Plattner, O.W. Westcott, I.V. Mitchell:
       Nucl. Instrum. Methods B15, 201 (1986)
4.176 K. Wittmaack, N. Menzel: Appl. Phys. Lett. 50, 815 (1987)
4.177 K. Wittmaack: Nucl. Instrum. Methods B7/8, 750 (1985)
4.178 K. Wittmaack: J. Vac. Sci. Technol. A4, 1662 (1986)
4.179 Z.L. Liau, J.W. Mayer, W.L. Brown, J.M. Poate: J. Appl. Phys. 49, 5295 (1978)
4.180 K. Wittmaack: J. Vac. Sci. Technol. A3, 1350 (1985)
4.181 E. Gillam: J. Phys. Chem. Solids 11, 55 (1959)
4.182 W.L. Patterson, G.A. Shirn: J. Vac. Sci. Technol. 4, 343 (1967)
4.183 H.F. Winters, J.W. Coburn: Appl. Phys. Lett. 28, 176 (1976)
4.184 N.J. Chou, N.W. Shafer: Surface Sci. 92, 601 (1980)
4.185 R.A. Kubiak, E.H.C. Parker, R.M. King, K. Wittmaack: J. Vac. Sci. Technol. A1, 34 (1983)
4.186 D. Briggs, J.C. Revière: In Practical Surface Analysis by Auger and X-ray Photoelectron
       Spectroscopy, ed. by D. Briggs, M.P. Seah (Wiley, Chichester 1983) p. 87
4.187 M.P. Seah, W.A. Dench: Surface Interf. Anal. 1, 2 (1979)
4.188 C.J. Powell: J. Vac. Sci. Technol. A4, 1532 (1986)
4.189 J. Powell: Surface Sci. 44, 29 (1974)
4.190 C.D. Wagner, L.E. Davis, W.M. Riggs: Surface Interf. Anal. 2, 53 (1980)
4.191 C.J. Powell: Surface Interf. Anal. 7, 256 (1985)
4.192 T.J. Chuang, K. Wandelt: IBM J. Res. Dev. 22, 277 (1978)
4.193 J. Kirschner, H.W. Etzkorn: In Thin Film and Depth Profile Analysis, ed. by H. Oechsner
       (Springer, Berlin, Heidelberg 1984) p. 103
4.194 H. Iwasaki, S. Nakamura: Surface Sci. 57, 779 (1976)
4.195 R.P. Frankenthal, D.J. Siconolfi: Surface Interf. Anal. 7, 223 (1985)

4.196 S.A. Schwarz, C.R. Helms, W.E. Spicer, N.J. Taylor: J. Vac. Sci. Technol. **15**, 227 (1978)
4.197 C.R. Helms, N.M. Johnson, S.A. Schwarz, W.E. Spicer: J. Appl. Phys. **50**, 7007 (1979)
4.198 J. Kirschner, H.W. Etzkorn: Scanning Electr. Microsc. **1982/I**, 93 (1982)
4.199 J. Kirschner, H.W. Etzkorn: Appl. Surface Sci. **14**, 221 (1982/83)
4.200 H.W. Etzkorn, J. Kirschner: Nucl. Instrum. Methods **168**, 395 (1980)
4.201 W. Heiland, H.G. Schäffler, E. Taglauer: Surface Sci. **35**, 381 (1973)
4.202 H.H. Brongersma, P.M. Mul: Surface Sci. **35**, 393 (1973)
4.203 H. Niehus, E. Bauer: Surface Sci. **47**, 222 (1975)
4.204 T.M. Buck: In *Methods of Surface Analysis*, ed. by A.W. Czanderna (Elsevier, Amsterdam 1975) p. 75
4.205 E. Taglauer, W. Heiland: Appl. Phys. **9**, 261 (1976)
4.206 T.W. Rusch, R.L. Erickson: J. Vac. Sci. Technol. **13**, 374 (1976)
4.207 E. Taglauer, W. Englert, W. Heiland, D.P. Jackson: Phys. Rev. Lett. **45**, 740 (1980)
4.208 W. Eckstein: In *Inelastic Particle-Surface Collisions*, ed. by E. Taglauer, W. Heiland (Springer, Berlin, Heidelberg 1981) p. 157
4.209 J.A. Van den Berg, D.G. Armour: Vacuum **31**, 259 (1981)
4.210 A. Zartner: IPP Report 9/31 (Max-Planck-Institut für Plasmaphysik, Garching 1979)
4.211 W.K. Chu, J.W. Mayer, M.-A. Nicolet: *Blackscattering Spectrometry* (Academic, New York 1978)
4.212 J. L'Ecuyer, J.A. Davies, N. Matsunami: Nucl. Instrum. Methods **160**, 337 (1979)
4.213 M. Hautala, M. Luomajärvi: Radiat. Eff. **45**, 159 (1980)
4.214 J.F. Ziegler: *Helium Stopping Powers and Ranges in All Elemental Matter* (Pergamon, New York 1977)
4.215 R.G. Smeenk, R.M. Tromp, F.W. Saris: Surface Sci. **112**, 261 (1982)
4.216 E.J. van Loenen, J.F. van der Veen, F.K. Legoues: Surface Sci. **157**, 1 (1985)
4.217 A. Feuerstein, H. Grahmann, S. Kalbitzer, H. Oetzmann: In *Ion Beam Surface Layer Analysis*, Vol. 1, ed. by O. Meyer, G. Linker, F. Käppeler (Plenum, New York 1976) p. 471
4.218 J.S. Williams: Nucl. Instrum. Methods **126**, 205 (1975)
4.219 K. Wittmaack: Appl. Phys. **12**, 149 (1975)
4.220 D.S. McPhail, M.G. Dowsett, H. Fox, R. Houghton, W.Y. Leong, E.H.C. Parker, G.K. Patel: Surface Interf. Anal. **11**, 80 (1988)
4.221 J. Kempf. Surface Interf. Anal. **4**, 116 (1982)
4.222 J.E. Kempf, H.H. Wagner: In *Thin Film and Depth Profile Analysis*, ed. by H. Oechsner (Springer, Berlin, Heidelberg 1984) p. 87
4.223 H.H. Andersen, H.L. Bay: J. Appl. Phys. **46**, 1919 (1975)
4.224 P. Blank, K. Wittmaack: J. Appl. Phys. **50**, 1519 (1979)
4.225 J. Kirschner, H.W. Etzkorn: Appl. Surface Sci. **3**, 251 (1979)
4.226 K. Wittmaack, W. Wach: Nucl. Instrum. Methods **191**, 327 (1981)
4.227 J. Kirschner, H.W. Etzkorn: Appl. Phys. A**29**, 133 (1982)
4.228 K. Wittmaack: Nucl. Instrum. Methods B**2**, 569 (1984)
4.229 W. Vandervorst, F.R. Shepherd: J. Vac. Sci. Technol. A**5**, 313 (1987)
4.230 W. Vandervorst, F.R. Shepherd, J. Newman, B.F. Phillips, J. Remmerie: J. Vac. Sci. Technol. A**3**, 1359 (1985)
4.231 J.B. Clegg: Surface Interf. Anal. **10**, 332 (1987)
4.232 H.S. Fox, M.G. Dowsett, R.F. Houghton: In *Secondary Ion Mass Spectrometry SIMS VI*, ed. by A. Benninghoven, A.M. Huber, H.W. Werner (Wiley, Chichester, New York 1988) p. 445
4.233 R.E. Whan, G.W. Arnold: Appl. Phys. Lett. **17**, 378 (1970)
4.234 K. Wittmaack, G. Staudenmaier: J. Nucl. Mat. **93 & 94**, 581 (1980)
4.235 J. Kempf, M. Nonnenmacher, H.H. Wagner: Appl. Phys. A**47**, 137 (1988)
4.236 R. Butz, H. Wagner: Phys. Stat. Sol. (a) **3**, 325 (1970)
4.237 W.O. Hofer: Thin Solid Films **29**, 223 (1975)
4.238 L.C. Feldman, J.M. Poate, F. Ermanis, B. Schwartz: Thin Solid Films **19**, 81 (1973)
4.239 B.D. Sartwell: J. Appl. Phys. **50**, 7887 (1979)

4.240 F. Folkmann: In *Ion Beam Surface Layer Analysis*, Vol. 2, ed. by O. Meyer, G. Linker, F. Käppeler (Plenum, New York 1976) p. 747

4.241 R. Behrisch, R. Weißmann: Phys: Lett. **30A**, 506 (1969)

4.242 R.R. Hart, H.L. Dunlap, O.J. Marsh: J. Appl. Phys. **46**, 1947 (1975)

4.243 P. Blank, K. Wittmaack: Radiat, Eff. **27**, 29 (1975)

4.244 Z.L. Liau, W.L. Brown, R. Homer, J.M. Poate: Appl. Lett. **30**, 626 (1977)

4.245 W. Wach, K. Wittmaack: Nucl. Instrum. Methods **149**, 259 (1978)

4.246 W.L. Brown, C.T. Reimann, R.E. Johnson: Nucl. Instrum. Methods **B19/20**, 9 (1987)

4.247 A.C. Hall: In *Characterization of Solid Surfaces*, ed. by P.F. Kane, G.B. Larrabee (Plenum, New York 1974) p. 33

4.248 D.J. Whitehouse: ibid, p. 49

4.249 A.E. Morgan, H.A.M. de Grefte, H.J. Tolle: J. Vac. Sci. Technol. **18**, 164 (1981)

4.250 A.E. Morgan, H.A.M. de Grefte, N. Warmoltz, H.W. Werner, H.J. Tolle: Appl. Surface Sci. **7**, 372 (1981)

4.251 K. Wittmaack: Nucl. Instrum. Methods **218**, 307 (1983)

4.252 Ch. Meyer, M. Maier, D. Bimberg: J. Appl. Phys. **54**, 2672 (1983)

4.253 M. Meuris, W. Vandervorst, G. Borghs, H.E. Maes: In *Secondary Ion Mass Spectrometry SIMS VI*, ed. by A. Benninghoven, A.M. Huber, H.W. Werner (Wiley, Chichester, New York 1988) p. 277

4.254 W. Katz, P. Williams, C.A. Evans, Jr., Surface Interf. Anal. **2**, 120 (1980)

4.255 R. Voigtmann, W. Moldenhauer: Surface Interf. Anal. **13**, 167 (1988)

4.256 G. Slodzian: Scanning Microsc. Suppl. **1**, 1 (1987)

4.257 H.N. Migeon, A.E. Morgan: In *Secondary Ion Mass Spectrometry SIMS IV*, ed. by A. Benninghoven, J. Okano, R. Shimizu, H.W. Werner (Springer, Berlin, Heidelberg 1984) p. 299

4.258 K. Wittmaack: Rev. Sci. Instrum. **47**, 157 (1976); In *Ion Beam Surface Layer Analysis*, Vol. 2, ed. by O. Meyer, G. Linker, F. Käppeler (Plenum, New York 1976) p. 649

4.259 K. Wittmaack: Scanning **3**, 133 (1980)

4.260 T.A. Whatley, C.B. Slack, E. Davidson: In *Proc. Sixth Int. Conf. X-Ray Optics and Micro-analysis*, ed. by G. Shinoda, K. Kohra, T. Ichinokawa (University of Tokyo Press 1971) p. 417

4.261 W.O. Hofer, H. Liebl, G. Roos, G. Staudenmaier: Int. J. Mass Spectrom. Ion Phys. **19**, 327 (1976)

4.262 P. Williams, C.A. Evans, Jr.: Int. J. Mass Spectrom. Ion Phys. **22**, 327 (1976)

4.263 R.v. Criegern, I. Weitzel, J. Fottner: In *Secondary Ion Mass Spectrometry SIMS IV*, ed. by A. Benninghoven, J. Okano, R. Shimizu, H.W. Werner (Springer, Berlin, Heidelberg 1984) p. 308

4.264 H. Frenzel, J.L. Maul, H. Mertens, R. Raab, Ch. Scholze: In *Secondary Ion Mass Spectrometry SIMS VI*, ed. by A. Benninghoven, A.M. Huber, H.W. Werner (Wiley, Chichester, New York 1988) p. 219

4.265 P.R. Boudewijn, H.W.P. Akerboom, C.W.T. Belle-Lieuwma, J. Haisma: Surface Interf. Anal. **7**, 49 (1985)

4.266 S.G. Puranik, B.V. King: Appl. Surface Sci. **28**, 180 (1987)

4.267 N.Q. Lam, H.A. Hoff: Surface Sci. **193**, 353 (1988)

4.268 H.J. Mathieu, D. Landolt: Surface Interf. Anal. **5**, 77 (1983)

4.269 R.R. Olson, P.W. Palmberg, C.T. Hovland, T.E. Brady: In *Practical Surface Analysis by Auger and X-ray Photoelectron Spectroscopy*, ed. by D. Briggs, M.P. Seah (Wiley, Chichester 1983) p. 217

4.270 S. Thomas: J. Appl. Phys. **45**, 161 (1974); Surface Sci. **55**, 754 (1976)

4.271 S. Hofmann, A. Zalar: Thin Solid Films **56**, 337 (1979)

4.272 K. Röll, W. Losch, C. Achete: J. Appl. Phys. **50**, 4422 (1979)

4.273 C.G. Pantano, T. Madey: Appl. Surface Sci. **7**, 115 (1981)

4.274 H.J. Mathieu: In *Thin Film and Depth Profile Analysis*, ed. by H. Oechsner (Springer, Berlin, Heidelberg 1984) p. 39

4.275 R.E. Honig: Int. J. Mass Spectrom. Ion Proc. **66**, 31 (1985)
4.276 A. Benninghoven, F.G. Rüdenauer, H.W. Werner: *Secondary Ion Mass Spectrometry* (Wiley, New York, Chichester 1987)
4.277 Y. Homma, Y. Ishii: J. Vac. Sci. Technol. A3, 356 (1985)
4.278 M. Lepareur: Rev. Techn. Thomson-CSF **12**, 225 (1980)
4.279 J.J. Le Goux, N.N. Migeon: In *Secondary Ion Mass Spectrometry SIMS III*, ed. by A. Benninghoven, J. Giber, J. Laszlo, M. Riedel, H.W. Werner (Springer, Berlin, Heidelberg 1982) p. 52
4.280 H.A. Storms, K.F. Brown, H.D. Stein: Anal. Chem. **49**, 2023 (1977)
4.281 V.R. Deline, C.A. Evans, Jr., P. Williams: Appl. Phys. Lett. **33**, 578 (1978)
4.282 V.R. Deline, W. Katz, C.A. Evans, Jr., P. Williams: Appl. Phys. Lett. **33**, 832 (1978)
4.283 K. Wittmaack: J. Appl. Phys. **52**, 527 (1981)
4.284 K. Wittmaack: Surface Sci. **112**, 168 (1981)
4.285 K. Wittmaack: Surface Sci. **126**, 573 (1983)
4.286 P.R. Boudewijn, M.R. Leys, F. Roozeboom: Surface Interf. Anal. **9**, 303 (1986)
4.287 T. Ambridge: Scanning El. Microsc. **1983**, 31 (1983)
4.288 A.M. Huber, G. Morillot, A. Friederich: In *Secondary Ion Mass Spectrometry SIMS IV*, ed. by A. Benninghoven, J. Okano, R. Shimizu, H.W. Werner (Springer, Berlin, Heidelberg 1984) p. 278
4.289 Y. Homma, Y. Ishii: In *Secondary Ion Mass Spectrometry SIMS V*, ed. by A. Benninghoven, R.J. Colton, D.S. Simons, H.W. Werner (Springer, Berlin, Heidelberg 1986) p. 161
4.290 M. Gauneau, R. Chaplain, A. Regreny, M. Salvi, C. Guillemot, R. Azoulay, N. Duhamel: Surface Interf. Anal. **11**, 545 (1988)
4.291 A.E. Morgan, P. Maillot: In *Secondary Ion Mass Spectrometry SIMS VI*, ed. by A. Benninghoven, A.M. Huber, H.W. Werner (Wiley, Chichester, New York 1988) p. 709
4.292 V.R. Deline: Nucl. Instrum. Methods **218**, 316 (1983)
4.293 W. Szymczak, K. Wittmaack: In *Secondary Ion Mass Spectrometry SIMS VI*, ed. by A. Benninghoven, A.M. Huber, H.W. Werner (Wiley, Chichester, New York 1988) p. 243
4.294 W. Vandervorst, F.R. Shepherd: Appl. Surface Sci. **21**, 230 (1985). Note that the abscissa in Fig. 3 is in error. The numbers for the final primary ion energy should be multiplied by a factor of 2. With reference to O_2^+ the quoted numbers may be considered an energy per oxygen atom (keV/atom).
4.295 K. Wittmaack: Nucl. Instrum. Methods **143**, 1 (1977)
4.296 R. Raab: Private communication (1988)
4.297 F. Simondet, P. Staib: In *Secondary Ion Mass Spectrometry SIMS IV*, ed. by A. Benninghoven, J. Okano, R. Shimizu, H.W. Werner (Springer, Berlin, Heidelberg 1984) p. 144
4.298 D.G. Welkie, R.L. Gerlach: ibid, p. 317
4.299 R. Levi-Setti, Y.L. Wang, G. Crow: J. Phys. (Paris) **45**, C9–197 (1984)
4.300 K. Wittmaack, M.G. Dowsett, J.B. Clegg: Int. J. Mass Spectrom. Ion Phys. **43**, 31 (1982)
4.301 K. Wittmaack: Appl. Phys. Lett. **29**, 552 (1976)
4.302 K. Wittmaack, J. Maul, F. Schulz: In *Electron and Ion Beam Science and Technology* (Sixth Int. Conf.), ed. by R. Bakish (The Electrochem. Soc., Princeton, N.J. 1974) p. 164
4.303 G. Müller: Appl. Phys. **10**, 317 (1976)
4.304 C.W. Magee, W.L. Harrington: Appl. Phys. Lett. **33**, 193 (1978)
4.305 K. Wittmaack: J. Appl. Phys. **50**, 493 (1979)
4.306 W. Reuter, M.L. Yu, M.A. Frisch, M.B. Small: J. Appl. Phys. **51**, 850 (1980)
4.307 C.P. Hunt, C.T.H. Stoddart, M.P. Seah: Surface Interf. Anal. **3**, 157 (1981)
4.308 A. Brown, J.C. Vickerman: Surface Interf. Anal. **8**, 75 (1986)
4.309 D.S. McPhail, M.G. Dowsett, E.H.C. Parker: J. Appl. Phys. **60**, 2573 (1986)
4.310 K. Wittmaack: Surface Interf. Anal. **10**, 311 (1987)
4.311 G. Slodzian: Optik **77**, 148 (1987)
4.312 Y. Homma, Y. Ishii, M. Oshima: Mass Spectrosc. **32**, 345 (1984)
4.313 S.D. Littlewood, J.A. Kilner, S. Biswas: In *Secondary Ion Mass Spectrometry SIMS VI*, ed. by A. Benninghoven, A.M. Huber, H.W. Werner (Wiley, Chichester, New York 1988) p. 89

4.314 W. Paul, H.P. Reinhard, U. von Zahn: Z. Phys. **152**, 143 (1958)
4.315 K. Wittmaack: In *Proc. 7th Int. Vacuum Congress & 3rd Int. Conf. Solid Surfaces*, Vol. III, ed. by R. Dobrozemsky, F. Rüdenauer, F.P. Viehböck, A. Breth (IVC & ICSS, Vienna 1977) p. 2573
4.316 L.J. Kiefer, G.H. Dunn: Rev. Mod. Phys. **38**, 1 (1966)
4.317 J. Tümpner: In *Secondary Ion Mass Spectrometry SIMS VI*, ed. by A. Benninghoven, A.M. Huber, H.W. Werner (Wiley, Chichester, New York 1988) p. 797
4.318 J. Kiko, H.W. Müller, K. Büchler, S. Kalbitzer, T. Kirsten, M. Warhaut: Int. J. Mass Spectrom. Ion Phys. **29**, 87 (1979)
4.319 H. Gnaser, H.L. Bay, W.O. Hofer: Nucl. Instrum. Methods **B15**, 49 (1986)
4.320 G. Blaise, R. Castaing: C.R. Acad. Sci. **B284**, 449 (1977)
4.321 R. Castaing, G. Blaise: In *X-ray Optics and Microanalysis*, ed. by D.R. Beamann, R.E. Ogilvie, D.B. Wittry (Pendel, Midland 1980) p. 3
4.322 G. Blaise: Scanning Electron Microscopy **1985/I**, 31 (1985)
4.323 E. Darque-Ceretti, F. Delamare, G. Blaise: Surface Interf. Anal. **7**, 141 (1985)
4.324 H. Oechsner, H. Schoof, E. Stumpe: Surface Sci. **76**, 343 (1978)
4.325 E. Stumpe, H. Oechsner, H. Schoof: Appl. Phys. **20**, 55 (1979)
4.326 H. Oechsner, H. Paulus, P. Beckmann: J. Vac. Sci. Technol. **A3**, 1403 (1985)
4.327 H. Oechsner: In *Secondary Ion Mass Spectrometry SIMS V*, ed. by A. Benninghoven, R.J. Colton, D.S. Simons, H.W. Werner (Springer, Berlin, Heidelberg 1986) p. 70
4.328 H. Oechsner, W. Rühe, E. Stumpe: Surface Sci. **85**, 289 (1979)
4.329 A. Wucher, H. Oechsner: Nucl. Instrum. Methods **B18**, 458 (1987)
4.330 A. Wucher, H. Oechsner: Surface Sci. **199**, 567 (1988)
4.331 A. Wucher, H. Oechsner: In *Secondary Ion Mass Spectrometry SIMS VI*, ed. by A. Benninghoven, A.M. Huber, H.W. Werner (Wiley, Chichester, New York 1988) p. 143
4.332 A. Wucher, W. Reuter: J. Vac. Sci. Technol. **A6**, 2316 (1988)
4.333 R. Jede, H. Peters, G. Dünnebier, O. Ganschow, U. Kaiser, K. Seifert: J. Vac. Sci. Technol. **A6**, 2271 (1988)
4.334 W. Reuter: In *Secondary Ion Mass Spectrometry SIMS V*, ed. by A. Benninghoven, R.J. Colton, D.S. Simons, H.W. Werner (Springer, Berlin, Heidelberg 1986) p. 94
4.335 A. Wucher, F. Novak, W. Reuter: J. Vac. Sci. Technol. **A6**, 2265 (1988)
4.336 J.W. Coburn, E. Kay: Appl. Phys. Lett. **18**, 435 (1971)
4.337 W.W. Harrison, K.R. Hess, R.K. Marcus, F.L. King: Anal. Chem. **58**, 341A (1986)
4.338 D.J. Hall, N.E. Sanderson: Surface Interf. Anal. **11**, 40 (1988)
4.339 J.B. Hasted: *Physics of Atomic Collisions* (Butterworths, London 1964)
4.340 J.-F. Hennequin, G. Blaise, G. Slodzian: C.R. Acad. Sci. (Paris) **268**, B 1507 (1969)
4.341 M. Bernheim, G. Blaise, G. Slodzian: Int. J. Mass Spectrom. Ion Phys. **10**, 293 (1972/73)
4.342 K. Wittmaack: Nucl. Instrum. Methods **132**, 381 (1976)
4.343 K. Wittmaack: In *Advances in Mass Spectrometry*, Vol. 8A, ed. by A. Quale (Heyden London 1980) p. 503
4.344 R. Holland, G.W. Blackmore: Surface Interf. Anal. **4**, 174 (1982); Int. J. Mass Spectrom. Ion Phys. **46**, 527 (1983)
4.345 P. Williams, L.A. Streit: Nucl. Instrum. Methods **B15**, 159 (1986)
4.346 P. De Bisschop, W. Vandervorst: In *Secondary Ion Mass Spectrometry SIMS VI*, ed. by A. Benninghoven, A.M. Huber, H.W. Werner (Wiley, Chichester, New York 1988) p. 809
4.347 M.W. Thompson: Phil. Mag. **18**, 377 (1968)
4.348 G. Betz: Nucl. Instrum. Methods **B27**, 107 (1987)
4.349 C.H. Becker, K.T. Gillen: J. Opt. Soc. Am. **B2**, 1438 (1984)
4.350 C.H. Becker, K.T. Gillen: In *Secondary Ion Mass Spectrometry SIMS V*, ed. by A. Benninghoven, R.J. Colton, D.S. Simons, H.W. Werner (Springer, Berlin, Heidelberg 1986) p. 85
4.351 G.E. Thomas: Surface Sci. **90**, 381 (1979)
4.352 M. Braun, B. Emmoth, R. Buchta: Radiat. Eff. **28**, 77 (1976)
4.353 N.H. Tolk, I.S.T. Tsong, C.W. White: Anal. Chem. **49**, 16A (1977)
4.354 R. Shimizu, T. Okutani, T. Ishitani, H. Tamura: Surface Sci. **69**, 349 (1977)

4.355 I.S. Tsong, G.L. Power, D.W. Hoffman, C.W. Magee: Nucl. Instrum. Methods **168,** 399 (1980)
4.356 R. Goutte, C. Guillaud, R. Javelas, J.P. Meriaux: Optik **26,** 574 (1967)
4.357 J.P. Meriaux, R. Goutte, C. Guillaud: J. Radioanal. Chem. **12,** 53 (1972)
4.358 J.E. Greene, F. Sequeda-Osorio, B.R. Natarajan: J. Appl. Phys. **46,** 2701 (1975)
4.359 G.T. Marcyk, B.G. Streetman: J. Electrochem. Soc. **123,** 1388 (1976)
4.360 P.W.J.M. Boumans, Anal. Chem. **44,** 1219 (1972)
4.361 J.E. Greene, F. Sequeda-Osorio, B.G. Streetman, J.R. Noonan, C.G. Kirkpatrick, Appl. Phys. Lett. **25,** 435 (1974)
4.362 B. Wannberg, U. Gelius, K. Siegbahn: J. Phys. E: Sci. Instrum. **7,** 149 (1974)
4.363 W.M. Riggs, M.J. Parker: In *Methods of Surface Analysis*, ed. by A.W. Czanderna (Elsevier, Amsterdam 1975) p. 103
4.364 P.W. Palmberg: J. Vac. Sci. Technol. **12,** 379 (1975)
4.365 E. Gisler, E.B. Bas: Vacuum **36,** 715 (1986)
4.366 E.L. Barish, D.J. Vitkavage, T.M. Mayer: J. Appl. Phys. **57,** 1336 (1985)
4.367 M. Grundner, W. Heiland, E. Taglauer: Appl. Phys. **4,** 243 (1974)
4.368 M. Bernheim, G. Slodzian: Nucl. Instrum. Methods **132,** 615 (1976)
4.369 D.A. Reed, J.E. Baker: Nucl. Instrum. Methods **218,** 324 (1983)
4.370 K. Wittmaack: Seventh Int. Workshop on *Inelastic Ion Surface Collisions*, Krakow (1988), unpublished
4.371 W.L. Harrington, R.E. Honig, A.M. Goodman, R. Williams: Appl. Phys. Lett. **27,** 644 (1975)
4.372 D.E. Carlson, C.W. Magee: Appl. Phys. Lett. **33,** 81 (1978)
4.373 C.W. Magee, C.P. Wu: Nucl. Instrum. Methods **149,** 529 (1978); C.W. Magee, S.A. Cohen, D.E. Voss, D.K. Brice: Nucl. Instrum. Methods **168,** 383 (1980)
4.374 M.-P. Macht, V. Naundorf: J. Appl. Phys. **53,** 7551 (1982)
4.375 E. Zinner: J. Electrochem. Soc. **130,** 199C (1983)
4.376 W. Vandervorst, H.W. Werner: In *Secondary Ion Mass Spectrometry SIMS VI*, ed. by A. Benninghoven, A.M. Huber, H.W. Werner (Wiley, Chichester, New York 1988) p. 409
4.377 R. Levi-Setti, J. Chabala, Y.L. Wang: Scanning Ion Microsc. Suppl. **1,** 13 (1987)
4.378 J.M. Chabala, R. Levi-Setti, Y.L. Wang: J. Vac. Sci. Technol. **B6,** 910 (1988)
4.379 A.R. Bayly, A.R. Waugh, P. Vohralik: Spectrochim. Acta **40B,** 717 (1985)
4.380 F.G. Rüdenauer: Surface Interf. Anal. **6,** 132 (1984)
4.381 W. Steiger, F.G. Rüdenauer, G. Ernst: Anal. Chem. **58,** 2037 (1986)
4.382 M.E. Kargacin, B.R. Kowalski: Anal. Chem. **58,** 2301 (1986)
4.383 J.D. Brown: Scanning Microsc. **2,** 653 (1988)
4.384 R. Schubert, J.C. Tracy: Rev. Sci. Instrum. **44,** 487 (1973)
4.385 C.W. Magee: Nucl. Instrum. Methods **191,** 297 (1981)
4.386 J.B. Pallix, C.H. Becker, N. Newman: J. Vac. Sci. Technol. **A6,** 1049 (1988)
4.387 H.A. Hoff, N.Q. Lam: Surface Sci. **204,** 233 (1988)
4.388 R.v. Criegern, T. Hillmer, V. Huber, H. Oppolzer, I. Weitzel: Fres. Z. Anal. Chem. **319,** 861 (1984)
4.389 W. Hösler, Summer course on *Characterization Techniques for VLSI and Advanced Semiconductor Devices*, Leuven (1987), unpublished, and private communication (1988)
4.390 H.H. Andersen, H.L. Bay: In *Sputtering by Particle Bombardment I*, Topics Appl. Phys., Vol. 47, ed. by R. Behrisch (Springer, Berlin, Heidelberg 1981) p. 145
4.391 H. Oechsner: Nucl. Instrum. Methods **B33,** 918 (1988)
4.392 N. Laegreid, G.K. Wehner: J. Appl. Phys. **32,** 365 (1961)
4.393 P.C. Zalm: J. Appl. Phys. **54,** 2660 (1983)
4.394 Ch. Steinbrüchel: Appl. Phys. **A36,** 37 (1985)
4.395 W. Wach, K. Wittmaack: J. Appl. Phys. **52,** 3341 (1981)
4.396 T. Ishitani, R. Shimizu, H. Tamura: Appl. Phys. **6,** 277 (1975)
4.397 S.A. Schwarz, C.R. Helms: J. Vac. Sci. Technol. **16,** 781 (1979)
4.398 C.W. Magee, R.E. Honig: Surface Interf. Anal. **4,** 35 (1982)
4.399 J.M. Sanz, S. Hofmann: Surface Interf. Anal. **5,** 210 (1983)
4.400 J. Fine, P.A. Lindfors, M.E. Gorman, R.L. Gerlach, B. Navinsek, D.F. Mitchell, G.P. Chambers: J. Vac. Sci. Technol. **A3,** 1413 (1985)

4.401 A. Zalar: Surface Interf. Anal. **9,** 41 (1986)
4.402 A. Zalar, S. Hofmann: Nucl. Instrum. Methods **B18,** 655 (1987)
4.403 F. Matsunaga, H. Kakibayashi, T. Mishima, S. Kawase: Jap. J. Appl. Phys. **27,** 149 (1988)
4.404 P. Williams, J.E. Baker: Appl. Phys. Lett. **36,** 842 (1980)
4.405 P. Williams, J.E. Baker: Nucl. Instrum. Methods **182/183,** 15 (1981)
4.406 S. Matteson: Appl. Surface Sci. **9,** 335 (1981)
4.407 K. Wittmaack: In *Symposium on Surface Science 3S' 85*, ed. by G. Betz, H. Störi, W. Husinsky, P. Varga. (Inst. f. Allg. Physik, Tu Wien 1985) p. 211
4.408 K. Wittmaack, D.B. Poker: Nucl. Instrum. Methods **B47,** 224 (1990)
4.409 S. Hofmann: Surface Interf. Anal. **2,** 148 (1980)
4.410 K. Tsunoyama, T. Suzuki, Y. Ohashi, H. Kishidaka: Surface Interf. Anal. **2,** 212 (1980)
4.411 W. Reuter, G.J. Scilla: Anal. Chem. **60,** 1401 (1988)
4.412 W. Reuter: Anal. Chem. **59,** 2081 (1987)
4.413 E. Hechtl, J. Bohdansky, J. Roth: J. Nucl. Mater. **103 & 104,** 333 (1981)
4.414 E. Hechtl, J. Bohdansky: J. Nucl. Mater. **122 & 123,** 1431 (1984)
4.415 E. Hechtl, J. Bohdansky: J Nucl. Mater. **133 & 134,** 301 (1985)
4.416 S. Nagata, H. Bergsaker, B. Emmoth, L. Ilyinsky: Nucl. Instrum. Methods **B18,** 515 (1987)
4.417 S. Hues, P. Williams: Nucl. Instrum. Methods **B15,** 206 (1986)
4.418 K. Wittmaack: Appl. Phys. Lett. **48,** 1400 (1986)
4.419 K. Wittmaack: Nucl. Instrum. Methods **B19/20,** 484 (1987)
4.420 W. Reuter: Nucl. Instrum. Methods **B15,** 173 (1986)
4.421 F.A. Stevie, P.H. Kohara, D.S. Simons, P. Chi: J. Vac. Sci. Technol. A6, 76 (1988)
4.422 K. Miethe, W.H. Gries, A. Pöcker: In *Secondary Ion Mass Spectrometry SIMS V*, ed. by A. Benninghoven, R.J. Colton, D.S. Simons, H.W. Werner (Springer, Berlin, Heidelberg 1986) p. 347
4.423 Y. Homma, H. Okamoto, Y. Ishii: Jap. J. Appl. Phys. **24,** 934 (1985)
4.424 J.F. Wager, C.W. Wilmsen: J. Appl. Phys. **50,** 874 (1979)
4.425 C.P. Hunt, M.P. Seah: Surface Interf. Anal. **5,** 199 (1983)
4.426 M.P. Seah, H.J. Mathieu, C.P. Hunt: Surface Sci. **139,** 549 (1984)
4.427 M.P. Seah, C.P. Hunt: Surface Interf. Anal. **5,** 33 (1983)
4.428 J.R. Parsons: Phil. Mag. **12,** 1159 (1965)
4.429 D.J. Mazey, R.S. Nelson, R.S. Barnes: Phil. Mag. **17,** 1145 (1968)
4.430 E.P. EerNisse: Appl. Phys. Lett. **18,** 581 (1971)
4.431 E.C. Baranova, V.M. Gusev, Yu. V. Martynenko, C.V. Starinin, I.B. Hailbullin: Radiat. Eff. **18,** 21 (1973)
4.432 A. Zalar, S. Hofmann: Vacuum **37,** 169 (1987)
4.433 N.Q. Lam: Surface Interf. Anal. **12,** 65 (1988)
4.434 K. Ishikawa, S. Hamasaki, K. Goto: Jap. J. Appl. Phys. **22,** L 547 (1983)
4.435 N. Bouadama, P. Devoldere, B. Jusserand, P. Ossart: Appl. Phys. Lett. **48,** 1285 (1986)
4.436 M.G. Dowsett, R.M. King, E.H. Parker: Appl. Phys. Lett. **31,** 529 (1977)
4.437 D.K. Skinner, J.G. Swanson, C.V. Haynes: Surface Interf. Anal. **5,** 38 (1983)
4.438 W.H. Gries, K. Miethe: Mikrochim. Acta I, 169 (1987)
4.439 M.-P. Macht, V. Naundorf: Nucl. Instrum. Methods **B15,** 189 (1986)
4.440 F. Schulte, M. Maier: Nucl. Instrum. Methods **B15,** 198 (1986)
4.441 A.E. Morgan, P. Maillot: Appl. Phys. Lett. **50,** 959 (1987)
4.442 B.V. King, D.G. Tonn, I.S.T. Tsong: Nucl. Instrum. Methods **B7/8,** 607 (1985)
4.443 D.G. Tonn, O.F. Sankey, I.S.T. Tsong: Nucl. Instrum. Methods **B15,** 193 (1986)
4.444 M. Kühlein, I. Jäger: Surface Interf. Anal. **6,** 129 (1984)
4.445 Y.-T. Cheng, A.A. Dow, B.M. Clemens: Appl. Phys. Lett. **53,** 1346 (1988)
4.446 W.L. Johnson, Y.-T. Cheng, M. Van Rossum, M.-A. Nicolet: Nucl. Instrum. Methods **B7/8,** 657 (1985)
4.447 A. Many, Y. Goldstein, S.Z. Weisz, O. Resto: Appl. Phys. Lett. **53,** 192 (1988)
4.448 H.D. Hagstrum: In *Inelastic Ion-Surface Collisions*, ed. by N.H. Tolk, J.C. Tully, W. Heiland, C.W. White (Academic, New York 1977) p. 1
4.449 K. Wittmaack: unpublished results (1983)

4.450 N. Menzel, K. Wittmaack: Nucl. Instrum. Methods **191**, 235 (1981)
4.451 W. Reuter, K. Wittmaack: Appl. Surface Sci. **5**, 221 (1980)
4.452 Y. Homma, K. Wittmaack: J. Appl. Phys. **65**, 5061 (1989)
4.453 Y. Gao, J.C. Harmand: J. Vac. Sci. Technol. A6, 2243 (1988)
4.454 F.A. Stevie, P.M. Kohara, S. Singh, L. Kroko: In *Secondary Ion Mass Spectrometry SIMS VI*, ed. by A. Benninghoven, A.M. Huber, H.W. Werner (Wiley, Chichester, New York 1988) p. 319
4.455 Y. Gao: J. Appl. Phys. **64**, 3760 (1988)
4.456 M.A. Ray, J.E. Baker, C.M. Loxton, J.E. Greene: J. Vac. Sci. Technol. A6, 44 (1988)
4.457 P. Williams: In *Secondary Ion Mass Spectrometry SIMS VI*, ed. by A. Benninghoven, A.M. Huber, H.W. Werner (Wiley, Chichester, New York 1988) p. 261
4.458 H.E. Smith, G.H. Morrison: Anal. Chem. **57**, 2663 (1985)
4.459 H. Gnaser: J. Appl. Phys. **60**, 1212 (1986)
4.460 L.A. Streit, P. Williams: J. Vac. Sci. Technol. A5, 1979 (1987)
4.461 S.D. Littlewood, J.A. Kilner: J. Appl. Phys. **63**, 2173 (1988)
4.462 L.A. Streit, P. Williams: In *Secondary Ion Mass Spectrometry SIMS VI*, ed. by A. Benninghoven, A.M. Huber, H.W. Werner (Wiley, Chichester, New York 1988) p. 201
4.463 H.W. Werner, A.E. Morgan: J. Appl. Phys. **47**, 1232 (1976)
4.464 H.W. Werner, N. Warmoltz: J. Vac. Sci. Technol. A2, 726 (1984)
4.465 J.A. van den Berg: Vacuum **36**, 981 (1986)
4.466 R. Pantel, D. Levy, D. Nicolas: J. Vac. Sci. Technol. A6, 2953 (1988)
4.467 W. Pamler: Surface Interf. Anal. **13**, 55 (1988)
4.468 R.v. Criegern, Th. Hillmer, I. Weitzel: Fresenius Z. Anal. Chem. **314**, 293 (1983)
4.469 J.B. Clegg: in *Secondary Ion Mass Spectrometry SIMS V*, ed. by A. Benninghoven, R.J. Colton, D.S. Simons, H.W. Werner (Springer, Berlin, Heidelberg 1986) p. 112 and private communication
4.470 K. Wittmaack: Appl. Phys. A38, 235 (1985)
4.471 J.P. Coad, H.E. Bishop, J.C. Revière: Surface Sci. **21**, 253 (1970)
4.472 B.A. Joyce, J.H. Neave: Surface Sci. **34**, 401 (1973)
4.473 P. Williams, R.K. Lewis, C.A. Evans, Jr., P.R. Hanley: Anal. Chem. **49**, 1399 (1977)
4.474 C.W. Magee, E.M. Botnick: J. Vac. Sci. Technol. **19**, 47 (1981)
4.475 K. Wittmaack: Nucl. Instrum. Methods **218**, 327 (1983)
4.476 Y. Homma, Y. Ishii, T. Kobayashi, J. Osaka: J. Appl. Phys. **57**, 2931 (1985)

5. Desorption of Organic Molecules from Solid and Liquid Surfaces Induced by Particle Impact

Bo U.R. Sundqvist

With 28 Figures

This chapter describes the use of particle-induced secondary ion mass spectrometry to analyze films of organic molecules. Because of the large demand in the biosciences for new analysis methods, this overview is devoted largely to the analysis of layers of biomolecules. Most of these molecules are very large.

First the various instrumental techniques – including ion source arrangements and analyzers – are described. Sample preparation is covered in some detail. Methods for detecting large molecular ions are described. The analytical applications are illustrated with some recent examples. Currently available ideas concerning desorption mechanisms are discussed. The present knowledge, which is still very incomplete, is summarized, and the main issues and problems in the field are formulated and discussed.

5.1 Historical Survey

For many years the mass spectrometry community has tried to develop new techniques to deal with the problem of producing gaseous ions of large organic molecules. Fairly successful methods, like chemical ionization [5.1] and field desorption [5.2], were introduced in the late sixties and early seventies. Despite this progress, rather small peptides like lyzine-bradykinin were still difficult, if not impossible, to study by mass spectrometric techniques. In 1974 *Macfarlane* and coworkers at the Texas Agricultural and Mining University were involved in the time-of-flight analysis of recoiling (β-emitting) nuclei when they realized that fission fragments passing through a sample layer of biomolecules can cause desorption and ionization of these molecules [5.3]. Fission fragments are high-velocity ions interacting mainly with the electrons in the medium. In the early seventies they were considered very "exotic" primary ions in the field of secondary ion mass spectrometry (SIMS). Soon after *Macfarlane*'s discovery, *Benninghoven* and coworkers at the University of Münster [5.4] used well-known SIMS probes such as 2 keV Ar ions and a quadrupole mass analyzer to obtain results very similar to those of *Macfarlane*. The latter primaries are low-velocity particles that lose energy mainly by elastic collisions with target atoms. While the Münster group headed toward developing the surface analysis aspects of their technique, the Texas group soon showed that many thermally labile biomolecules can be studied by what they called plasma desorption mass spectrometry (PDMS) [5.3].

The impact of *Macfarlane*'s and *Benninghoven*'s contributions on the development of mass spectrometry was rather limited until about 1981 when *Surman* and *Vickerman* [5.5] introduced the idea of using a liquid matrix. *Barber* and coworkers [5.6] have often been associated with this innovation; however, their original paper on FAB ("fast atom bombardment") dealt only with solid samples. The reason for the success of the liquid matrix technique is still debated, and the problem will be discussed in Sect. 5.5.1. The impact on the field of mass spectrometry was enormous. Since FAB sources soon became commercially available, it took less than a year for the method to be applied in a large number of laboratories.

By the time FAB was introduced, the PDMS method had already been demonstrated to be feasible for analyzing molecules larger than ever before in the history of mass spectrometry. The detection of bovine insulin was reported in 1981 [5.7]. In 1983, the Uppsala group showed that intact molecular ions of proteins with masses up to 23 500 u can be transferred into the gas phase, mass analyzed, and detected [5.8]. Recently the protein porcine pepsin (34 636 u) was successfully studied [5.9]. Using FAB and magnetic sector instruments it has been demonstrated that molecules of masses up to about 20 000 u can be studied [5.10]. The upper mass limits for these techniques are expected to be pushed even further in the next few years. We can safely say that the introduction of particle-induced desorption methods in mass spectrometry has revolutionized the field and has widened the applicability of mass spectrometry by increasing the mass range by about one order of magnitude.

The applications of organic mass spectrometry are numerous, and we cannot cover the field in any detail here. One recent compilation of studies reviews mass spectrometry in general [5.11]; another covers the use of SIMS [5.12]. SIMS as a means of surface analysis has been reviewed by *Wittmaack* [5.13]. *Cooks* et al. [5.14, 15] have discussed high-molecular-weight mass spectrometry, both applications and mechanisms. Applications in biomedical mass spectrometry have been reviewed by *Morris* [5.16]. Applications of LLS-SIMS (FAB) and HSF-SIMS (PDMS) are reviewed in articles by *Rinehart* [5.17], *Macfarlane* [5.18], and *Sundqvist* and *Macfarlane* [5.19].

While the emphasis in this review will be on biomolecules, it should be noted that SIMS has also been used successfully to analyze polymer surfaces. The field has been described by *Briggs* [5.20].

Although mass spectrometry is the dominating application of sputtering in organic overlayer analysis, interest is growing in the use of ion-impact-induced sample erosion in other bioscience fields. With remarkable success sputtering with low-velocity particles has been used to erode virus [5.21, 5.22] and pollen [5.23]. In a dynamic sense it has also been possible to analyze what is left as a function of erosion time, and thereby to derive structural information on the systems studied. This new application of ion sputtering in biosciences holds great promise.

5.2 Nomenclature

The introduction of particle impact methods in mass spectrometry has given rise to new names and acronyms. Names have been chosen to reflect either the analytical use of the method or the inventor's opinion of the underlying mechanism. Once a name is in use, it is often very difficult to introduce more appropriate notations, and the old ones tend to stay around. One could therefore argue that suggesting new names is futile. *Wittmaack's* [5.13] attempt to replace the notation "static" SIMS with the more appropriate notation "low-fluence" SIMS is a discouraging illustration. On the other hand, the present nomenclature is very confusing, and a more precise terminology is desirable.

SIMS (secondary ion mass spectrometry) is an old and concise name for methods to study the mass distribution of (secondary) ions ejected from a surface bombarded with energetic (primary) ions. Here the notation SIMS will be used in a general sense as the common denominator in the name of all methods by which secondary ions are studied. To define a SIMS experiment sufficiently well the following parameters have to be specified: the energy or velocity of the primary particle, the sample state (matrix), and the speed of secondary ion extraction. (The method of mass analysis may be added for completeness.)

The particle velocity can be classified according to the scheme commonly used in radiation damage and sputtering theory [5.24, 25]. Particles transferring energy to target atoms mostly via elastic collisions (nuclear stopping regime) are considered low in velocity. High-velocity particles, on the other hand, lose energy predominantly via inelastic collisions with target electrons (electronic stopping regime). The transition from one regime to the other is roughly defined by the Bohr velocity ($v_0 = 2 \times 10^8$ cm/s). In terms of energy per atomic mass

Table 5.1. New terminology for old methods

Old name	New name
SIMS (secondary ion mass spectrometry)	LSS-SIMS: Low-velocity impact, solid sample, slow extraction SIMS (with quadrupole mass analyzer, QMA, or magnetic sector instrument, MSI). LSF-SIMS: Low-velocity impact, solid sample, fast extraction SIMS (with time-of-flight system, TOF).
PDMS (plasma desorption mass spectrometry)	HSF-SIMS: High-velocity (or fission fragment) impact, solid sample, fast extraction SIMS (with TOF).
FAB (fast atom bombardment)	LLS-SIMS: Low-velocity impact, liquid sample, slow extraction SIMS (with MSI).
LDMS (laser desorption mass spectrometry)	PSF-SIMS: Photon (laser) impact, solid sample, fast extraction SIMS (with TOF).
ESD (electron stimulated desorption)	ESS-SIMS: Electron impact, solid sample, slow extraction SIMS (with QMA, TOF, or MSI).

this number corresponds to $E(\text{eV})/M(\text{u}) = 2 \times 10^4$. (This means that the notation "fast atom bombardment" [5.6] used in conjunction with keV particles of mass 40 u is not very appropriate.)

Another important parameter is the state of the sample, either solid or liquid (Sect. 5.4.1). Last but not least it is appropriate to specify the time required for secondary ion extraction and transfer through the mass spectrometer (Sect. 5.3.3). Somewhat arbitrarily we consider extraction (and transport) to be "slow" if the electric field used has a strength $|E| < 1$ kV/mm, whereas the case $|E| \geq 1$ kV/mm is referred to as "fast" extraction.

Based on this convention the new names for old methods will be as in Table 5.1 (note that in this scheme photons or electrons may also be included as primaries). In addition to the primary particle used, we may also add information on the intensity of the incident particle beam, i.e., whether the analysis is run under "dynamic" (high-fluence) or "static" (low-fluence) conditions. In this chapter the suggested new terminology will be used, and the old names will be given in parentheses.

5.3 Basic Aspects of Large Molecule Analysis

5.3.1 Quantification of Ion Yields

Secondary ions are produced when an energetic primary ion (or a neutral particle) hits a sample surface (Fig. 5.1). The sample can be in either liquid or solid form. With low-energy particles (slow particles) the sample must be bombarded from the front, because the range of these particles is usually small compared with the sample thickness. When fast ions are used, it is often convenient to use a thin backing and let the primary particles enter the sample from the back. The details of the technical arrangement for ion production also depend very much on the type of mass analyzer used.

In the field of sputtering, the yield Y is defined as the mean number of particles (charged and neutral) emitted per incident primary particle. A subscript M indicates the yield of a particular particle of mass M, $Y_M = \beta_M Y$, where β_M is the branching ratio, i.e., the fraction of particles with mass M relative to the total number of particles emitted. If the molecules (M) are mixed into a matrix of other molecules, β_M can be very small. In this chapter the ion yield Y^q (q = charge state) is more often discussed and defined as the mean number of secondary ions produced per incident primary particle. Sometimes it is also convenient to consider the yield of a particular positive or negative secondary ion of mass M, $Y_M^q = P_M^q Y_M$, where P_M^q is the ionization probability. In time-of-flight (TOF) mass spectrometry with multiple stop detectors, the total number of secondary ions (of either charge) per incident particle is often referred to as the multiplicity, which is the same as the total ion yield of either charge. For slow heavy particle impact on organic solids, the yield of positive ions is about

(a)   IEI ⪆ 1 kV/mm

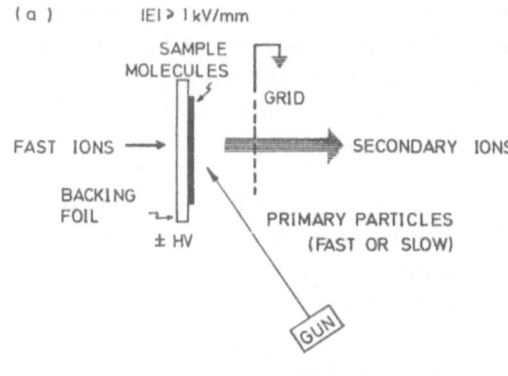

SAMPLE
MOLECULES

GRID

FAST IONS →   → SECONDARY IONS

BACKING
FOIL

± HV   PRIMARY PARTICLES
             (FAST OR SLOW)

GUN

Fig. 5.1. The ion source arrangement in a system for (a) fast extraction, as in many time-of-flight systems; (b) slow extraction, as in many magnetic field sector instruments

(b) $|\vec{E}| \approx 0.01 - 1$ kV/mm

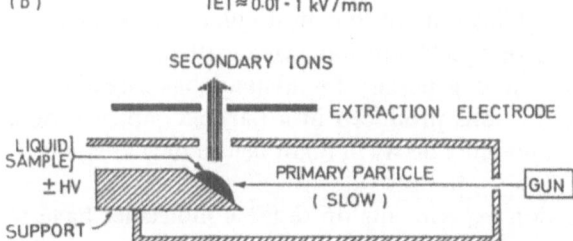

SECONDARY IONS

EXTRACTION ELECTRODE

LIQUID
SAMPLE
± HV   PRIMARY PARTICLE   GUN
                (SLOW)
SUPPORT

1 or often less, while for fast heavy ion impact the corresponding number is often 10–100.

The number I_M^q of molecular ions detected in a SIMS-type mass spectrometric analysis can be written as

$$I_M^q(\Phi) = P_M^q c_M Y_M \eta A \Phi f_M(\Phi) g_M(\tau_d) \ , \tag{5.1}$$

where c_M is the concentration of the molecules of interest in the sample, $\Sigma c_M = 1$, η is the transmission and detection efficiency of the employed instrument, A is the probed area, and Φ is the primary particle fluence (ions/cm^2 or atoms/cm^2). The functions $f(\Phi)$ (Sect. 5.3.2) and $g(\tau_d)$ describe the effect of bombardment-induced damage and molecular stability on the measured intensity,

$$f_M(\Phi) = \exp(-\tilde{\sigma}_M \Phi) \quad \text{and} \tag{5.2}$$

$$g_M(\tau_d) = \exp\left(-\frac{\tau_d}{\tau_M}\right) \ , \tag{5.3}$$

where $\tilde{\sigma}_M$ is the damage cross section, τ_M the average lifetime of the ejected molecular ion, and τ_d the time required to transport the ion from the sample to the detector. Equation (5.2) is valid for $\Phi \tilde{\sigma}_M \leq 1$. The importance of bombardment-induced damage and fragmentation of molecular ions is discussed in some detail in Sects. 5.3.2, 3.

Note that (5.1) relates to a thick sample. For monolayer or submonolayer samples, the product $c_M Y_M$ has to be replaced by $N_M \sigma_M$, where N_M is the areal density of the molecular species on the backing used, and σ_M the cross section for ion-induced desorption. In general $\tilde{\sigma}_M \neq \sigma_M$. The ionization efficiency P_M^q in (5.1) is often a function of the concentration c_M, particularly in samples containing mixtures of molecules.

In sputtering the number of neutrals ejected is generally much larger than the number of ions produced (Chap. 3 and Sect. 5.5.2); i.e., the ionization efficiency P_M^q is much less than 1. Therefore, various postionization methods have been tried in SIMS (Chap. 4). For bioorganic molecules *Campana* and *Freas* [5.26] have shown that the addition of a reactive gas (chemical ionization) to a fast atom bombardment source may increase the ion yield considerably. The same study also indicates that the increased pressure in the extraction region results in collisional stabilization of the molecular ion, which also increases the measured molecular-ion yield. Substantiated further, this would be a very important finding, since it is generally found (and has already been pointed out) that large molecular ions produced in a particle impact process acquire a large internal energy and often decay in flight before they are detected (Sect. 5.3.3) [5.27].

Very promising postionization experiments on organic molecules have recently been performed by *Grotemeyer* and collaborators [5.28]. The desorbed neutrals were cooled in a supersonic argon jet and then ionized by resonantly enhanced multiphoton ionization (REMPI) using an eximer-dye-laser system. Although a laser was also used for desorption, the basic postionization principle should apply equally well to particle impact ion sources. Such a development is now under way in several laboratories.

5.3.2 Bombardment-Induced Damage

When an ion hits a sample of organic molecules a certain area $\tilde{\sigma}_M$ around the point of a single particle impact may be modified or destroyed. Therefore, the original molecules in that area will not appear as molecular ions when this area is hit by another primary particle (Fig. 5.2). The area $\tilde{\sigma}_M$ is called the disappearance or damage cross section [5.29]. Strictly speaking a certain volume of the sample will be destroyed. For low-velocity impact that volume will have the approximate shape of a half-sphere, for a high-velocity ion with a long range the volume is more like a cylinder. In the following, only damage on the sample surface is taken into account. If $Y_M^q(0)$ is the yield of molecular ions from a fresh sample, the yield $Y_M^q(\Phi)$ can be written

$$Y_M^q(\Phi) = Y_M^q(0)\left\{1 - \frac{\delta A_M}{A}\right\} , \qquad (5.4)$$

where $\delta A_M = \delta A(\Phi, \tilde{\sigma}_M)$ is the fraction of A damaged by ΦA incident ions. In the expression for δA_M we must include a correction for overlapping areas. This

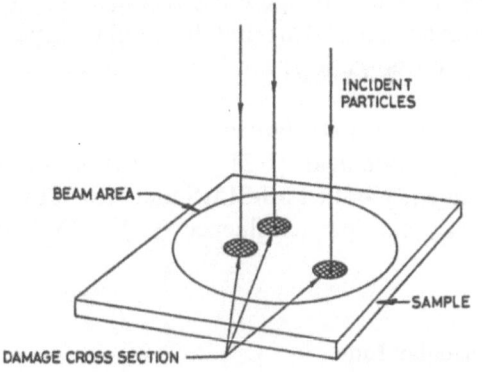

Fig. 5.2. Schematic illustration of the damage cross-section concept

INCIDENT PARTICLES

BEAM AREA

SAMPLE

DAMAGE CROSS SECTION

can be done in an approximate way as shown by *Salehpour* [5.30] to give

$$\frac{\delta A_M}{A} = 1 - \exp(-\tilde{\sigma}_M \Phi) \ . \tag{5.5}$$

Comparison of (5.4, 5) with (5.1, 2) shows that

$$f_M(\Phi) = 1 - \frac{\delta A_M}{A} \ . \tag{5.6}$$

In Fig. 5.3 [5.30] experimental data from a damage cross-section determination of 90 MeV $^{127}\text{I}^{14+}$ on valine ($M = 117$ u) are shown. Equation (5.6) was used to fit the dashed line to the data, and a value of $\tilde{\sigma}_M = 6.8 \pm 1.8 \times 10^{-13}$ cm^2 was derived. Similar results were found for other amino acids and peptides [5.30]. Note that the damage cross section will also reflect damage of the sample molecules caused by "late" effects of the radiation. The term "late" denotes times

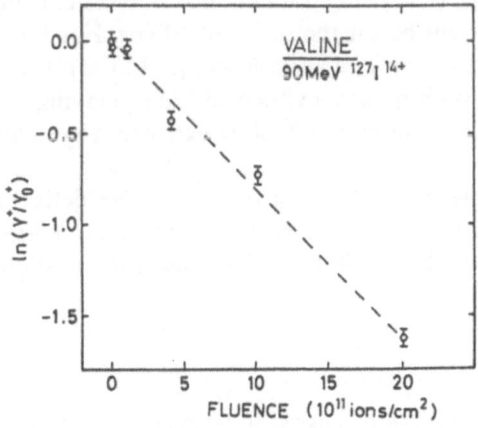

VALINE
90 MeV $^{127}\text{I}^{14+}$

$\ln(Y^*/Y_0^*)$

FLUENCE (10^{11} ions/cm^2)

Fig. 5.3. Normalized intensities of molecular ions of valine ($M = 117$ u) as a function of the primary ion fluence for 90 MeV $^{127}\text{I}^{14+}$. The dashed line is a least-squares fit of (5.5) to the data. The damage cross section extracted from the graph is $(6.8 \pm 1.8) \, 10^{-13}$ cm^2 [5.30]

long after the ejection and ionization. In fact damage cross sections are not directly correlated with the area of removed material around the point of impact (crater formation). In general, damage cross sections are larger than areas of sample removal.

Sichtermann and *Benninghoven* have measured damage cross sections for low-energy primaries and molecules like amino acids [5.31]. The cross sections are on the order of 10^{-14} cm². Consequently both LSS-SIMS and HSF-SIMS on such surfaces can be run only at low primary particle fluences (static SIMS or low-fluence SIMS).

5.3.3 Fragmentation of Desorbed Molecular Ions

The ion extraction technique used affects the mass spectra of secondary ions in many ways. Large molecular ions produced by particle impact appear to have a relatively high internal energy [5.27]. The molecular ions are produced with a distribution of excitation energies that will cause fragmentation with a distribution of rates (and a distribution of lifetimes), so that the number of unfragmented molecular ions rapidly decrease with time. To make the measured molecular ion intensity as large as possible, it will always be an advantage, therefore, to keep the total flight time τ_d, in the mass analyzer as short as possible, cf. (5.3). The time τ_d is the sum of extraction time τ_{ex} and the drift time τ_{dr}. The extraction time τ_{ex} required to transport an ion of mass M and charge q from one plane electrode at voltage V, where the ion is initially at rest, to an adjacent plane parallel grounded electrode at a distance D is

$$\tau_{ex} = 2D\sqrt{\frac{M}{2E}} \qquad \text{or} \tag{5.7}$$

$$\tau_{ex}(s) = 2 \times 10^{-6} D(\text{cm}) \sqrt{\frac{M(u)}{2E(\text{eV})}} . \tag{5.8}$$

Here $E = qV$ is the kinetic energy of the ion after extraction. For $D = 0.1$ cm and $M(u)/E(\text{eV}) = 1$, the extraction time will be on the order of 100 ns. For large molecules ($> 10\,000$ u) the extraction time may be as large as 1 μs. In the present discussion (Fig. 5.1) fast extraction systems are defined as those having an extraction field in excess of 1 kV/mm. Consequently slow extraction systems have lower extraction fields.

After extraction the ions are allowed to drift a distance d to the detector before they are detected in an array of channel electron multiplier plates (Fig. 5.4). In a TOF system, d is typically 20–300 cm. The total time-of-flight τ_d is

$$\tau_d = \sqrt{\frac{M}{2E}}(2D + d) . \tag{5.9}$$

In mass spectrometry, fragmentation is often discussed in terms of prompt and

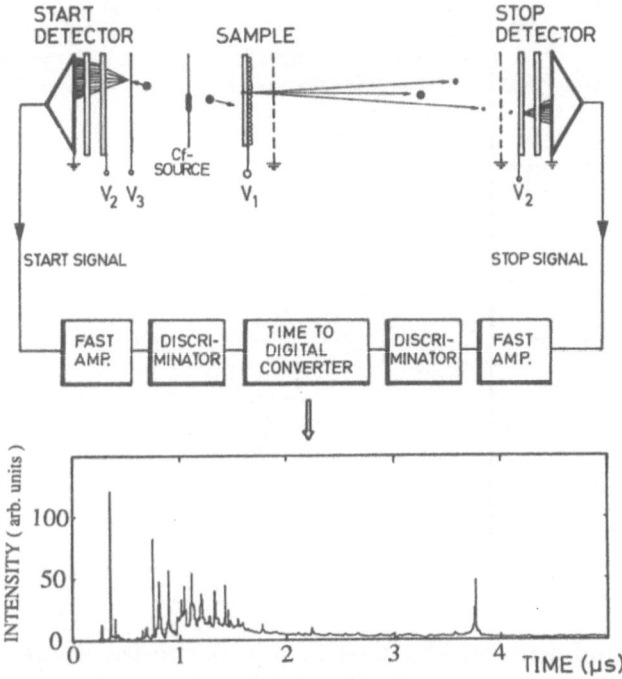

Fig. 5.4. Principle of a HSF-SIMS (PDMS) TOF. Time spectra, $I(t)$, can be converted to mass spectra, $I(M)$, through (5.9). The voltages V_1, V_2, V_3 are on the order of $\pm (10–20)$ kV, -2 kV, and -2.2 kV, respectively

metastable processes. Prompt fragmentation is mainly associated with sharp peaks in the mass spectra. Those peaks will arise if the decay happens before the parent ion has obtained significant kinetic energy and before the ion is extracted from the sample. Metastable fragmentations, associated with broad peaks and appearing at masses not normally expected for the fragments, are due to decays in the mass analyzer (the field-free region in a TOF system). Of course, these definitions depend on the experimental conditions. *Chait* [5.27] has considered the time intervals involved in a time-of-flight system and a magnetic sector system for analysis of a molecule of mass 1 000 u. In such systems prompt decompositions will occur at times less than $10^{-8}–10^{-9}$ s and $10^{-6}–5 \times 10^{-5}$ s, respectively, while the time intervals for metastable decompositions are $10^{-8}–10^{-4}$ s and $10^{-6}–5 \times 10^{-5}$ s.

Figure 5.5 illustrates the effect of metastable decay in the field-free region of an ion in a straight time-of-flight spectrometer. In Fig. 5.5a the spectrum of a stable ion is shown. In Fig. 5.5b some of the stable ions have decayed, the shaded area corresponds to flight times of the charged fragments due to the decay. As internal energy is released in the decay, fragments gain or lose their velocity in the direction of the spectrometer. Figure 5.5c shows the effect of

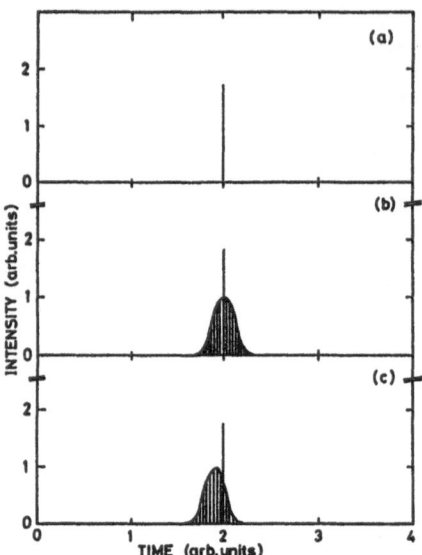

Fig. 5.5. The effect of metastable decay on line shapes in a time-of-flight mass spectrometer: (a) stable ion peak (b) stable ion peak plus metastable peak (c) effect of postacceleration on the peak shape

postacceleration at the stop detector. The flight time in the postacceleration region of the fragments will be shorter than the flight time for the heavier parent ion. Accordingly, the broad peak due to metastable decays will shift toward flight times shorter than for the parent ion. In a straight time-of-flight system this results in rather complicated line shapes for large molecules. In addition, many decays will happen in the acceleration gap before the parent ion has acquired the full acceleration energy.

The effect of metastable decays in the field-free region of a TOF spectrometer can be studied by using a deceleration grid in front of the stop detector (Fig. 5.6) [5.27]. The parent ions that survive the acceleration time will acquire the full acceleration energy, while charged fragments from a decay have a lower energy and can be filtered out with the deceleration grid. *Chait* [5.32] recently used this technique and illustrated the difference in stability of molecular ions desorbed with fast ions from multilayer samples and a monolayer of sample molecules adsorbed to nitrocellulose (Sect. 5.4.1). The intensity of molecular ions of porcine insulin is strongly reduced in the case of multilayer samples; i.e., almost all molecular ions have decayed in the field-free region. With porcine insulin adsorbed to nitrocellulose almost all molecular ions survive the field-free region. This result proves that molecular ions, desorbed from a monolayer adsorbed to nitrocellulose, have considerably lower internal energies and therefore longer lifetimes than ions desorbed from multilayer samples produced using the electrospray method.

Figure 5.7 shows another way of studying the effect of metastable decays in a time-of-flight analyzer. The breakup of a molecular ion into neutral and charged fragments contributes to the spectrum, and the fragments can be

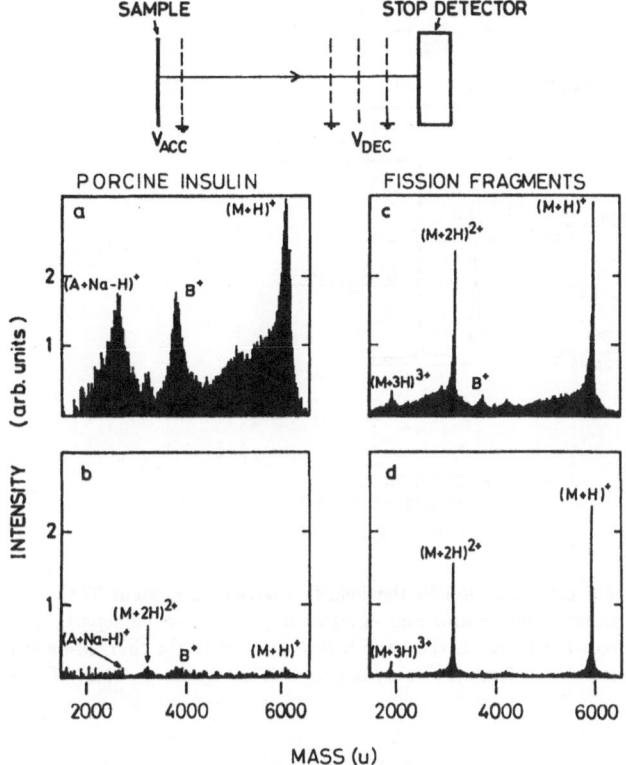

Fig. 5.6. Partial time-of-flight mass spectra of porcine insulin: (**a**) electrosprayed sample ($V_{acc} = 10\,\text{kV}$; $V_{dec} = 0\,\text{kV}$), (**b**) electrosprayed sample ($V_{acc} = 10\,\text{kV}$; $V_{dec} = 8.4\,\text{kV}$), (**c**) nitrocellulose-bound sample ($V_{acc} = 10\,\text{kV}$; $V_{dec} = 0\,\text{kV}$), (**d**) nitrocellulose-bound sample ($V_{acc} = 10\,\text{kV}$; $V_{dec} = 9\,\text{kV}$). Above the spectra, the sample, acceleration grid, field-free path, deceleration grid, and stop detector are schematically shown. After [5.32]

separated and studied simultaneously by adding an electrostatic mirror, as in this setup [5.33].

5.3.4 Optimization of Mass Resolution

Once the yield of secondary ions has been maximized, the next important consideration is the design of proper ion collection optics before the ions enter the mass analyzer. Three types of instruments are used for mass analysis of large molecular ions: magnetic sector field instruments, quadrupole-based instruments, and time-of-flight systems. Magnetic sector field instruments can offer the best mass resolution, and such systems are the only ones with which separation of neighboring mass lines ($\Delta M = 1\,\text{u}$) has already been demonstrated in the mass region around $10\,000\,\text{u}$ [5.34]. However, achieving this excellent performance requires a very efficient ion production process. With the ion sources used

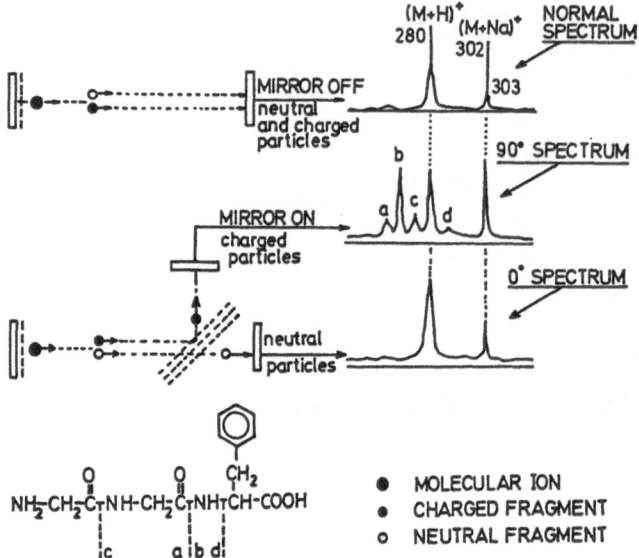

Fig. 5.7. The effect of decay of a molecular ion in the field-free part of a straight TOF mass spectrometer. The figure also shows how neutral and charged fragments can be separated by introducing an electrostatic mirror. The molecule is a peptide (gly-gly-phe). In the charged-particle spectrum (90°), the peaks a, b, and c arise from the decay of $(M + H)^+$, d from $(M + Na)^+$ [5.33]

so far, and taking into account the mass scanning procedure, the limited molecular ion yields (sensitivity) become a serious problem at mass numbers around 20 000 u. Also the cost of the hardware, as well as the fairly complicated operation of the apparatus, makes it hard to go further up in mass with such systems. Quadrupole instruments are also difficult to push to high mass while preserving high mass resolution and appropriate transmission [5.35]. The simplest technique, which has the advantage of almost unlimited mass range and the disadvantage of low mass resolution, is the time-of-flight technique. In principle, another important advantage is sensitivity. The transmission through such a mass analyzer may be as high as 50%–90% [5.36].

With mass analysis of high-mass molecular ions keep in mind that, because elements are usually composed of different isotopes, an analyzed molecular beam gives rise to a distribution of mass lines rather than to a single mass peak. In biomolecules it is mainly carbon (98.90% ^{12}C, 1.1% ^{13}C) and sulfur (95.02% ^{32}S, 0.75% ^{33}S, 4.21% ^{34}S, 0.04% ^{36}S) isotopes that cause this effect. Most mass spectrometric studies performed so far have covered molecules only up to masses of about 2000–3000 u. Figure 5.8 plots the calculated isotopic distributions for a homologue of phospholipase A_2 from Australian tiger snake (averaged molecular weight $M = 13\,309.9$ u) and bovine insulin ($M = 5\,733.5$ u). Also indicated is the resolving power, $M/\Delta M$ (FWHM) for the envelopes of these distributions. These distributions, based on binomial distributions, are cal-

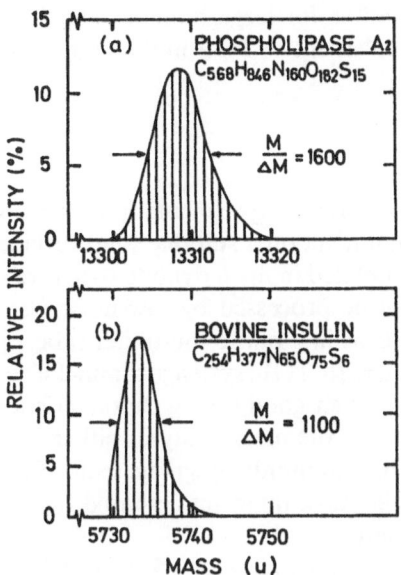

culated from the natural isotopic abundance of isotopes for the elements and their mass, and the chemical formula of the compound. For very heavy molecules the distribution will be almost Gaussian, while for the lighter molecules there will be a "cutoff" on the low mass wing corresponding to a molecule where all carbon nuclei are ^{12}C isotopes (bovine insulin in Fig. 5.8). If the isotopic distribution in itself does not contain important information, the isotopically averaged mass, as extracted from a peak distribution corresponding to the envelope of the isotopic distribution, should give the best confirmation of the molecular weight [5.37].

The consequence for the design of a convenient mass analyzer for molecular weight determination is that we need either a very good resolving power $(M/\Delta M = 20000$ or better) so that the individual mass peaks can be resolved or a comparatively low resolution (2 000), which matches the width of the envelope of the isotopic distributions. For low-mass work, available magnetic sector field and quadrupole-based instruments, with well-developed ion sources for liquid and solid samples, are preferable. At masses below 10 000 u very powerful sector instruments are now commercially available. If new developments drastically improve ion production for large molecular ions, these instruments will continue to be very useful. In the high-mass region, TOF systems will be preferable. However, much more development is needed before standard analytical problems can be well handled with TOF systems.

Another technique not directly associated with sputtered-ion analysis but which should be mentioned here is ion cyclotron resonance (ICR) in combination with Fourier-transform techniques [5.38]. With that method very high mass resolution is possible, but it requires longtime storage of ions (on the order of 1 ms). This has been demonstrated as possible for large biomolecular ions [5.39], and the technique may become a complement to TOF in the high-mass region when high mass resolution is needed.

5.3.5 Detection of Large Molecular Ions

The detection of secondary ions produced in SIMS experiments is based on secondary electron emission produced by ion impact on a suitable solid surface (Fig. 5.9a). The secondary electron signal is amplified in open dynode structures, so that a current pulse is produced that can be processed by low-noise electronics. The dynode structures can be discrete, as in a photo multiplier tube, or continuous, as in a channel electron multiplier. In TOF systems channel electron multiplier plate arrays are used because of their good timing characteristics (geometrically and electronically). Depending on the mass analyzer either current integration or single-ion counting is used. Commonly magnetic sector field instruments are run in the former mode, while most quadrupole-based instruments and TOF instruments are run in the latter mode.

In studies on large water clusters, *Beuhler* and *Friedman* [5.40–42] have shown that there is a "threshold" velocity (v_{th}) for the production of secondary electrons. They found that this threshold was independent of the mass of the cluster, $v_{th} = 1.8 \times 10^6$ cm/s. The Uppsala group found the same threshold velocity for bovine insulin, a peptide with a mean molecular weight of 5 733.5 u [5.43], and for other large molecules [5.44]. Tentatively generalizing these results, we can conclude that for efficient detection of a large molecule of mass M(u) > 1 000, it must have an energy E (eV) satisfying the condition

$$E/M > 1.7 \, \text{eV/u} \ . \tag{5.10}$$

Accordingly, very high acceleration voltages (or postacceleration voltages) will have to be used if very large molecules like proteins are studied.

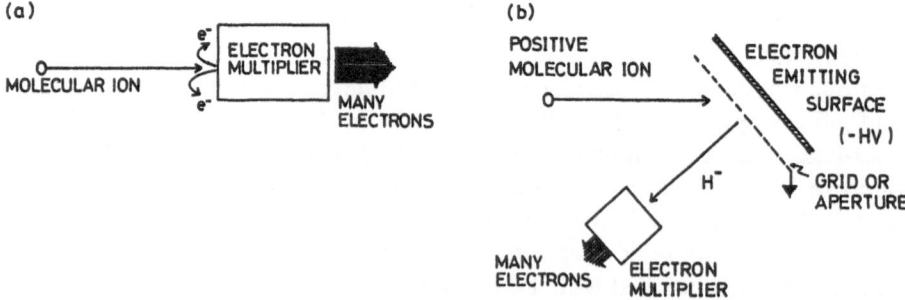

Fig. 5.9. (a) Simple stop detector based on secondary electron production. (b) A more complicated arrangement may be required for the detection of very large molecules. A small fragment like H^- can be produced when the large molecular ion hits the converter surface with a very low velocity. The H^- is used to generate secondary electrons as in (a)

In straight time-of-flight spectrometers large molecular ions are often identified via a postaccelerated charged fragment originating from a metastable decay [5.44]. Accordingly, the detection problem encountered in other types of spectrometers could be solved by first fragmenting the molecule and then postaccelerating a light fragment [5.45]. A light fragment like an H^+ or H^- emitted from a large molecule can be detected already at an energy of a few hundred electron volts (Fig. 5.9b). Fragmentation may sometimes be easy because the molecule often has a large amount of internal energy (Sect. 5.3.3).

5.4 Experimental Procedure

5.4.1 Sample Preparation

In general, the chemical environment of the molecule in the sample is very important for the desorption-ionization process. Many different sample preparation techniques are in use, so experimental data are very difficult to compare. In publications this problem is often deliberately passed over and sample preparation is not described in detail. Since complete recipes are not available yet, only an outline of the main principles will be given here.

a) Solid Samples

Figure 5.10 illustrates methods for the preparation of solid samples. The first three methods have the first stage in common, namely the preparation of a solution of the sample molecules. The simplest method is to deposit a droplet of the solution with a microsyringe onto a backing and to allow the solvent to evaporate. In the spin-casting technique [5.46, 47] the backing rotates at a high speed and then a droplet is deposited at the center of the backing surface. This results in a rather homogeneous layer of molecules. In the electrospray method [5.48–50] the solvent should be easy to evaporate. The solution is electrosprayed onto a backing (usually in air). A voltage of a few kV is applied to an electrode in a capillary tube filled with the molecule solution. The tube is placed 1–2 cm from the backing foil, which is held at ground potential. The small droplets formed in the spray are converted to a dry layer on the backing. The layer consists of μm-sized particles. This spray technique is less sample consuming than spin casting, but the uniformity of the films is not as good as in spin casting.

In all three methods the material concentration in the solution is an important parameter. Small amounts of the solvent are always left in the sample layer. In the limit of very low concentrations the films prepared with these methods become very thin and the choice of backing becomes critical.

In contrast to the first three methods, molecular-beam deposition is a vacuum technique [5.51]. Its advantage is the high purity of the prepared samples.

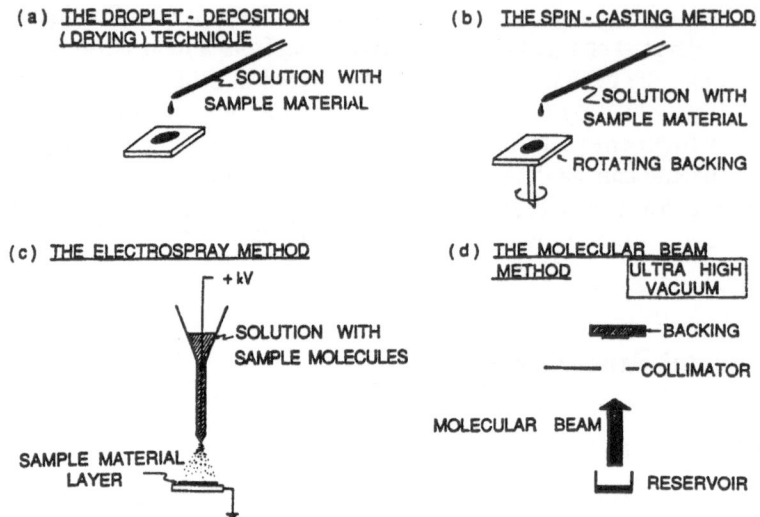

Fig. 5.10. Different sample preparation techniques: (**a**) droplet deposition, (**b**) spin casting, (**c**) electrospraying, and (**d**) evaporation

The requirement is, of course, that the molecules can be made to evaporate (or sublimate) unfragmented. Therefore, the method is of limited use for large, thermally unstable molecules, i.e., most biomolecules. Another disadvantage is that large quantities of material are needed.

A promising development in the field is based on the idea of preparing sample surfaces with active sites, possibly with molecules with a distinct specificity. Molecules are then adsorbed from a solution onto the sample backing. The analysis of such films will, however, require a very efficient ionization process and an efficient secondary ion analyzer, because the total amount of sample available is very small (a monolayer or less). *Jordan* et al. [5.52] have demonstrated that this approach may become very important to increase the analytical potential of SIMS methods. In a first study Nafion, an ion-containing polymer, was used as backing. Later *Macfarlane*'s group studied samples with monolayer and submonolayer coverage of rhodamine G on mylar films [5.53]. *Jonsson* et al. [5.54] for the first time adsorbed small proteins on nitrocellulose films and used these samples in HSF-SIMS (PDMS) studies. With their sample preparation technique they observed molecular ions of lower internal energy than with the electrospray technique (Fig. 5.6). Moreover, multiply charged molecular ions are more frequently observed when samples are prepared by adsorption of molecules to polymer films. The use of nitrocellulose for sample preparation has considerably increased the usefulness of HSF-SIMS (PDMS) in applications to peptides and proteins.

A general problem with all preparation techniques is sample purity. Pure means that the sample does not contain components that impair desorption (increased bonding to the surface) or ionization (lowered ionization probability).

Such compounds are often introduced in the biochemical separation procedures used during sample production. Therefore, mixture analysis is generally difficult. The problem has not been studied in detail, but demonstrations have shown that even small amounts of Na may quench yields of biomolecular ions [5.54]. For successful analysis of "unknown" molecules, it is often important to purify the sample in advance. With samples prepared by adsorbing the molecules on a backing, salts and low-molecular-weight compounds can often be effectively removed by rinsing the sample with agents like water or methanol [5.54].

b) Liquid Samples

The most important analytical procedure today for magnetic sector instruments is the solution of the sample material in a liquid matrix. This liquid must have a low vapor pressure ($p \leq 10^{-3}$ Pa). Commonly used liquids are glycerol or thioglycerol. Thioglycerol has a slightly higher vapor pressure than glycerol, but often gives higher yields of protonated molecules, probably because it is more acidic. The lifetime of a glycerol droplet (a few microliters) in vacuum under ion bombardment of typical current densities of $10^{13}-10^{14}$ particles/cm^2 s is 10–20 minutes. Hence more sample and/or glycerol is sometimes added in the ion source to achieve useful mass spectra. One disadvantage with liquid matrices is that the spectra of secondary ions will, in addition to peaks due to the matrix, also contain peaks due to mixed clusters, i.e., ions consisting of combinations of sample and matrix molecules.

The operation principle of a high mass resolution magnetic field instrument based on scanning implies a low sensitivity. Therefore many primary particles must be used (Sect. 5.3.1) to get a sufficiently intense signal throughout the spectrum. As discussed earlier, radiation damage is a problem with solid samples, and the use of a liquid matrix is mandatory, particularly if large molecules are to be studied (the branching ratio β_M is low). So far liquid matrices have been used only with slow primary ions. In principle, a liquid matrix should also help with fast primary ions, but because of the long range of fast ions, damage in the bulk of the liquid sample may cause additional problems.

An original motivation for *Barber* and coworkers [5.6] to introduce the LLS-SIMS (FAB) technique was to reduce sample charging by the use of neutral primaries. This idea is reflected by the name chosen for the method, disregarding the fact that energetic neutral particles also cause charging of an insulating sample through secondary electron production and secondary ion production. Sample charging is a general problem with all techniques discussed in this chapter. It is more or less significant, depending on the techniques used, but it certainly affects the ultimate performance of all instruments in terms of mass precision and mass resolution.

Soon after the introduction of LLS-SIMS (FAB), *Aberth* et al. [5.55] showed that a Cs$^+$ beam, replacing the neutral particle beam in a FAB source, gave even better results in terms of efficiency, partly because the charged primary ion beam could be focused and directed toward the desired spot on the sample. The authors argued that the effect of the liquid matrix is mainly to replenish the surface layer with intact sample molecules and thereby to overcome radiation damage (Sect. 5.3.2) in solid-sample SIMS.

5.4.2 Instrumentation for Mass Analysis

To understand previously reported experimental data we should know about the essential differences between the two approaches used for mass analysis of large molecular ions. This may also be a starting point for future innovations, since various combinations of the two techniques are conceivable.

a) Magnetic Sector Field Instruments

Figure 5.11 shows the principle of a magnetic sector field instrument. Details of the design depend on the application. *Wollnik* provides a detailed description of the field [5.56].

Some problems encountered when going to high masses in a magnetic sector instrument can be discussed with reference to the basic equation for ion transport through such a device,

$$\frac{M}{q} = \frac{R^2 B^2}{2E} \; , \tag{5.11}$$

where M is the mass of a particle of charge q and energy E, moving with a radius of curvature R at a right angle to a magnetic field of strength B.

For a given maximum field strength and a fixed ion energy, the linear dimensions of the system – i.e., the radius of the magnetic sector – increases with mass. Since the minimum energy needed for ion detection by secondary electron production increases with increasing mass of the molecule (Sect. 5.3.5), the dimensions of the system will scale with the maximum mass of ions to be analyzed. In principle this problem can be circumvented by using low acceleration voltages and postacceleration. However, with low acceleration voltages

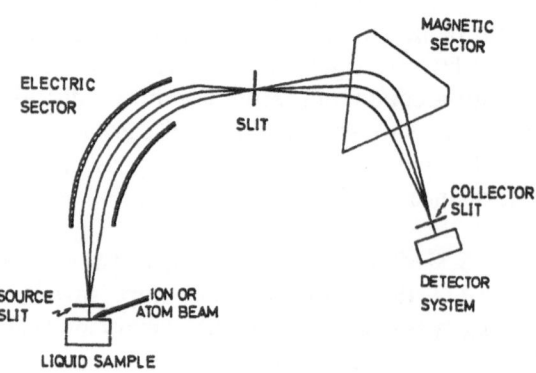

Fig. 5.11. Sector field mass analyzer. The arrangement of the electric and magnetic fields is the one used in a double-focusing mass spectrometer of Nier-Johnson geometry

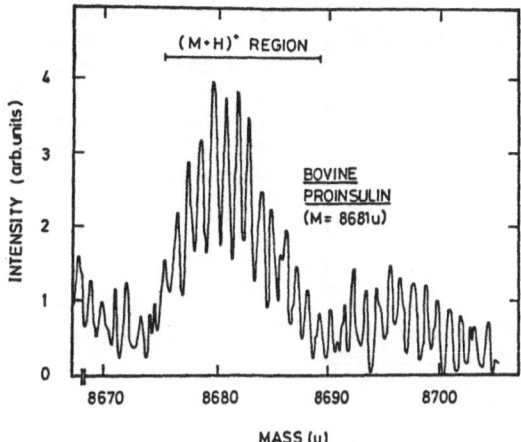

Fig. 5.12. Part of the mass spectrum of bovine proinsulin measured with a magnetic sector field instrument, LLS-SIMS (FAB) [5.34]

the flight time of the molecules through the mass analyzer will be long, a disadvantage since many molecular ions decay already at short times after emission (Sect. 5.3.3).

To increase the mass resolution the slits defining object and image must be reduced, and consequently the transmission decreases. If we want to resolve the isotopic pattern characteristic for a large molecule, an intense secondary ion beam is therefore needed. To save bombardment time the magnetic field is often scanned only over a selected mass region. Figure 5.12 illustrates the excellent mass resolution achievable with present technology in sector instrument design.

b) Time-of-Flight Systems

In most time-of-flight (TOF) systems fast extraction is preferred (Sect. 5.3.3). Therefore the sample is deposited on a plane surface (Fig. 5.1), making it rather difficult to use a liquid sample. This has been partly overcome by *Olthoff* and *Cotter* [5.57], who used the time-lag focusing principle of *Wiley* and *McLaren* [5.58] to study the effect of a glycerol matrix in ion-induced desorption of organic molecules. In magnetic sector field and quadrupole-based instruments, where low extraction fields are used, the requirement of a plane surface is usually not so strict and more "exotic" sample geometries can be tolerated (Fig. 5.1).

The advantage of TOF systems for very large molecules has been demonstrated in several studies. Figure 5.13 shows the spectrum for a large biomolecular ion, i.e., porcine trypsin [5.8]. This spectrum was collected for a sample where the trypsin was adsorbed on a nitrocellulose backing with the technique described by *Jonsson* et al. [5.54]. Cluster ions of biomolecules up to mass 70 000 u have been observed with HSF-SIMS (PDMS) [5.59].

In most modern TOF systems, the single secondary ion counting mode is used. For fast ion impact, the multiplicity of secondary ions is larger than one. Hence time measuring instruments must be able to register many stop events per

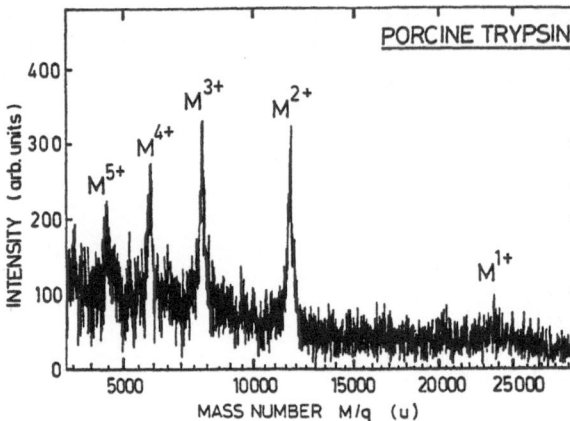

Fig. 5.13. TOF spectrum of porcine trypsin (molecular weight 23 463) adsorbed on nitrocellulose, produced by bombardment with fission fragments, HSF-SIMS (PDMS), background subtracted

start event. Most often so-called multistop time-to-digital converters (TDC) are used. Even if the multiplicity is lower for slow particle impact, multistop TDCs are necessary also in such TOF systems because the sample is bombarded with pulsed primary ion beams with 10^2–10^4 primary particles per pulse.

Problems in TOF systems may arise from time-of-flight differences, grid effects (variations in D and d, cf. (5.9)), and initial velocity distributions (variations in V). Therefore, a time resolution $(t/\Delta t)$ better than 3000 $(M/\Delta M \approx 1\,500)$ can hardly be achieved [5.60] in a straight TOF spectrometer $(\Delta t$ is the full width at half maximum). For a large molecule it is not possible to achieve unit mass resolution in a straight TOF system. On the other hand, the resolution figure mentioned fits very well with the width of the envelope of the isotopic distribution for a large molecule (Fig. 5.8), and that resolution should be appropriate for extracting the isotopically averaged mass.

A TOF system is very efficient because of the high transmission (i.e., high collection efficiency). However, intimately connected with this feature is the lack of selectivity, i.e., everything is integrated. Besides stable molecular ions and fragments, fast neutrals as well as charged particles from metastable decays are detected. This results in slightly more complex spectra than with the sector field instruments (see Fig. 5.20 below). Also, the analyzer sets no limit to the mass region covered. In fact the long flight times for large molecules pose neither a theoretical nor a practical problem. The limit on mass range at present is in the bombardment-induced desorption and the detection of large molecular ions. A problem connected with the TOF technique is the limited dynamic range in the intensities that can be studied. The time-of-flight spectra (in straight TOF spectrometers) contain neutral and charged particles from metastable decays. Therefore, in the acceleration region and the field-free drift path a "background" limits the sensitivity for detection of low-intensity secondary ions. A sector instrument suppresses such a background.

In practice the mass resolution in a straight TOF system is determined by the precision with which sample grids and the stop detector are aligned and by field inhomogeneities due to the grids [5.61]. If these problems are treated carefully, the initial velocity distribution of the secondary ions becomes the principal limitation. *Mamyrin* and coworkers [5.62] have used the reflector principle to solve that problem. The secondary ions are reflected in an electrostatic mirror, and a time-of-flight focus can be created by a proper design of flight distances and mirror geometry. *DellaNegra* and *LeBeyec* [5.63] have demonstrated that this is feasible in an HSF-SIMS (PDMS) system. For molecules with masses up to 3 000 u, a mass resolution $M/\Delta M$ (FWHM) of 4 000 could be obtained. Recently *Niehuis* and coworkers presented even more impressive results [5.64]. With an LSF-SIMS setup and for a small molecule they achieved a mass resolution of 13 000.

The inclusion of the reflector principle in a TOF system will always lead to longer flight times than in a straight system. From an intensity point of view (5.3.1) the application of this technique to the analysis of molecular ions with short lifetimes is usually not advisable.

5.5 Experimental Observations

5.5.1 Low-Velocity Particle Impact

So far secondary ion yields have been studied mainly as a function of primary particle mass and energy. *Standing* et al. [5.65] have measured the yield of protonated alanine (and fragments) for different primary particles (Li, Na, K, and Cs) in the energy regime 1–14 keV. Based on a comparison of stopping powers and absolute molecular ion yields for the different primary particles, the authors' conclusion is that the yields are directly related to the nuclear stopping power. Their conclusion seems convincing if only the experimental results for the heavier primaries are considered. However, if the authors' data for Li are plotted as a function of the nuclear and electronic stopping powers, the yields seem directly correlated to the electronic stopping power (Fig. 5.14).

The most abundant secondary ions emitted from small polar organic molecules are protonated $(M + H)^+$ and deprotonated $(M - H)^-$ species. These molecules are known to appear as zwitterions in solution. Occasionally dimer ions are observed. Other parentlike positive or negative ions (for example $(M + Na)^+$ or $(M + Cl)^-$) often appear in the spectra. Sometimes M^+ and M^- ions, probably formed in electron-transfer processes, are also detected. In addition to parentlike ions there are characteristic molecule fragments in the high-mass range and intense more unspecific fragments in the low-mass region [5.29].

Figure 5.15 gives spectra from the Münster group's work as examples of LSS-SIMS spectra [5.66]. In a review on this subject *Benninghoven* [5.29] claims that the primary ion parameters in general have a minor influence on the relative peak heights in the spectra. The absolute ion yields, however, depend strongly on the primary ion energy and mass.

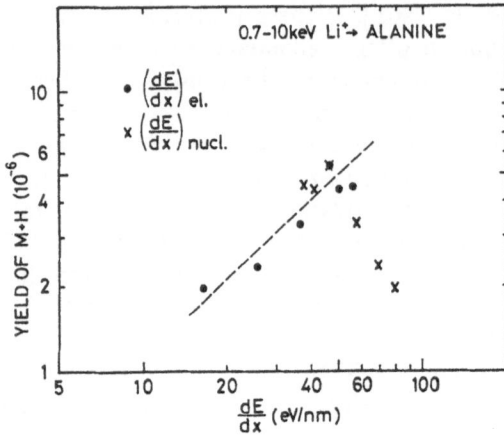

Fig. 5.14. Yield of protonated alanine $(M + H)^+$ as a function of the electronic and nuclear stopping powers of Li ions incident on the alanine sample [5.65]

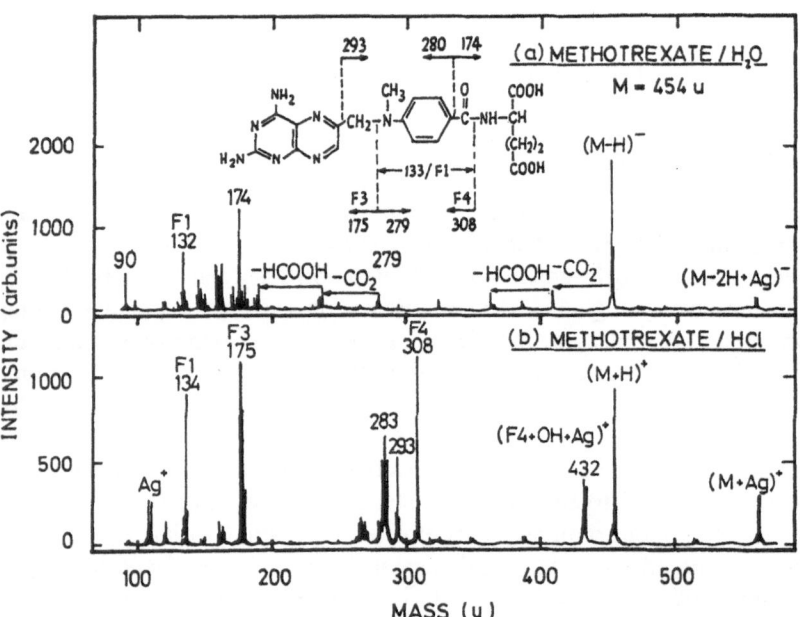

Fig. 5.15. (a) Negative and (b) positive secondary ion spectrum of methotrexate on an Ag backing: 2 nmol dissolved in 1 ml of distilled water and in 0.1 normal HCl, respectively, were deposited on 1 cm². Primary ions: Ar⁺, 3 keV; 2.5×10^{-10} A (DC current) on 0.1 cm² [5.66]

The observed ions strongly reflect the chemical nature of the analyzed molecules and their environment in the sample [5.29]. It has been observed that varying the acidity of thioglycerol influences the relative intensities of multiply charged molecular ions of bovine insulin in LLS-SIMS (FAB) spectra [5.67].

The Münster group has also studied secondary ion yields from monolayers of amino acids [5.51]. The layers were prepared under UHV conditions and deposited by the molecular-beam method on top of cleaned metal surfaces. The main findings were the following: Deposition and reevaporation experiments at various substrate temperatures showed that on reactive metals like Cu the first monolayer is considerably more strongly bound than the second and third layer, and so on. On noble metals like Au the first monolayer is only weakly bound. For molecules on a Cu surface, the $(M + H)^+$ emission starts only when a monolayer is completed while the $(M - H)^-$ intensity grows monotonically with coverage in the submonolayer region. On a gold surface, however, both peaks grow monotonically in the submonolayer region. These findings indicate, according to the authors, that the first monolayer on a Cu surface is $(M - H)^-$, supporting *Benninghoven*'s precursor model [5.29] (see Sect. 5.6.3 below).

Michl [5.68] has performed extensive studies on secondary ions (in particular clusters) emitted from low-temperature solids of condensed gases (like Ar, Kr, Xe, N_2, and O_2) when bombarded with keV ions. The clusters produced from these samples are often built on units of molecular composition different from that of the solid, and spectra of mixtures are often quite different from the summed spectra of the different components. *Michl*'s results therefore show that in ejection and ionization "new" systems form, particularly with samples that consist of a mixture of reactive compounds and for which sputtering yields are high. In the "plume" of sputtered material the pressure may initially be high and gas-phase reactions are likely to take place.

Wong et al. [5.69, 70] and *Barofsky* et al. [5.71] have performed studies on the role of the liquid matrix in SIMS. They have measured the sputtering yield for 6 keV Xe impact on glycerol at room temperature. The sputtering yield was found to be about 1 000 glycerol molecules per primary ion. *Williams* and *Gillen* [5.72] have studied the sputtering yield from frozen glycerol and found a yield of 54 ± 9 for 8 keV Ar^+. They argue that this sputtering yield does not agree with the results above, because there should only be a small difference between sputtering yields from a solid and a liquid, assuming that only knockon sputtering contributes. For condensed gas sputtering it is known, however, that the sputtering yield is temperature dependent [5.73]. If we take this into account and also the fact that the previous authors used 6 keV Xe as primary particles, there is probably no contradiction between the measured values.

In another study, *Wong* and *Röllgen* examined [5.74] the renewal of the surface layer from samples of a pentapeptide and stachyose distributed in glycerol (Fig. 5.16). Stachyose ($M = 666$ u) and the pentapeptide ($M = 777$ u) were chosen because stachyose is homogeneously miscible with glycerol, even at high concentrations, and the pentapeptide has surface-active properties; i.e., it segregates to the surface in a glycerol solution. Remember that a glycerol layer evaporates at about 50 nm/s even without ion bombardment. This implies that the near-surface concentration of molecules dissolved in glycerol is higher than in the bulk. For stachyose the sample used was a 1:1:6 mixture of stachyose, ammonium chloride (NH_4Cl), and glycerol. For the pentapeptide a mixture of

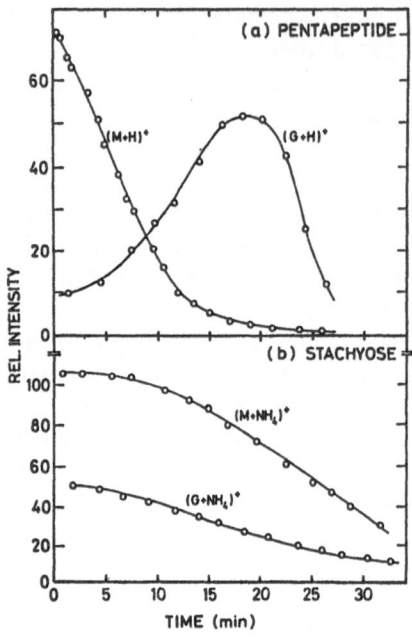

Fig. 5.16. Variation of molecular-ion yield from (a) pentapeptide $(M + H)^+$ and (b) stachyose $(M + NH_4)^+$ dissolved in glycerol as a function of time of irradiation with 6 keV Xe ions at a flux of 10^{13} Xe atoms/cm^2 s [5.74]

1:24 of pentapeptide and glycerol was used. The stachyose is positively ionized by NH_4^+ attachment and the pentapeptide by protonization. The molecular-ion yield of stachyose, the pentapeptide, and glycerol was studied as a function of the irradiation time, i.e., the fluence of 6 keV Xe (Fig. 5.16). For the stachyose sample (Fig. 5.16b) the molecular-ion yields of stachyose and glycerol decrease slowly as a function of irradiation time. The conclusion is that the surface renewal is due to the bulk sputtering of the radiation-damaged stachyose. With no glycerol, no molecular-ion yield of stachyose was observed, supporting the idea of glycerol-assisted elimination of radiation damage.

With the pentapeptide a quite different behavior was found (Fig. 5.16a). The molecular-ion intensity of the peptide rapidly decreases when most of the peptide is sputtered away, and the glycerol molecular-ion intensity increases before all the glycerol is gone. This is, of course, due to the expected surface enrichment of the peptide. Also irradiation time-scale effects on molecular-ion yields were studied during the first seconds of bombardment. The results indicate that the diffusion times from the bulk to the surface of sample molecules are not important. However, ion migration by the field imposed on the sample seems to be important.

Olthoff and *Cotter* [5.57] recently studied the effect of the glycerol matrix on the molecular-ion yield of a peptide gramicidin *S* using a low-energy pulsed ion gun and the time-of-flight technique. By varying the instantaneous primary ion current and the pulse repetition rate they found that at the highest rates the molecular-ion intensity decreases, in particular when a low-concentration (gramidin *S*/glycerol − 0.1 nmol/µl) sample was used. This indicates, according to the authors, a finite recovery time to repair radiation damage. When discussing experiments on the role of the glycerol matrix, we should stress that it is very difficult to maintain full control of a mixed liquid sample. In the *Olthoff* and *Cotter* study [5.57] the gramicidin *S* was mixed into methanol before mixed with glycerol. When such a sample is put in a vacuum system the peptide may be surface enriched simply because the methanol evaporates faster than glycerol.

The following explanations for damage removal with liquid glycerol have been proposed: bulk diffusion processes, large sputtering yields, ion migration, rapid evaporation of the liquid containing damage, and gross liquid flow due to beam effects. Of these, only bulk diffusion seems to be ruled out at present [5.75]. More research is needed on the role of liquid matrices like glycerol.

The molecular ions produced in sputtering from a liquid matrix may also be less excited internally than when a solid matrix is used, as demonstrated for sputtering of salt-cluster ions from a glycerol solution [5.70].

5.5.2 High-Velocity Particle Impact

The sputter erosion of metal surfaces with low-velocity heavy particles was a well-established phenomenon at the time *Benninghoven* and coworkers made the first experiments on bioorganic solids and *Barber* later introduced LLS-SIMS (FAB). On the other hand, sputtering of insulators due to electronic energy deposition became an interesting subject, largely because of studies involving high-velocity ion bombardment of biomolecules. The direct connection to electronic energy deposition was demonstrated independently and at about the same time, around 1980, by experiments at Erlangen and Uppsala (Fig. 5.17) [5.76, 77]. The first international meeting discussing electronic sputtering of biomolecules was held at Uppsala in 1981 [5.78]. Today there are much experimental data on secondary ion yields of biomolecules as a function of primary ion parameters such as velocity [5.36, 76, 79], angle of incidence [5.80, 81], and charge state [5.82–85]. In the electronic stopping regime there are probably more experimental yield data on biomolecules available than in the nuclear stopping regime.

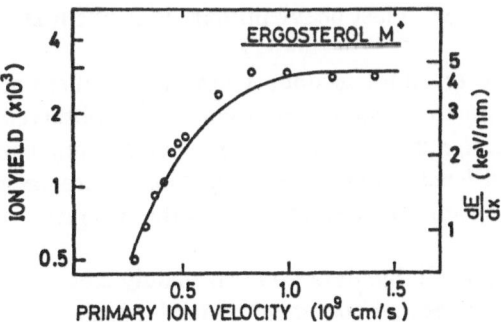

Fig. 5.17. Yields of molecular ions of ergosterol ($C_{28}H_{44}O$) as a function of the velocity of ^{63}Cu ions in the electronic stopping regime [5.74] ($v = \sqrt{2E/M}$ cm/ns where E is in MeV and M in u). The electronic stopping powers (full line) were taken from data tables [5.98]

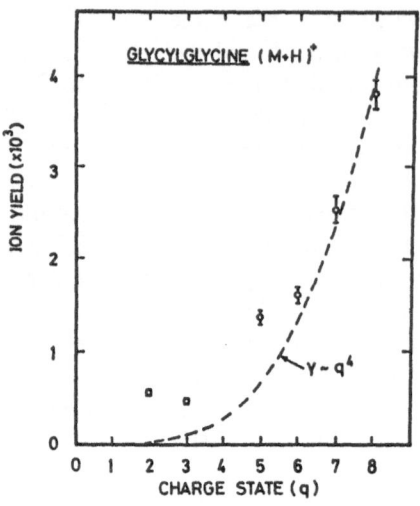

Fig. 5.18. Yield of molecular ions $(M + H)^+$ of glycylglycine $(C_4H_8O_3N_2)$ as a function of the charge state of 20 MeV ^{16}O ions [5.82]

As observed in the work with low-velocity primaries, high-velocity primary ion parameters are in general found to have a minor influence on the relative peak heights in mass spectra as well as on the types of molecular ions observed. The main effect of the primary particle is to influence the absolute yield of secondary ions. As illustrated in Fig. 5.17, the ion yields are closely related to the electronic stopping power. Ion yield also depends strongly on the charge state of the incident ion, as can be seen in Fig. 5.18. For the case in Fig. 5.18 the yield for the highest charge states (q) was found to be proportional to q^4 [5.82].

Guthier et al. [5.83] reported on the production of hydrocarbon fragments of the type C_nH_m from biomolecular films. At the same primary ion velocity (1 MeV/u) the relative intensities of these fragments were higher for ^{238}U ions than for ^{14}N primary ions. The interpretation was that in the core of the ion track very high energy densities are produced in the case of the ^{238}U beam, so that violent fragmentation occurs. The Uppsala group has found a strong nonlinearity of yields for large molecular ions when studied as a function of energy density in the ion track [5.86]. Multilayer samples were used. It was argued that the "excitation area," which defines a certain minimum energy density, becomes critical when large molecules are to be desorbed. Molecular ions of the protein porcine phospholipase $A_2(M = 13\,980\ u)$, for example, could be observed only under bombardment with fast heavy primary ions like fission fragments [5.43].

A similar "size effect" was also observed for secondary ion clusters of amino acids. For 90 MeV ^{127}I primaries, up to 16 valines in a cluster are observed, whereas for 8.5 MeV ^{12}C the maximum number of molecules per cluster that can be identified above the background level is four (Fig. 5.19). This difference is again attributed to the difference in excitation area (volume) for the two primary ions.

As for low-energy primaries, the available data concern mainly secondary ions (rather than neutrals). However, a recent experiment by the Uppsala group

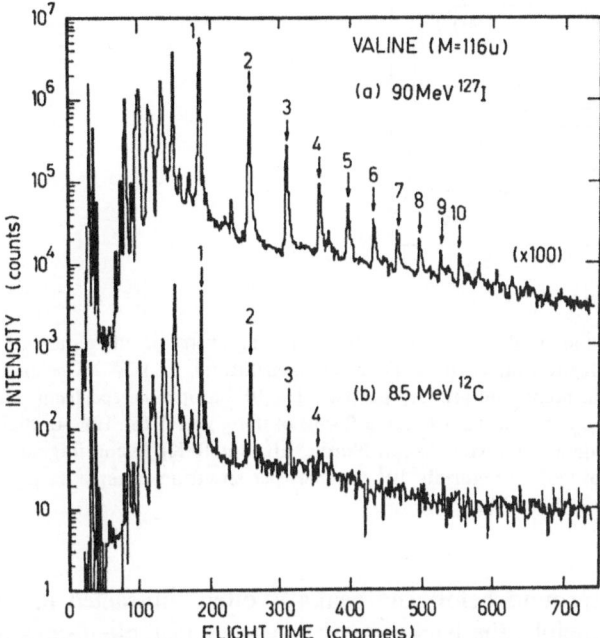

Fig. 5.19. TOF mass spectra obtained by bombarding a layer of the amino acid valine ($M = 116$ u) with **(a)** 90 MeV $^{127}I^{14+}$ and **(b)** 8.5 MeV $^{12}C^{4+}$ ions. The labels 1, 2, 3, . . . indicate the positions of the peaks $(M + H)^+$, $(2M + H)^+$, $(3M + H)^+$, etc. The lower spectrum was recorded with twice as many primary ions as the upper spectrum

[5.87] showed that one fission fragment can desorb more than one thousand intact molecules of the amino acid leucine.

Säve et al. [5.88, 89] have performed HSF-SIMS experiments on Langmuir-Blodgett films of fatty acids. Some of these experiments used a marker molecule layer positioned at various depths in the film. Marker-specific molecular ions were found to originate from layers deep below the surface (10–20 nm). This finding can be best explained by crater formation.

5.5.3 Comparison of Mass Spectra and Ion Yields for Low- and High-Velocity Particle Impact

With the data outlined above it is interesting to compare the mass spectra for the same molecule using either low- or high-velocity primary particles. The spectra shown in Fig. 5.20 were taken in the LLS-SIMS (FAB) (Fig. 5.20a) and in the HSF-SIMS (PDMS) (Fig. 5.20b) modes, i.e., with two different instruments and with two different types of matrices. The molecule studied was bovine insulin with a mean molecular weight of 5 733.5 u.

Evidently the TOF spectrum (Fig. 5.20b) contains a large contribution from fast neutrals (about 20%) and charged particles resulting from metastable

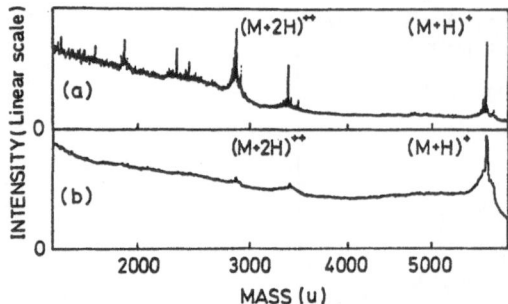

Fig. 5.20. Mass spectra of bovine insulin taken with the following instrumental parameters: (a) Magnetic sector field instrument with ZAB-SE (Vacuum Generators), 6.8 keV Xe beam of neutrals, 2.2×10^{13} atoms/s, total number of primary particles 3×10^{15} atoms per spectrum. The sample consisted of 10 µg of bovine insulin dissolved in a few µl of thioglycerol. (b) Time-of-flight instrument with ^{252}Cf fission fragment source (Bio-Ion Nordic AB, Bin 10 K), fission fragments, 1000 particles/s, total number of fission fragments 10^7 particles per spectrum, Sample: 50 µg of bovine insulin electrosprayed on a thin Al backing

decays. The smooth background below the peaks is often subtracted (as in Figs. 5.13, 5.24–26). In principle, the background due to metastable decays in TOF spectra may be eliminated by adding an electrostatic filter. Nevertheless, a background similar to the one in Fig. 5.20a will probably be left. The background is often called the "chemical noise" and most likely results from the certain intensity at every mass due to prompt fragmentation.

The similarities of the two spectra in Fig. 5.20, produced with very different instruments (mass analyzers), different basic ion impact processes (slow and fast ions), and different sample states (liquid and solid) may indicate a large degree of isomerization of energy [5.15] before the actual ejection of ions (desorption-ionization). The ejection and formation of ions take place a "long" time after the primary ion energy deposition. The actual desorption-ionization process involved in the above two cases may very well be the same. The main difference is the yield with which the molecular ions are produced in the two primary ion velocity regimes. (The intensity of the doubly charged molecular-ion peak in the low-velocity primary ion spectrum in Fig. 5.20a should be divided by two before comparison, because the integrated ion charge was measured.) A detailed and careful comparison using the same analyzer and sample state may, however, substantiate the idea that not only the yield but also the relative intensity of fragment ions to molecular ions is different for fast- and slow-ion impact.

Such experiments have been made. *Kamensky* et al. [5.90] compared the molecular-ion yields for electrosprayed samples of four molecules in the mass region 100–2000 u in the same vacuum system with a TOF system using 4 keV Cs and 60 MeV Cu as primary ions. This comparison was limited to molecular secondary ion yields because of the experimental technique. The finding was that the relative yield with high- over low-velocity ions was higher for all molecules, and more so the heavier the molecule. Originally the Manitoba

(*Standing* et al.) and Rockefeller (*Field* et al.) groups [5.91] used LSF- and HSF-SIMS with TOF to compare mass spectra of a few small organic molecules and found very small differences in the fragmentation patterns (samples were prepared in one laboratory and sent by mail to the other laboratory). Later the Manitoba-Rockefeller collaboration extended experiments [5.92] to include larger molecules and molecular-ion yields. The findings were qualitatively the same as those of *Kamensky* et al. [5.90].

Recently *Ens* et al. [5.93] completed a system at Uppsala where fission fragments from a ^{252}Cf source and 18 keV Cs$^+$ ions can be used alternatively to study TOF mass spectra from a compound without exposing the sample to air in between. Figure 5.21 shows the results of an experiment in which a set of peptides and small proteins were studied. Two types of sample preparation, namely electrospray and adsorption to nitrocellulose, were used. The figure shows relative molecular-ion yields normalized to the yield for the lightest molecule studied. The absolute yield for this molecule was estimated to be about ten times larger for fission fragments than for 18 keV Cs$^+$ ions. With multilayer samples fast ion impact is relatively more efficient. With low-velocity ion impact, beyond mass 5 777.6 u (porcine insulin) the molecular-ion peak falls below the limits of detectability. Although not studied in this experiment, molecular ions of porcine trypsin ($M = 23\,463$ u) have been observed from an electrosprayed sample using HSF-SIMS (PDMS) [5.8]. For adsorption to nitrocellulose the relative difference in yields is less pronounced, and here insulin and even larger molecules can be observed for slow ion impact. If we assume that the sputtered volume of material is larger for fast ion impact, this may explain why the fast ion

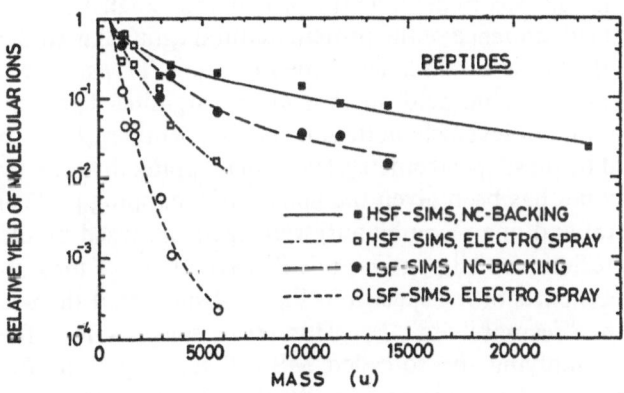

Fig. 5.21. A comparison of relative molecular-ion yields of peptides bombarded with fission fragments and 18 keV Cs$^+$ ions as a function of peptide mass. The fission fragments enter the sample surface from the back at 90° and the Cs$^+$ ions from the front at a 66° angle of incidence. The mass analyzer used was a TOF system. Samples were prepared with two different techniques, namely the electrospray method (multilayer) and adsorption to nitrocellulose backings (monolayer). The yields are normalized to the yield for the lightest peptide studied, arbitrarily set to 1 [5.93]. The lines are drawn to guide the eye

is relatively more efficient for large molecules and for multilayer samples. For monolayer samples the difference is less dramatic, and the difference for the largest molecules may be due to a slightly larger excitation area for fast ion impact.

5.5.4 Application Highlights

In this section we discuss a few applications to show how problems in biosciences can be solved using the techniques described above. The first example, a study by *Richter* et al. [5.94], illustrates a problem supposed to be of growing importance. The study tried to verify the primary structure of two types of eglin C. One occurs naturally and the other is prepared by recombinant DNA technology. Eglin C is a polypeptide of isotopically averaged mass 8 091 u and with an atomic composition $C_{373}H_{550}N_{96}O_{107}$. In nature it is produced by leech and is a strong and specific inhibitor of the human granulocytic proteinases elastase and cathepsin G. The recombinant eglin C, an *E. coli* expression product, was found to be indistinguishable from natural eglin C from a biological, immunological, and amino acid compositional point of view. However, it turned out to be different when analyzed with reversed-phase high-performance liquid chromatography (HPLC). Use of the LLS-SIMS (FAB) technique and a large sector instrument showed the mass of the molecular ion of the recombinant eglin C to be 42 u larger than for natural eglin C. The measured masses of the dominant peaks in the final analysis were $8\,091.79 \pm 0.28$ and $8\,133.74 \pm 0.4$, respectively. The value $8\,091.79 \pm 0.28$ fits well to the expected molecular ion, i.e., a protonated molecule $(M + H^+)$ for natural eglin C. Does the mass difference indicate a chemical modification of the recombinant eglin C?

The next step in the analysis was to degrade the protein into smaller pieces. Table 5.2 gives the amino acid sequence of the protein (natural eglin C) with the amino acids labeled with the accepted three-letter identification. Trypsin gives cleavages in all lysine-A_n (A_n = amino acid number n) and arginine-A_n amide bonds. This results in a mixture of seven chain fragments – i.e., peptides $(P)P_1$ to P_7 – which can be analyzed by mass spectrometry. (Protein mapping this way by mass spectrometric techniques has been given the name FAB mapping [5.95].) These mixtures (one natural and one recombinant) were again analyzed by the same mass spectrometric technique. All peptides but P_1 gave the same mass in the two cases. The low-mass parts of the spectra in Fig. 5.22 show that there is a 42-mass-unit shift in the "recombinant" P_1. This result suggests that the modification is in P_1. By applying the so-called MS-MS technique[1] to P_1, *Richter* and coworkers [5.94] were able to identify the modification as an acetylation (addition of CH_3CO) of the *N*-terminus of the peptide. Thus it was possible to detect a minor divergence from natural eglin C, namely a small

---

[1] A molecular ion, selected in one mass spectrometer, is excited in some way (e.g., collisionally activated), and the resulting fragment ions are studied in a second mass spectrometer.

Table 5.2. Structure of natural eglin C

	10
H-Thr-Glu-Phe-Gly-Ser-Glu-Leu-Lys-Ser-Phe-Pro-Glu-Val-Val-Gly-	
20	30
Lys-Thr-Val-Asp-Gln-Ala-Arg-Glu-Tyr-Phe-Thr-Leu-His-Tyr-Pro-	
	40
Gln-Tyr-Asp-Val-Tyr-Phe-Leu-Pro-Glu-Gly-Ser-Pro-Val-Thr-Leu-	
50	60
Asp-Leu-Arg-Tyr-Asn-Arg-Val-Arg-Phe-Tyr-Asn-Pro-Gly-Thr-Asn-	
	70
Val-Val-Asn-His-Val-Pro-His-Val-Gly-OH	

Amino acid		Atom composition	Molecular weight[a]
Alanine	Ala	C_3H_5ON	71.080
Arginine	Arg	$C_6H_{12}ON_4$	156.189
Aspartic acid	Asp	$C_4H_5O_3N$	115.090
Asparagine	Asn	$C_4H_6O_2N_2$	114.105
Cysteine	Cys	C_3H_5ONS	103.144
$\frac{1}{2}$ Cystine	Cys	C_3H_4ONS	102.136
Glutamic acid	Glu	$C_5H_7O_3N$	129.117
Glutamine	Gln	$C_5H_8O_2N_2$	128.132
Glycine	Gly	C_2H_3ON	57.052
Histidine	His	$C_6H_7ON_3$	137.143
Leucine/Isoleucine	Leu/Ile	$C_6H_{11}ON$	113.161
Lysine	Lys	$C_6H_{12}ON_2$	128.176
Methionine	Met	C_5H_9ONS	131.198
Phenylalanine	Phe	C_9H_9ON	147.179
Proline	Pro	C_5H_7ON	97.118
Serine	Ser	$C_3H_5O_2N$	87.079
Threonine	Thr	$C_4H_7O_2N$	101.106
Tryptophan	Trp	$C_{11}H_{10}ON_2$	186.216
Tyrosine	Tyr	$C_9H_9O_2N$	163.178
Valine	Val	C_5H_9ON	99.134

[a] Isotopically averaged masses of amino acid residues (C = 12.0112, O = 15.9994, H = 1.0080, N = 14.0067, S = 32.064)

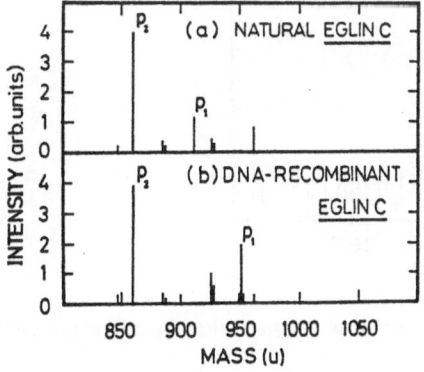

Fig. 5.22. LLS-SIMS (FAB) mass spectra (low-mass region) of the peptide mixtures obtained by tryptic digestion of (a) authentic eglin C and (b) recombinant DNA eglin C [5.94]

acylsubstituent in a posttranslational event at a distinct site in a large bio-molecule, by mass spectrometric techniques in combination with chemical degradation.

Before the LLS-SIMS technique (FAB) was introduced in 1981, *Macfarlane*'s HSF-SIMS system (PDMS) was one of the few systems that could be used to analyze large thermally unstable biomolecules. One famous application during that period dealt with the palytoxin problem. Palytoxin is a marine toxin whose structure had been investigated for a decade. A Japanese group had come very close to putting the complex structure together. They needed the molecular weight for final assembly of the structure. The peak of the spectrum in Fig. 5.23 is labeled $(M + Na)^+$, and therefore the molecular weight deduced was 2651 u. The molecular ion was verified to be a sodium attachment by adding a Cs salt to the sample: the molecular ion peak shifted by 110 u. Within a few months the structure was known because of the HFS-SIMS measurement on the native toxin and its derivatives [5.96].

The third application is again related to biotechnology. A Danish company markets human insulin manufactured from porcine (pig) insulin. Insulin consists of two peptide chains *A* and *B* coupled together via disulfur bridges. Figure 5.24a shows the mass spectrum of porcine insulin collected with an HSF-SIMS system [5.8]. The second spectrum (Fig. 5.24b) shows the result when an alanine residue was taken away from the *B* chain (note the shift in the positions of the molecular-ion peak and *B*-chain peak). In the next step a threonine residue was added at the same position in the molecule; the peaks shifted again but now to higher masses. The spectrum (Fig. 5.24d) for natural human insulin can finally be seen to agree with that in Fig. 5.24c. In the future mass spectrometric techniques may be applied for batch control in biotechnical industry processes, as discussed with reference to Fig. 5.24.

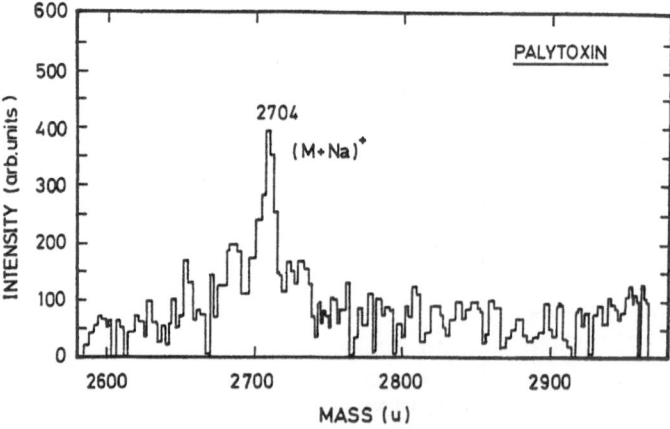

Fig. 5.23. HSF-SIMS (PDMS) positive-ion mass spectrum of palytoxin (background subtracted) [5.96]

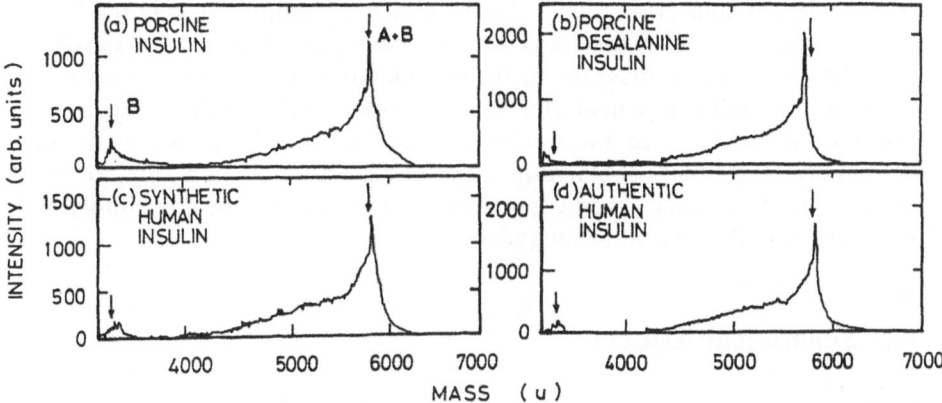

Fig. 5.24. High-mass region of HSF-SIMS (PDMS) spectra of (a) porcine insulin, (b) desalanine insulin, (c) semisynthetic human insulin, and (d) human insulin (background subtracted). The mass differences observed for the molecular-ion peaks in (b–d) relative to that of porcine insulin are -71.5 ± 3.0, $+31.9 \pm 3.0$, and 27.4 ± 3.0 u. The corresponding calculated values are -71, $+30$, and $+30$ u [5.8]

Figure 5.25a shows how HFS-SIMS can also be used to determine molecular weights in mixtures [5.8]. A 1:1:1:1 molar mixture of four peptides and proteins – the releasing hormone for luteinizing hormone (L, MW 1 182), porcine insulin (I, MW 5 778), pancreatic spasmolytic peptide (P, MW 11 711), and porcine phospholipase A_2 (A, MW 13 980) – was prepared and analyzed using HFS-SIMS. The relative peak heights reflect the different yields with which

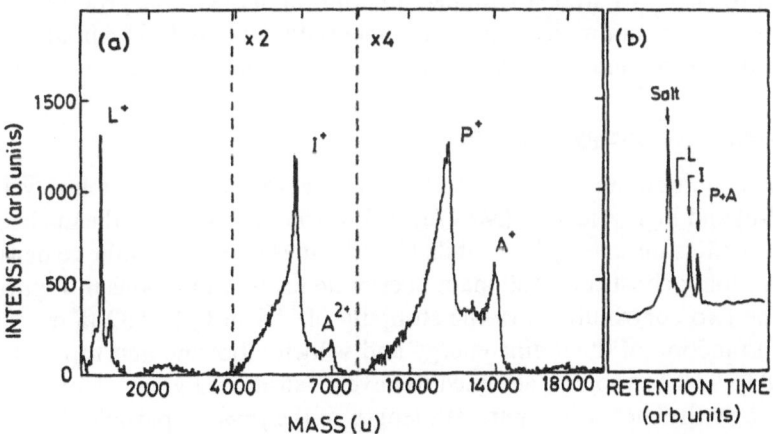

Fig. 5.25. (a) Positive-ion HSF-SIMS (PDMS) spectrum of a 1:1:1:1 molar mixture of LHRH, i.e., the releasing hormone for luteinizing hormone (L, MW 1182), porcine insulin (I, MW 5778), pancreatic spasmolytic peptide (P, MW 11 711), and porcine phospholipase A_2 (A, MW 13 980) (background subtracted). (b) HPLC (high performance liquid chromatography) diagram of the same mixture [5.8]

these molecular ions are desorbed. The mass determinations could be made with the same precision as in spectra of pure samples [5.8]. Figure 5.25b shows the reversed-phase HPLC diagram for the same mixture. Besides the large disturbing peak from salt components, the heaviest components of this mixture are not resolved. In addition to the poorer resolution here, the precision in mass determination is orders of magnitude worse. This example shows that for small proteins, particle-induced mass spectrometry can now compete favorably with conventional biochemical techniques.

5.6 Theoretical Aspects

5.6.1 General

With the large range of conceivable applications, the search for a deeper understanding of sputtering mechanisms in large organic molecules is well motivated. Even though the initial basic ion-solid interactions are well known [5.97], we soon encounter severe problems if we try to describe how the deposited energy is converted into molecular motion. The processes under consideration take place in a time interval ranging from 10^{-14} s to 10^{-9} s. Time-of-flight experiments provide information only about what is going on at times after more than 10^{-9} s and outside the solid. Details of the energy deposition and sputtering process are hidden.

At present we are left with many unanswered questions. How are bonds broken? How is momentum transferred to a large molecule? How do the molecules get ionized? Are the ions formed at the surface or in the gas phase? Since it is far too early to present a unified description of the whole process, our ambition here is to summarize the experimentally known facts about the processes and then discuss the various mechanisms proposed so far.

5.6.2 Mechanisms of Energy Deposition

The dominant (initial) energy-loss mechanisms are very different at low velocities (keV/u) and high velocities (MeV/u). Below the Bohr velocity the nuclear stopping dominates the energy loss, while for velocities above the Bohr velocity electronic stopping constitutes the main energy-deposition mechanism. Figure 5.26 plots the two contributions to the stopping of ^{127}I in C_3H_5NO (a model protein) as a function of the iodine energy and velocity (bottom and top scale, respectively) [5.98]. The solid is assumed to have a density of 1 g/cm^3. Note that the graph shows the energy loss per path length of the *primary* particle. As can be seen, the total energy loss per unit length is roughly 15 times larger in a fission fragment experiment (100 MeV ^{127}I) than in a typical slow particle experiment (10 keV ^{127}I).

At low incident velocities the projectile loses its energy predominantly via elastic binary collisions with the atoms in the solid. The energy transfer may

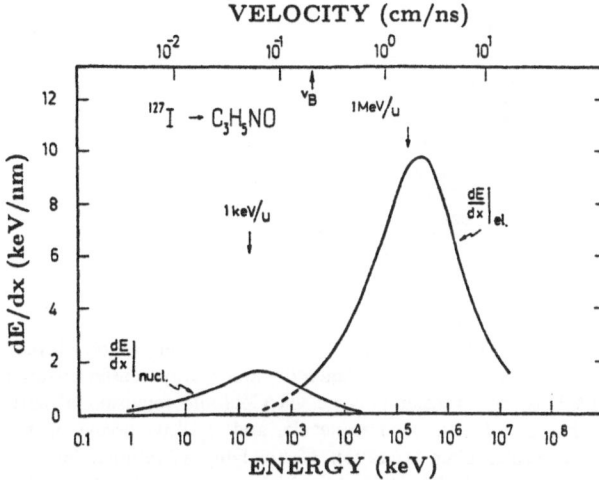

Fig. 5.26. Nuclear and electronic energy loss per unit length for ^{127}I incident on a solid of a model protein C_3H_5NO with a density of $\rho = 1 \text{ g/cm}^3$. The nuclear energy loss in the organic material was calculated using Ref. 5.24, Eq. (2.2.1) and a Lenz-Jensen potential. The electronic stopping powers were taken from data tables [5.98]

suffice to move an atom from its lattice site. If the recoiling atom is energetic enough, it may cause other atoms to recoil and a collision cascade to develop. Such a cascade may cross the surface and cause single-atom sputtering (nuclear sputtering). Three different regimes of sputtering events can be identified: the single-knockon regime, the linear-cascade regime, and the spike regime [5.24]. In the first regime the recoiling target atoms cause a small number of further recoils. In the linear-cascade region the density of moving atoms is still low, but in the spike regime most atoms are moving in a certain (spike) volume [5.99].

At higher energies, the incident particle loses energy via ion-electron collisions. In insulators the lifetimes of excited electronic states may be long enough to allow the excitation energy to be transferred to atomic motion and sputtering (electronic sputtering) [5.100]. Again, three regimes may be identified, depending on the density of ionizing and dissociating events. As electronic energy is also deposited in the nuclear stopping regime, due to both the primary particle and the recoils, there can in principle also be electronic sputtering in this regime. Figure 5.27 schematically illustrates the two different initial energy-deposition modes.

5.6.3 Nuclear Stopping Regime

The ejection of a large thermally labile molecule due to the impact of a slow particle is difficult to understand within the classification scheme for sputtering events given above. A single-collision ejection mechanism may work for small molecules. For larger molecules, however, a more collective momentum-transfer

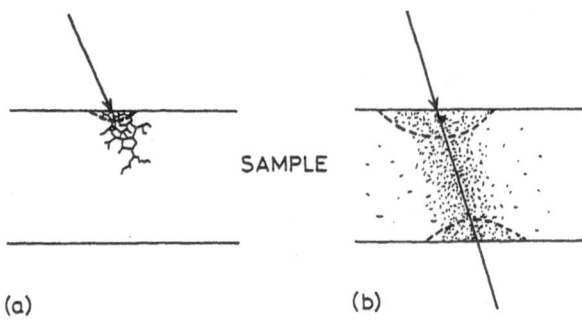

Fig. 5.27. Schematic illustration of the effects of low- and high-velocity particle impact on a layer of organic molecules. For a slow particle (**a**) there is a direct momentum and energy transfer in atomic collisions. For a fast particle (**b**) this energy transfer may go via Coulomb repulsion of partly screened nuclei in the infratrack, electronic excitation of molecules and repulsive decays of these excited states, and expansion of molecules after excitation of low-lying vibrational levels by low-energy secondary electrons. The areas marked by dashed lines indicate the regions from which atoms and molecules may be ejected

mechanism via surrounding atoms and molecules is needed to explain why the molecule leaves the surface intact. Dense linear cascades or spikes where atomic collisions terminate in weak molecular collisions are more likely to be involved. *Sigmund* [5.101] has proposed and *Sigmund* and *Claussen* have developed [5.102] an elastic collision spike model. The atoms in the spike volume are described as ideal gas atoms, the energy transport in the gas is described as heat conduction, and the initial temperature in the spike is determined from the energy deposition in elastic collision cascades. This model describes the experimental data for metals [5.96] better than the linear-cascade model. The problem is, however, that LLS-SIMS experiments typically are performed under conditions far from spike conditions, e.g., 10 keV Xe hitting an organic sample. Even at low velocities it may be important to consider the role of the electronic stopping, which may constitute an appreciable fraction of the total energy deposition if the recoil interactions are taken into account. In the example above, 10 keV Xe on an organic sample, the energy deposition due to electronic stopping amounts to about one fourth of the total energy deposition.

From the calculations of *Winterbon* et al. [5.103] the mean transverse width for the region of 10 keV Xe atoms moving in an organic solid can be estimated to be on the order of 10 nm. Accordingly, the volume of atoms set in motion is large compared with the volume of quite large molecules.

An important aspect in this discussion relates to the fact that the bioorganic layers are insulators. Insulating molecular solids exhibit a variety of different properties *relevant* for the discussion of desorption-ionization mechanisms. Unlike metals, molecular solids have an internal chemical structure. Moreover, long-lived localized charges and electronic excitation are possible in molecular solids, but not in metals. Finally, the heat of vaporization is high for metals, compared with molecular solids.

In the discussion of cluster-ion ejection mechanisms in low-temperature solids bombarded by keV particles, *Michl* [5.68] has stressed the importance of secondary ionization and electronic excitation at the surface caused by moving atoms and molecules in the collisional cascade. An illustration is that in 70 eV electron ionization of neutral clusters of $(NO)_n$, the solvating molecule of the cluster ions is N_2O_3 [5.104], similar to what was found in *Michl*'s SSL-SIMS experiments. Local charging in the impact region in an insulator should also be taken into account. For molecular solids he therefore questions the relevance of *Garrison*'s calculations, [5.105] which used the classical trajectory approach to study the sputtering of large molecules from a surface. According to the simulations, quite large molecules can be sputtered intact. In the calculations ionization mechanisms were not taken into account, and no difference was made between an insulator and a metal sample.

5.6.4 Electronic Stopping Regime

If we assume that the energy deposited into electronic excitations causes the sputtering, we have to know how the electronic energy is converted into atomic and molecular motion. When the ion traverses a layer of material, it produces a cylindrical region of intense ionization and excitation because of the direct Coulomb interaction of the swift ion, the "infratrack" [5.106]. The radius of the infratrack is the Bohr adiabatic radius, which is proportional to the velocity of the fast ions and is typically on the order of a few tenths of a nanometer. In addition, high-energy (keV) electrons (δ electrons) are produced in "hard" ion-electron collisions. These δ electrons have fairly large ranges and form a cylinder of comparatively low energy density around the incident ion path called the "ultratrack" [5.106]. The ultratrack radius is determined by the maximum projected δ-electron range, roughly proportional to the square of the ion velocity [5.107]. The ultratrack radius may range from tens to hundreds of nonometers for the particles of interest.

During the last few years, various mechanisms for the conversion of the electronic energy to nuclear motion have been proposed. For a long time it has been known that fast particles like fission fragments [5.108] cause radiation damage in certain insulators, so-called track-forming materials. The tentative explanation for this effect is that the positive-ion core, formed initially after the passage of a fast ion, "explodes" because of the Coulomb forces. In agreement with experimental observations [5.108], this track formation was supposed to be effective only in materials with slow relaxation such as insulators. In 1976 *Haff* predicted sputtering effects in the electronic stopping regime on the basis of the Coulomb explosion model [5.109]. The question of track formation was resurrected recently with the discovery of electronic sputtering. Around 1980 several studies were published on sputtering yields in the electronic sputtering regime for samples of alkali halides [5.110], ices [5.111–113], condensed noble gases [5.114, 115], metal halides [5.116], and layers of biomolecules [5.76, 77].

Models for fast ion erosion of inorganics have been reviewed by *Johnson* and *Brown* [5.117] and more recently by *Schou* [5.118]. *Brown* et al. [5.111] also considered the Coulomb explosion mechanism for sputtering [5.109] in connection with work on fast ion sputtering of ices. The model was later questioned by *Ritchie* and *Claussen* [5.119], who argued that in molecular solids there is probably not enough time for the positive ions to become accelerated appreciably before the electrons return to the core and neutralize it. Therefore they proposed that the returning electrons create a plasma fed with energy from the excited states produced in the region around the core. In the plasma there would be enough time for the ions to be heated by the electrons and for the core to expand, causing momentum and energy transfer to the surrounding region. In the *Watson-Tombrello* model [5.120] the concept of an expanding lattice driven by residual Coulomb forces is also used. In this high-excitation-density regime *Itoh* and *Nakayama* [5.121] have discussed the local lattice distortion caused by the vicinity of two excited sites.

A more recent model, based on the repulsive decay of localized excited states, is used to explain the sample-thickness dependence of ion and electron sputtering of condensed layers of Ar [5.122–124] and the sputtering of molecular solids in the low-ionization-density regime [5.125]. Most excitations of fast particles produce Ar^+, which diffuses rapidly (it may diffuse tens of nanometers before becoming localized), but can gain energy by pairing with a neutral Ar to form Ar_2^+. As electrons are available, this system can repulsively decay to $Ar^* + Ar$ with a release of 1 eV in kinetic energy to the two Ar atoms. If this happens at a surface it can lead to sputtering.

Also, Ar^* has an attractive state formed with a ground-state Ar, i.e., the eximer, Ar_2^*. The only way out of this potential minimum is a vertical transition, which gives rise to the dominant eximer luminescent band. As the Ar atom separation in the eximer state is much less than the equilibrium lattice spacing in solid Ar, potential energy is again released into kinetic energy of Ar atoms. This is a second chance for sputtering if the decay takes place close to the surface. In such a decay, energy is released via recoils, and a minicascade is produced over a small volume, i.e., a few atomic distances [5.126]. For ionized molecules in a solid such a process will always occur with electronic recombination [5.117, 123]. For lower lying states the process is always in competition with photon emission.

Exciton models for ion and electron sputtering of alkalihalides were discussed earlier by *Townsend* [5.100]. A mechanism of this type in organic molecular solids – i.e., electron relaxation via repulsive excited states – has been proposed by *Johnson* and *Sundqvist* [5.127] as a possible energizing process for the desorption of large molecules. Molecular solids are known to luminesce upon ion impact [5.121]. As yet, however, studies on the relationship of the luminescence spectra to desorption efficiency have not been carried out for molecules of biological interest.

With reference to the so-called ion track model, *Hedin* et al. [5.128] have examined the strong nonlinear dependence of the yields of large molecular ions on the energy deposited by the fast primary particle (Sect. 5.6.2). The secondary electron shower creates a distribution of excitations ("hits") over a well-defined radial distribution. For desorption to occur at least m bonds must be broken by secondary electrons generated by the passing ion. A possibility rarely considered so far is that low-energy electrons in the track may efficiently excite low-lying vibrational levels in the molecular solid, causing increased pressure in the track. This is the basis for a recent model by *Williams* and *Sundqvist* [5.129] discussed below. Mechanisms for conversion of electronic energy to atomic and molecular motion in organic solids have recently been reviewed by *Johnson* [5.130].

5.6.5 Ejection Mechanisms

At the time of desorption and ionization the memory of the initial energy supply mechanism is probably lost; i.e., the degree of energy isomerization may be high. The initial energy transfer takes place in about 10^{-13} s, i.e., in a time shorter than typical molecular vibration periods. This probably holds for both primary particle velocity regimes. Particularly for low-energy heavy particles, nuclear collision cascades can overlap and cooperative phenomena may become important (spike effects). Electronic stopping may contribute even at low energies, depending on the incident ion and its energy. For fast particles three main routes have been considered for converting the electronic energy to atomic and molecular motion. The direct Coulomb explosion of the infratrack is one mechanism. Second, the incident ion causes excitations of bonds directly through its Coulomb field and via secondary electrons. The bonds break in a repulsive decay of the excited state. The recoiling nuclei in a repulsive decay can initiate further recoils and a collisional minicascade develops. The third possibility is the grand-scale excitation of low-lying vibrational levels in a large molecule by low-energy secondary electrons, leading to expansion of the molecular volume. If the energy transfer is fast enough, the likelihood for breaking of internal bonds in the molecule may be small. In fact, slow heating of such samples is known to produce only fragments of large molecules [5.131].

The idea of rapid heating [5.131] was adopted by *Macfarlane* and *Torgerson* [5.132] and by *Lucchese* [5.133] in attempts to explain the ejection in HSF-SIMS (PDMS). The concept of temperature is not generally accepted, nor is it particularly useful with the short time scales involved in such nonequilibrium processes. Available thermal models do not address the mechanism of electron-atom coupling or the microscopic details of ejection.

The bonding of a large molecule, like a protein to a substrate, can be estimated to amount to several electron volts [5.129]. An ejection model must explain how these surface bonds are selectively broken and stay broken for an extended period, and leave structural bonds in the molecule intact. The molecule must be given enough kinetic energy to overcome the surface binding and escape with a few electron volts of excess energy. This translational energy may seem trivial compared with the energy deposited by a fast heavy primary ion, but for a large molecule it represents an enormous momentum.

As an example, consider the momentum of a 10 000 u molecule with 1 eV kinetic energy, which correponds to 200 hydrogen atoms with 50 eV each. This aspect was recently addressed for the first time. A model by *Williams* and *Sundqvist* [5.129] suggested that the very low (eV) energy electrons in the ultratrack excite almost all intramolecular bonds to the first vibrational level or higher. This excitation leads to a rapid expansion of the molecules in part of the ultratrack. Expanding molecules pushing against neighbors on a rigid substrate can then generate enough momentum for molecules to leave the surface. The expansion is estimated from thermal expansion coefficients of high-molecular-weight polymers. To achieve the kinetic energies in question (a few electron volts) 3–5% linear expansion is needed. The excitation required turns out to be equivalent to a temperature increase of 1 000 K. The time scale of the expansion $(10^{-13}–10^{-12}$ s) is estimated from what is known about the fundamental vibrational frequency of a large molecule, like a protein. If the amount and duration of the expansion are combined, the corresponding kinetic energy of a molecule of mass 10 000 u turns out to be a few electron volts.

If we assume that about one quarter of the energy lost by the incident ion is deposited as secondary electrons in the ultratrack [5.134], we can estimate the sputtering yield in this model. From the studies of *Säve* et al. [5.89] of film thickness effects in electron sputtering, we can assume that energy deposition down to about 10 nm contributes to sputtering. The total energy deposition is about 10 keV/nm for 90 MeV ^{127}I in an organic solid; i.e., 25 keV (10 nm × 10 keV/nm × 0.25) is converted to sputtering of organic material. *Hedin* et al. [5.87] have measured a sputtering yield of 1 200 intact molecules for 78.2 MeV ^{127}I incident on a multilayer of the amino acid leucine ($M = 131$ u). The sublimation energy for leucine is 1–2 eV, and the estimated number of ejected molecules – i.e., 2 500–1 250 – is in reasonable agreement with experiments.

Local expansion of a solid by ion impact caused either by a low-velocity ion (collision cascades) or a fast ion (an expanding track core) may create a mechanical shock wave. Such a model has been treated by *Bitensky* and *Parilis* [5.135]. *Urbassek* and *Michl* [5.136] have proposed a gas-flow model for sputtering of condensed gases. They argue that when the deposited energy density surpasses the critical temperature of the medium, the gas formed expands into vacuum, increases the sputtering yield, and shifts the energy spectrum of ejected molecules to lower energies.

Recently the Uppsala group showed that molecular ions of large molecules, like bovine insulin, are ejected at a nonnormal angle to the surface and at an angle related to the direction of the incident fast ion. If the incidence angle of the fast ion is 45°, the ejection angle of molecular ions of insulin is almost at right angles to the direction of the fast ion [5.137]. This finding strongly indicates that the ejection of intact organic molecules is caused by a compressional wave or pressure pulse initiated by the expanding cylindrical fast ion track. The observed ejection-angle effects, scaling of sputtering yields with electronic stopping power [5.87] and crater formation [5.89], have been successfully simulated by *Fenyö* et al. [5.138] using a molecular dynamics calculation of electronic sputtering. In the model the fast ion causes an expansion of the cylindrical ion track. The parameters in the Lennard-Jones potential for the molecule–molecule interaction are chosen to reproduce the sound velocity and the cohesive energy in an organic solid. Inspired by the molecular dynamics simulation of *Fenyö* et al. [5.138], *Johnson* et al. [5.139] give an analytical description of sputtering by fast ions at high energy densities. They describe a volume ejection mechanism (linearized "pressure pulse") in which the sputtering yield scales with the third power of the stopping power, when the particle penetration depth is large (cylindrical geometry). This accounts for the measured yield dependence on stopping power [5.87] and roughly describes the measured ejection angles for intact molecular ions [5.137].

Although the details of the ejection mechanisms in organic solids and the two primary particle velocity regimes are not known in detail, it is now well established from neutral-yield measurements that more material is generally ejected and a larger surface area is excited in the electronic stopping regime. (Fig. 5.27). The molecules may be ionized directly in the ejection process or at a later stage. Figure 5.28 schematically illustrates the whole ejection/ionization process [5.15, 140]. It also shows that several layers of the sample can be ejected and that depending on local energy density around the track damage, fragmentation or intact ejection dominates.

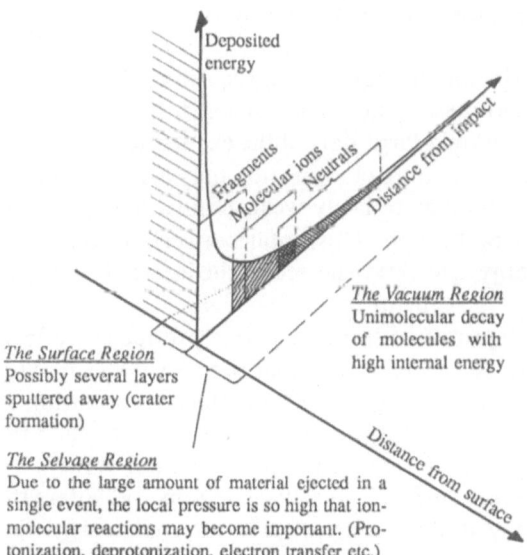

The Surface Region
Possibly several layers sputtered away (crater formation)

The Selvage Region
Due to the large amount of material ejected in a single event, the local pressure is so high that ion-molecular reactions may become important. (Protonization, deprotonization, electron transfer etc.)

Fig. 5.28. Schematic illustration of the sputtering ionization process. The low- and high-velocity particles have different distributions of deposited energy and therefore affect different sample volumes. After [5.15]

5.6.6 Ionization Mechanisms

The next step to consider is the ionization process. The precursor model of *Benninghoven* [5.29] assumed that the ionized molecules (or clusters of molecules) exist preformed on the surface. In the extreme version of this model even very large molecular ions are assumed to be preformed. Several groups have reported strong experimental evidence for the model. An example is the case when multiply charged ions are desorbed from very acidic thioglycerol [5.67]. It is not clear why the charge sign is conserved so well in such a complicated process. In other models, recombination and ionization are assumed to happen in the gas phase outside the surface [5.139]. The argument here is that because of the assumed large number of neutral molecules sputtered in one event, an appreciable local pressure may exist close to the surface. This region is often called the selvage (Fig. 5.28). Some of the charged fragments and alkali ions produced close to the center of the ion track may react with a neutral in a chemical sense (ion molecules reaction), thus producing an ionized molecule. These processes are frequently referred to as protonization (deprotonization) or cationization.

It is often argued that there is a velocity mismatch between ejected fast light fragments and large molecules. However, as the ejection process seems to involve several layers, that is not necessarily true. Furthermore, collisional stabilization of the excited molecular ions may occur in the selvage region. The same considerations are relevant concerning the point of origin of cluster ions. They can be ejected as clusters or formed by recombination in the gas phase.

The experiments by *Säve* et al. [5.88, 89] show that for electronic sputtering with fast heavy ions, crater formation takes place and material from several monolayers is ejected. The investigations also showed that molecules from several molecular layers below the surface may end up as secondary ions. This indicates that evaporation of neutrals from a large charged ejected cluster may be an important route for ion formation. *Derrick* has proposed such an ionization mechanism [5.141].

Another aspect is that in experiments on liquid samples, run under dynamic (high-fluence) SIMS conditions, processes like chemical ionization and collisional stabilization may be important in a large part of the excited region. In the selvage (Fig. 5.28), processes like electron impact ionization may also play a role. To desorb a big molecule a large number of fairly weak bonds (hydrophobic, electrostatic, and hydrogen) must be broken. This implies a certain minimum energy density, and part of the energy is also transferred to the internal energy of the desorbed molecule.

As indicated in Sect. 5.5.3 the spectra of secondary ions produced by low-energy and high-energy primary particles are very similar in many respects. This is the basis for a model developed by *Lin* et al. [5.142] for the interpretation of spectra of large molecular ions using the reaction kinetics concept, in which the deposited energy is shared statistically among all the possible final states.

Once the molecular ion has left the surface and selvage region and enters the vacuum region, it may still dissociate and release part of its internal energy. The mass spectra of large molecules reflect considerable fragmentation. The decay processes are to a large extent governed by the laws of unimolecular dissociations [5.27]. After the internally excited molecule leaves the surface, the internal energy is released in successive two-body decays (Fig. 5.28).

5.7 Summary and Prospects

We have examined the two main approaches for mass spectrometric analysis of large organic molecules. The high mass resolution that can be achieved with magnetic sector field instruments allows the isotopic "pattern" of large organic molecules ($M \approx 10\,000$ u) to be resolved. A necessary prerequisite for the analysis of very large organic molecules based on low-velocity particle impact is a liquid sample matrix like glycerol. Apparently the liquid matrix overcomes radiation damage. Time-of-flight systems for mass analysis used with fast ion impact on a solid sample have a modest mass resolution, but can be used to study larger molecules than magnetic sector field instruments.

At present our understanding of the mechanisms for ejection and ionization of large intact molecular ions is poor, both in the nuclear and the electronic stopping regions. Even the basic mechanism for conversion of electronic energy to nuclear motion is not yet established.

Despite instrumental limitations and a lack of understanding of mechanisms, ion impact techniques have already been successfully applied to problems in biosciences. Most biomolecules of interest have masses higher than those that can presently be studied by other mass spectrometric techniques. Improved sample-preparation procedures should increase the available mass range considerably. The large neutral yields of intact molecules, as measured in electronic sputtering, indicate that it is very important to develop new postionization methods. Concerning the basic mechanisms, experiments comparing fast and slow primary particle impact on the same sample and in the same mass analyzer might improve our understanding of ejection and ionization.

Acknowledgements. I would like to thank Dr. K. Wittmaack for his continuous and patient support and for his constructive criticism of the different versions of this chapter. I also gratefully acknowledge valuable comments and suggestions by Professors R.E. Johnson, P.J. Derrick, P. Roepstorff, F. Röllgen, and P. Sigmund. Many thanks also go to Dr. P. Håkansson, Mrs. I. Ericson, and Miss S. Persson for their help and patience. I am also very grateful to B. Green, VG Analytical, Manchester, UK, for permission to use his LLS-SIMS (FAB) data.

References

5.1   M.S. Munson, F.H. Field: J. Am. Chem. Soc. **88**, 2621 (1966)
5.2   H.D. Beckey: Int. J. Mass. Spectrom. Ion Phys. **2**, 500 (1969)
5.3   D.F. Torgerson, R.P. Skowronski, R.D. Macfarlane: Biophys. Res. Commun. **60**, 616 (1974)

5.4   A. Benninghoven, D. Jaspers, W. Sichtermann: Appl. Phys. **11**, 35 (1976)
5.5   D.J. Surman, J.C. Vickerman: J. Chem. Res. **170** (1981)
5.6   M. Barber, R.S. Bordoli, R.D. Sedgwick, A.N. Tyler: J. Chem. Soc. Chem. Commun. **325** (1981)
5.7   P. Håkansson, I. Kamensky, B. Sundqvist, J. Fohlman, P.A. Peterson, C.J. McNeal, R.D. Macfarlane: J. Am. Chem. Soc. **104**, 2948 (1982)
5.8   B. Sundqvist, P. Roepstorff, J. Fohlman, A. Hedin, P. Håkansson, I. Kamensky, M. Lindberg, M. Salehpour, G. Säve: Science **226**, 696 (1984)
5.9   A.C. Craig, Å. Engström, H. Bennich, I. Kamensky: 35th Ann. Conf. Mass Spectrometry and Allied Topics, Denver, 1987, Book of abstracts, p. 528
5.10 M. Barber, B.N. Green: Rap. Commun. Mass Spectrom. **1**, 80 (1987)
5.11 A.L. Burlingame, J.O. Whitney, D.H. Russel: Anal. Chem. **56**, 417R (1984)
5.12 N.H. Turner, B.I. Dunlop, R.J. Colton: Anal. Chem. **56**, 373R (1984)
5.13 K. Wittmaack: Surf. Sci. **89**, 668 (1979)
5.14 R.G. Cooks, K.L. Busch, G.L. Glish: Science **222**, 273 (1983)
5.15 K.L. Busch, R.G. Cooks: Science **218**, 247 (1982)
5.16 H.R. Morris: Int. J. Mass Spectrom. Ion Phys. **45**, 331 (1982)
5.17 K.L. Rinehart, Jr.: Science **218**, 254 (1982)
5.18 R.D. Macfarlane: Phys. Scripta **T6**, 110 (1983)
5.19 B. Sundqvist, R.D. Macfarlane: Mass Spec. Rev. **4**, 421 (1985)
5.20 D. Briggs: *Proc. 2nd Int. Conf. on Ion Formation from Organic Solids*, ed. by A. Benninghoven, Springer Ser. Chem. Phys., Vol. 25 (Springer, Berlin, Heidelberg 1983) p. 156
5.21 I. Bendet, N. Rizk: Biophys. J. **16**, 357 (1976)
5.22 W.W. Newcomb, J.W. Boring, J.C. Brown: J. Virology **51**, 52 (1984)
5.23 S. Blackmore, D. Clougher: Grana **23**, 85 (1984)
5.24 P. Sigmund: In *Sputtering by Particle Bombardment I*, ed. by R. Behrisch, Topics Appl. Phys., Vol. 47 (Springer, Berlin, Heidelberg 1981) p. 9
5.25 P. Sigmund: Nucl. Instrum. Methods **B27**, 1 (1987)
5.26 J.E. Campana, R. Freas: J. Chem. Soc. Chem. Commun. **1414** (1984)
5.27 B.T. Chait: Int. J. Mass Spectrom. Ion Phys. **53**, 227 (1983)
5.28 J. Grotemeyer, U. Boesl, K. Walter, E. Schlag: Org. Mass Spectrom. **21**, 645 (1986)
5.29 A. Benninghoven: *Ion Formation from Organic Solids* (IFOS II), ed. by A. Benninghoven, Springer Ser. Chem. Phys. Vol. 25 (Springer, Berlin, Heidelberg 1983) p. 64
5.30 M. Salehpour, P. Håkansson, B. Sundqvist: Nucl. Instrum. Methods **B2**, 752 (1984)
5.31 W. Sichtermann, A. Benninghoven: Int. J. Mass Spectrom. Ion Phys **40**, 177 (1981)
5.32 B.T. Chait: Int. J. Mass Spectrom. Ion Proc. **78**, 237 (1987)
5.33 K.G. Standing, W. Ens, R. Beavis, G. Bolbach, D. Main, B. Schueler, J.B. Westmore: *Ion Formation from Organic Solids* (IFOS III), ed. by A. Benninghoven, Springer Ser. Chem. Phys. Vol. 9 (Springer, Berlin, Heidelberg 1986) p. 37
5.34 Kratos News Letter 25 (1984), Kratos Ltd, Manchester, UK
5.35 K. Wittmaack: Vacuum **32**, 65 (1982)
5.36 P. Håkansson, B. Sundqvist: Radiat. Eff. **61**, 179 (1982)
5.37 J. Yergey, D. Heller, G. Hansen, R.J. Cotter, C. Fenselau: Anal. Chem. **55**, 353 (1983)
5.38 M. Comisarow, A.L. Marchall: Anal. Chem. **47**, 491A (1975)
5.39 D.F. Hunt, J. Shabanowitz, J.R. Yates III, N.Z. Zhu, D.H. Russel, M.E. Castro: Proc. Natl. Acad. Sci. USA **84**, 620 (1987)
5.40 R.J. Beuhler, L. Friedman: J. Appl. Phys. **48**, 3928 (1977)
5.41 R.J. Beuhler, L. Friedman: Int. J. Mass Spectrom. Ion Phys. **23**, 81 (1977)
5.42 R.J. Beuhler, L. Friedman: Nucl. Instrum. Methods **170**, 309 (1980)
5.43 B. Sundqvist, A. Hedin, P. Håkansson, I. Kamensky, M. Salehpour, G. Säve: Int. J. Mass Spectrom. Ion Proc. **65**, 69 (1985)
5.44 A. Hedin, P. Håkansson, B.U.R. Sundqvist: Int. J. Mass Spectrom. Ion Proc. **75**, 275 (1987)
5.45 H. Haberland, M. Winterer: Rev. Sci. Instrum. **54**, 764 (1983)

5.46 L.F. Thompson, M.J. Bowden (eds.): *Introduction to Microlitography* (Am. Chem. Soc., Washington D.C., 1983), p. 187–193
5.47 G. Säve, P. Håkansson, B.U.R. Sundqvist, U. Jönsson, G. Olofson, M. Malmquist: Anal. Chem. **59**, 2059 (1987)
5.48 J. Zeleny: Phys. Rev. **10**, 1 (1917)
5.49 E. Brunnix, G. Rudstam: Nucl. Instrum. Methods **131**, 131 (1961)
5.50 C.J. McNeal, RD. Macfarlane, E.L. Thurston: Anal. Chem. **51**, 2036 (1979)
5.51 A. Benninghoven, W. Lange, M. Jirikowsky, D. Holtkamp: Surf. Sci. **123**, L721 (1982)
5.52 E.A. Jordan, R.D. Macfarlane, C.R. Martin, C.J. McNeal: Int. J. Mass Spectrom. Ion Phys. **53**, 345 (1983)
5.53 R.D. Macfarlane: J. Trace Micropr. Techn. **2**, 267 (1984)
5.54 G. Jonsson, P. Håkansson, B. Sundqvist, P. Roepstorff, P. Carlsen, K.E. Johansson, I. Kamensky, M. Lindberg: Anal. Chem. **58**, 1084 (1986)
5.55 W. Aberth, K.M. Strauts, A.L. Burlingame: Anal. Chem. **54**, 2089 (1982)
5.56 H. Wollnik (Ed.): *Charged Particle Optics* (North Holland, Amsterdam 1981)
5.57 K. Olthoff, R.J. Cotter: Nucl. Instrum. Methods B**26**, 566 (1987)
5.58 W.C. Wiley, I.H. McLaren: Rev. Sci. Instrum. **26**, 1150 (1955)
5.59 B. Sundqvist, A. Hedin, P. Håkansson, I. Kamensky, J. Kjellberg, M. Salehpour, G. Säve, S. Widdiyasekera: Int. J. Mass Spectrom. Ion Phys. **53**, 167 (1983)
5.60 B.T. Chait, W.C. Agosta, F.H. Field: Int. J. Mass Spectrom. Ion Phys. **39**, 339 (1981)
5.61 R. Riggi: Nucl. Instrum. Methods B**22**, 588 (1987)
5.62 B.A. Mamyrin, V.I. Karataev, D.V. Shmikk, V.A. Zagulin: Sov. Phys. JETP **37**, 45 (1973)
5.63 S. DellaNegra, Y. LeBeyec: Int. J. Mass Spectrom. Ion Proc. **61**, 21 (1984)
5.64 E. Niehuis, T. Heller, H. Feld, A. Benninghoven: *Ion Formation from Organic Solids* (IFOS III), ed. by A. Benninghoven, Springer Ser. Chem. Phys., Vol. 9 (Springer, Berlin, Heidelberg 1986) p. 198
5.65 K.G. Standing, B.T. Chait, W. Ens, G. McIntosh, R. Beavis: Nucl. Instrum. Methods **198**, 33 (1982)
5.66 A. Eicke, V. Anders, M. Junack, W. Sichtermann, A. Benninghoven: Int. J. Mass Spectrom. Ion Phys. **46**, 479 (1983)
5.67 L.R. Schronk, R.J. Cotter: 33rd Ann. Conf. Mass Spectrometry and Allied Topics, San Diego, 1985, Book of abstracts, p. 177
5.68 J. Michl: Int. J. Mass Spectrom. Ion Phys. **53**, 255 (1983)
5.69 S.S. Wong, F.W. Röllgen, I. Manz, M. Przybylski: Biomed. Mass. Spectrom. **12**, 43 (1985)
5.70 S.S. Wong, U. Giessmann, F.W. Röllgen: 32nd Ann. Conf. Mass Spectrometry and Allied Topics, San Antonio, 1984, Book of abstracts, p. 186
5.71 D.F. Barofsky, A.M. Ilias, E. Barofsky, JH. Murphy: 32nd Ann. Conf. Mass Spectrometry and Allied Topics, San Antonio, 1984, Book of abstracts, p. 182
5.72 P. Williams, G. Gillen: *Ion Formation from Organic Solids* (IFOS IV), ed. by A. Benninghoven (Wiley, Chichester 1989) p. 15
5.73 W.L. Brown, W.M. Augustyniak, E. Simmons, K.J. Marcantonio, L.J. Lanzerotti, R.E. Johnson, J.W. Boring, C.T. Reimann, G. Foti, V. Pironello: Nucl. Instrum. Methods **198**, 1 (1982)
5.74 S.S. Wong, F.W. Röllgen: Nucl. Instrum. Methods B**14**, 436 (1986)
5.75 W.V. Ligon: Int. J. Mass Spectrom. Ion Proc. **61**, 11 (1984)
5.76 P. Dück, W. Treu, H. Fröhlich, W. Galster, H. Voit: Surf. Sci. **95**, 603 (1980)
5.77 P. Håkansson, A. Johansson, I. Kamensky, B. Sundqvist, J. Fohlman, P.A. Peterson: IEEE Trans. Nucl. Sci. NS-**28**, 1776 (1981)
5.78 B. Sundqvist (ed.): *Proc. of the Nordic Symp. on Ion Induced Desorption of Molecules from Bioorganic Solids*, Nucl. Instrum. Methods **198** (1982)
5.79 S. DellaNegra, D. Jacquet, I. Lorthiois, Y. LeBeyec, O. Becker, K. Wien: Int. J. Mass Spectrom. Ion Phys. **53**, 215 (1983)
5.80 P. Håkansson, I. Kamensky, B. Sundqvist: Surf. Sci. **116**, 302 (1982)

5.81   E. Nieschler, B. Nees, N. Bischof, H. Fröhlich, W. Tiereth, H. Voit: Surf. Sci. **145**, 294 (1984)

5.82   P. Håkansson, E. Jayasinghe, A. Johansson, I. Kamensky, B. Sundqvist: Phys. Rev. Lett. **47**, 1227 (1981)

5.83   W. Guthier, O. Becker, S. DellaNegra, W. Knippelberg, Y. LeBeyec, U. Weikert, K. Wien, P. Wieser, R. Wurster: Int. J. Mass Spectrom. Ion Phys. **53**, 185 (1983)

5.84   E. Nieschler, B. Nees, N. Bischof, H. Fröhlich, W. Tiereth, H. Voit: Radiat. Eff. **83**, 121 (1984)

5.85   K. Wien, O. Becker, W. Guthier, S. DellaNegra, Y. LeBeyec, B. Monart, K. Standing, G. Maynard, C. Deutsch: Int. J. Mass Spectrom. Ion Proc. **78**, 273 (1987)

5.86   P. Håkansson, I. Kamensky, M. Salehpour, B. Sundqvist, S. Widdiyasekera: Radiat. Eff. **80**, 141 (1984)

5.87   A. Hedin, P. Håkansson, M. Salehpour, B.U.R. Sundqvist: Phys. Rev. B**35**, 7377 (1987)

5.88   G. Säve, P. Håkansson, B.U.R. Sundqvist, E. Söderström, S.E. Lindquist, J. Berg: Int. J. Mass Spectrom. Ion Proc. **78**, 259 (1987)

5.89   G. Säve, P. Håkansson, B.U.R. Sundqvist, R.E. Johnson, E. Söderström, S.E. Lindquist, J. Berg: Appl. Phys. Lett. **51**, 1379 (1987)

5.90   I. Kamensky, P. Håkansson, B. Sundqvist, C.J. McNeal, R.D. Macfarlane: Nucl. Instrum. Methods **198**, 65 (1982)

5.91   W. Ens, K.G. Standing, B.T. Chait, F.H. Field: Anal. Chem. **53**, 1241 (1981)

5.92   W. Ens, D.E. Main, K.G. Standing, B.T. Chait: Accepted for publication in Anal. Chem. (April, 1988)

5.93   W. Ens, P. Håkansson, B.U.R. Sundqvist: in *Proceedings of SIMS VI*, Versailles, 1987, ed. by AM. Huber, A. Benninghoven, H.W. Werner and G. Slodzian (Wiley, Chichester 1988) p. 623

5.94   W.J. Richter, F. Raschdorf, W. Maerki: in *Mass Spectrometry in the Health and Life Sciences*, ed. by A.L. Burlingame and N. Castagnoli, Jr., Anal. Chem. Symp. Ser. 24 (Elsevier, Oxford, New York, Tokyo, 1985) p. 193

5.95   H.R. Morris, A. Dell, A.T. Etienne, M. Judkins, R.A. McDowell, M. Panico, G.W. Taylor: Pure Appl. Chem. **54**, 267 (1982)

5.96   R.D. Macfarlane, D. Uemura, K. Ueda, Y. Hirata: J. Am. Chem. Soc. **102**, 875 (1980)

5.97   J. Lindhard, M. Scharff, H.E. Schiott: Kgl. Dan. Vidensk. Selsk. Mat. Fys. Medd. **33**, No. 14 (1963)

5.98   J.F. Ziegler (ed.): *Handbook of Stopping Cross-Sections for Energetic Ions in All Elements*, (Pergamon, New York 1980)

5.99   H.H. Anderssen, H.L. Bay: In *Sputtering by Particle Bombardment I*, ed. by R. Behrisch, Topics Appl. Phys., Vol. 47 (Springer, Berlin, Heidelberg 1981) p. 145

5.100   P. Townsend: in *Sputtering by Particle Bombardment II*, ed. by R. Behrisch, Topics App. Phys., Vol. 52 (Springer, Berlin, Heidelberg, New York, Tokyo 1983) p. 147

5.101   P. Sigmund: Appl. Phys. Lett. **25**, 169 (1974)

5.102   P. Sigmund, C. Claussen: J. Appl. Phys. **52**, 990 (1981)

5.103   K. Winterbon, P. Sigmund, J. Sanders: Kgl. Dan. Vidensk. Selsk. Mat. Fys. Medd. **37**, No. 14 (1970)

5.104   D. Galomb, R.E. Good: J. Chem. Phys. **49**, 4176 (1968)

5.105   B.J. Garrison: Int. J. Mass Spectrom. Ion Phys. **53**, 243 (1983)

5.106   W. Brandt, R.H. Ritchie: In *Physical Mechanisms in Radiation Biology*, ed. by R.D. Cooper, R.W. Wood (USEC Technical Inf. Center, Oak Ridge, Tennesse 1974) p. 20

5.107   E.J. Kobetich, R. Katz: Phys. Rev. **170**, 1391 (1968)

5.108   R.L. Fleischer, P. Price, R.M. Walker: *Nuclear Tracks in Solids* (University of California Press, Berkeley, 1975)

5.109   P. Haff: Appl. Phys. Lett. **29**, 473 (1976)

5.110   J.P. Biersack, E. Santner: Nucl. Instrum. Methods **132**, 229 (1976)

5.111   W.L. Brown, L.J. Lanzerotti, J.M. Poate, W.M. Augustyniak: Phys. Rev. Lett. **40**, 1027 (1978)

5.112   W.L. Brown, W.M. Augustyniak, E. Brady, B. Cooper, L.J. Lanzerotti, A. Ramirez, B. Evatt, R.E. Johnson: Nucl. Instrum. Methods **170**, 321 (1980)

5.113   W.L. Brown, W.M. Augustyniak, L.J. Lanzerotti, R.E. Johnson, R. Evatt: Phys. Rev. Lett. **45**, 1632 (1980)

5.114 R.W. Ollerhead, J. Böttiger, J.A. Davies, J. L'ecuyer, H.K. Haugen, N. Matsumani: Radiat. Eff. **49**, 203 (1980)

5.115 F. Besenbacher, J. Böttiger, O. Graversen, J.L. Hansen, H. Sörensen: Nucl. Instrum. Methods **191**, 221 (1981)

5.116 J.E. Griffith, R.A. Weller, L.E. Seiberling, T.A. Tombrello: Radiat. Eff. **51**, 223 (1980)

5.117 R.E. Johnson, W.L. Brown: Nucl. Instrum. Methods **198**, 103 (1982); Nucl. Instrum. Methods **209/210**, 469 (1983)

5.118 J. Schou: Nucl. Instrum. Methods B27, 188 (1987)

5.119 R.H. Ritchie, C. Claussen: Nucl. Instrum. Methods **198**, 133 (1982)

5.120 C.C. Watson, T.A. Tombrello: Radiat. Eff. **89**, 263 (1985)

5.121 N. Itoh, T. Nakayama: Phys. Lett. A**92**, 471 (1982)

5.122 C.T. Reimann, R.E. Johnson, W.L. Brown: Phys. Rev. Lett. **53**, 600 (1984)

5.123 W.L. Brown, C.T. Reimann, R.E. Johnson: Nucl. Instrum. Methods B19/20, 9 (1987)

5.124 P. Borgesen, J. Schou, H. Sörensen, C. Claussen: Appl. Phys. **29**, 222 (1982)

5.125 F.L. Rook, R.E. Johnson, W.L. Brown: Surf. Science **164**, 625 (1985)

5.126 B.J. Garrison, B.E. Johnson: Surf. Sci. **148**, 388 (1984)

5.127 R.E. Johnson, B. Sundqvist: Int. J. Mass Spectrom. Ion Phys. **53**, 337 (1983)

5.128 A. Hedin, P. Håkansson, B. Sundqvist, R.E. Johnson: Phys. Rev. B**31**, 1780 (1985)

5.129 P. Williams, B.U.R. Sundqvist: Phys. Rev. Lett. **58**, 1031 (1987)

5.130 R.E. Johnson: Int. J. Mass Spectrom. Ion Proc. **78**, 357 (1987)

5.131 R.J. Beuhler, E. Flanigan, L.J. Greene, L. Friedman: J. Am. Chem. Soc. **96**, 3990 (1974)

05.132 R.D. Macfarlane, D.F. Torgerson: Science **191**, 920 (1976)

5.133 R.R. Lucchese: J. Chem. Phys. **86**, 443 (1987)

5.134 A. Mozumder: *Advances in Radiation Chemistry*, ed. by M. Burton, I. Magee (Wiley, New York 1969) Vol. 1, p. 1

5.135 I.S. Bitensky, E.S. Parilis: Nucl. Instrum. Methods B21, 26 (1987)

5.136 H.M. Urbassek, J. Michl: Nucl. Instrum. Methods B22, 480 (1987)

5.137 W. Ens, B.U.R. Sundqvist, P. Håkansson, A. Hedin, G. Jonsson: Phys. Rev. B**29**, 763 (1989)

5.138 D. Fenyö, B.U.R. Sundqvist, B. Karlsson, R.E. Johnson: Phys. Rev. B42 (1990) 1895

5.139 R.E. Johnson, B.U.R. Sundqvist, A. Hedin, D. Fenyö: Phys. Rev. B40 (1989) 49

5.140 R.G. Cooks, K.L. Busch: Int. J. Mass Spectrom. Ion Phys. **53**, 111 (1983)

5.141 P.J. Derrick: Fresenius Z. Anal. Chem. **324**, 486 (1986)

5.142 S.H. Lin, I.S. Tsong, A.R. Ziv, M. Szymonski, C.M. Loxton: Phys. Scripta T6, 106 (1983)

6. Production of Microstructures by Ion Beam Sputtering

Wolfgang Hauffe

With 30 Figures

Controlled ion beam sputtering has found widespread application for producing topographical microstructures on solids. Depending on the task and material, various methods have been developed to etch or to machine the desired microstructure. Ion beam thinning is used for producing freestanding thin foils with nanometer thickness (Sect. 6.2). By material-induced selective ion beam etching, microtopographical features are developed, revealing the internal grain and phase structure or generating a defined surface roughness (Sect. 6.3). Seeding with foreign atoms during ion beam sputtering can texture surfaces that otherwise would remain smooth (Sect. 6.4). Masking generates well-defined patterns to submicrometer dimensions laterally, and reactive ion beam etching gives a high depth-to-width ratio (Sect. 6.5). Ion beam slope cutting reveals selected microareas in heterogeneous solids (Sect. 6.6). Microfiguring by sputtering with movable screens and workpieces produces defined three-dimensional shapes (Sect. 6.7). Finally, scanned microfocused ion beams are used to modify the locally removed material thickness with high resolution and program-controlled contour change (Sect. 6.8).

6.1 General Remarks

During the last two decades ion beams have become an important tool for producing topographical structures with micrometer dimensions in a large variety of geometrical shapes and for very different purposes. For example, ion beams can be used to make components for microelectronics, integrated optics, and surface acoustic wave techniques or to texture materials to increase their absorption or adhesion. On the other hand, modern microscopic and microanalytical methods require that microstructures and microregions of heterogeneous samples will be revealed.

Microstructure generation is based on controlled sputtering by energetic ions that remove surface atoms in the small region of the collision cascade diameter, on the order of nanometers. The sputtering effect depends on and can be influenced by the properties of the bombarded material, the ion beam parameters, and environmental conditions. Additionally, the bombardment regime and sample manipulation can be modified in various process steps. The advantages of the ion beam sputtering process compared with other etching or

Topics in Applied Physics, Vol. 64
Sputtering by Particle Bombardment III Eds.: R. Behrisch · K. Wittmaack
© Springer-Verlag Berlin Heidelberg 1991

machining methods are the universal applicability to all substances, the controllable high accuracy of attack vertically down to monolayers, laterally with nanometer resolution and the fact that thermal effects can be kept small. Beyond this, undesired chemical effects can be avoided, but reactive ion beam etching is also useful in some applications (Sect. 6.5.2).

The methods of microstructure production are mostly connected with selective removal of material from flat and smooth samples. In some special cases, samples preformed by other means are finished by ion beam sputtering to their final shapes. Selective attack can be achieved by sputtering with broad uniform ion beams due to local differences in the sputtering velocity of the bombarded material. Nevertheless, also on homogeneous materials a microrelief (e.g., a faceting structure) can be developed according to the bombardment conditions, or this roughening can be stimulated by foreign atoms impinging simultaneously with the bombarding ions and modifying the sputtering mechanism. For producing defined pattern arrays on homogeneous substances or in layer systems the ion beam intensity has to be varied across the surface. Intensity can be varied with contact or projected masks if broad uniform ion beams are used, or by modifying the local ion dose with other techniques, e.g., movable screens and workpieces or scanned focused or specially shaped ion beams.

The problems to be solved to optimize these etching and machining processes are first determined by the specific properties of the physical (and in some applications also chemical) sputtering mechanism in the given material. They concern sputtering yield, selectivity, relief formation, and contour change of an existing profile. Depending on the special microstructure, additional sputtering effects have to be considered. For example, in all concave relief structures, sputtering by reflected ions and redeposition of sputtered atoms have to be taken into account. In multicomponent systems, the sputtering process can be modified by the presence of different elements. Second, the possibilities and limitations of ion beam process control have to be considered, for example, control of dose and density distribution, mass and energy homogeneity, direction, divergence, positioning, and stability.

The quality of a produced microstructure has to be assessed not only by its shape or resolution, but also by such other parameters as radiation damage or cross contamination of material, which can be important for electrical properties. In most cases, resolution is not limited by the sputtering mechanism itself, but determined by other processes, for example, mask preparation or ion beam forming. Microfocused ion beam equipment with today's highest available performance has a spot size approaching the order of the collision cascade diameter, and such machines are primarily not designed for sputtering of microstructures, but for ion beam lithography, ion implantation, or ion microscopy.

In microstructure generation by ion beam sputtering we also have to consider the relation between expenditure on ion beam equipment and desired results, as well as productivity. Some of the widely used sputtering applications can be carried out with simple ion guns; however, most processes require highly sophisticated ion beam and sample manipulation. The following sections describe the various processes that apply ion beam sputtering to produce microstructures. We discuss aims, sample properties, ion beam processing, and resulting microstructure, including applications, achievable quality, and limitations.

Some of the topics have been reviewed recently. Here we outline new and special results and further developments. This concerns ion beam thinning [6.1, 2], texturing [6.3], and the wide field of mask etching, which has been summarized in several books and reviews [6.4]. Basic considerations of the sputtering process underlying these ion beam applications are discussed in the preceding volumes of this series [6.5, 6] and earlier reviews [6.7]. Publications on ion beam etching up to 1978 are collected in *Hawkins* bibliography [6.8, 9].

6.2 Ion Beam Thinning

For such applications as transmission electron microscopy (TEM) and for piezoelectric transducers or infrared filters, samples have to be prepared with freestanding thin foils with a thickness down to some nanometers. Generally materials with different compositions have to be thinned, and they are mostly inhomogeneous in the case of TEM samples, but even small areas are sufficient for the investigations. Further applications require high-quality thinning over larger areas but only for some selected homogeneous and single crystalline materials. Ion beam thinning has many advantages compared with mechanical ultramicrotomy or chemical/electrolytic methods:

i) Applicability to a large variety of substances
ii) Removal of material without mechanical deformation or chemical influences
iii) Controlled accuracy of material attack, often combined with smoothing of the surface

More than twenty years' experience in optimizing the working conditions for ion thinning has been reported in *Goodhew*'s recent monograph [6.1], which includes a list of ion beam milling apparatus manufacturers.

6.2.1 Standard Ion Beam Thinning

The standard arrangement of an ion beam thinning apparatus consists of two ion guns directed from opposite sides under oblique ion incidence onto the sample disk, which rotates around its axis Fig. 6.1. The typical bombardment conditions are [6.1, 2, 10]:

Ion species: Ar^+, Kr^+ (mostly from saddle-field ion guns)
Ion energies: 1 . . . 10 keV
Ion beam densities: 1 . . . 200 $\mu A/mm^2$

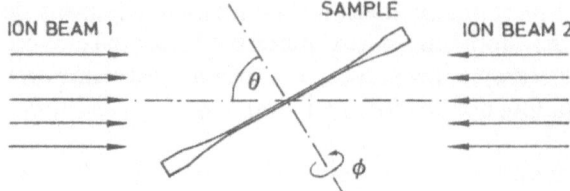

Fig. 6.1. Ion beam thinning arrangement for producing samples for transmission electron miscroscopy (TEM): θ is the ion incidence angle; ϕ, the sample rotation

Depending on the sputtering yield and initial sample thickness, several hours bombardment time are necessary to thin a sample to the perforation normally used for TEM objects for observing the thin borders of a hole by transmitted electrons. The incidence angle can be 60° to the normal at the beginning and can be changed to grazing incidence of 80° ... 88° to reduce remaining surface roughness and the thickness of the radiation-damaged layer [6.11]. With this standard ion beam thinning process cross sections through surface layer systems have also been prepared [6.11, 12], and different initial sample shapes could be thinned successfully [6.13].

6.2.2 Modified Ion Beam Thinning

To enlarge the area of possibly thin regions, ion milling can be combined with preformed deepenings in the sample, which can be elaborated also by ion beam machining [6.14]. A mask (for example, an Mo grid with 25 μm diameter wire and a 250 μm wide mesh) is placed above the sample disk Fig. 6.2a With ion incidence perpendicular to the surface, deepenings are produced. After this machining step the grid is removed and the bombardment continues in the conventional mode (Fig. 6.2b). The crossbar structure gives rise to more than one hole with thin borders and stabilizes the sample (Fig. 6.2c). Figure 6.3 shows

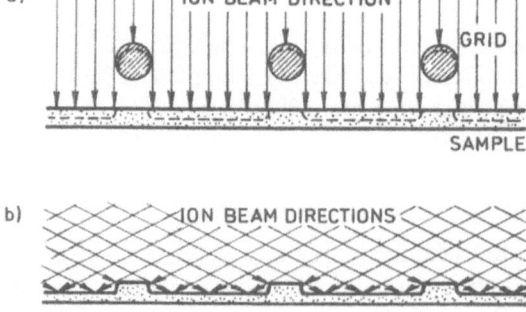

Fig. 6.2. Two-step ion milling of a thin sample with an integrated grid for TEM [6.14]: (a) grid projection onto the sample by a normally incident ion beam, (b) ion milling under oblique ion incidence and sample rotation without the grid, (c) final TEM object

Fig. 6.3. Scanning electron micrograph of a glass sample to be thinned for TEM after the grid projection process. (The grid is still in place.) [6.14]

a glass object after the machining step of Fig. 6.2a. The Mo grid is still in place. In other preparation tasks this crossbar structure with thin regions can be used as substrate for investigating thin films [6.14].

Two other special techniques of thin-film production by ion beam machining are based on mask etching (Sect. 6.5) or ion beam slope cutting (Sect. 6.6).

6.2.3 Applications

As early as 1962 *Paulus* and *Reverchon* [6.15] prepared TEM specimens of spinel with ion beam equipment fully corresponding to today's techniques. Metals, semiconductors, ceramics, and various heterogeneous materials have been successfully prepared by ion beam thinning [6.16–24], and also organic and biological substances such as human enamel were thinned by ion beam milling [6.18]. Thin foils of Si, GaP, and GaAs have been prepared for high-resolution TEM lattice imaging [6.23, 24]. Thin samples with uniform thickness in larger areas were fabricated for infrared detectors and piezoelectric devices also using Ar^+ ions in the keV range [6.25, 26, 27]. *Castellano* and *Hokanson* [6.28] described ion milling of piezoelectric crystals with angular correction by modifying the ion beam density.

6.3 Material-Induced Ion Beam Etching

The material-induced microtopography development on solids during ion bombardment is of essential interest for many practical applications. Under homogeneous unidirectional ion bombardment sputtering reveals the internal grain and phase structure of the solid. This was one of the earliest applications of *ion beam etching in the* ion bombardment emission electron microscope (EEM), and with replica techniques in TEM, and in the scanning electron microscope (SEM) [6.29, 30, 31]. Also now ion etching is a valuable preparation technique, not only for substances that cannot be etched by chemical methods, and it can be used to get additional information in all applications of ion beams for such analytical approaches as secondary ion mass spectrometry (SIMS), emission ion microscopy (EIM), scanning ion microscopy (SIM), or sputter depth profiling. Unidirectional ion bombardment with higher fluence on many substances develops a microroughness that is applied for changing surface properties, e.g., the absorption coefficient of solar absorbers [6.32] or the adhesion of biomedical implants [6.33].

6.3.1 Polycrystalline Materials

Figure 6.4 shows polycrystalline iron after bombardment with 10 keV Kr$^+$ ions [6.34]. Figure 6.5 shows the development of slopes at twin grain boundaries on polycrystalline copper [6.35, 36]. Compared with chemical grain boundary etching, ion beam sputtering develops single slopes at most grain boundaries and grooves occur only in some cases, preferentially at normal ion incidence [6.35, 36]. Figure 6.6 will help us understand the development of single slopes. We know from sputtering yield measurements at single crystals that the grains are sputtered at different velocities, according to their orientation to the ion beam direction [6.37]. Neglecting the small relief structure on the grains, sputtering effects of reflected ions, and redeposition of sputtered material (which

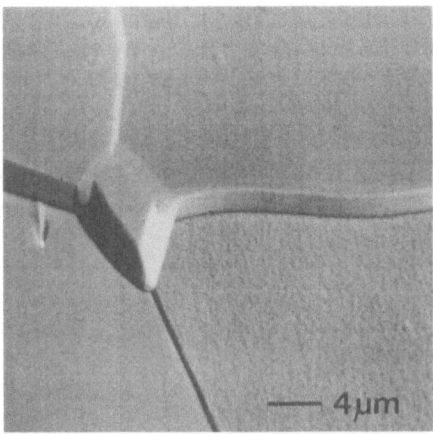
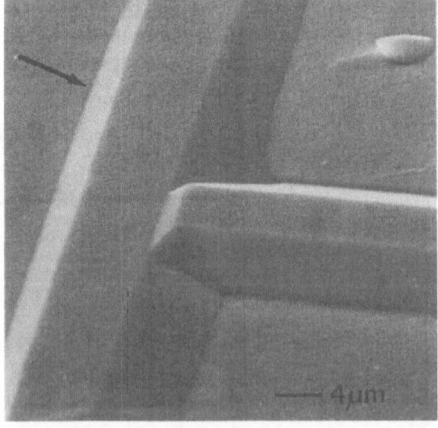

Fig. 6.4. Selective grain etching of polycrystalline Fe after bombardment with 10 keV Kr$^+$ ions under 50° (to the normal) [6.34]

Fig. 6.5. Microtopography at twin grain boundaries on Cu after bombardment with 10 keV Ar$^+$ ions (3 μA/mm^2), $\theta_0 = 50°$, $t_B = 40$ min) [6.35, 36] Arrow: ion beam direction

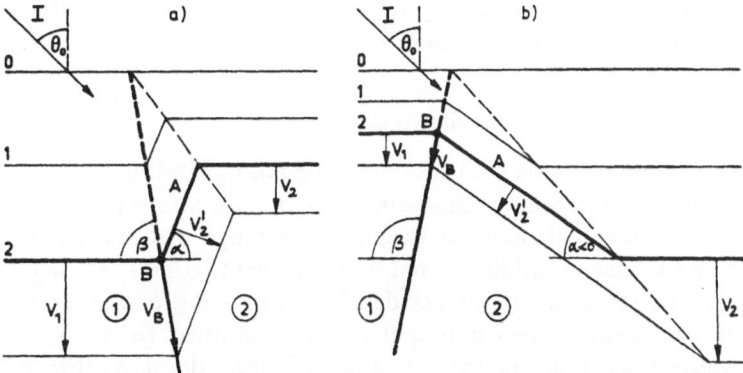

Fig. 6.6. Model of slope development at a boundary between two grains with (a) $v_1 > v_2$ and (b) $v_1 < v_2$. Here A is the slope; I the ion beam direction; θ_0 the angle of incidence; B the intersection line of the grain boundary; β the inclination of the grain boundary; v_1, v_2, v_2', and v_B are velocities; and α is the slope angle [6.35, 36]

all occur but do not prevent our basic discussion), the erosion velocity v_i of an area i of a surface relief is

$$v_i = \frac{1}{N_S} J_i Y_i \,, \tag{6.1}$$

$$v_i = \frac{1}{N_S} J_0 \cos \theta_i \, Y_i(\theta_i) \,, \tag{6.2}$$

with the particle density N_S of the solid, sputtering yield Y, incidence angle θ, and the ion beam density J. The inclination α of the resulting slope A is due to the geometric relation at point B in Fig. 6.6:

$$\frac{\sin(\alpha + \beta)}{\sin \beta} = \frac{v_2'(\alpha)}{v_1} \,, \tag{6.3}$$

and to the function

$$v_2' = v_2'[Y(\theta(\alpha))] \,, \tag{6.4}$$

where $Y(\theta)$ represents the dependence of the sputtering yield on the incidence angle on the slope A [6.36]. Proceeding from both standard cases (Fig. 6.6), this model has been applied and experimentally confirmed for different three-dimensional arrangements of grain boundaries and ion beam directions from normal to grazing incidence, including also the development of grooves under special conditions [6.35, 36]. The general properties mentioned here for poly-crystalline single-element solids are also expected for multielement systems and heterogeneous materials, but modified by effects resulting from the additional elements or different materials (Sects. 6.3.2–5).

6.3.2 Single Crystals and Amorphous Substances

Single crystalline and amorphous surfaces (including amorphization by the ion bombardment itself) of numerous single-element solids show an electron microscopically smooth removal of material over a large range of bombardment conditions, also in the case of unidirectional bombardment [6.6, 11, 35, 38]. In many other cases, however, a microrelief develops consisting of one or two families of facets and more or less nonregular transition areas [6.35, 38, 39]. Sometimes a well-defined relief of three families of facets develops (Fig. 6.7) [6.35, 38, 39]. At ion incidence in symmetric crystal planes and on amorphous surfaces at smaller angles of incidence, furrows develop, oriented perpendicular to the ion beam direction and at larger incidence angles parallel to the projection of the ion beam on the surface [6.40]. The microstructure grows with the bombarding ion fluence showing the following properties [6.38, 41]. The average and maximum lengths of the structure elements increase, whereas their number per unit length decreases. Also small structure elements always exist. This behavior can be explained by the contribution of reflected ions to the sputtering process [6.38, 41]. In all concave surface structures, and therefore in most microstructure development processes (including mask etching, Sect. 6.5), the erosion velocity v_n of a surface element (in the direction of the normal) is given by [6.38, 42]

$$v_n = \frac{1}{N_S} \left[J_D Y_D + \iint_{(E_R, \Omega)} dJ_R Y_R - \iint_{(E_S, \Omega)} dJ_S(K - Y_S) \right] , \tag{6.5}$$

$$v_n = v_D + v_R - v_S , \tag{6.6}$$

with particle density N_S of the sample, the density of directly impinging ions J_D,

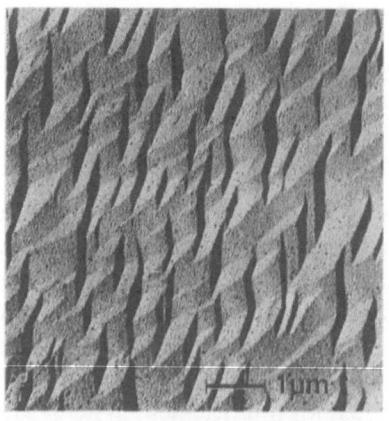

Fig. 6.7. Electron micrograph (replica) of the faceted structure on a Cu single-crystal surface bombarded with 10 keV Ar$^+$ ions, (3 μA/mm^2, 60 min, 55° to the normal) [6.35]

the sputtering yield Y_D for directly impinging ions, the density of scattered ions dJ_R (impinging with energy E_R on the regarded surface element), the sputtering yield Y_R for scattered ions, the density of sputtered atoms dJ_S (impinging with energy E_S on the regarded surface element), the sticking coefficient K of sputtered atoms, and the sputtering yield Y_S for atoms impinging on the regarded surface element from the environment Ω. In (6.6) v_D represents the erosion velocity caused by directly impinging ions, v_R the influence of reflected ions, and v_S the redeposition and sputtering by sputtered atoms. With sufficiently high contributions of reflected (scattered) ions (v_R) to the facet displacement velocity, a "coarsening" mechanism leads to the successive decrease and disappearance of small relief structures next to larger facets, and so on [6.41]. With these observations and the model, other general properties of the microstructure can also be explained [6.38]. The roughness is always small compared with the removed material thickness. There is no difference for the general mechanism in amorphous or crystalline solids, but the crystal structure will modify the process. For practical applications this microrelief can be developed to the desired roughness by ion dose control.

6.3.3 Heterogeneous and Organic Materials

The main microtopography phenomena reported above can also be observed on heterogeneous samples, but accompanied by additional effects. On perturbation-free prepared initial surfaces after low dose bombardment, a microrelief develops, revealing the internal grain and phase structure. Figure 6.8 shows a heterogeneous ceramic (based on ZnO) after bombardment with 10 keV Kr^+ ions (10 µA/mm²). Here both orientation differences and the diffferent materials are responsible for variations of the components' sputtering velocities. To determine the slopes at grain and phase boundaries, (6.3) can also be applied

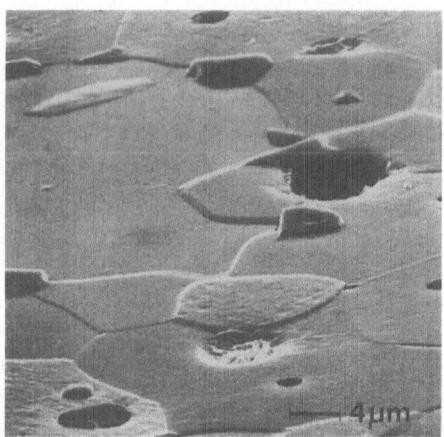

Fig. 6.8. Multicomponent ceramic surface (based on ZnO) after etching with 10 keV Kr^+ ions [6.43]

———— 4 µm

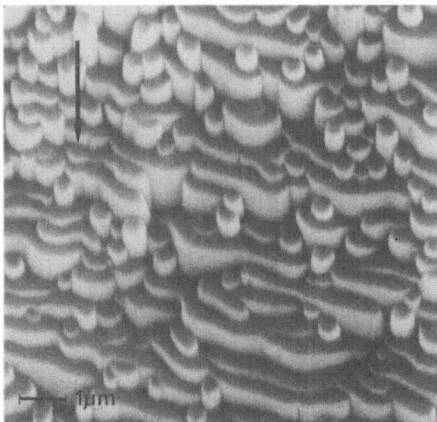

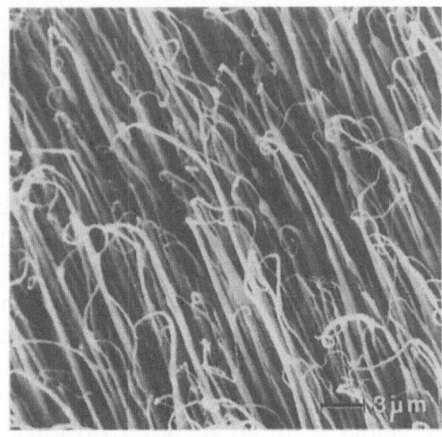

Fig. 6.9. Human tooth enamel after ion bombardment with 10 keV Kr$^+$ ions. Arrow: ion beam direction [6.47]

Fig. 6.10. Ion etch structure of an organic material, triglycinsulfate, after bombardment with 10 keV Kr$^+$ ions [6.46]

[6.44]. In addition, with increasing step height between the components, redeposition of one species of atom on a neighboring surface element of another material can modify the sputtering process [6.45]. On organic and biological materials characteristic relief structures develop with ion beam sputtering. The microtopography on human tooth enamel after etching with 10 keV Kr$^+$ ions (Fig. 6.9) is comparable with that on amorphous insulators [6.46, 47]. Another typical relief structure for a large group of organic substances is seen on triglycinsulfate (Fig. 6.10) after bombardment with 10 keV Kr$^+$ ions [6.46]. Such a fiberlike structure has been observed also on polyfluorethylene and many other polymers [6.48]. Here, besides physical sputtering, thermal and other effects probably are involved.

6.3.4 Influence of the Initial Surface State

The production of microtopographical structures can be drastically influenced by the initial surface state. Surfaces from the melting and manufacturing process and mechanically polished surfaces usually contain many defects. Even electrolytically well-prepared surfaces are often covered with a layer of contaminants originating from manipulation and inspection. Such layers have to be avoided or can be used to develop desired microstructures by subsequent ion bombardment.

Another topographical structure caused by the initial surface state is the well-known cone or pyramid. Figure 6.11 shows pyramids on silver (852) grain, initiated by a contamination layer in SEM and developed under oblique ion incidence of 51°. There are different cone-formation mechanisms [6.49–53]. One type of cone is formed under screening particles that protect the underlying

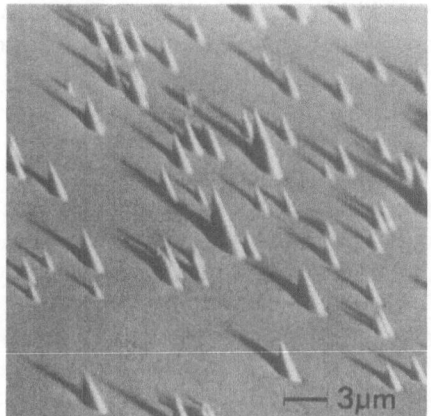

Fig. 6.11. Cone (pyramid) formation on an Ag (852) surface during oblique bombardment with 10 keV Ar⁺ ions (SEM image taken with back-scattered electrons) [6.49]

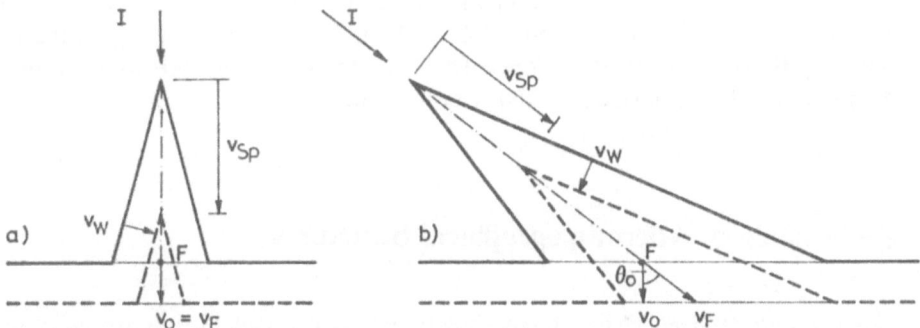

Fig. 6.12. Comparing the removal conditions for single cones at (**a**) normal and (**b**) oblique ion incidence under simplified conditions: The cones are drawn symmetric to and displaced in the beam direction. The displacement velocities v_w and v_0 are assumed to be equal in (**a**) and (**b**) [6.49]. Here I is the ion beam direction; F is the foot; v_{sp}, v_w, v_0, and v_F are velocities; and θ_0 is the inclination angle

material. This effect, applied with lithographic masks [6.54, 55], can be used to produce well-defined arrangements of cones (Sect. 6.5). These cones developed at normal ion incidence decrease and finally disappear (Fig. 6.12a) [6.49]. This phenomenon can be used to control their size. However, at oblique ion incidence cones can grow too. The difference between cones developed at normal and oblique ion incidence can be examined in Fig. 6.12, where their development is compared under some simplifying conditions [6.38, 49]. With oblique cones a small amount of redeposited material can cause them to grow relative to the surrounding area. Also this process can be stopped at the desired cone height. For nucleation of such cones only a short-term sputter inhibition is necessary. The development conditions of cones can be more complicated, taking into consideration the presence of more than one kind of atom and the influence of surface diffusion, especially the ion-bombardment-enhanced mobility on the surface, including such mechanisms as whisker growth [6.3, 6, 50, 56].

6.3.5 Applications

The manifold applications of material-induced ion-etched surface microstructures can be divided into two main groups. In the first group, the internal structure of the solid is revealed by a small ion fluence and then investigated by electron or ion microscopy [6.10, 30, 35]. These composition-dependent relief formations have to be considered in microanalytical investigations based on ion bombardment [6.34, 57].

In the second group of applications the ion beam texturing of solid surfaces is used for different purposes. High-fluence unidirectional bombardment of polycrystalline, organic, or heterogeneous materials leads in many cases without additional means to surface roughening. *Mirtich* and *Sovey* [6.58] applied this "natural texturing" to produce conelike structures on metals and fluoropolymers with $0.5 \ldots 1.0$ keV Ar^+ ions ($0.2 \ldots 1.5$ mA/cm^2) to get a high adhesive bonding. By the same method *Banks* [6.33] textured biopolymers and orthopedic implant alloys consisting of Ni, Co, Cr, and Mo.

6.4 Seeding of Microtopographical Structures

Wehner and *Hajicek* [6.60] have shown that cones can be intentionally developed on metals at normal ion incidence, if together with the bombarding ions a small amount of foreign atoms hit the target surface. However, not only conelike structures can be produced by "seeding" with foreign atoms. Also faceting structures were initiated during sputtering and simultaneous exposure of the sample to foreign atoms [6.61]. This process expands the variety of materials that can be textured by ion bombardment. Note that this seeding mechanism differs from the deposition of larger amounts of material, which forms screening islands protecting the underlying material, similar to mask etching (Sect. 6.5) [6.50].

6.4.1 Facets and Conelike Structures

Figure 6.13 shows the basic experimental arrangement for seeding atoms during sputtering [6.38, 61]. Here an Ar^+ ion beam is directed onto an Si(111) surface and onto a cylinder of the seeding material. In this way, simultaneously with the high-energy inert-gas ions, slow seed atoms hit the silicon surface. Depending on the atom-flux density, a faceting structure is developed (Fig. 6.14). Without exposure to foreign atoms, silicon remains smooth at 10 keV Ar^+ ion bombard-

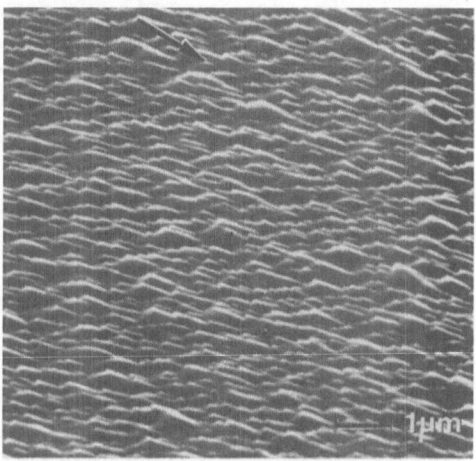

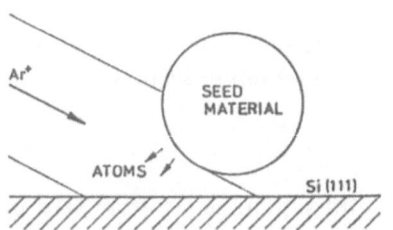

Fig. 6.13. Sputtering of an Si(111) surface with Ar⁺ ions and seeding with foreign atoms [6.61]

Fig. 6.14. Faceting structure on Si(111) developed under 10 keV Ar⁺ ion bombardment at 63° to the normal and simultaneous impinging of foreign atoms (as in Fig. 6.13) [6.38]

ment [6.38]. Already a certain amount of foreign atoms in beams of simple ion sources can produce the seeding effect [6.59, 61]. When the exposure to foreign atoms is switched off, the developed relief diminishes and disappears [6.38]. Figure 6.15 shows the direct evidence for the seeding effect. The bombarding Ar⁺ ions and the metal atoms fall onto the Si surface from different directions

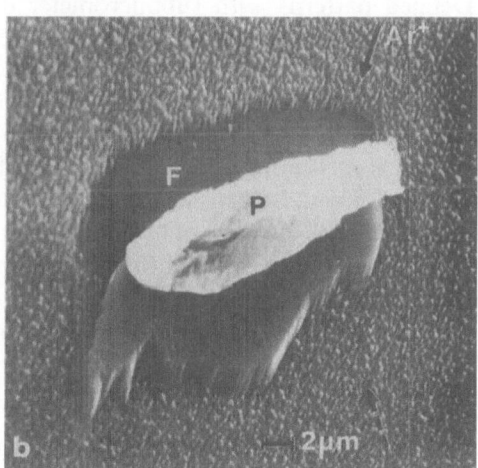

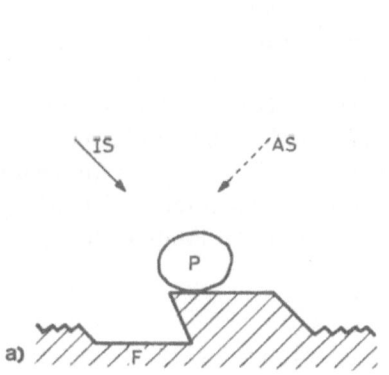

Fig. 6.15. Evidence for the seeding effect: (a) schematic drawing, (b) SEM image. Microtopography on Si in the environment of a dust particle P under ion bombardment according to Figs. 6.13, 14 [6.38]: IS is the ion beam; AS, the seeding atoms; and F, the shadow region for seed atoms

(Fig. 6.13). Therefore, around a dust particle, regions without exposure to metal atoms exist, which are sputtered without microtopographical features [6.61]. The flux of atoms can also be delivered from such sources as evaporation of the seed material [6.38]. *Wehner* [6.50] tested a lot of target/seed atom combinations at bombardment with Hg ions and found a relation between the melting points of both materials to be responsible for the cone formation. From this data a suitable seeding material can be chosen. *Robinson* and *Rossnagel* [6.56] investigated the influence of the diffusion process on seeding-induced cone formation. *Hudson* [6.62] textured a large number of elements with $0.5 \ldots 2.0$ keV Xe^+ ions of 2 mA/cm^2 density with tantalum seeding.

6.4.2 Applications

Surface modification by ion beam sputtering and simultaneous atom seeding has been developed to a technology for producing surface roughness on materials [6.32, 33, 56, 62–64]. The changing of optical properties to get high absorptance over the solar spectrum is one of the important applications. During the last years much work has been done in altering the surface morphology of materials for biomedical implants [6.33, 63]. Polymer and metal surfaces have been modified by ion beam processing with natural texturing (Sect. 6.3), seeding, or masking techniques (Sect. 6.5). *Weigand* and *Banks* [6.63] applied Ta seeding for texturing orthopedic materials by Ar^+ ions.

6.5 Ion Beam Mask Etching

Defined patterns with submicrometer dimensions can be produced only by anisotropic attack and therefore by dry-etching methods using directed beams of ions and atoms. To get higher efficiency and selectivity in etching of special materials and material combinations, which are relevant in microelectronics, reactive methods are applied using chemical effects combined with ion impact (chemical sputtering [6.6]). In this way the anisotropy of ion beam attack and chemical reactivity can be applied simultaneously for considerable advantages in applications. Because of its importance in the microelectronics industry, mask etching has been described in several reviews [6.3, 4, 65–76]. In the following, after basic considerations, some nonconventional applications of masking are described, opening new possibilities and expanding the limits of this technique.

6.5.1 Physical Sputtering Through Masks

Mask etching requires the uniform removal of all nonscreened regions of a defined pattern array on a surface or layer system. Mask forming – generally carried out by exposure of the pattern in a resist layer to electrons, ions, or photons, followed by its development – is not a subject of this chapter, with the exception that the mask itself is produced by ion beam etching steps.

The ideal etching process should transfer the mask pattern array without lateral alterations in the substrate down to the desired depth. On the basis of the properties of real physical sputtering (Sect. 6.3 [6.5, 6]), the following problems have to be considered:

i) The minimum mask thickness is determined by the relation of the sputtering rates of mask and substrate, together with the desired etching depth. Etching rates are well known for a great variety of substances under typical bombardment conditions [6.65–76].
ii) The ion attack leads to lateral shrinking of the mask dimensions, resulting in specially shaped sidewalls of the transferred patterns [6.65, 68, 69, 73, 74, 77, 78, 79].
iii) Redeposition and cross contamination of sputtered material, as well as sputtering by reflected ions, lead to additional sidewall contour changes [6.65, 67, 72, 75, 80–82].
iv) Mask ion etching can develop a microroughness in the sputtered regions, depending on the bombardment conditions and seeding effects [6.67, 72, 73].
v) The ion-beam-induced thermal loading may influence the resist and the resulting structure, and radiation damage can alter a thin surface region up to some nanometers [6.66, 76].

These characteristics require a careful choice, compromising bombardment conditions concerning the kind of ions, ion energy, ion beam density, ion beam direction, and the bombardment regime, as well as the mask design with respect to material, pattern dimension, and sidewall shape. In most applications Ar^+ ions up to some keV and ion beam densities up to some mA/cm^2 are used. To reduce thermal loading and radiation damage, often energies below 1 keV are applied with ion beam densities below $1\ mA/cm^2$ [6.74, 76].

The geometric transfer of the mask structure into the layer/substrate system has been investigated experimentally and theoretically by computer simulation, taking into consideration also redeposition effects and additional sputtering by reflected ions [6.77, 81, 83, 84]. Lateral alterations and sidewall shapes can be controlled by oblique ion incidence and sample rotation [6.65, 73, 75, 76, 78, 84–87]. Figure 6.16 gives an example of a test pattern transferred from resist into an SiO_2 layer on Si [6.78]. The standard process of pattern transfer has been modified for different purposes. To use the higher stability of metallic or carbon masks against sputtering, the "lift-off" technique is applied to covert the resist pattern into a stable material before the ion etching step [6.68, 88].

Mask etching can be combined with additional ion milling steps to produce specially shaped structures. Figure 6.17 shows how a regular array of sharp cones or ridges is produced starting with mask etching. After the resist is removed, the ion bombardment is continued to change the surface profile up to

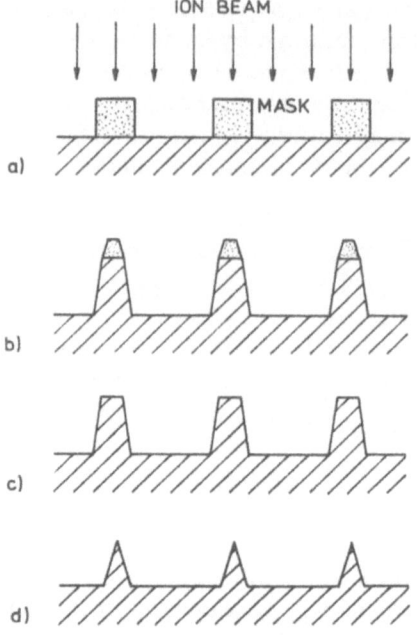

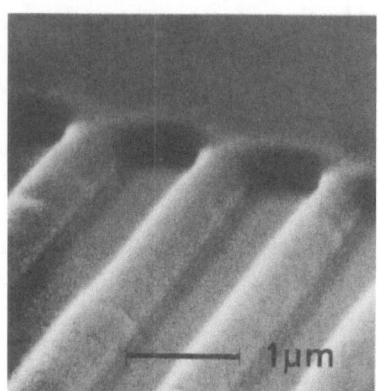

Fig. 6.16. Ion beam etching through a resist mask on SiO_2/Si under 30° (to the normal) with sample rotation [6.78]

Fig. 6.17. Development of a regular cone arrangement by masking [6.38]: **(a)** resist structure, **(b)** ion etching through the mask, **(c)** removing of resist, **(d)** cone development by ion etching

the desired shape [6.38, 54, 55]. With oblique ion incidence in the second etch step, a triangular-profiled periodical microstructure was produced [6.89, 90]. The spatial frequency of this grating can be doubled by a third ion beam etching step [6.91]. The mask need not necessarily be in contact with the material to be patterned. With ion optical systems the mask pattern array can be demagnified onto the sample [6.92–94]. However, these techniques are generally used for lithographic purposes and seldom for direct sputtering. In the present state of mask etching methods, the sputtering mechanism itself is not the resolution-limiting factor. However, with some special masking techniques using a writing electron or an ion beam or STM, mask structures near the theoretical limit of the collision cascade diameter can be produced [6.95–97].

6.5.2 Pattern Generation by Reactive Ion Beam Etching

The physical sputtering process with inert-gas ions is universally applicable to any material. However, introducing chemical reactivity in ion beam processing lessens some difficulties especially connected with microstructure production in semiconductors, insulators, and their combinations with metals. The main advantages are the considerable change (increase or reduction) of the sputtering rates of the mask, the layer, and the substrate materials up to several orders of magnitude, the higher selectivity, and the mostly volatile reaction products (and therefore low redeposition effects) [6.76, 79, 98–104]. The chemical reactivity can modify the process in different ways, e.g., by a reactive ion beam or a reactive component in the ion beam or by the residual gas atmosphere near the sample [7.105–110]. With reactive ion beam etching (RIBE) by XeF_2 the sputtering efficiency of Si can be enhanced to a factor of 25 [6.101, 102]. On the other hand, a portion of reactive O_2^+ ions in an Ar^+ ion beam reduces drastically the sputtering rate of metals (Cr, V, T, Al) by forming stable oxides with lower sputtering yields [6.98]. In several processes a reactive gas current is directed onto the sample together with the bombarding ion beam. The ion-bombard-ment-induced sputtering and chemical etching act together as ion-beam-assisted etching (IBAE) [6.109, 111, 112]. In this way the advantages of anisotropy are combined with the high efficiency and volatility of the reaction products. If the velocity is determined by the chemical etch step, such methods are also de-scribed as chemically assisted ion beam etching (CAIBE) [6.113–115]. Unlike with physical sputtering, the mechanisms that limit or promote the etch process have to be found out for each special system. For example, the enhanced etching of Al by Ar^+ ions in a Cl_2 atmosphere consists in continuous removal of the always present Al_2O_3 oxide layer by physical sputtering [6.112], or the chemical process forms reaction products weakly bonded and quickly sputtered [6.105–108]. New combinations of such mechanisms are being sought. Pro-cesses have been developed for Si, SiO_2, Si_3N_4, GaAs, GaP, InP, silicides, and various metals [6.76, 107]. The volatility of the reaction products allows etching of deep channels determined only by the accuracy of the ion beam direction and requiring a low divergence [6.106].

6.5.3 Applications

Active and passive components in Si and GaAs planar technology have been produced by ion beam etching [6.66, 116, 117]. Magnetic bubble memories with submicrometer resolution are patterned in permalloy [6.65, 118–120]. Reflective array filter configurations have been fabricated in $LiNbO_3$ by reactive sputtering methods [6.121]. In several applications microstructures with differ-ent depth are developed by moving the sample behind an aperture (Sect. 6.7.1) [6.121]. Micropatterns for integrated optics components and optical grating were fabricated with a large variety of design and spacing [6.121–124]. A special process to produce high-resolution patterns for X-ray masks was carried out with an additional oblique evaporation of a metallization layer on a resist profile. This metallization layer has been ion milled to remove selected regions [6.125–127].

Mask etching has also been applied to produce specially shaped surface textures on polymers for biomedical implants. For such purposes relatively large macroscopic areas were masked successively with crossbar structures in micro-meter dimensions. To form pillars this crossbar structure was replicated [6.33].

Small particles (latex) as "natural lithographic masks" were used to form surface textures by reactive ion beam etching [6.128]. Larger holes in diamond of millimeter thickness with diameters greater than 50 μm were fabricated by masking techniques [6.129], and small holes with 100 nm diameter have been developed in 100 nm thin foils [6.130]. Special particle or wire masking techniques were used to produce thin material sections for TEM samples [6.2, 131].

6.6 Ion Beam Slope Cutting

Another technique based on ion beam sputtering is ion beam slope cutting (IBSC). The method produces representative microcuts through heterogeneous materials [6.38, 132, 133]. This method has essential advantages in comparison with classical mechanical grinding methods and can be applied universally to most substances, especially to materials with very different components, porous materials, fibers, and biological samples (if they are stable under vacuum) [6.133, 134]. Additional advantages are achieved if the ion bombardment is carried out in situ in a scanning electron microscope (SEM) [6.38] to control cutting and to observe the plane which has been cut. The IBSC method is favored also for sample preparation for scanning Auger microscopy (SAM), ion microprobe microanalyzers (IMMA) and scanning ion microscopy (SIM). Based on ion beam slope cutting, a new technique for producing thin foils for TEM investigations has been developed [6.135, 136]. Furthermore the method can be used for special machining tasks, e.g., for producing sharp edges and for profile measurements.

6.6.1 Basic Method and Modifications

Figure 6.18a shows the principle of ion beam slope cutting [6.133]. The sample is sputtered by ions up to a sharp borderline, which is given by the shadow projection of the screen edge. High-fluence bombardment develops a slope area between the bombarded and unbombarded regions. This slope is a representative cut through the material with the following properties:

i) The slope is free of mechanical deformation.
ii) The slope position can be chosen with any wanted inclination and can be controlled by SEM.
iii) The slope is hit by grazing incident ions and therefore is not etched selectively, which means that also on heterogeneous samples smooth-cut planes develop.

The slope area can be observed in SEM in different positions. Figure 6.19 shows an example where the primary electron beam falls perpendicular on the slope, which is cut under 45° through a thick film layer and seen in material contrast in SEM [6.137]. However, not only surface layer systems can be cut by this

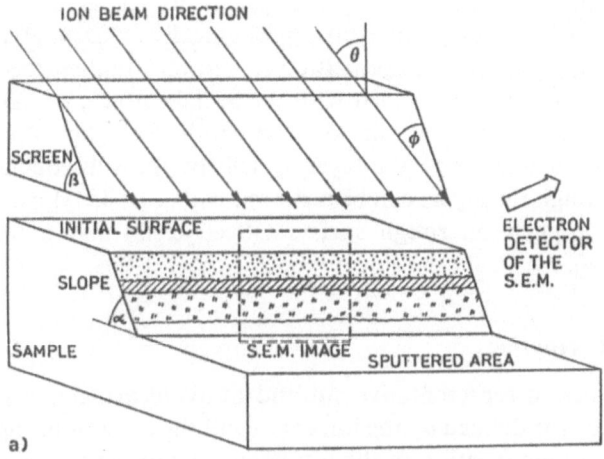

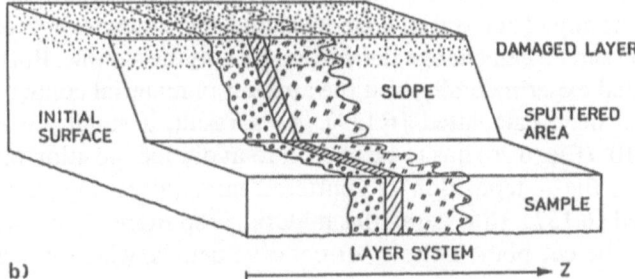

Fig. 6.18. Ion beam slope cutting (IBSC) [6.38, 132]: (a) The sample surface is sputtered up to a sharp border line. At this border line a slope has developed, representing a cut through the layer system. (b) Modification of IBSC for analyzing deeper layer systems. IBSC is started from a mechanically polished area through the damaged layer into disturbance-free regions. Beginning on the left at the initial surface we can investigate the material in direction z to any desired depth

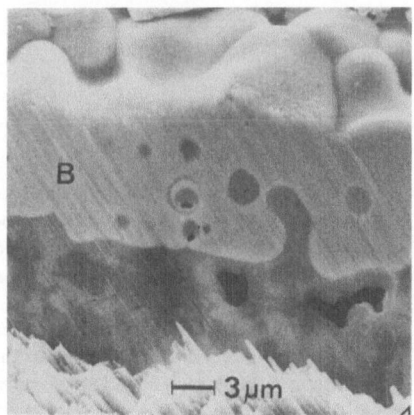

Fig. 6.19. Observation of an ion beam slope cut by material and topography contrast in SEM with the electron beam impinging normally to the slope [6.137]. Here B is the slope area. The sample is thick film based on AgPd

technique. Combining it with mechanical treatments reveals deeper regions of interest. Figure 6.18b shows a slope going out from a mechanically polished area and revealing not only the mechanically deformed layer, but also undisturbed material. Here, if we begin inspection on the left with the initial surface, we can investigate the sample in direction z up to any desired depth. With an other position of the mechanical cut relative to the layer system by the subsequent IBSC, very shallow cuts through the layers can be achieved with, e.g., 45° slopes. The method can also be applied on rough surfaces (Sect. 6.6.3) and more complicated samples [6.138].

6.6.2 Conditions of Slope Formation

A necessary condition to form a representative cut and to avoid artifacts is a stable incidence plane of the ions defined by the ion beam and the beam-limiting screen edge. However, this stable position of the ion beam shadow line on the sample can be influenced by different effects (Fig. 6.20). Assuming a mechanically and thermally stabilized screen position and a high-definition ion incidence plane, the sputtering of the screen edge itself and the condensation of sputtered material on the screen can change the cutting line on the sample. Both influences have been tested experimentally, and the amount of material condensed onto the screen has been calculated [6.139]. As a result, a small angle $(90° - \beta - \theta)$ of $1° \ldots 10°$ (Fig. 6.20) has to be chosen to avoid redeposition on the screen edge. Further, the redeposition of sputtered material on the slope itself has to be considered [6.137]. In general, this material is sputtered from the slope, but it can modify the cut plane in holes or concave details, where it can condense.

The ion bombardment is carried out by 10 keV Kr$^+$ ions with a density up to 0.5 mA/mm^2. The inclination of the screen edge and the slope are measured

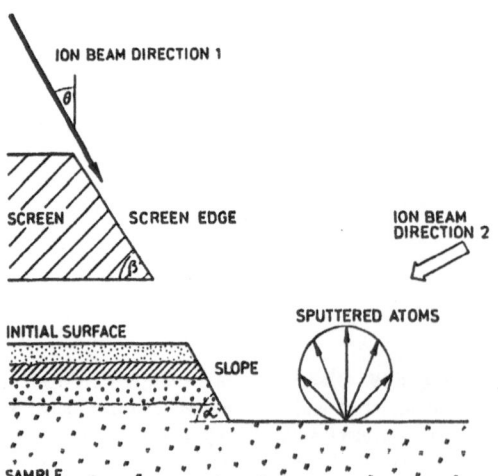

Fig. 6.20. Conditions of slope formation. The screen edge and the slope itself can be influenced by redeposition of sputtered material. A short-duration second ion bombardment in direction 2 is used for cleaning and selective ion etching of the slope. Here θ is the incidence angle of the ion beam; β, the screen edge angle; and α, the slope angle [6.137]

with grazing electron incidence in SEM [6.137]. The screen edge is prepared by mechanical grinding.

The possible artifacts can be avoided and additional advantages can be obtained by a short-duration second ion bombardment in direction 2 (Fig. 6.20) removing only some nanometers of the slope. This serves the following purposes:

i) The slope from redeposits is cleaned (important also for subsequent elemental analysis by scanning Auger microscopy).
ii) Ion-beam-induced furrows developed during cutting are smoothed (furrows can also be avoided by changing the angle ϕ in Fig. 6.18 during cutting).
iii) The internal structure in the slope is selectively ion etched for observation in topographical contrast in SEM, as well as in the composition mode. This also allows one to separate redeposition effects in holes, etc.

6.6.3 Applications

Ion beam slope cutting has been applied to various problems in materials science (metals, semiconductors, insulators, ceramics, fibers, powders, polymers), microelectronics, and biomedicine [6.137, 138, 140–145]. Moreover, basic experiments on the sputtering mechanism have been performed [6.78]. Thick-film systems based on NiP and AgPd (Fig. 6.19) were cut by ions to get information on thickness, homogeneity, grain and phase structure, porosity, interface, and defect structure [6.137, 140]. Figure 6.21 shows an example of cutting very rough surface regions of an Fe sample deformed by macroparticle bombardment erosion [6.141]. The surface can be directly correlated with the underlying structure, which is impossible when preparing embedded materials. Also, laser-modified hard metals have been cut successfully, showing structural changes in a

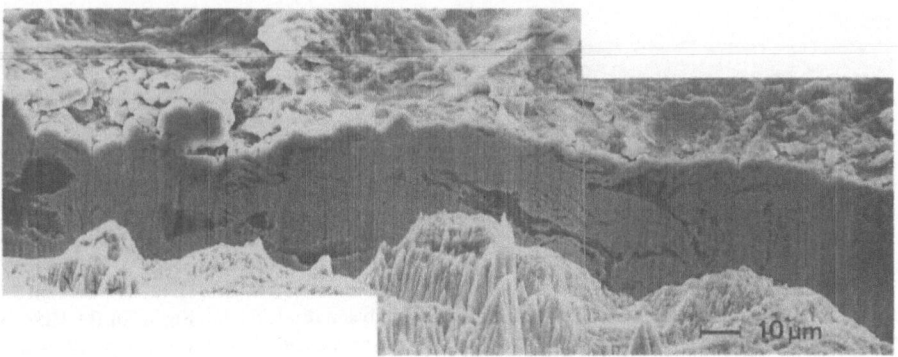

Fig. 6.21. IBSC through an Fe surface region deformed by long-duration macroparticle bombardment [6.141]

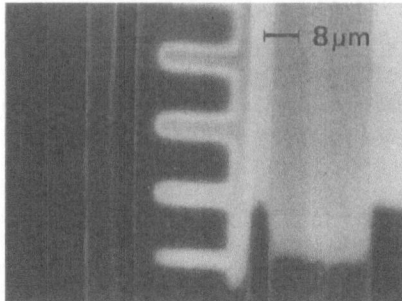

Fig. 6.22. MOS FET structure cut by IBSC under a small slope angle: superposition of SEM signals in EBIC and secondary electron mode [6.38]

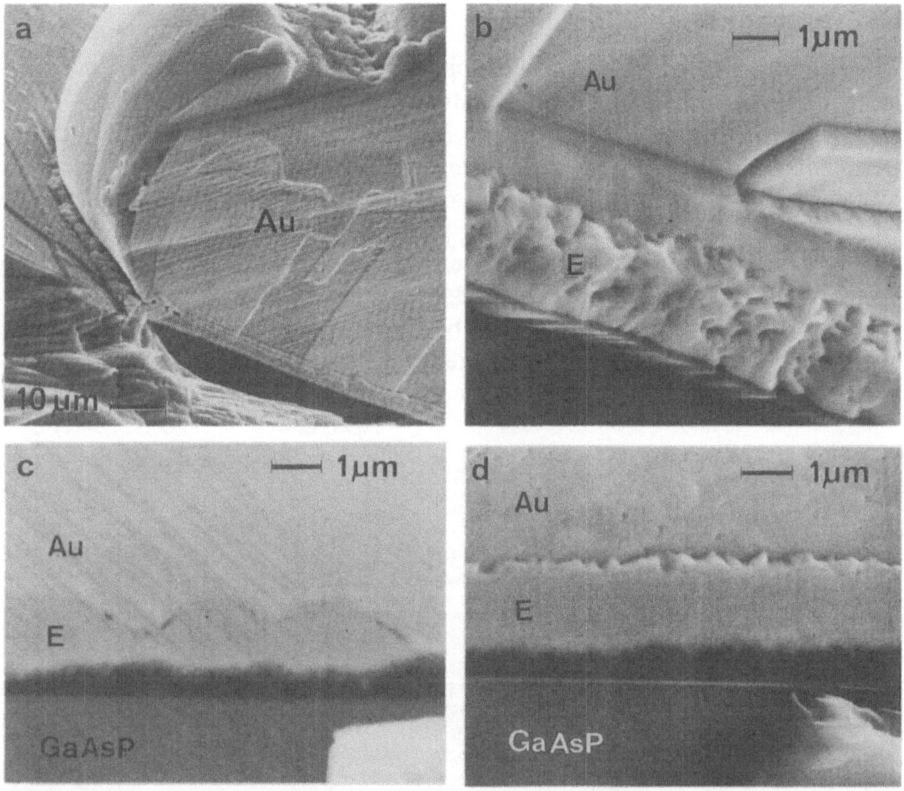

Fig. 6.23. Ion beam slope cutting of microelectronic bonding contacts [6.138]: (**a**) IBSC of an Au sphere contact on Al on GaAsP (selectively etched by ion beam direction 2 in Fig. 6.20). (**b**) IBSC of the bonding zone selectively etched: E is the eutectic layer on a small Si_3N_4 layer on GaAsP (black). (**c**) IBSC of a bonding zone without selective etching. Material contrast in SEM. The black layer below the eutectic layer E consists of Al. (**d**) Homogeneous layer system in a bonding zone (selectively etched): E is the eutectic layer; the black is Al; and the bright line is Si_3N_4

2 μm surface layer [6.142]. Figure 6.22 shows a MOSFET structure cut under a small angle, revealing contrast of electron-beam-induced current (EBIC mode) superimposed by secondary electron contrast [6.38]. The additional selective ion etching of the slope described in Sect. 6.6.2 (Fig. 6.20) is demonstrated in Fig. 6.23 with slope cutting of microelectronic bonding contacts [6.134, 138]. In Fig. 6.23a an Au sphere contact is cut, and the cut plane shows the selectively etched grain structure of Au and the zone below between Au and GaAsP. Figures 6.23b–d shows layer systems of the bonding zone with higher resolution. In Fig. 6.23c only material contrast is seen, whereas Figs. 6.23b and d demonstrate the superposition of material and ion-etched topographical contrast of different strengths. Also organic materials (polyethylene) and biomedical samples (human tooth enamel) were cut by IBSC [6.143, 144].

6.6.4 Thin-Foil Cutting by the Ion Beam Microtome

A new technique for producing thin foils for TEM investigations based on ion beam slope cutting is applicable to a large group of samples that cannot be prepared by conventional methods [6.134–136]. The initial form of the sample must not be a disk. Figure 6.24a shows the arrangement with the special task to produce a longitudinal section from the axis of an Al microwire with 25 μm diameter. A mounting that allows the object to be handled for the ion beam machining process, for inspection in the SEM, and for insertion into the TEM is necessary. In a first cutting step, one half of the microwire is fully removed by

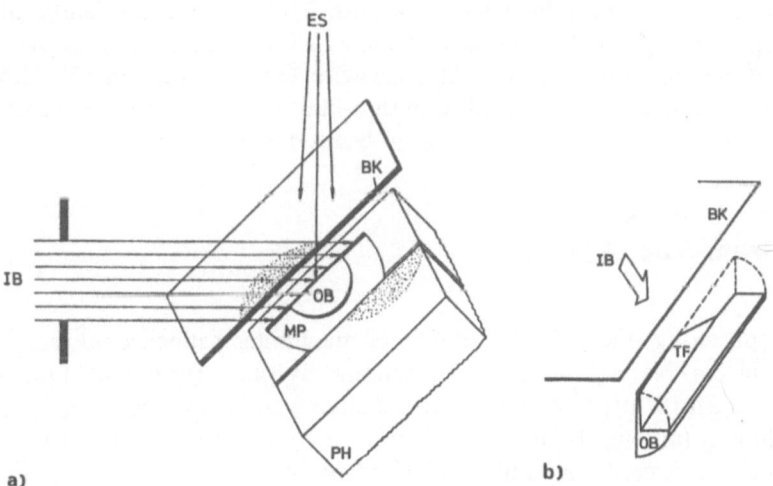

Fig. 6.24. Ion beam microtomy of a longitudinal section from the axis of a microwire for TEM investigation [6.136]. Here PH is the sample holder; MP, the mounting plate; OB, the object (microwire); BK, the screen edge; IB, the ion beam; and ES, the electron beam. (**a**) Arrangement of object, ion beam, and screen edge for ion beam cutting under SEM control. (**b**) Second ion beam cutting; TF is a thin section

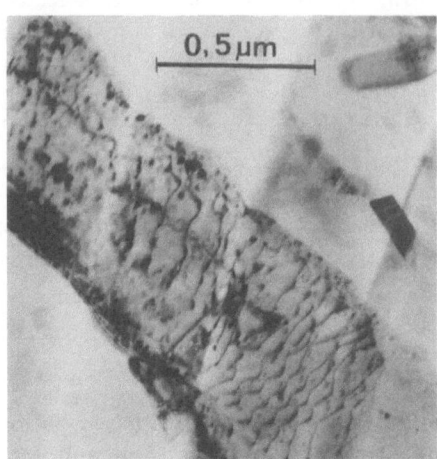

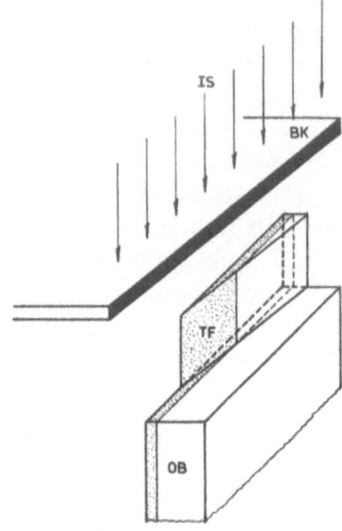

Fig. 6.25. Transmitted electron micrograph with 1 MeV from the microwire cut by IBSC according to Fig. 6.24 [6.136]. (TEM image courtesy of Dr. Häußler, IFE Halle)

Fig. 6.26. Ion beam microtomy of a thin film on a substrate [6.145]. Here OB is the object; TF, the thin film; IS, the ion beam; and BK, the screen edge

IBSC. Afterward the holder is rotated and adjusted as shown in Fig. 6.24b for the second cutting step. To obtain a stable object, the second cutting should not be carried out over the total diameter of the wire. To achieve a sufficiently thin section, a small wedge angle is applied, as in Fig. 6.24b. Figure 6.25 shows a transmission electron image of the Al microwire taken in the 1 MeV TEM (JEOL). The method can also be applied in thin-film studies. A thin-film/substrate system can be prepared by only one ion beam cut (Fig. 6.26) [6.145].

6.7 Ion Beam Microfiguring

Ion beam sputtering offers the possibility of machining defined contours on single workpieces. Natural samples or mechanically prefabricated workpieces can be formed into a large variety of desired shapes in analogy to mechanical macromachining (milling, lathing), if the total amount of the material to be removed does not exceed one millimeter. The amount of material removed can be controlled by the local ion dose. In broad homogeneous ion beams of inert-gas ions in the keV range with densities to $50 \, mA/cm^2$ and diameters of some centimeters, movable screens and workpieces are applied to modify the local ion fluence. By ion beam lathing, rotation-symmetric workpieces can be formed into the desired shape [6.146–151]. The microfiguring processes have to be inspected by scanning electron microscopy. Therefore, in special cases machining should be carried out in situ with SEM [6.38]. Ion beam sputtering is now being introduced in common high-precision mechanical technologies as an ultrafine finishing process [6.146].

6.7.1 Sputtering Through Movable Screens

To modify the local dose of a broad uniform ion beam, movable screens with more or less complicated openings have been used. To analyze depth profiles, a simple translation of a screen with a straight border and constant speed has been applied to generate a ramp section through surface region of silica glass with a very small inclination [6.148]. With specially shaped apertures, arrays of ion beams were formed, allowing batch fabrication of micromechanical elements through programmed movement of the screen or workpiece [6.151].

To form a defined crater, the ion fluence distribution of an ion gun can be modified by lenses and apertures. However, *Karger* fabricated a much more precise profile with a rotating screen containing a special hole contour cut by a programmed laser beam (Fig. 6.27) [6.149].

6.7.2 Ion Beam Microlathing

Spencer and *Schmidt* have described ion beam lathing using contact masks on the rotating workpiece [6.65]. More complicated shapes can be machined with the arrangement shown in Fig. 6.28 [6.150]. The ions fall through a slit with variable width above the rotating workpiece, which can be translated. The amount of material removed is controlled by the slit opening as well as by workpiece movement. By ion beam microlathing tips for FEM and STM have

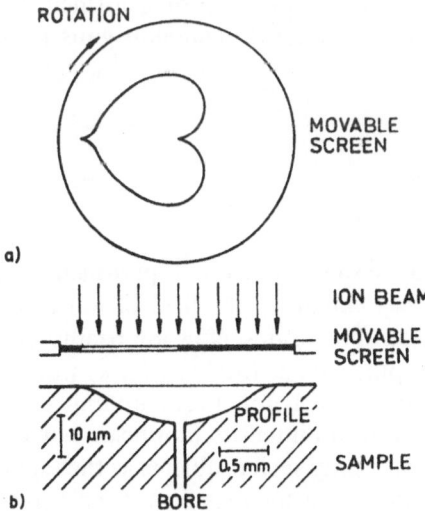

Fig. 6.27. Karger's method for generating a given three-dimensional shape on glass. [6.149] (a) movable mask with a hole (b) arrangement of sample, movable screen, and ion beam

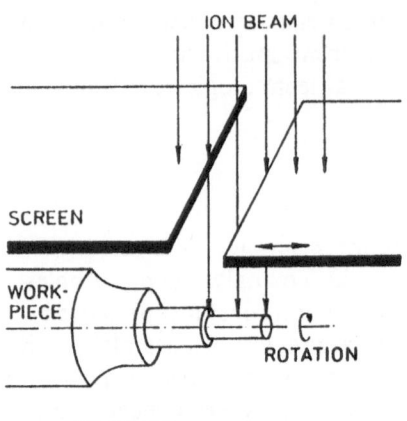

Fig. 6.28. Ion beam microlathing of a rotating workpiece [6.150]. The ion beam is limited by two screens forming a slit. The workpiece can be moved under the slit

been produced and *Davies* and *Bowen* [6.151] reduced a 3-mm-long brass cylinder with a 300 μm diameter down to a 200 μm diameter with high-precision computer-controlled correction of roundness, using a 1 keV Ar$^+$ ion beam of 2.5 mA/cm^2 for 90 min.

6.7.3 Edge Sharpening

During three-dimensional shaping by ion beam sputtering the contour change of an existing profile by the sputtering mechanism has to be considered. The resulting profile depends on the initial shape of the workpiece, as well as on the sputtering behavior of the material. The machining process can be predetermined with a known $Y(\theta)$ function. *Taniguchi* transformed a diamond tool with trapezoidal initial cross section into a sharp edge with a radius smaller than 10 nm to use as a microtome knife [6.146]. In the same way, sharp cones can be created for a cone arrangement, starting from an initial profile formed, for example, by mask etching (Fig. 6.17). Sharp tips for field emission guns have been produced with this method [6.147].

6.8 Microfocused Ion Beam Sputtering

The development of microfocused ion beam systems based on liquid-metal ion sources with a spot size down to 10 nm has expanded the field of ion beam processing to microstructures with nanometer dimensions. Now ion beams with some 10 nm in diameter are used for such purposes as localized implantation, lithography, microanalysis, microscopy, and special ion beam milling tasks [6.152–157]. However, because the small ion beam intensity (ion beam current in the pA range) removes material at a low rate, only such problems as the repair of microstructures produced by other sputtering techniques have been solved effectively. Microfocused inert-gas ion beams available to micrometer spot size [6.158, 159], have been applied for program-controlled production of three-dimensional optical surfaces [6.160]. With chemical reactive methods in micro-focused ion beam processing, higher efficiency together with other advantages of this technique can be expected.

6.8.1 Scanning Ion Beam Sputtering

Figure 6.29a schematically shows the machining process in microfocused ion beam sputtering [6.161]. The beam is scanned across the sample in two orthogonal directions. This is the typical feature in all microfocused ion beam equipment used for ion beam microanalyzers or scanning ion microscopes, which can thus be applied also for microstructure production as well as surface analysis or microscopy [6.152–158, 162–164]. This is of essential interest, if ion beam sputtering is applied for removing or depositing material in microareas, e.g., for repairing integrated circuits in selected regions with high-precision positioning and resolution [6.165–168]. In situ ion imaging of the produced microstructure requires only small intensities. With frame storage only one monolayer has to be removed to get a picture, whereas sputter machining requires higher intensities to remove the desired material quickly [6.157].

If microstructures are formed with small lateral dimensions compared with the removed depth, ion reflection and redeposition effects become important. The removed material thickness during one scan should be smaller than the beam diameter (Fig. 6.29a: $h < d$) [6.161, 164, 169]. Mostly the microbeam is computer-controlled and can be addressed to each position with individual ion fluence determined by intensity and dwelling time. These parameters of the scanning mode are important for producing a desired surface relief, as Fig. 6.29b shows with an inclined facet [6.161]. Smooth area removal requires overlapping of subsequent scans.

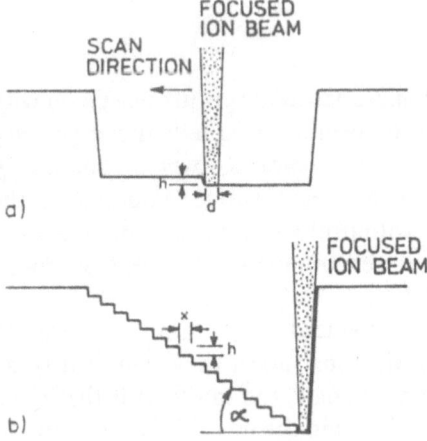

Fig. 6.29. Microfocused ion beam machining [6.161]: (a) formation of a uniform hole, (b) formation of an inclined facet (α: slope angle)

Typical parameters for high-resolution machining and imaging apparatus using liquid-metal ion sources are as follows [7.156, 157]:

Ion species: Ga, In, Au, Be, Pd, . . .
Ion energies: 20–100 keV
Spot size: 10 nm–3 μm
Ion density: 1.5 A/cm^2

The ion beam divergence is always very small (10^{-3} r), because lens aberrations are minimized in the ion beam forming system [6.156]. Characteristic sputtering yields for a 70 keV Ga$^+$ ion beam are $Y = 4.2$ for Al and $Y = 2.3$ for Cr [6.157].

With special deflection programs, different machining problems have been solved [6.169, 170]. *Seliger* et al. [6.162, 163] produced a pattern with 38 nm line-width in a 40 nm Au film on Si. To get higher productivity in microstructure production multibeam arrangements have been designed [6.94].

Focused ion beam milling has recently been successfully combined with reactive methods analogous to large-area chemically assisted ion beam etching (Sect. 6.5.2) by means of a reactive-gas flow directed onto the sample [6.171–175]. High-resolution etching effects by a microfocused ion beam have also been obtained on such special self-developing substances as nitrocellulose [6.176], but this ion-beam-induced etching belongs directly neither to physical nor to chemical sputtering.

For three-dimensional microanalysis with scanned microfocused ion beams, the repetitive scan mode is applied, corresponding to simultaneous irradiation of larger areas. With successive removal of surface layers of heterogeneous samples, material-dependent etch structures are developed (Sect. 6.3). Such structures give additional information and have to be considered when determining the fully three-dimensional elemental distribution [6.155, 177].

Using selected area irradiation (e.g., together with inclined samples), ion beam slope cuts (Sect. 6.6) should be possible with microfocused ion beams. This is a consequence of the high parallelism of the beam as shown, for example, at the right slope in Fig. 6.29b.

6.8.2 Applications

Yasuda has produced a three-dimensional microparabolic profile on silicon with a program-controlled focused 10 keV Ar$^+$ ion beam 30 μm in diameter [6.160]. The electrostatically deflected ion beam was scanned over a square of 200×200 μm^2, with a controlled ion fluence to each surface point. Thus the microrelief shown in Fig. 6.30 was formed with profiles in the x, y, and diagonal directions. *Harriott* et al. [6.161, 178] repaired optical circuits and produced microfacets for laser optics.

Wagner et al. [6.156, 167] used a 100 nm focused 57 keV Ga$^+$ ion beam to repair optical and X-ray masks, and as a demonstration they removed small squares of 0.5 μm–2 μm. For the X-ray mask repair, the mask with the intentional defects was imaged with the same ion beam in the normal scanning mode.

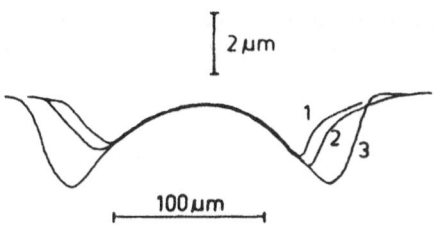

Fig. 6.30. Microparabolic profile machined on Si with a focused 10 keV Ar$^+$ ion beam by *Yasuda* [6.160]: surface profile along x axis (*1*), axis (*2*), and 45° to the x axis (*3*)

The edges of the pattern could be repaired by sputter machining of the defects using a sufficiently slow scanning speed; nevertheless, the repair proceeded rapidly (< 10 s/μm^2).

Photomasks could be repaired by removing Cr in small areas or by milling light scattering microstructures on glass in microregions [6.166]. *Harriott* et al. used a similar technique to cut conductors in integrated circuits [6.167]. An ion beam with 1 A/cm^2 and 0.1–0.2 μm diameter was applied for a cut 1 μm wide. This cut was much more precise than with laser cutting. *Shaver* and *Ward* repaired integrated circuits this way and could test the machining result by the voltage contrast mode in the ion beam equipment [6.165]. Ion beam machining and ion-beam-induced microdeposition of the sputtered material were applied to produce conducting connections in metal-insulator multilayer systems [6.168].

6.9 Conclusions

Although not all basic sputtering processes are fully understood, ion beam sputtering has been well established as a sensitive tool for producing topographical microstructures. The methods are rapidly developing to produce the more complicated structures necessary in microelectronics and related fields, and to provide higher resolution as well as greater productivity and reliability. The equipment has to be improved not only with stable high-performance ion guns, but also with high-precision sample operation and controlled environmental conditions, including defined beams of atoms directed onto the sample simultaneously with the ions for modifying the physical or chemical sputtering process. For micromachining the workpiece position, translation and rotation should be performed with nanometer accuracy in relation to the ion beam limiting screens. With the manipulation techniques provided by scanning tunneling microscopy, the resolution of ultraprecision machining can be expected to approach a near-atomic scale. At this level of ion beam processing, the size of the collision cascade becomes important and has to be minimized by the bombardment parameters. Parallel to these developments, the quality of the in situ process inspection by electron microscopy – mainly by scanning electron microscopy but also by the microfocused ion beam itself – has to be improved, along with computer control of the ion beam and workpiece.

References

6.1   P.J. Goodhew: Thin Foil Preparation for Electron Microscopy, Vol. 11 in *Practical Methods in Electron Microscopy*, ed. by A.M. Glauert (Elsevier, Amsterdam 1985)

6.2   H. Bethge, J. Heydenreich (eds.): *Electron Microscopy in Solid State Physics* (Elsevier, Amsterdam 1987)

6.3   O. Auciello, R. Kelly (eds.): *Ion Bombardment Modification of Surfaces* (Elsevier, Amsterdam 1984)

6.4   I. Brodie, J.J. Muray: *The Physics of Microfabrication* (Plenum, New York 1984)

6.5   R. Behrisch (ed.): *Sputtering by Particle Bombardment I*, Topics Appl. Phys., Vol. 47 (Springer, Berlin, Heidelberg 1981)

6.6   R. Behrisch (ed.): *Sputtering by Particle Bombardment II*, Topics Appl. Phys, Vol. 52 (Springer, Berlin, Heidelberg 1983)

6.7   P.D. Townsend, J.C. Kelly, N.E. Hartley: *Ion Implantation, Sputtering and their Applications* (Academic, London 1976)

6.8   D.T. Hawkins: Ion Milling (Ion Beam Etching) I, A bibliography 1954–1975, J. Vac. Sci. Technol. **12**, 1389 (1975)

6.9   D.T. Hawkins: Ion Milling (Ion Beam Etching) II, A Bibliography 1975–1978, J. Vac. Sci. Technol. **16**, 1051 (1979)

6.10   L.E. Murr: *Electron optical applications in materials science* (McGraw-Hill, New York 1970)

6.11   H. Bach: J. Non-Crystalline Solids **3**, 1 (1970)

6.12   H. Bach, H. Schröder: Z. Phys. **224**, 122 (1969)

6.13   P.J. Goodhew: J. Mater. Sci. **8**, 581 (1973)

6.14   W. Hauffe: to be published

6.15   M. Paulus, F. Reverchon: Proc. Coll. Int. C.N.R.S. "Le Bombardement Ionique" (Bellevue 1962) p. 223

6.16   D.J. Barber: J. Mater. Sci. **5**, 1 (1970)

6.17   M. Molčik: Prakt. Metallographie **7**, 361 (1970) and **12**, 299 (1975)

6.18   W.J. Hamilton, G. Judd, G.S. Ansell: J. Dental Res. **52**, 703 (1973)

6.19   J. Franks: Microscopy (GB) **25**, 227 (1977)

6.20   K. Hojou, T. Oikawa, K. Kanaya, T. Kimura, K. Adachi: Micron (GB) **8**, 151 (1977)

6.21   P.H. Holloway: Surf. Interf. Anal. **3**, 118 (1981)

6.22   A. Barna: Proc. VIIIth Europ. Congr. Electron Microscopy, Budapest 1984, p. 107

6.23   N. Baba, K. Adachi, Y. Muranaka, K. Suzuki, K. Kanaya: Proc. XIth Int. Congr. Electron Microscopy, Kyoto 1986, p. 355

6.24   C.W.T. Bulle-Lieuwma, P.C. Zalm: Proc. XIth Int. Congr. Electron Microscopy, Kyoto 1986, p. 351

6.25   D. Beecham: J. Appl. Phys. **40**, 4357 (1969)

6.26   H. Bach, R. Haspel, N. Neuroth: J. Phys. E **9**, 557 (1976)

6.27   H.C. Huang, J.D. Knox, Z. Turski, R. Wargo, J.J. Hanak: Appl. Phys. Lett. **24**, 109 (1974)

6.28   R.M. Castellano, J.L. Hokanson: Proc. 29th Ann. Symp. Frequ. Contr. 1975, p. 128

6.29   Ch. Fert, N. Colombie, B. Fagot, Phan van Chuong: Proc. Coll. Int. C.N.R.S. "Le Bombardement Ionique" (Bellevue, 1962) p. 67

6.30   L. Wegmann: Schweiz. Arch. Wiss. Techn. **30**, 143 (1964)

6.31   A.D.G. Stewart, Proc. Vth Int. Congr. Electron Microscopy, Philadelphia 1962, D 12

6.32   H.R. Kaufman, R.S. Robinson: J. Vac. Sci. Technol. **16**, 175 (1979)

6.33   B.A. Banks: Chap. 10 in Ref. [6.3] p. 399

6.34 W. Hauffe: Proc. IIIrd Int. Conf. *Secondary Ion Mass Spectrometry SIMS III*, Springer Ser. Chem. Phys., Vol. 19 (Springer, Berlin, Heidelberg 1981) p. 82
6.35 W. Hauffe: Thesis A, Dresden Technical University 1971
6.36 W. Hauffe: Physica Status Solidi (a) **4**, 111 (1971)
6.37 D. Onderdelinden: Canad. J. Phys. **46**, 739 (1968)
6.38 W. Hauffe: Thesis B, Dresden Technical University 1978
6.39 P. Haymann, C. Waldburger: Proc. Coll. Int. C.N.R.S. "Le Bombardement Ionique", Bellevue 1962, p. 294
6.40 M. Navez, C. Sella, D. Chaperot: Proc. Coll. Int. C.N.R.S. "Le Bombardement Ionique" (Bellevue, 1962) p. 233
6.41 W. Hauffe: Physica Status Solidi (a) **35**, K 93 (1976)
6.42 W. Hauffe: Proc. VIIIth Int. Conf. Atomic Coll. Solids, Moscow 1977, C 21
6.43 W. Hauffe: unpublished
6.44 K. Wetzig, H. Wittig: Kristall und Technik **5**, 561 (1970)
6.45 M. Tanemura, F. Okuyama: J. Vac. Sci. Technol. A **4**, 2369 (1986)
6.46 W. Hauffe: Reprint "Ionenätzung-Grundlagen und Anwendungen" Phys. Soc. GDR, Berlin 1979
6.47 W. Hauffe: Spectrum (Acad. Sci. GDR) No. 1, p. 18 (1983)
6.48 C. Weismantel, M. Rost, O. Fiedler, H.J. Erler, H. Giegengack, J. Horn: Proc. 6th Int. Vac. Congr., Kyoto 1974, p. 439
6.49 W. Hauffe: Proc. Int. Conf. Ion Beam Modification of Materials, Budapest 1978, p. 1079
6.50 G.K. Wehner: J. Vac. Sci. Technol A **3**, 1821 (1985)
6.51 A.D.G. Stewart, M.W. Thompson: J. Mat. Sci. **4**, 56 (1969)
6.52 I.H. Wilson, Rad. Effects **18**, 95 (1973)
6.53 R.S. Gvosdover, V.M. Efremenkova, L.B. Shelyakin, V.E. Yurasova: Radiat. Eff. **27**, 237 (1976)
6.54 J. Linders, H. Niedrig, N. Ram, H. Koch: Rad. Eff. **88**, 105 (1986)
6.55 J. Linders, H. Koch, H. Niedrig: Proc. XIth Int. Congr. Electron Microscopy, Kyoto 1986, p. 667
6.56 R.S. Robinson, S.M. Rossnagel: J. Vac. Sci. Technol. **21**, 790 (1982)
6.57 R. Castaing, G. Slodzian: J. Phys. E: Sci. Instr. **14**, 1119 (1981)
6.58 M.J. Mirtich, J.S. Sovey: J. Vac. Technol. **16**, 809 (1979)
6.59 G. Carter, M.J. Nobes, I. Abril, R. Garcia-Molina: Surf. Interf. Analysis **7**, 41 (1985)
6.60 G.K. Wehner, D.J. Hajicek, J. Appl. Phys. **42**, 1145 (1971)
6.61 J. Punzel, W. Hauffe: Phys. Status Solidi (a)**14**, K 97 (1972)
6.62 W.R. Hudson: J. Vac. Sci. Technol. **14**, 286 (1977)
6.63 A.J. Weigand, B.A. Banks: J. Vac. Sci. Technol. **14**, 326 (1977)
6.64 Z.W. Kowalski: J. Mat. Sci. **20**, 1521 (1985)
6.65 E.G. Spencer, P.H. Schmidt: J. Vac. Sci. Technol. **8**, 52 (1971)
6.66 C.M. Milliar-Smith: J. Vac. Sci. Technol. **13**, 1008 (1976)
6.67 P.G. Gloersen: J. Vac. Sci. Technol. **12**, 28 (1975)
6.68 H.I. Smith: Proc. IEEE **62**, 1361 (1974)
6.69 R. Wechsung: Vakuum-Technik **26**, 227 (1977)
6.70 L.D. Bollinger: Solid State Technol., Nov. 1977, p. 66
6.71 S. Somekh: J. Vac. Sci. Technol. **13**, 1003 (1976)
6.72 P.G. Gloersen: Solid State Technol., April 1976, p. 68
6.73 H. Dimigen, H. Lüthje: Philips techn. Rundschau, **35**, 217 (1976)
6.74 H. Dimigen, H. Lüthje, H. Hubsch, U. Convertini: J. Vac. Sci. Technol. **13**, 976 (1976)
6.75 R.E. Lee: J. Vac. Sci. Technol. **18**, 184 (1979)
6.76 L.D. Bollinger: Solid State Technol., Jan. 1983, p. 101
6.77 J.P. Ducommun, M. Cantagrel, M. Moulin: J. Mat. Sci. **10**, 52 (1975)
6.78 W. Hauffe, H.O. Prater: Proc. Nat. Conf. Physics of Semiconductor Surfaces, Acad. Sci. GDR, Binz 1979, p. 231

6.79 M. Cantagrel, M. Marchal: J. Mater. Sci. **8**, 1711 (1973)
6.80 R.E. Chapman: J. Mat. Sci. 12, 1125 (1977)
6.81 R. Smith, M.A. Tagg, J.M. Walls: Vacuum **34**, 175 (1984)
6.82 W. Hauffe: Proc. VIth Nat. Conf. Interaction of Atomic Particles with Solids, Minsk (USSR) 1981, p. 253
6.83 J.C. Moreno-Marin, J.A. Valles-Abarca, A. Gras-Marti: J. Vac. Sci. Technol. B **4**, 322 (1986)
6.84 I.W. Rangelow, P. Thoren, R. Kassing: Microelectron Eng. **3**, 631 (1985)
6.85 H. Gokan, S. Esho: J. Vac. Sci. Technol. **18**, 23 (1981)
6.86 S. Hosaka, S. Hasimoto: J. Vac. Sci. Technol. **15**, 1712 (1978)
6.87 A.R. Neureuther: J. Vac. Sci. Technol. **16**, 1767 (1979)
6.88 H. Aritome, S. Matsui, K. Moriwaki, S. Namba: J. Vac. Sci. Technol. **16**, 1939 (1979)
6.89 L.F. Johnson: Chap. 9 in [Ref. 6.3]
6.90 L.F. Johnson, K.A. Ingersoll: Appl. Opt. **20**, 2951 (1981)
6.91 L.F. Johnson, K.A. Ingersoll: Appl. Phys. Lett. **38**, 532 (1981)
6.92 B.A. Free, G.A. Meadows: J. Vac. Sci. Technol. **15**, 1028 (1978)
6.93 G. Stengl, R. Kaitna, H. Löschner, P. Wolf, R. Sacher: J. Vac. Sci. Technol. **16**, 1883 (1979)
6.94 G. Stengl, H. Löschner, W. Maurer, P. Wolf: J. Vac. Sci. Technol. B **4**, 194 (1986)
6.95 W.W. Molzen, A.N. Broers, J.J. Coumo, J.M.E. Harper, R.B. Laibowitz: J. Vac. Sci. **16**, 269 (1979)
6.96 A.N. Broers, W.W. Molzen, J.J. Coumo, N.P. Wittels: Appl. Phys. Lett. **29**, 596 (1976)
6.97 M.A. McCord, R.F. Pease: J. Vac. Sci. Technol. B **5**, 86 (1987)
6.98 M. Cantagrel: IEEE Transact. Electron. Devices **22**, 483 (1975)
6.99 J.W. Coburn, H.F. Winters, T.J. Chuang: J. Appl. Phys. **48**, 3532 (1977)
6.100 M.A. Bösch, L.A. Coldren, E. Good: Appl. Phys. Lett. **38**, 264 (1981)
6.101 U. Gerlach-Meyer, J.W. Coburn, E. Kay: Surf. Sci. **103**, 177 (1981)
6.102 U. Gerlach-Meyer: Surf. Sci. **103**, 524 (1981)
6.103 H.R. Kaufman, J.J. Coumo, J.M.E. Harper: J. Vac. Sci. Technol. **21**, 737 (1982)
6.104 B.A. Heath: Solid State Technol., Okt. 1981, p. 75
6.105 S.W. Pang, G.A. Lincoln, R.W. McClelland, P.D. DeGraff, M.W. Geis, W.J. Piacentini: J. Vac. Sci. Technol. B **1**, 1334 (1983)
6.106 G.A. Lincoln, M.W. Geis, S.W. Pang, N.N. Efremow: J. Vac. Sci. Technol. B **1**, 1043 (1983)
6.107 W. Katzschner, A. Steckenborn, R. Löffler, N. Grote: Appl. Phys. Lett. **44**, 352 (1984)
6.108 R.A. Barker, T.M. Mayer, R.H. Burton: Appl. Phys. Lett. **40**, 583 (1982)
6.109 R.A. Barker, T.M. Mayer, W.C. Pearson: J. Vac. Sci. Technol. B **1**, 37 (1983)
6.110 P.C. Zalm, L.J. Beckers, F.H.M. Sanders: Nucl. Instr. Meth **209/210**, 561 (1983)
6.111 M.W. Geis, G.A. Lincoln, N. Efremow, W.J. Piacentini: J. Vac. Sci. Technol. **19**, 1390 (1981)
6.112 N.L. DeMeo, J.P. Donnelly, F.J. O'Donnell, M.W. Geis, K.J. O'Connor: Nucl. Instr. Meth. Phys. Res. B **7/8**, 814 (1985)
6.113 J.v. Zwol, J.v. Laar, A.W. Kolfschoten, J. Dieleman: J. Vac. Sci. Technol. B **5**, 1410 (1987)
6.114 C. Garner: J. Vac. Sci. Technol. B **5**, 332 (1987)
6.115 I.W. Rangelow: Microelectronic Engineering **3**, 639 (1985)
6.116 I. Adesida: Nucl. Instr. Meth. Phys. Res. B **7/8**, 923 (1985)
6.117 D.B. Rensch, R.L. Seliger, G. Csanky, R.D. Olney, H.L. Stover: J. Vac. Sci. Technol. **16**, 1897 (1979)
6.118 M.J. Vasile, C.J. Mogab: J. Vac. Sci. Technol. A **4**, 1841 (1986)
6.119 E.G. Spencer, P.H. Schmidt, R.F. Fischer: Appl. Phys. Lett. **17**, 328 (1970)
6.120 R.K. Watts, H.M. Darley, J.B. Krüger, T.G. Blocker, D.C. Guterman, J.T. Carlo, D.C. Bullock, M.S. Shaikh: Appl. Phys. Lett. **28**, 355 (1976)
6.121 H.L. Garvin: Electronic Packaging and Production **18**, 90 (1978)
6.122 L.F. Johnson, K.A. Ingersoll: Appl. Phys. Lett. **35**, 500 (1979)
6.123 H.L. Garvin: Solid State Technol. **16**, 31 (1973)
6.124 H.L. Garvin, E. Garmire, S. Somekh, H. Stoll, A. Yariv: Appl. Optics **12**, 455 (1973)
6.125 D.C. Flanders, A.M. Hawryluk, H.I. Smith: J. Vac. Sci. Technol. **16**, 1949 (1979)
6.126 D.C. Flanders, A.E. White: J. Vac. Sci. Technol. **19**, 892 (1981)

6.127 D.C. Flanders, N.N. Efremov: J. Vac. Sci. Technol. B **1**, 1105 (1983)
6.128 H.W. Deckman, J.H. Dunsmuir: Appl. Phys. Lett. **41**, 377 (1982)
6.129 E.G. Spencer, P.H. Schmidt: J. Appl. Phys. **43**, 2956 (1972)
6.130 B.P. Beecken, W. Zimmermann: J. Vac. Sci. Technol. A **3**, 1839 (1985)
6.131 H.W. Deckman, J.H. Dunsmuir, B. Abeles: J. Vac. Sci. Technol. A **3**, 950 (1985)
6.132 W. Hauffe: Patent DD 139670 (1977)
6.133 W. Hauffe: Proc. IXth Nat. Conf. Electron Microscopy (GDR), Dresden 1978, p. 84 and Proc. Xth Nat. Conf. Electron Microscopy (GDR), Leipzig 1981, p. 307
6.134 W. Hauffe: Elektronika (Poland) **26**, 3 (1985)
6.135 W. Hauffe: Patent DD 218954 (1982)
6.136 W. Hauffe: Proc. VIIIth Europ. Congr. Electron Microscopy, Budapest 1984, p. 105
6.137 W. Hauffe: Proc. 10th Conf. of the Internat. Soc. for Hybrid Microelectronics – Poland Chapter, Krakow 1986, p. 27
6.138 W. Hauffe: Proc. Nat. Conf. Physics of Semiconductor Surfaces, Acad. Sci. GDR, Binz 1982, p. 177 and 1983, p. 133
6.139 W. Hauffe, T. Chudoba: Proc. Nat. Conf. Physics of Semiconductor Surfaces, Acad. Sci. GDR, Binz 1986, p. 235
6.140 B. Holodnik, A. Jakubovicz, M. Lukaszewicz, W. Hauffe: Active and Pass. Elec. Comp. **12**, 127 (1986)
6.141 W. Hübner, W. Hauffe: to be published
6.142 W. Hauffe, B. Schultrich: Praktische Metallogr. **25**, 517 (1988)
6.143 W. Berger, U. Ludwig, W. Hauffe, F.E. Karasz: J. of Appl. Polymer Science **34**, 919 (1987)
6.144 B. Böhm, W. Hauffe: Zahn-, Mund- u. Kieferheilk. **76**, 134 (1988)
6.145 W. Hauffe: Materials Science (Poland) **13**, 21 (1987)
6.146 N. Taniguchi: Annals of the CIRP **32**, 1 (1983)
6.147 J.A. Kubby, B.M. Siegel: J. Vac. Sci. Technol. B **4**, 120 (1986)
6.148 R.A. Chappel, C.T.H. Stoddart: Proc. 7th Int. Vac. Congr. and 3rd Int. Conf. Solid Surfaces, Vienna 1977, p. 2297
6.149 A.M. Karger: Applied Optics **12**, 451 (1973)
6.150 W. Hauffe: Patent DD 152232 (1980)
6.151 S.T. Davies, D.K. Bowen: J. Vac. Sci. Technol. B **5**, 337 (1987)
6.152 R. Levi-Setti, G. Crow, Y.L. Wang: Proc. SEM (USA) Chicago 1985, p. 335
6.153 R. Levi-Setti: Proc. SEM (USA), Chicago 1983, p. 1
6.154 A.R. Waugh, A.R. Bayly, K. Anderson: Proc. IVth Int. Conf. Secondary Ion Mass Spectrometry SIMS IV, Springer Ser. Chem. Phys., Vol. 36 (Springer, Berlin, Heidelberg 1984) p. 138
6.155 R.E. Thurstans, J. Wolstenholme: Vacuum **37**, 289 (1987)
6.156 A. Wagner: Nucl. Instr. Meth. Phys. Res. **218**, 355 (1985)
6.157 J. Melngailis: J. Vac. Sci. Technol. B **5**, 469 (1987)
6.158 H. Liebl: J. Appl. Phys. **38**, 5277 (1967)
6.159 H. Liebl: J. Vac. Sci. Technol. **12**, 385 (1975)
6.160 H. Yasuda, J. Appl. Phys. **45**, 484 (1974)
6.161 L.R. Harriott, R.E. Scotti, K.D. Cummings, A.F. Ambrose: J. Vac. Sci. Technol. B **5**, 207 (1987)
6.162 R.L. Kubena, R.L. Seliger, E.H. Stevens: Thin Solid Films **92**, 165 (1982)
6.163 R.L. Seliger, R.L. Kubena, R.D. Olney, J.W. Ward, V. Wang: J. Vac. Sci. Technol. **16**, 1610 (1979)
6.164 H. Morimoto, Y. Sasaki, Y. Watanabe, T. Kato: J. Appl. Phys. **57**, 159 (1985)
6.165 D.C. Shaver, B.W. Ward: J. Vac. Sci. Technol. B **4**, 185 (1986)
6.166 P.J. Heard, R.A. Cleaver, H. Ahmed: J. Vac. Sci. Technol. B **3**, 87 (1985)
6.167 L.R. Harriott, A. Wagner, F. Fritz: J. Vac. Sci. Technol. B **4**, 181 (1986)
6.168 J. Melngailis, C.R. Musil, E.H. Stevens, M. Utlaut, E.M. Kellogg, R.T. Post, M.W. Geis, R.W. Mountain: J. Vac. Sci. Technol. B **4**, 176 (1986)
6.169 H. Yamaguchi, A. Shimas, S. Haraichi, T. Miyauchi: J. Vac. Sci. Technol. B **3**, 71 (1985)
6.170 T. Kato, H. Morimoto, K. Saitoh, H. Nakata: J. Vac. Sci. Technol. B **3**, 50 (1985)

6.171 K. Asakawa, S. Sugata: J. Vac. Sci. Technol. B **3**, 402 (1985)
6.172 Y. Ochiai, K. Gamo, S. Namba: J. Vac. Sci. Technol. B **1**, 1047 (1983)
6.173 Y. Ochiai, K. Gamo, S. Namba: J. Vac. Sci. Technol. B **3**, 67 (1985)
6.174 Y. Ochiai, K. Shihoyama, T. Shiokawa, K. Toyoda, A. Masuyama, K. Gamo, S. Namba: J. Vac. Sci. Technol. B **4**, 333 (1986)
6.175 Y. Ochiai, K. Gamo, S. Namba, K. Shihoyama, A. Masuyama, T. Shiokawa, K. Toyoda: J. Vac. Sci. Technol. B **5**, 423 (1987)
6.176 Y. Yasuoka, K. Harakawa, K. Gamo, S. Namba: J. Vac. Sci. Technol. B **5**, 405 (1987)
6.177 F. Rüdenauer: Beitr. elektronenmikr. Direktabb. Oberfl. **18**, 25 (1985)
6.178 L.R. Harriott, R.E. Scotti, K.D. Cummings, A.F. Ambrose: Appl. Phys. Lett. **48**, 1704 (1986)

Additional References with Titles

Alexander, K.B., Angelini, P., Miller, M.K.: "Precision Ion Milling of Field-Ion Specimens", in Proc. 36. Int. Field Emiss. Symp., J. de Physique **50**, Suppl. Coll. C8 (1989)

Barna, A., Barna, P.B., Zalar, A.: "Analysis of the development of large area surface topography during ion etching", Vacuum **40**, 115 (1990)

Kirk, E.C.G., McMahon, R.A., Cleaver, J.R.A., Ahmed, H.: "Scanning ion microscopy and microsectioning of electron beam recrystallized silicon on insulator devices", J. Vac. Sci. Technol. **B6**, 1940 (1988)

Miyamoto, I., Ezawa, T., Nishimura, K.: "Ion beam machining of single point diamond tools for nano-precision turning", Nanotechnology **1**, 44 (1990)

Morishita, S., Okuyama F.: "Interplay of erosion and redeposition processes in seed cone formation", J. Vac. Sci. Techn. **9**, 3295 (1990)

Sudraug, R., Ben Assayag, G.: "Focused-ion-beam milling, scanning electron microscopy, and focused-droplet deposition in a single microcircuit surgery tool", J. Vac. Sci. Techn. **B6**, 234 (1988)

7. Production of Thin Films by Controlled Deposition of Sputtered Material

Edwin D. McClanahan and Nils Laegreid

With 16 Figures

Controlled deposition of sputtered material is presently the most versatile method of forming thin films on solid substrates. Compared with other widely used methods of thin-film formation – vacuum evaporation, electroplating, electroless plating, and chemical vapor deposition – sputter deposition permits a much wider selection of film materials, produces films with higher purity and better controlled composition, provides films with greater adhesive strength and homogeneity, and permits better control of deposit thickness.

In this chapter, we describe the different technological approaches currently used for sputter deposition. Unlike most other work described in this book, sputtering for thin-film production is performed using a plasma rather than a focused ion beam. After the fundamentals of sputter deposition, we discuss direct current (dc) glow discharge, dc-supported plasma discharge, radio-frequency (rf) discharge, and magnetically enhanced discharge. In each case, examples illustrate the power of the sputter deposition technique.

7.1 Historical Review

Sputtering was discovered over a hundred years ago and was first reported by *Grove* [7.1] in 1852 and independently by *Plücker* [7.2] in 1858. As long ago as 1877, sputtering was used to coat mirrors [7.3]. By 1930 it was used to coat such things as phonograph masters [7.4, 5]. When equipment improvements made evaporative coating possible, this less-complicated and better understood technique superseded sputter deposition [7.6]. As a result, the development of the sputter deposition process languished until the late 1950s. Coincidentally, improvements in the understanding of the sputtering phenomenon and the demand for high-performance thin films of a variety of materials (stimulated by advances in microelectronics and semiconductor technologies) prompted new developments in sputter deposition.

Thin films can be deposited by many different methods, including thermal evaporation [7.7], chemical vapor-phase deposition [7.8], plasma deposition [7.9], polymerization of monomers [7.10], oxidation of metals or semiconductors [7.11], and sputter deposition. Each method has advantages and disadvantages, and no one method is best for all applications. Sputter deposition is one of the more complex methods and, in many cases, is more expensive.

Topics in Applied Physics, Vol. 64
Sputtering by Particle Bombardment III Eds.: R. Behrisch · K. Wittmaack
© Springer-Verlag Berlin Heidelberg 1991

However, sputter deposition permits better control of the composition and dimensions of the deposited film and greater flexibility in the types of materials that may be deposited [7.12]. The unique features of sputter deposition have led to its being described as a "controlled method of coating almost anything with a thin film of almost anything to make it more resistant to anything (or more adherent, or more conductive, or provide better lubrication)" [7.13].

State-of-the-art sputter film deposition makes possible the successful formation of films of a multitude of substances including metals, alloys, insulators, semiconductors, mixtures of metals and ceramic insulators, ferroelectrics, and even organic polymers. Examples of laboratory-developed film deposition operations that could be adapted for production include the deposition of almost all metals plus oxides, sulfides, nitrides, hydrides, many alloys, compounds, and glasses. The layers can be laminates of different metals or compounds. The composition of the deposit can be graded [7.14]. Sputter deposition has also been used to produce pinhole-free chromium masks for photoetching, different types of corrosion- and abrasion-resistant coatings, piezoelectric and ferromagnetic films, film components for microcircuit elements, protective and passivating films, flexible circuitry, and decorative coatings for glass and plastics. High-rate sputtering has now been developed to the point that it can be used commercially to manufacture thick films for corrosion and abrasion resistance, braze metal layers for joining, and produce high-performance optical surfaces.

Sputtering is not limited to coating preparation. Bulk free-standing sputter-deposited materials with unique metallurgical properties have been produced – for example, high-coercivity permanent magnet alloys, free-standing turbine blades, special alloy sheet, closed-cell metal foams, and a bulk-property 1.25-cm-thick test specimen of CuZr alloy weighing 5.5 kg [7.14]. Other novel applications are being developed: the coating of prosthetic devices with bone [7.15]; the storage of ^{85}Kr, a fission product from spent nuclear fuel [7.16]; and the production of diamond-like carbon and high-temperature superconducting films [7.17, 18].

Originally, sputter deposition was achieved by the dc glow discharge method (also called diode or cold cathode sputtering). This method is still widely used because of its simplicity, but it has many shortcomings. The rf plasma discharge and magnetron methods of sputter deposition developed in recent years overcome some of the limitations of the glow discharge method.

7.2 Fundamentals of Sputter Deposition

The principles of sputtering are discussed in detail in Volumes I and II of this series [7.19, 20] and other sources [7.4, 6, 21–23]. However, a brief review will help explain the reasons for, and the differences between, various sputter deposition methods.

When an atom with an energy of more than about 30 eV hits a surface, a small fraction of the energy and momentum of the incoming ion will, through lattice collisions, be reversed and may cause ejection of surface atoms (i.e., sputtering). The sputtered atoms leave the target surface with relatively high energies ($\sim$ 10 eV) compared with evaporated atoms ($\sim$ 0.1 eV). The average number of the atoms ejected from the surface per incident ion is called the sputtering yield [7.19]. To achieve a practical deposition rate on a strategically located substrate, an ion source of sufficiently high flux and a means for accelerating the ions are necessary. The ion source is usually a plasma (i.e., an electrically neutral mixture of positive ions and electrons) generated by electron impact in a noble gas at subatmospheric pressures (typically 2–10 Pa). The ions are accelerated in an electric field obtained by applying a negative potential with respect to the plasma potential to an electrode immersed in that plasma.

Sputter deposition methods used today (such as dc glow discharge, supported plasma discharge, rf discharge, and magnetically enhanced discharge) have common simple goals: to generate and maintain a desired plasma and to establish a bias or electric field for the acceleration of ions to the electrode or target being bombarded (the deposition source). When the fast ion comes within a few angstroms of the target surface, it is neutralized by an electron from the surface [7.24]. The neutralized atom, upon interacting with the target surface, may cause the emission of not only target atoms but also secondary electrons [7.24]. These so-called γ electrons are essential for generating and maintaining a dc glow discharge. The number of secondary electrons emitted per incident atom depends mainly on the velocity of the atom and the material of the target.

The flow of electrons from the target (cathode) constitutes an electric current that, coupled with the flow of the energetic incoming ions, results in an appreciable power dissipation at the target (typically 1–10 W/cm^2). This power must be dissipated within the chamber since there is no increase in the potential energy of the substrate or the deposited film. The only permanent changes that occur in the chamber are a transfer of material from the target to the substrate and, in the case of reactive sputter deposition, a chemical reaction of the gas with the target metal, usually an exothermic process. Essentially all the electric power supplied to the sputter deposition chamber is converted into heat inside the chamber. Such heat must be controlled to provide stable, reproducible conditions for film deposition.

The sputtering yield varies with the target material, the kind of impinging ion, and the energy of that ion [7.25, 26]. The yield generally increases with the atomic weight of the impinging ion for ion energies in excess of 600 eV. At a given ion energy, the yield increases significantly with increasing obliqueness of the impinging ions up to about 60° away from normal [7.25, 27]. The effect must always be taken into account in the geometry and size of thin-film deposition equipment to achieve the required deposit properties and uniformity.

The erosion rate of the target $\dot{z}$ (cm/s) and, in turn, the deposition rate on the substrate are uniquely determined by the current density $j(A/cm^2)$ and the sputtering yield Y (atoms/ion), as shown in the following relationship:

$$\dot{z} = \frac{jY}{eN} \, , \tag{7.1}$$

where e is the elementary charge and N (atoms/cm^3) is the target density. Deposition equipment using supported plasma or magnetically enhanced discharges is therefore designed to obtain high ion current densities over the target surface.

The angular distribution of the sputtered atoms varies with the energy of the bombarding ion [7.28–30]. For normal beam incidence and amorphous or polycrystalline targets, the dependence of the angular differential yield on the polar angle θ can usually be written in the form $dY/d\theta = Y(0°)/\cos^n \theta$. It is under cosine ($n < 1$) at low ion energies, near cosine at medium energies (~ 1000 eV), and over cosine ($n > 1$) at high ion energies. This fact has to be taken into account in considering target edge losses and deposit thickness uniformity. In many microelectronic applications, a uniform deposit thickness may be required for metallization of semiconductor integrated circuits. The broad angular distribution of sputtered atoms provides beneficial effects. Since the atoms are arriving from a wide range of angles, depending on gas pressure and electrode spacing, the walls and the tops of the steps are fairly uniformly covered [7.31].

Some of the superior adhesion and structural properties of sputter-deposited films can be attributed to the relatively high energies with which the atoms impact on the substrate surface (Chap. 2). The high impact velocity theoretically implies, and actually results in, differences in the initial nucleation and subsequent growth of sputter-deposited films, as compared with films deposited by other means where the impact energies are much lower. Since the energy distribution of sputtered atoms can vary with bombardment and ejection angles [7.32], control of the properties of sputter-deposited film can become complicated [7.33]. If desired, sputtered atom velocity may be reduced by increasing the mass or pressure of the sputtering gas [7.34].

The principles discussed above can be used to explain many of the problems inherent in sputter deposition and to provide an understanding of the many different methods of sputter deposition currently used. In addition, they emphasize that the same basic principles apply to all types of sputter deposition.

In alloys, one element is usually sputtered more readily than the others (preferential or selective sputtering [7.35, 36]). Therefore, when sputtering begins from a new target, one kind of atom will be removed from the target more frequently and deposited on the substrate at a higher concentration than would be found in the target. This unbalanced rate will continue until the concentration of the preferentially sputtered element on the target surface is reduced to the point where the rate of removal of the different elements equals their numerical ratio in the alloy bulk. This holds true only when the target is sufficiently cooled to prevent bulk diffusion of the target atoms [7.37]. The result of such preferential sputtering is that the base layer of the sputter-deposited film will not have the same composition as the bulk of the alloy. In those instances where homogeneity throughout the film thickness is desired, a shutter has to be interposed between the target and the substrate for the initial film deposition period. When the composition of the deposited film has stabilized, the shutter is removed.

Another problem can occur during the deposition of an alloy film composed of constituents with atoms that vary in mass. Lighter atoms will be ejected more nearly perpendicular to the target surface [7.35, 38]. Heavier atoms will "spray" outward, resulting in a film (assuming a planar target and substrate) with a relatively lower concentration of more massive atoms in the center and a lower concentration of less massive atoms at the outer portions. This separation in space can be prevented by moving the substrate in reference to the target throughout the time of deposition or by sputtering from composite targets.

Since sputtering is caused by a collision cascade [7.39], the temperature of the target has essentially no influence on the ejection rate of atoms from the target surface [7.40]. If the temperature of the target is below the condensation temperature of the sputtering gas, energetic sputtered adsorbed atoms will also be added to the flux of atoms impinging on the substrate surface. This is particularly important when mercury is used as the sputtering gas [7.41]. At the other extreme, if the temperature of the target is high, sublimation of the target atoms can take place. The fact that sputtering is basically a collision process makes the sputter deposition of refractory metals and other high-melting-point materials relatively easier than evaporative coating processes.

The structure of the deposited material may be modified by controlling the deposition rate, the substrate temperature, and the particle flux [7.33, 42]. Deposit structures ranging from amorphous to essentially single crystalline have been obtained [7.43].

Although controlling many variables (gas composition, deposition rate, pressure, substrate temperature, and bias) can be tedious, it does make it possible to change dramatically the structure and therefore the physical properties of sputter-deposited films. For example, stainless steel can be deposited as ferrite, austenite, or a mixture of the two [7.44]. Similarly, TiO_2 can be deposited as rutile or a mixture of rutile and anatase [7.45]. The mechanical properties of sputter-deposited polytetrafluoroethylene (PTFE) indicate a shorter, highly cross-linked polymeric molecule [7.46]. A bias sputter-deposited film of MoS_2 may lose its lubricity if the bias voltage is too high [7.47]. The high negative bias results in the depletion of sulfur in the film.

7.3 dc Glow Discharge

Electrical breakdown in gases has been studied for over a century [7.48, 49]. Figure 7.1 illustrates the three fundamental types of continuous electric discharge: dark discharge, glow discharge, and arc discharge.

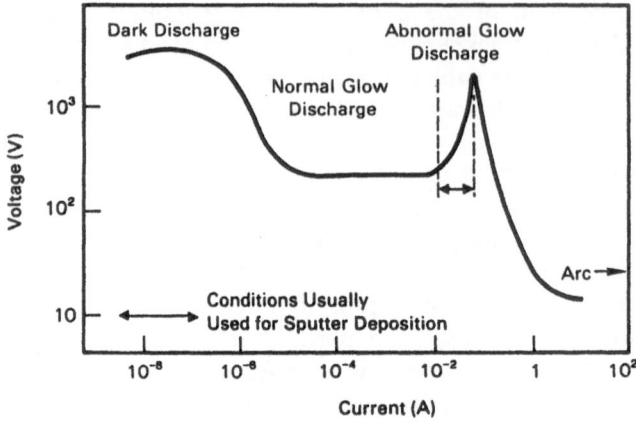

Fig. 7.1. Voltage/current characteristic for three types of self-sustained discharges ($p = 2$ to 30 Pa)

A continuous electric discharge is easily produced between an anode and a cathode mounted 1–10 cm apart in a vacuum chamber. If the chamber pressure is maintained constant between 2 and 30 Pa and a gradually increasing dc voltage is applied between the anode and the cathode, a glow discharge will occur in the gas and will stabilize when the voltage reaches a value called the normal cathode fall. The voltage will depend on the sputtering gas and the electrode material. The discharge will have five rather distinct regions (Fig. 7.2): cathode glow, cathode dark space, negative glow, Faraday dark space, and positive column.

At low power, the region of cathode glow will cover only a portion of the area of the cathode, and the discharge current will be small. As the applied power increases, the glow area and the current will increase, but the cathode current density and cathode voltage will remain constant (the normal glow discharge range). When the power is increased to the point where the glow covers the entire surface of the cathode, further power increases will result in an

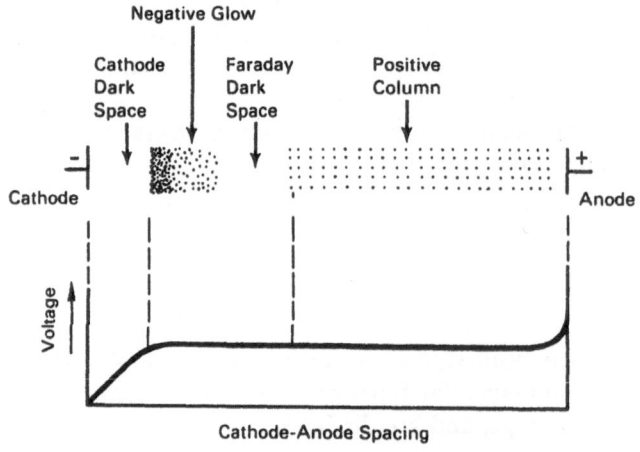

Fig. 7.2. Features of a dc glow discharge system

increase in current density and a rise in the discharge voltage. This condition is called the abnormal glow discharge and is used for sputter deposition.

If the power of a glow discharge is increased, a point on the cathode may become hot enough for thermal electron emission. This results in an arc with a low cathode fall (on the order of 10 V) accompanied by a corresponding increase in discharge current. This type of arc is a nuisance in sputtering and generates unwanted splattered particulate matter.

The cathode dark space established in a glow discharge (Fig. 7.2) is also called Crookes' dark space or the cathode fall region. The thickness of this space is approximately equal to the mean distance traveled by an electron emitted from the cathode surface before it makes an ionizing collision with a gas molecule (1–5 cm).

The negative glow region is just beyond the cathode dark space. It is composed of a mixture of ionized gas atoms and electrons in equal numbers, excited gas atoms, and neutral atoms. The negative glow is essentially electrically neutral. The decay of ionized and excited gas atoms is accompanied by photon emission.

In the Faraday dark space, the electrons lose energy by colliding with gas atoms as they travel through the negative glow. At the far end of that region, they no longer possess enough energy to either excite or ionize the gas atoms. Thus, no light is generated for a short distance.

The positive column and the Faraday dark space are not essential to discharge maintenance. In the positive column, the electric field gradient is weak, and the average charge density is near zero. Excitation and ionization occur here. While not perfectly understood, they are believed to result from fast electrons in the Maxwellian tail of the electrons ejected from the target [7.21].

7.3.1 Operation

Glow discharge begins when free electrons appear in the static electric field between the cathode and the anode. Such electrons may originate from cosmic radiation or from a very high potential momentarily applied to electrically floating vacuum chamber components. These electrons are accelerated in the electric field and gain enough energy to ionize gas atoms. The resulting ions are then attracted by the cathode. They bombard the surface of the cathode with sufficient energy to cause sputtering and the ejection of secondary electrons from the surface. These secondary or γ electrons are accelerated in the electric field to cause further ionization, at which point avalanche conditions take over, creating more ions. An ion space charge is formed in front of the cathode, resulting in a strong electric field gradient perpendicular to the cathode (cathode dark space, Fig. 7.2). The supply of electrons and the number of ionizing collisions thus stabilizes and the various discharge regions are well defined.

Attempts to find a simple relationship over the range of interest for the critical discharge parameters (cathode voltage, current density, and discharge pressure) for the commonly used sputtering gases have been unsuccessful [7.49].

The lack of success is due to an unreasonable number of additional variables (e.g., gas, gas purity, electrode material and geometry, and gas density) that can change with the discharge if the walls and electrodes are uncooled.

The current density in an H_2 or N_2 glow discharge is given by the following empirical formula [7.49]:

$$j = A_d V_c^b p^2 , \tag{7.2}$$

where V_c is the cathode potential and p is the gas pressure; A_d and b are empirical constants. This relation holds over a limited pressure and cathode voltage range and depends on the discharge gas, electrode materials, and electrode geometry.

In the glow discharge, the secondary electrons are accelerated through the cathode fall region and begin ionizing collisions at the edge of the negative glow region. Ions generated in that region are immediately accelerated toward the cathode. Ions created in the body of the negative glow move toward its edge by diffusion. When they arrive at the edge of the negative glow, they are also accelerated toward the cathode, where they produce sputtered atoms and secondary electrons.

The electrons drift through the Farady dark space and the positive column (if present) and are collected at the anode. The only regions active in producing a self-sustaining glow are the cathode dark space, and a portion of negative glow (Fig. 7.2).

In glow discharge sputtering, the substrate and/or substrate holder can serve as the anode during deposition. The role of the target (cathode) and substrate holder may be reversed for sputter cleaning of the anode, if a shutter is not available to serve as a temporary anode.

a) Target-to-Substrate Spacing

The target-to-substrate spacing is a critical parameter because of the target's dual role as a source of electrons and sputtered atoms. If we reduce the cathode-to-anode spacing under otherwise fixed glow discharge conditions, the length of the positive column shrinks, whereas the extension of all other characteristic discharge regions remains about the same. Accordingly, the anode-to-cathode spacing can be varied over a large range without affecting the main discharge features.

Usually placing the substrate as close to the target as possible is desirable to maximize the deposition rate and retain the energy of the sputtered atoms. In practice, the anode is moved well into the negative glow region to achieve a spacing that is at least two times the thickness of the cathode dark space. In conventional systems, this spacing is 1–10 cm, which permits maintenance of the self-sustained glow and provides the maximum feasible deposition rate.

For each sputter deposition application, the optimum cathode-to-anode spacing must be determined. In a planar arrangement, if the target and substrate are too close to one another, the discharge in the midportion may be reduced or even extinguished, even though there is a glow discharge around the edge. If the target and substrate are too far apart, sputtered atom scattering will become important, but the flux of atoms arriving at the substrate will be very uniform. Calculated values for the mean free path of sputtered atoms in argon show that nearly all the sputtered atoms will be thermalized before impinging on the substrate at a target-to-substrate spacing of 5 cm when the gas pressure is greater than 1 Pa [7.50].

b) Chamber Pressure

If, for a given power input, the chamber pressure is increased, the current density is increased (8.2). Hence the deposition rate increases, provided the pressure is not too high. The cathode dark space and the negative glow region contract toward the target as the pressure is increased, but expand slightly with increasing target voltage. At high pressures ($p \geq 7$ Pa), ions frequently collide with neutral atoms, and very few ions reach the target with the full energy defined by the cathode fall [7.51]. Therefore, the sputtering yield and, in turn, the deposition rates are reduced. The sputtered atoms will also collide with a very large number of gas atoms and lose energy rapidly. Some may even be returned to the target after several collisions. The number of collisions and thus the loss of energy are roughly proportional to the distance traveled. A prime reason for preserving a sizable fraction of the sputtered atom energy is derived from the increased mobility of the atoms, which has a beneficial effect on nucleation and growth.

A rather important relationship in determining the operating pressure (p) and the electrode spacings (d) in sputtering hardware geometry is the experimental fact that the breakdown voltage (V_g) for plane parallel electrodes is a unique function of the product of the gas density (n) and the spacing. This is Paschen's law [7.52], which may be written

$$V_g(p, d) = V_g(pd) . \tag{7.3}$$

Thus, the areal density of gas atoms (atoms/cm^2) between the electrodes at which breakdown occurs is a constant (Nd = constant). Figure 7.3 shows the breakdown voltage curve for argon.

To maintain a stable discharge or plasma between two electrodes at a given spacing, a minimum pressure (gas density) must be equaled or exceeded. Conversely, to exclude the plasma to prevent unwanted sputtering, there is a spacing between the target and the target shield that cannot be exceeded for a given pressure.

7.3.2 Sputter Deposition System

Figure 7.4 illustrates the principal parts of a glow discharge sputter deposition system, which are discussed in the following paragraphs.

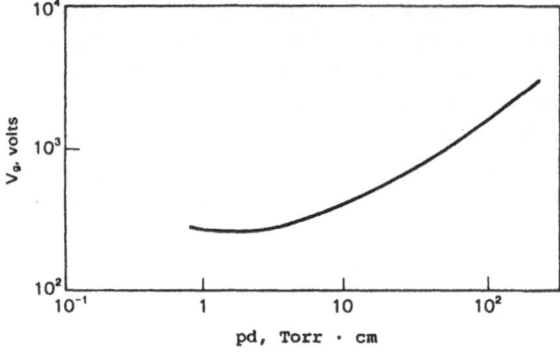

Fig. 7.3. Breakdown voltage V_g as a function of the reduced electrode spacing pd in argon

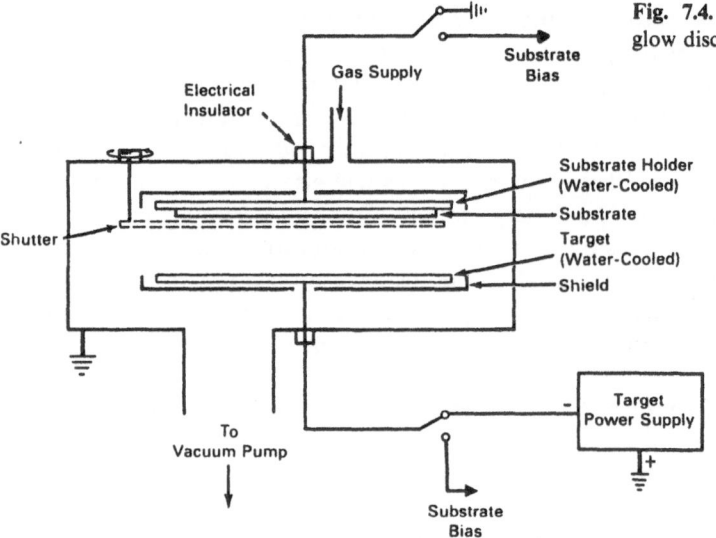

Fig. 7.4. Schematic of a dc glow discharge system

a) Deposition Chamber

Most modern chambers for research or production applications are made from stainless steel. Both in-line and batch processing are used. Reliable load locks for system entry and exit provide continuous operation for extended periods. Semicontinuous processes such as loading and unloading from cassettes are also used. The cassettes can be introduced and removed through load locks with minimal impact on the sputtering atmosphere [7.53].

The inside of the chamber should be easy to clean and have little hardware. The total surface area inside the chamber should be as small as possible, and only construction materials with low vapor pressure and low outgassing rates should be used. Corrosion-resistant steels are recommended, wherever feasible. The vacuum chamber sealing materials must be chosen on the basis of expected operating temperatures and/or the required bakeout temperature; elastomer

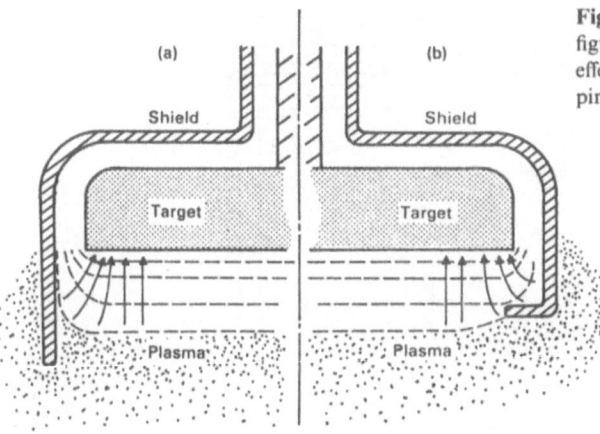

Fig. 7.5. Two different shield configurations for reducing target edge effects: (a) extended, (b) overlapping

seals and a bakeout temperature of 100°–150°C are typical. For special cases, a bakeout temperature of 400°C may be desired; metal seals must be used at such temperatures. To minimize the outgassing of the walls during sputter deposition, the chamber is often cooled externally.

After the initial pump down – and bakeout, if necessary – the pressure in the chamber is increased to 1 Pa–15 Pa by admitting an inert gas, usually argon (except with reactive sputtering). Usually a steady flow of clean gas is maintained through the chamber to flush out contaminants. The pumping capacity of the vacuum system must be adequate to maintain the desired pressure.

b) Target, Substrate, and Shutter

Sputter deposition hardware consists of the target, the shutter, and the substrate. The target is generally larger in area than the substrate, and the edges extend beyond those of the substrate if uniform deposit thickness is required over the substrate. The target surface should be smooth with no sharp points that could result in localized arcing. The target must be shielded on the back and around the edges to exclude the penetration of the plasma and prevent sputtering. The shield, which is grounded electrically, should conform to the shape of the back side of the target and extend around the edges (Fig. 7.5).

The flux of ejected atoms is nearly constant over a large planar surface except at the edges, where there is a definite increase in the ion current density caused by the ion-focusing effect of the curved equipotential surfaces around the target border. Moreover, the electric field configuration at the target edge causes the impinging ions to strike the target surface at an oblique angle, resulting in an increased sputtering yield. These effects combine to cause a greater erosion rate (7.1) of the target near the edge, creating a groove (Fig. 7.6). If the target material consists of only a thin layer on a backing plate, erosion through the target material may occur near the edge, resulting in contamination of the deposited

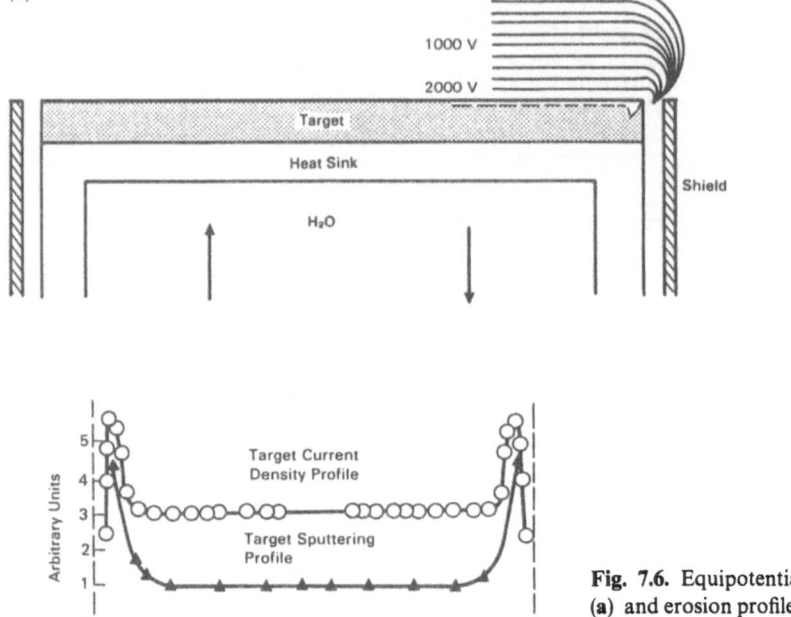

(a)

(b)

Fig. 7.6. Equipotential curves (**a**) and erosion profile (**b**) profile of dc glow discharge shielded target

film with backing plate material. To avoid this, the target material must be thicker around the edges.

The edge effect (also called the rim effect) can be reduced considerably by extending the target shield (Fig. 7.5a). An overlapping extension (Fig. 7.5b) is preferred when the target retainer ring is a different material. The shield extensions will block the ions that would otherwise strike the edge of the target at an oblique angle. This setup works well at relatively high pressures because the space-charge region (cathode dark space) does not extend beyond the edge of the shield.

To ensure a uniform ion current density with resulting thickness uniformity, the geometric relationship of the target, substrate, and shield should be as axially symmetric as feasible. If there is a lack of symmetry and some portion of the target is too close to the chamber wall, the internal jigging, or some other grounded surface, the plasma density may be distorted at that location.

The spacing between the shield and the target is important and must be carefully maintained over the entire surface (normally 1–3 mm). If, at any point, the spacing is greater than the length of the cathode dark space, a local plasma will form and enhanced sputtering will take place. Conversely, if the shield is too close to the target, arcing can result from mechanical vibration or bridging by conductive "flitters." Proper shield design and positioning will confine sputtering to the face of the target opposite the substrate.

Perhaps the most common difficulty with improper hardware spacing in the sputtering chamber is the "wall effect." As a general rule, all chamber and hardware surfaces not at the same potential as the cathode (except the target and substrate shields) should be located at a distance from the target equal to more than five times the length of the cathode dark space. If the chamber wall is too close to the target (usually the result of selecting a vacuum chamber too small for the size of the target) a serious reduction of the plasma density around the outside of the target can result from wall recombination phenomena. In addition, fewer sputtered atoms are deposited on the substrate near the edge, because of the loss of gas-scattered target atoms to the wall. The combined effect is that markedly fewer atoms are available for deposition from the space around the target edge. In extreme cases, too close a wall spacing can result in nonuniform film deposition over an area equal to 94% or more of the target area [7.54]. With proper wall spacing, uniform film deposition (within the range of 1.5%) can be obtained over the whole substrate area except for a narrow strip around the perimeter.

The chamber should be equipped with a shutter for cleaning contaminants off the surfaces of the target and the substrate before deposition of the desired film. The shutter is usually held at anode potential and therefore must be capable of dissipating the requisite power. During the initial phase of sputter deposition, any contaminants on the surface of the target will be sputtered off. After the target surface has been cleaned, the shutter is removed and the desired material deposited. The same procedure, but with inverted polarity, can be used to clean the surface of the substrate.

Three shutter arrangements can be used. The first consists of two shutters, one close to the surface of the substrate and the other close to the surface of the target. The spacing between a shutter and the surface to be cleaned must be greater than the length of the cathode dark space; if it is not, sputtering will not occur. The second arrangement consists of a single shutter that can be moved linearly so that it can be positioned close to either the target or the substrate.

A third arrangement can be used with the proper material combinations. A single shutter is positioned close to the surface of the substrate. With the shutter retracted, the surface of the substrate can be sputtered and the contaminants deposited on the surface of the target. When the substrate has been cleaned, the shutter can be interposed between the two surfaces, and the face of the target is sputtered until it is clean. In all cases, the shutter must be moved out of the line of sight between the target and the substrate after the cleaning operation. When the shutter is retracted, no part of it should be closer to the target than five times the length of the cathode dark space, as pointed out before (general rule). To reduce the space required for retraction, a shutter is sometimes made in the form of a folding fan.

A shutter may also be used as a substrate for a deposited getter if the target material is a good getter [7.55]. In such a case, a sufficiently thick layer of getter material is deposited on the shutter to reduce the partial pressure of reactive impurity gases in the system.

The substrate should be positioned at the top of the chamber above the target to minimize the possibility of dust or debris settling on the substrate, especially during the initial stages of pump down when considerable turbulence exists in the chamber. Impurities in the target could also cause defects in the deposited film. As an alternative to locating the substrate above the target, it can be placed vertically so that sputtered particles travel horizontally. However, when the substrate holder is located above the target, it is frequently cumbersome to attach the substrate. Locating the target above the substrate holder permits simple gravity-assisted placement of the substrates on the surface of the substrate holder. Usually this convenience takes precedence in system designs.

c) Vacuum Pumping and Gas Supply Systems

The vacuum system should pump the system down to 1×10^{-5} Pa or less during the initial evacuation. This preconditioning operation removes contaminants from the chamber before sputter deposition. The desired inert-gas flow is conventionally regulated by adjusting a leak valve located between the gas inlet to the sputter deposition chamber and the regulatory valve on the supply cylinder. In critical applications, high-purity gas should be used and continuously analyzed with a differentially pumped mass spectrometer. With a diffusion-pumped system, throttling the inlet to the pump with a valve or orifice is necessary to keep the pressure low enough to prevent breakdown of the top jet, which would adversely affect the pressure control and markedly increase backstreaming of oil molecules into the sputtering chamber. Vacuum systems equipped with cold traps are often provided with a throttle valve between the trap and the deposition chamber. A better scheme, suggested by *Hoffman* [7.56], places a fixed orifice between the cold trap and diffusion pump, thus retaining the full pumping capacity of the cold trap.

A bakeable, completely static system with no inert-gas flow during film deposition has also been used. The system's merit was demonstrated by producing very pure films of $Nb_{12}Al_3Ge$ with superconducting critical current densities of 4.4×10^5 A/cm^2, measured with an applied transverse-magnetic field of 10 tesla [7.57].

d) Power Supply

The power supply for a dc glow discharge system should provide a potential of 2–5 kV with a current equal to at least 1 mA/cm^2 of the target surface area. The output need not be filtered or regulated, but it should have a high reactance to prevent overloading of the power supply in case of electrical breakdown inside the chamber (arcing).

e) Electrode Temperature Control

Less than a few percent of the electrical energy supplied to the system is used to eject sputtered atoms and secondary electrons [7.13, 25]. The remainder of the energy appears as heat; about three quarters of the heat is absorbed by the target, and most of the rest is absorbed by the substrate. The target is heated primarily by ion impact. The substrate is heated primarily by electrons, the heat of condensation of deposited atoms, and radiation [7.58].

The low-pressure gaseous environment is an excellent thermal insulator; therefore, heat is dissipated from the target and substrate almost exclusively by radiation and conduction through the support structures. As a result, the temperatures of these two components can rise to harmfully high levels, allowing diffusion and/or melting unless the system design provides cooling. Overheating can occur, especially at higher deposition and power rates or when sputter deposition is used to produce thick coatings.

The amount of material removed from the target face during a typical single-film deposition cycle in microelectronic applications is very small. Thus, a target will last through many deposition cycles. The back of the target may be metallurgically bonded to the target holder, making cooling of the target relatively easy and efficient. Keep in mind, however, that the target is at a considerable negative potential. Therefore, the tubing used for circulating the coolant and the coolant itself must be of some electrically nonconductive material.

Cooling the substrate, or preventing its heating up, is often just as important as cooling the target. Since the substrate is a specially prepared item to be coated only with a film and not otherwise altered, achieving the desired cooling can be difficult. It is relatively easy to cool the substrate holder, but if there is any gap between the substrate and the holder, the low-gas-pressure environment will provide an effective heat-transfer barrier. This barrier can be overcome by coating or "wetting" the mating surfaces with a thin layer of a low-melting-point metal or alloy, for example, gallium (melting point 29.8°C) or a gallium-indium alloy (melting point 15.7°C) [7.59]. If gallium is used as a thermal bridge, the substrate and substrate holder must not react with the wetting material. Where the substrate temperature has been kept low enough (about 150°C), it is reported that a silicone-based vacuum grease can be used [7.4]. A thin sheet of a soft metal, such as indium or tin, will also provide a satisfactory heat bridge, if sufficient contact pressure can be exerted to provide surface contact on the two faces of the foil by deforming the filler metal [7.60].

Both substrates and targets can also be effectively cooled by sealing them to backing plates using gaskets or O-rings and passing a cooling fluid through the space between the two surfaces [7.61].

An alternative to cooling a substrate is reducing the amount of heating to which it is subjected. The bulk of the energy imparted to the substrate comes from plasma electrons; if the incidence of the electrons is reduced, the heating rate will also be reduced. A weak magnetic field can be applied parallel to the substrate surface to reduce the heating rate.

If the target is not cooled, the surface may be vaporized, increasing the rate of film growth considerably. With sputter deposition of alloys, one component may be preferentially sputtered. If the temperature of the target rises high enough, the preferentially depleted component could be continually replenished by diffusion of the component to the surface. As a result, the sputter-deposited film would have a composition considerably different from that of the target material [7.62].

Low target and substrate temperatures are advisable to simplify the control process because the gas density and hence the deposition rate could be affected. However, for the substrate, each application will probably dictate its desired temperature, because the microstructure, composition, and stress in the deposit produced is usually very temperature dependent.

7.3.3 Biased Sputter Deposition

In ordinary glow discharge sputter deposition, the deposited film may be contaminated by incorporation of gas atoms. Such contamination can occur even though the sticking coefficient of argon and other inert gases having thermal energies is theoretically zero at ambient temperature [7.63]. *Kornelsen* showed that if monoenergetic inert-gas ions or atoms arrive at the substrate with energies of 40 eV or more, they may penetrate solid surfaces and become trapped [7.64, 65].

The structure and composition of sputter-deposited films can be markedly altered by negatively biasing the substrate with respect to the anode [7.66, 67]. If the substrate is negatively biased with respect to the anode, it becomes a secondary cathode and is surrounded by a dark space, which is similar to the cathode dark space. The film will then be continuously bombarded by ions. The impinging ions may sputter contaminants from or be incorporated in the growing film. The magnitude of the substrate bias potential is important. For sputter deposition of a nickel film in argon, the optimum bias voltage was found to be -75 V with respect to the anode [7.60].

Contaminants originating in the residual gas may also be removed by bias sputtering. A more important one is oxygen, the major contaminant in sputter deposition. Such impurities can sometimes be removed by preferential sputtering, depending on the relative strengths of the bonds [7.4]. With oxygen, for example, the oxygen-to-metal bond is weaker than the metal-to-metal bond for metals such as tantalum, molybdenum, and niobium. In the deposition of these metals, biased sputtering is very effective in producing high-purity films. With aluminum and magnesium, however, the oxide bond is stronger than the metallic bond, and biased sputtering is of little help in producing oxygen-free films.

Another feature of biased sputter deposition is that atoms that adhere weakly to the substrate will be sputtered off and atoms that adhere tightly will not. As a result, a denser, more tightly bonded film can be obtained when the substrate is properly biased. Biased sputtering can also exert an influence on the density and crystalline structure of the deposited film [7.42]. It has also been found to have a very definite effect on the magnetic properties of the deposited film [7.68, 69]. Titanium dioxide deposits have been made for potential coatings for high-energy laser components. Only by using bias sputtering could a smooth nonscattering surface be obtained [7.45]. Bias sputtered deposits did not have the usual columnar grain structure with the associated cobblestone effect at the surface that is usually responsible for light scattering.

7.3.4 Sputter Deposition of Alloys and Compounds

Often the film to be deposited is an alloy or a compound. Many such materials can be successfully sputter deposited from targets of the desired film material. However, with some alloys and compounds satisfactory film quality can be achieved only with considerable difficulty or not at all. Sputtering of compounds is complicated by large differences in (i) the sputtering yields of the components and their sticking coefficients, (ii) their chemical reactivity, and (iii) their physical properties (vapor pressure and electrical and thermal conductivity). If one element of the compound is a gas, it may be necessary to add the element to the sputtering atmosphere to ensure stoichiometry, further complicating the process since its concentration must be carefully controlled. Note that some materials cannot be obtained – or can be obtained only at prohibitive cost – in the proper size and shape to form targets with the desired purity throughout. For these and similar cases, multiple targets, multicomponent targets, or reactive sputter deposition may be the answer.

a) Sputter Deposition of Alloys

Ordinarily, a metal alloy film can be sputter deposited from a target of the desired composition. Because different constituents usually have different sputtering rates, target preconditioning is necessary. After initial deposition of a film rich in the high-yield component onto the shutter, a balance in the composition of the target surface material will be reached. The shutter can then be removed and a film that duplicates the composition of the target material will be deposited on the substrate [7.66]. However, conditions that can affect the film composition include diffusion in the target and surface segregation [7.19], different sticking coefficients of the alloy components [7.70], and preferential sputtering if the substrate is biased negatively. Another problem not solved by preconditioning is the different ejection pattern produced when an alloy consisting of a blend of light and heavy atoms is sputtered (Sect. 7.2).

One alternative to an alloy target is a group of targets for each component. The target area for each alloy component will determine the composition of the deposit. The target/substrate arrangement must also be designed to give the desired alloy composition on the substrate. Another approach is to rotate the substrate past each target in turn [7.71, 72]. This method can also be used to produce films in which the alloy content can be varied in a predetermined pattern [7.73]. If the rotation speed is too low, the substrate may be coated with discrete layers of different materials. Thus, if a laminated film is desired, it can be obtained by adjusting the speed, target areas, and target voltages [7.74].

Another way to achieve a homogeneous film is to use a single multicomponent target rather than a group of targets. The target material may be composed of a carefully blended mixture of powders of the different materials pressed into a compact mass by very high pressures. Or the target may be fabricated from the principal material to be sputter deposited; plugs or strips of alloying elements can then be inserted into close-fitting holes or grooves. This method avoids the problem of impurities and particulate contamination associated with pressed powder targets. For parallel strip sources, *Dahlgren* [7.75] found that a strip spacing-to-target/substrate distance ratio of ≤ 0.5 would produce a composition variation of less than 0.01% in a binary alloy.

b) Deposition of Compounds by Reactive Sputtering

Sputter deposition has been used successfully to produce thin films of many compounds. Much of the work has been devoted to oxides, nitrides, and sulfides, with the bulk of the effort on oxide films. A few examples reported in the literature are the deposition of alumina films [7.76], zirconia films [7.77], zinc oxide films [7.78], vanadium oxide films [7.79], nitride films [7.80, 81], and sulfide films [7.47, 82, 83]. Figure 7.7 illustrates the variety of materials (glass, crystals, plastic), shapes, and sizes that can be coated by "reactive sputtering" of metal targets. The notation "reactive sputtering" refers to the case where neutral, excited, or ionized gaseous species react with the target, sputtered particles, or substrate. Reactive sputtering of elemental targets has the advantage that multilayer compound coatings can be produced simply by changing the sputtering gas [7.84].

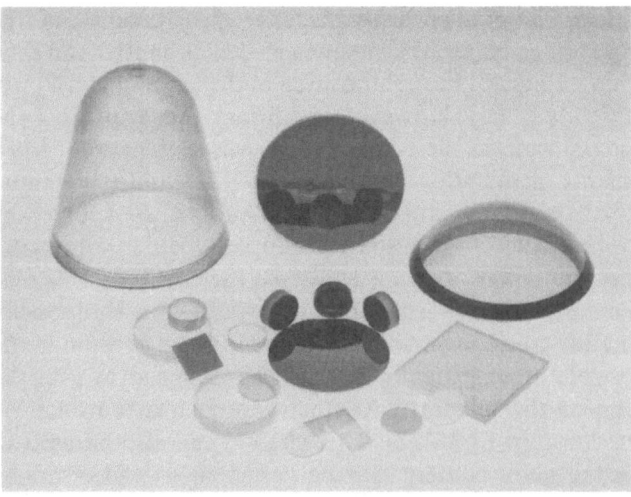

Fig. 7.7. Optical coatings prepared by reactive sputtering of metal targets at Pacific Northwest Laboratory

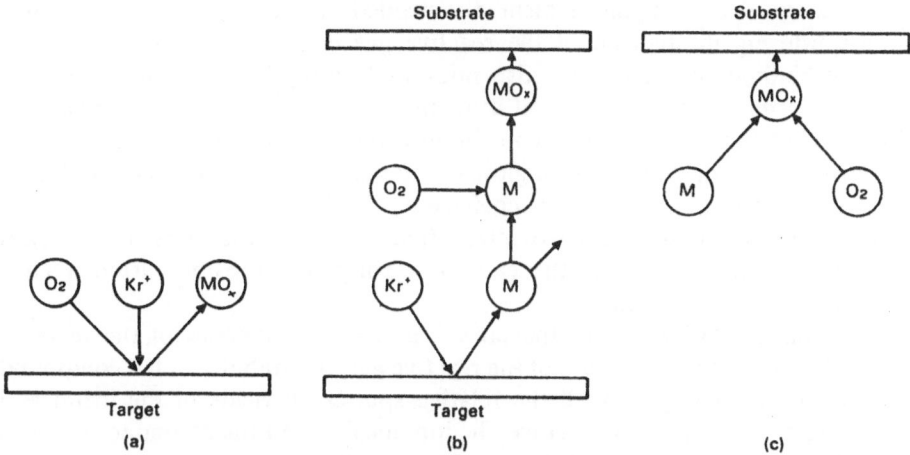

Fig. 7.8. Three different mechanisms for reactive sputter deposition [7.86]: (a) at the target, (b) in the plasma volume, and (c) at the substrate

Nonconducting compounds are usually deposited by rf sputtering. It is possible to deposit very thin oxide films from anodized metallic targets [7.85] in a dc glow discharge when the anodized layer is a photoconductor. Thin films of oxides, nitrides, and sulfides can also be deposited by reactive sputtering in a chamber atmosphere consisting of reactive gas (O_2, N_2, or H_2S) mixed with inert gas. The target material may be the desired film compound or the metal of which the desired compound is to be formed.

Holland [7.86] listed three ways the active gas could combine with the metal to form a sputtered gas-metal film (Fig. 7.8): (a) at the surface of the target, (b) in the space between the target and the substrate, or (c) on the surface of the substrate. *Burkhardt* and *Gregor* compared the properties and stoichiometry with the sputter deposition parameters for Si_3N_4 produced by reactive sputtering and rf sputtering in Ar and N_2 [7.87]. They concluded that some of the N_2 in the film had to come from the target; thus, the first mechanism might predominate.

To get a better picture of the reactive sputtering mechanism, *Bomchil* et al. [7.88] studied the sputtering of In in Ar–O_2 mixtures in a dc triode configuration that allowed variation of certain sputtering parameters such as target voltage and current independently of one another. For a sputtering system operating at a fixed target voltage, they concluded that the initial O_2 partial pressure could be varied from zero to a characteristic pressure (p^*), below which only metal atoms are sputtered. The pressure is related to the abrupt steplike drop in the sputtering rate and can be estimated from a semiempirical equation derived by *Abe* and *Yamashina* [7.89]. *Shinoki* and *Itoh* [7.90] proposed a model that takes into account the reaction at the target and the gettering action of the sputtered atom deposit.

By measuring the Ar^+ sputtering yields for Ti, Ta, Mo, and W as a function of the partial pressures of reactive gas (O_2 or N_2), *Hrbek* was able to show that "the reactive sputtering mechanism is controlled simultaneously by the adsorption-sputtering mechanism of the reactive gas adsorbed, by changes in the surface binding energy due to adsorption and finally by collision mechanics" [7.91]. *Goranchev* et al. proposed a qualitative physical model to explain the influence of the oxygen content in the gas flow on the discharge current for reactive sputtering of Si in a dc glow discharge [7.92]. Their model took into account the abrupt change in the coefficient of secondary ion-electron emission when the target surface became oxidized. It also included the effect of the oxygen content on ionization near the cathode during the transition from a glow discharge in Ar to one in O_2.

Chemisorption occurs at the target surface. The kinetics of the reaction depend on the partial pressure of the reactive gas, the stability of the compound or compounds being formed, the relative sputtering yields of the metal and compound, the target temperature, the ion density, and the secondary electron emission coefficient [7.88, 92]. For low reactive gas pressures and high sputtering rates, the reaction takes place at the substrate. For high reactive gas partial pressures, the ratio of the compound formed at the substrate to that sputtered from the target needs to be measured experimentally.

The second mechanism (Fig. 7.8b) can be ruled out if the pressure is low and the target-to-substrate spacing is small. For pressures in the 5–10 Pa range, the compound could be formed in the target/substrate gap, depending on the activated species.

Schwartz [7.93] and *Hollands* and *Campbell* [7.94] believe that the third mechanism is the most likely. *Winters* et al. [7.95] proposed a simple model based on incorporation of the species at the substrate for predicting the composition of sputtered multicomponent thin films. Those results cannot be compared with experiments because the uncertainty in such parameters as the sticking and sputtering coefficients is too large.

Kelly and *Lam* [7.85] discussed the sputtering behavior of 33 oxides. They found that the wide variation of sputtering yields could be explained if they divided the compounds into three groups according to whether the yield was most dependent on collisional, thermal, or oxygen sputtering. *Schiller* et al. [7.96] have shown that for high target power densities obtained with magnetron sputtering, target surfaces remain unoxidized even at high O_2 partial pressures. To produce stoichiometric oxide films under these conditions, it is necessary to maintain the proper oxygen-metal atom arrival ratio and substrate temperature. *Betz* and *Wehner* [7.36] conclude that, despite the large amount of experimental data, multicomponent sputtering is far from being completely understood.

7.4 dc-Supported Plasma Discharge

The gas pressure required to maintain a conventional dc glow discharge with the required supply of electrons and ions can be 2–4 Pa or greater. If electrons are supplied from a separate source, the glow can be maintained at a pressure more than two orders of magnitude lower, down to 0.013 Pa [7.97]. The plasma density can be increased by more than a factor of 10 over that for a glow discharge, and it can be made independent of the voltage applied to the sputtering target.

7.4.1 General Description

In dc-supported plasma discharge, a cathode is heated to a temperature at which the required electron emission takes place (Fig. 7.9). The external discharge power supply maintains the current flow between the cathode and the anode. The electronic space charge usually associated with thermionic emitters in vacuum tubes is neutralized by plasma production, thus reducing the voltage required to produce the full emission current. The positive-ion space charge close to the cathode acts as a virtual anode and permits full emission current to flow with a cathode-plasma voltage as low as 20 V. Higher voltages can produce even greater currents due to the effect of secondary electron emission resulting from increased cathode heating by ion impact.

As Fig. 7.9 shows, the plasma will extend from near the cathode to the anode. A secondary cathode (target) immersed in the plasma column and maintained at a relatively large negative potential will attract ions from the plasma. The ions will strike the target surface with energies determined by the target potential. Since the potential relative to the target anode has essentially no effect on plasma generation and maintenance, it may be adjusted independently to vary the energy of the incident ions. The substrate in a supported plasma discharge is not the anode, so we have greater freedom of design and control. In addition, the substrate may be placed closer to the target than the length of the mean free path of the gas molecules or the sputtered atoms [7.98]. Therefore, the sputtered atoms will arrive at the substrate with higher average energy, having been subjected to fewer collisions.

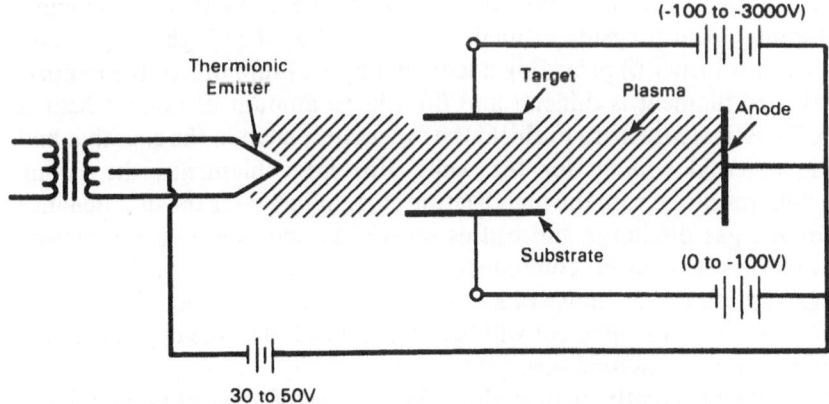

Fig. 7.9. *Thermionic dc-supported plasma discharge (triode sputter deposition system)*

These conditions provide a cleaner film and one that may be more adherent. However, note that thermionic emitters can be another source of impurities [7.99].

Because of the third electrode in the form of the target, supported plasma discharge systems are often called triode systems [7.100]. Sometimes an additional positively charged electrode, or grid, is added a short distance in front of the cathode. The electrode increases the ionization efficiency, permitting an even further lowering of the chamber pressure than is possible with the triode system. This approach has been called tetrode sputtering [7.101].

There are advantages and disadvantages in lowering the pressure or the concentration of gas atoms in the chamber. On the plus side, there will be fewer energy-robbing impacts and fewer deflected atoms. On the minus side, there will be fewer gas atoms to ionize and more energetic neutral atoms. With the longer mean free path that accompanies the reduced pressure, a larger number of energetic neutral atoms will strike the surface of the substrate with energies equal to the target potential [7.102]. Such energetic impacts can result in sputtering of the deposited film, produce surface defects, and entrap gas atoms. Thus, a balance must be determined experimentally between the extremes of chamber pressure to provide the best combination for a particular film deposition application.

7.4.2 Refractory Metal Hot Cathodes

Refractory metals may be used as the emitter for supported plasma discharges. The temperature at which appreciable electron emission occurs is different for different metals, and only tungsten has been found to have the physical properties to last for a reasonable length of time when used as a thermionic emitter. Since the threshold energies for tungsten sputtering fall in the range 35–50 eV, where the yields are 10^{-3} atoms/ion or less [7.25], tungsten can tolerate ion bombardment in this range for periods in excess of 50 h for a filament with practical dimensions.

Refractory metal cathodes have limitations. Satisfactory thermionic emission is obtained from tungsten around 2300°C [7.103]. The high temperature produces two problems: (i) providing adequate physical support for the temperature-weakened filament is difficult and (ii) a large amount of radiant heat is produced. Thus, it is necessary to shield the other equipment in the chamber and to provide external cooling for the walls. For large filaments, the strong magnetic field resulting from the high heating current increases the maintenance voltage for the gas discharge and causes severe inhomogeneity in the plasma distribution. These two effects combine to increase the erosion of the filament by sputtering. Another disadvantage of a refractory metal hot cathode is that some atoms sputtered from the filament will become embedded in the deposited film, adding to any other contaminants.

These problems greatly reduce the effectiveness of refractory metal hot cathodes. The problems are complex and satisfactory solutions have not yet been discovered. As a result, such systems are not widely used in sputter film deposition.

7.4.3 Oxide-Coated Hot Cathodes

Vacuum tubes use heated oxide-coated emitters because they produce electrons more efficiently than bare metal. The thermionic work functions of the oxide materials are considerably lower than those for refractory metals. As a result, the temperatures required to produce thermionic emission are much lower. However, the oxide coatings are easily poisoned and require special care.

There has been some success with a special oxide-coated cathode (from a commercial thyratron) [7.104]. The oxide emitter was at all times protected from contact with air during system downtime by carefully containing it in a dry inert atmosphere. In addition, the emitter also had to be thermally reactivated in vacuum before each use. Because of the special care required and many difficulties, oxide-coated hot cathodes are not used in commercial sputter deposition systems.

7.4.4 Thin-Film Hot Cathodes

A thin-film hot cathode is composed of a filament coated with a thin film of a material with a low thermionic work function. Under ordinary conditions, the thin-film coating would be depleted by evaporation and/or ion bombardment.

However, a recently developed method greatly extends the life of the thin film, continuously replenishing it [7.105, 106]. The new method incorporates an auxiliary electrode (target) of the low work function material located adjacent to the filament and maintained at a potential negative to that of the anode (Fig. 7.10). Atoms are sputtered from the secondary target surface and deposited on the hot filament surface at the same rate as atoms are sputtered and evaporated from the cathode surface. This procedure results in a continuously maintained electron-emitting film on the hot cathode surface.

The base metal of the filament is usually tungsten; the coating material may be uranium, thorium, cerium, lanthanum, or some other low thermionic work function material that has a low vapor pressure at the operating temperature. The combination of the low work function material and the tungsten base metal may produce a material with a work function even lower than either of the individual materials. For example, the work functions of tungsten and thorium are 4.5 eV and 3.35 eV, respectively [7.107], but the work function of W-Th is only 2.7 eV [7.108].

The continuous deposition of a thin-film emitter on a tungsten filament produces a cathode superior to either a bare or a thoriated tungsten emitter in a gas discharge. The peak electron emission density for thoriated tungsten is $\sim 3 \text{A/cm}^2$ at 2000 K in a vacuum of less than 1.3×10^{-4} Pa; this value drops to a few mA/cm^2 (about that for pure tungsten) at 2000 K in a rare-gas plasma with a 30–40 V negative bias applied. Using a replenished thin-film uranium-tung-

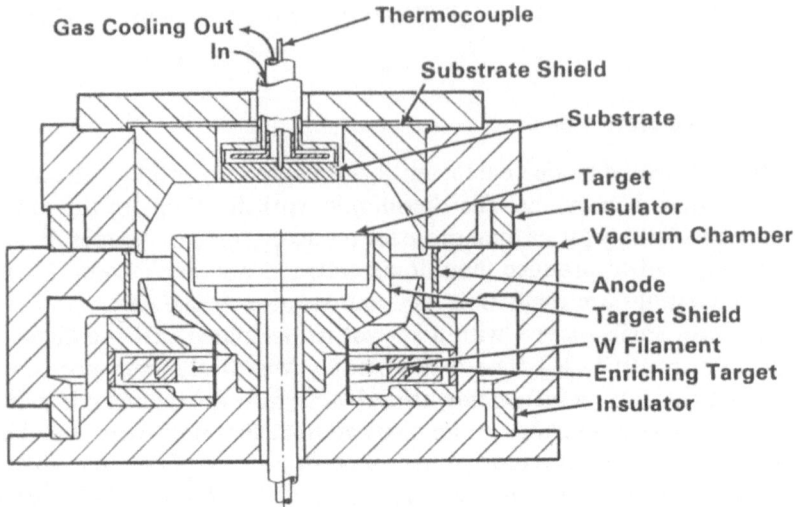

Fig. 7.10. Cross section of a thermionically supported discharge sputtering apparatus with a replenished thin-film hot cathode

sten cathode, electron emission densities of 5 A/cm^2 have been obtained with a thermal efficiency of ~100 mA of electron emission per watt (the power includes filament heating and atom replenishment) [7.109].

Use of a replenished thin-film hot cathode introduces contaminants into the chamber from the sputtering and evaporation of the thin filament coating, as well as stray sputtered material from the auxiliary cathode. However, most of the material can be controlled by the location of the cathode and strategically placed shielding (Fig. 7.11). Many of the low work function materials used to form thin films on hot cathodes are good gettering agents and reduce the amount of other contaminants, especially oxygen, inside the chamber.

The advantages of the thin-film hot cathode (high emission density, high thermal efficiency, long life, low induced magnetic field, and high gettering capacity) outweigh the complication of an extra electrode/power supply combination to make this the most effective method of supported plasma sputter deposition.

The freestanding hollow turbine blades shown in Fig. 7.12 were made in sputtering equipment using a thin-film cathode, permitting a deposition rate of greater than 50 µm/min. Coating mandrels were prepared by brazing airfoil sections to modified commercial turbine blades. The mandrels were polished and rotated over a sputtering target until ~1 mm of stainless steel had been applied. The blades were completed by removing the mandrels with nitric acid. One of the blades shown in Fig. 7.12 was sectioned to reveal its structure.

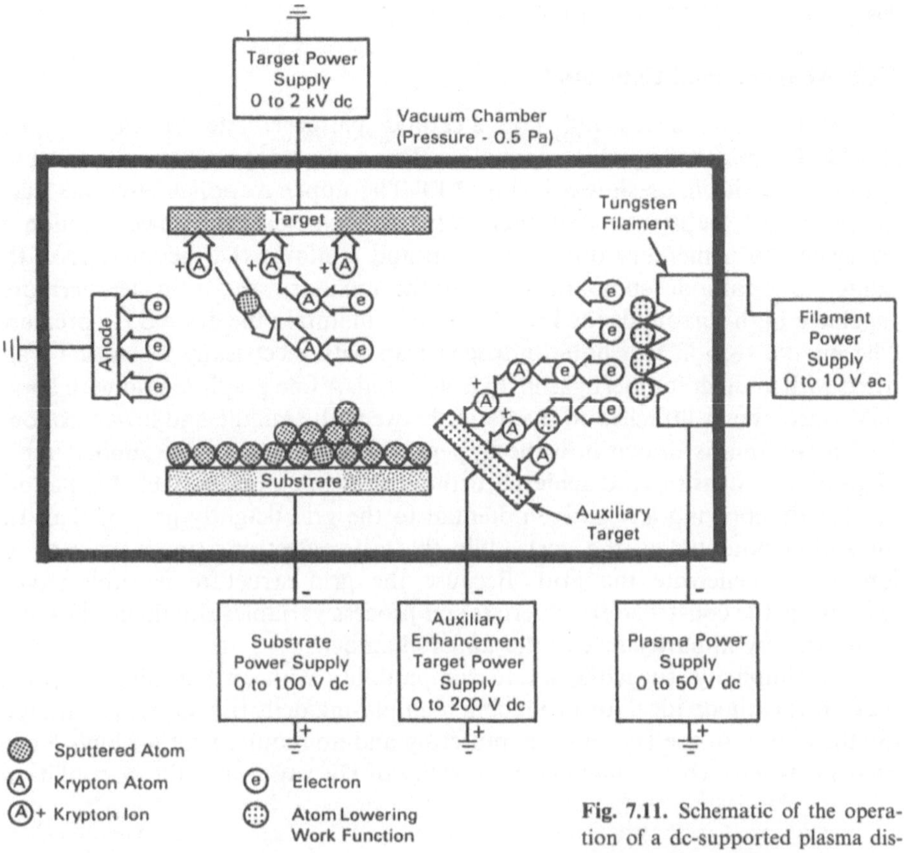

Target Power
Supply
0 to 2 kV dc

Vacuum Chamber
(Pressure - 0.5 Pa)

Target

Tungsten
Filament

Filament
Power
Supply
0 to 10 V ac

Anode

Substrate

Auxiliary
Target

Substrate
Power Supply
0 to 100 V dc

Auxiliary
Enhancement
Target Power
Supply
0 to 200 V dc

Plasma Power
Supply
0 to 50 V dc

Sputtered Atom

Krypton Atom

Krypton Ion

Electron

Atom Lowering
Work Function

Fig. 7.11. Schematic of the operation of a dc-supported plasma discharge with a continuously replenished thin-film hot cathode

Fig. 7.12. Turbine blades fabricated by the soluble mandrel method

7.4.5 Mercury-Pool Cathodes

Wehner [7.41] has successfully used a cathode patterned after the mercury-pool rectifier for sputtering yield experiments. This equipment could be modified for sputter deposition, as shown in Fig. 7.13. The upper section houses the main anode, target, substrate, and their respective shields. The lower section is connected to a mercury diffusion pump and contains the mercury pool, the igniter, the cathode spot anchor, and the auxiliary anode. A temperature-regulated bath surrounds the lower section to maintain the desired Hg pressure. The middle section is a liquid-nitrogen trap with electrically insulating walls fitted with an inlet for inert gas if Hg is not used. A fine graphite grid with about 50% open area is attached to the flange between the middle and lower sections.

The plasma is drawn into the upper section with the main anode supply. High plasma densities are achieved without drawing an excessively high anode current by applying a negative potential to the grid (slightly greater than the ionization potential of the gas). Only the faster electrons (more efficient for ionization) penetrate the grid. Because the grid structure is subjected to sputtering, the construction materials and process variables should be chosen to minimize any impurity flux to the upper chamber.

The simplicity, ruggedness, and compatibility with an Hg diffusion pump make this cathode ideal for producing high plasma densities. Other advantages are the option to use Hg ions for sputtering and no requirement for high-purity inert gases. Drawbacks include the toxicity of Hg vapor and the possibility of

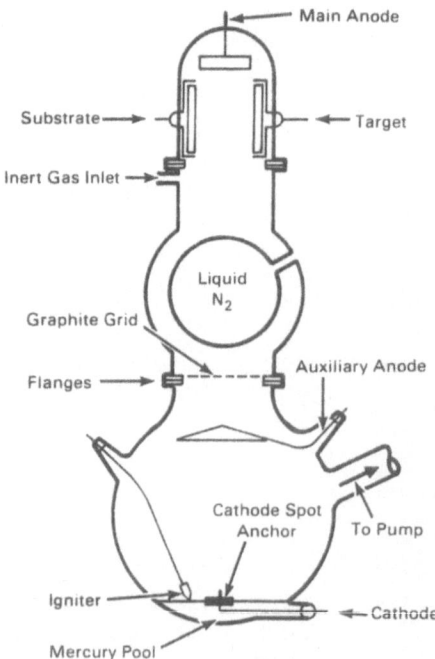

Fig. 7.13. Mercury-pool cathode sputtering apparatus

forming the very explosive mercury azide, should N_2 be inadvertently introduced during operation.

In general, the mercury-pool cathode is a very useful research tool because of the ease of operation and short cycle times. It is not used, however, in any commercially available sputtering equipment.

7.5 Radio-Frequency Discharge

Perhaps the greatest restriction in the use of a dc glow discharge sputter deposition system is the very limited ability to deposit films of electrically nonconducting materials. Some nonconducting films can be sputter deposited by dc reactive sputtering, and very thin films can be sputter deposited from targets thin enough to permit charges to be drained off. However, neither method is satisfactory for depositing relatively thick films of dielectric or most other insulating materials.

The electrically insulating layer of target material can be considered as a sheet of dielectric between the two plates of a parallel-plate capacitor. The metal backing of the target acts as one plate and the plasma acts as the other. If the insulating layer is immersed in a dc plasma, it will float at a negative potential because the electrons have a much higher velocity than the slow-moving positive ions. As the surface charges up negatively, however, plasma ions will arrive at a rate compensating for the electrons, since the net current to the surface must be zero. The surface assumes a potential called the floating or wall potential. The incoming ions will have a very low energy arriving at the surface ($\sim$ 3–10 eV) [7.110], i.e., below the sputtering threshold [7.25]. The energy can be increased only by increasing the electronic charge on the insulating surface. This can be achieved by taking advantage of the different mobilities of the electrons and ions. If an alternating current is applied across the capacitor, a supply of electrons will flow alternately from one surface of the target to the other. When the electrons flood the side of the target facing the plasma, the surface will become negatively charged, but the charge will be quickly compensated for by incoming ions. However, if the frequency of the applied current is high enough, it will take nearly all of the cycle for the electronic charge to be neutralized by the ions, and the small remaining portion of the cycle will recharge the surface with electrons. The amount of charge accumulated on the surface depends on the peak-to-peak rf potential applied to the target.

7.5.1 Technological Features

The rf method for obtaining and maintaining an electronic charge on an insulating surface was first proposed by *Anderson* et al. [7.111] and was later developed by *Davidse* and *Maissel* [7.112, 113]. Calculations suggested that the frequency should be 10 MHz or more, because at lower frequencies the average energy of the bombarding ions would be markedly reduced as a result of positive charge accumulation [7.111]. If the frequency becomes too high (for example, > 100 MHz), the ions cannot respond to the rf field and again the sputtering rate diminishes. In practice, the permitted industrial frequency of 13.56 MHz is usually used.

This rf sputter deposition method is also widely used for producing films of metals, alloys, and semiconductors. By capacitively coupling the power supply to a metal target electrode, the metal surface may be negatively charged and therefore sputtered in the same manner as the insulating surface. The capacitor in series with the target serves the same function as the dielectric plate target discussed above.

The mechanism whereby ions are formed in the rf discharge system is different from that in the dc glow discharge system. In the rf system, the plasma is generated primarily by electrons oscillating in the rf field. As a result, the high cathode voltage and secondary electrons required for plasma maintenance in the dc glow discharge are not needed in an rf system. The average energy of the incoming ions is comparable with the energies obtained in glow discharge sputtering. Therefore, secondary electrons still contribute to ion production, but not in a major way. The ionization mechanism also permits the discharge to be maintained down to a chamber pressure of 0.1 Pa.

Although an rf sputtering system allows more flexibility in the choice of materials that can be sputtered and can be operated at lower chamber pressures, it is generally more complicated than a dc glow discharge system. The system components have to be carefully designed with attention to eliminating stray capacitances and inductances that may generate fields and thus produce unwanted sputtering of chamber components other than the discharge electrodes. Furthermore, it is frequently difficult to match the impedance of the highly reactive load presented by the discharge system to the rf generator. An impedance-matching network, usually a tunable LC network, is used between the rf generator and the discharge system. Information regarding component dimensions and layout of impedance-matching networks is available in the literature [7.114, 115].

7.5.2 Applications

The rf sputter deposition technique is universally applicable to the deposition of films of metals, semiconductors, and insulators, and is widely used in industry [7.116]. It has also been used to deposit thin films on surfaces of such small articles as special screws [7.117]. An rf-powered flat rectangular target was positioned on the axis of a horizontally positioned barrel [7.118], and the parts were placed in the bottom of the barrel, which was then rotated. The tumbling action presented all surfaces to the sputtered target, and the metal film coatings varied no more than ± 10% in thickness.

7.6 Magnetically Enhanced Discharge

Most of the more recently developed deposition approaches deal with various ways to obtain high plasma densities. Plasma density can be increased by applying magnetic fields in selected configurations to the various discharge arrangements already discussed. For a given method of sputter film deposition, the rate of ionization can be increased by creating an environment such that the number of electron-atom collisions will be greater than it would otherwise be. Two methods can be used: (i) The distance that an electron travels in its passage from the cathode to the anode can be increased, thus enhancing the probability of an ionizing collision; or (ii) the electrons can be constrained from collision with the internal surfaces of the apparatus. Magnetic fields help with both methods.

If an electron moves in a path perpendicular to a magnetic field, it will experience a force in a direction perpendicular to both the field and the velocity. The acceleration perpendicular to the velocity does not cause a change in the magnitude of the velocity but merely changes its direction; i.e., the electron will move in a circular path rather than in a straight line. This movement will greatly increase the total length of travel and the probability of an ionizing collision. If the magnetic field is not perpendicular to the direction of motion, the electrons are forced to spiral around the magnetic flux lines. Simultaneously, they will be prevented from moving very far in any direction perpendicular to the flux lines. This method of containment has been termed the magnetic bottle effect [7.51].

The radius of the helix will decrease and the total distance traveled by an electron will increase as the strength of the magnetic field is increased. In addition, since the radius of the helix is inversely proportional to the mass of the charged particle, only the electrons will be noticeably affected.

Penning first proposed the use of a magnetic field to increase the plasma density and thus the rate of sputter deposition of thin films [7.119]. Since then, magnetic fields have been applied to various modes of sputter deposition.

7.6.1 Applications

The proper application of a magnetic field to a dc glow discharge system has produced a number of beneficial results. It has been reported that the ionization rate has been increased by as much as a factor of 50 [7.54] and that the rate of film deposition has been increased by as much as 30 times [7.120].

A magnetic field can be superimposed on a glow discharge system either longitudinally (axially) or transversely. A longitudinal magnetic field will have flux lines parallel to the path of the electrons as they pass through the cathode fall and will not affect the direction of travel nor the path length of the electrons in that region. However, when the electrons reach the negative glow region, where there is no electric field, the electron motion becomes random due to the many impacts with atoms and other electrons. Electrons with velocity components transverse to the axis will then be affected by the magnetic field. The path lengths of these electrons will increase, and the electrons will be contained within the negative glow space (magnetic bottle effect). This increases the ionization rate and, consequently, the ion density in the portion of the negative glow near the cathode dark space. Moreover, the thickness of the cathode dark space is reduced. The net effect is an increased sputtering rate and an increased film deposition rate.

A transverse magnetic field, normal to the cathode fall electric field, will act on the electrons passing through the cathode dark space as well as those in the negative glow region. The longer electron paths will produce corresponding increases in the rates of ionization and sputtering, significantly greater than the increase obtained with the longitudinal magnetic field. However, a transverse magnetic field will concentrate the discharge at one side of a planar electrode (target), producing a drastic imbalance in the sputtering rate over the planar target surface. This imbalance will show up in a lack of uniformity in the deposited film [7.121].

The disadvantages of a unidirectional transverse mangetic field can be overcome by using a radially directed magnetic field. Such a field can be produced by using two solenoids connected to oppose one another. The resulting field has been termed a quadrupole magnetic field (Fig. 7.14). This system has been tested with a reported 30-fold increase in the rate of film deposition and relatively uniform thickness [7.120]. However, an adequate magnetic field cannot be obtained in any but very small chambers without the use of excessively large solenoids, and this technique has not seen broad application.

Since the supported plasma discharge has all the advantages of the magnetic-field-enhanced plasma, magnetic fields are not normally used to increase plasma density. Generally, any applied magnetic field will distort the plasma and usually increase the discharge voltage drop. Target erosion rates become nonuniform for practical target shapes and sizes. A magnetic field parallel to the discharge axis may be used to reduce wall recombination, thus increasing discharge efficiency and plasma density near the deposition electrodes. Such fields are usually generated by two magnetic coils on the outside of the sputtering apparatus described by *Kay* [7.54].

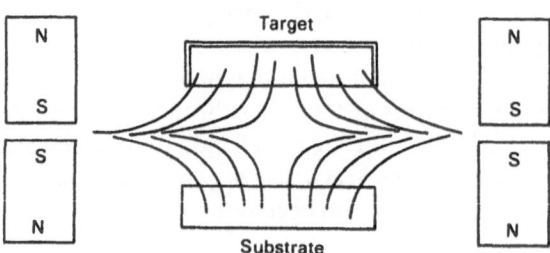

Fig. 7.14. Sputter deposition in a quadrupole magnetic field

As to the application of a magnetic field to an rf discharge (Sect. 7.5), electrons in an rf field can gain ionization energy only when they remain in the rf field. The field exists primarily between the two electrodes. If an electron escapes from the space between the two electrodes because of diffusion or a random collision, it will no longer gain energy from the rf field and will, therefore, be lost to the glow. A magnetic field with its flux lines parallel to the rf field can constrain the electrons and keep them in the rf field. As a result, a magnetic field can be an important adjunct to an rf discharge system, although the magnetic field may interact with the rf electric field. Therefore, system geometry and materials must be considered to guard against possible detrimental plasma distortion.

7.6.2 Magnetron Mode of Sputter Deposition

In ordinary glow discharge with planar electrodes, most of the ions generated in the negative glow impinge on the cathode and produce secondary electrons. However, there are also a number of metastable atoms and photons in the negative glow that can produce secondary electrons when they impinge on the target surface. These components have no charge and are not attracted to the cathode. As a result, they travel in random directions, and most never strike the target. However, if two planar cathodes are brought close enough together so that the two negative glows merge, almost all of the photons and metastable atoms, as well as the ions, will impinge upon the cathodes, and the rate of emission of secondary electrons will markedly increase. In addition, each electron will also make more ionizing and exciting collisions, thus multiplying the effectiveness of the discharge and causing a considerable increase in the current density (by orders of magnitude with the same voltage). Results are similar whenever the negative glow from one part of a cathode surface overlaps that from another, increasing the effective use of the uncharged particles. This phenomenon is known as the hollow-cathode effect [7.122].

Theoretically, the greatest benefit from the hollow-cathode effect is obtained by sputtering inside a hollow cylinder so that no particles can be lost in a radial direction. However, in actual usage, the performance of a hollow cathode is limited by end losses. The end losses can be almost eliminated by applying magnetic fields or by using properly designed electron reflecting surfaces. Doing this achieves high currents and high sputter deposition rates [7.123]. Such systems are called magnetron sputtering sources and are defined as diode devices in which magnetic fields are used with the cathode surface to form electron traps. These are configured so that the $E \times B$ electron drift currents can close on themselves [7.124].

Magnetron sputtering has been applied to two general types of target geometries: cylindrical and planar. The cylindrical mode of sputter deposition was first used by *Penning* and *Moubis* [7.125, 126]. Increased plasma densities were achieved by using two magnetic coils in a Helmholtz arrangement outside the vacuum vessel. Incentives include increased sputtering rates, large-area deposition, uniform deposits on more complex substrate geometries, lower substrate heating rates, and improved deposit properties. This approach is restricted to applications where targets can be easily fabricated.

Much recent work has centered on dc operation of the configurations above. Investigators have reported on the basic post configuration [7.127–129], the alternative post assembly [7.128–131], the basic hollow-cathode configuration [7.127, 129], and the modified hollow cathode [7.129]. In addition, investigative work has been reported on the rf operation of post and hollow-cathode configurations [7.66, 129].

a) Cylindrical Magnetron Sputter Deposition

Cylindrical magnetron sputtering has been described by *Wasa* and *Hayakawa* [7.132], *Thornton* [7.123], and *Thornton* and *Penfold* [7.124]. Cylindrical magnetron sputtering has two principal embodiments. The so-called post configuration consists of a tubular anode with the cathode (target) in the shape of a cylinder located concentrically on the axis of the cylinder (Fig. 7.15a). The hollow-cathode configuration consists of a concentric rod-shaped anode at the axis of symmetry of a tubular-shaped cathode (Fig. 7.15b). This configuration is also called the inverted magnetron. In both cases, alternative configurations have ring-shaped anodes located at the cylinder ends. In the post configuration, the cylinder surrounding the post is used as a substrate holder only; in the hollow-cathode configuration, the inside of the cathode cylinder remains open for strategic location of the substrate.

The dc cylindrical magnetron mode of sputter deposition permits the use of low chamber pressure, typically 0.1 Pa, with a magnetic field strength of 0.02 tesla. The current density may run as high as 200 mA/cm^2 with erosion rates as high as 333 nm/s (with copper) [7.123]. Uniform sputtering over the target surface is obtained with sufficiently uniform magnetic fields. Intrinsic stress in

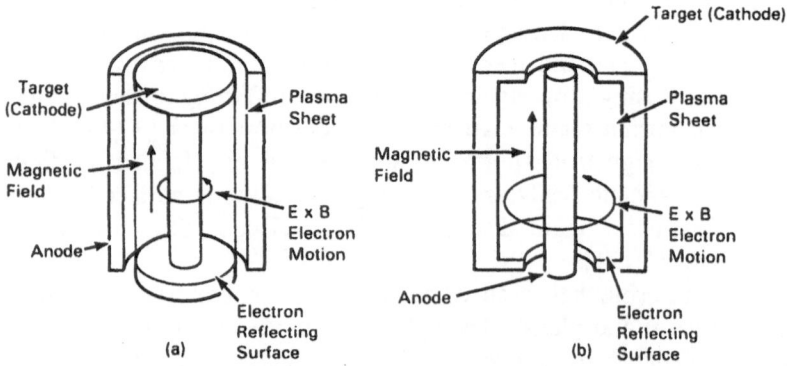

Fig. 7.15. Cylindrical magnetron arrangements: (**a**) post configuration and (**b**) inverted magnetron or hollow-cathode configuration

sputtered films produced by this technique may be controlled by the sputtering gas and its pressure [7.133, 134].

Hollow cathodes are effective for coating substrates of complex shape, because all surfaces having an unobstructed view of the cathode receive a uniform coating. For example, wires and rods can be coated uniformly without being rotated.

b) Planar Magnetron Sputter Deposition

The planar magnetron (ring-gap plasmatron) configuration uses a planar cathode with permanent magnets attached to the back side to provide a ring-shaped magnetic tunnel field at the target surface (Fig. 7.16) [7.135]. The magnetic field confines the electrons within the space covered by the field.

The lines of magnetic flux form a semicircular pattern in a plane perpendicular to the target surface. The lines of flux enter the space above the surface nearly perpendicular to the surface, curve through a semicircular path, and return in a direction nearly perpendicular to that surface. Electrons will be confined in the space where the magnetic field is perpendicular to the electric field. More than 95% of the electrons generally remain in the plasma region. Stray electrons can be caught with an anode trap (a supplementary anode located near the target edge). This type of sputter deposition system can increase film deposition rates to more than 1.5 μm/min at a chamber pressure in the range 0.1–0.7 Pa [7.135]. The ion current density on portions of the cathode can be as high as 1 A/cm^2.

Additional advantages of the planar magnetron method of sputter deposition include simplicity of design, especially if permanent magnets are used; restriction of plasma to regions near the cathode; superior adaptability to

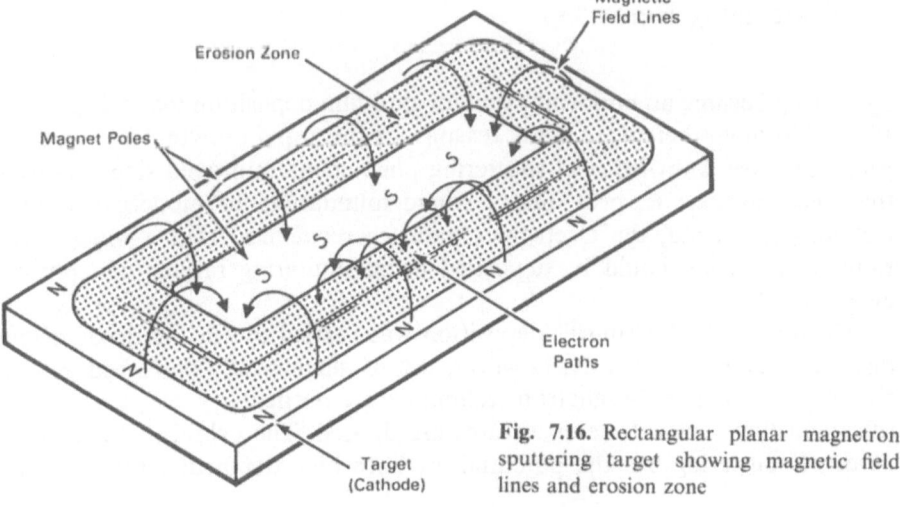

Fig. 7.16. Rectangular planar magnetron sputtering target showing magnetic field lines and erosion zone

coating arrangements by adequate selection of target configuration; and adaptability to the use of rf power so the method can deposit nonconducting coatings [7.136].

The target can be circular or rectangular. The sputtering pattern from the target surface will be irregular because of the configuration of the ring-shaped magnetic field. To achieve uniformly deposited films, the substrate must be moved in reference to the target. The effective target area is limited. Even though the rate of sputtering is relatively high on a unit area basis, the total amount of deposited material is limited by the area of the ring section. As a result, the rate of uniform film growth, averaged over the area of a moving substrate, is not noticeably greater than that obtained by other methods.

The planar magnetron has had much greater commercial use than the cylindrical magnetron. It is widely used in the semiconductor industry for metallization. More recently, however, it has been demonstrated by *Schiller* et al. to be very useful for deposition of oxides using reactive sputtering [7.96, 137]. Stoichiometric TiO_2 and Ta_2O_5 films have been deposited at 1 050 nm/min and 680 nm/min, respectively. *Waits* has also described the planar magnetron process in detail [7.138].

The advantages of the planar magnetron mode include an energy requirement only about one third that required for diode sputter deposition, a low chamber pressure, and a high rate of sputtering from the target. However, the inhomogeneity of the sputtering pattern results in a planar target use of only about 30%. Attempts to improve the target material use have involved periodic shifting of the magnetic field [7.139, 140]. In addition, it is difficult to provide controlled bias and ion etching of the substrates during deposition. In the planar magnetron, that can be done only with proper design and placement of magnetic fields in addition to the target magnetic field.

7.7 Concluding Remarks

Sputtering became an important tool for thin-film deposition technology in the 1960s. Although there were many reasons for this rapid growth, two stand out. First, the basic knowledge of sputtering phenomena improved significantly in the preceding decade, providing a sound foundation for modern sputtering technology. Second, the electronics industry recognized that a much larger range of materials could be deposited using sputtering rather than physical evaporation.

During the 1960s, sputter deposition was carried out in the dc or rf glow discharge regime. Currently, however, use of magnetically enhanced gas discharge is on the rise. Magnetron techniques are particularly useful where high rates and low substrate temperatures are desired. The cylindrical magnetron mode of sputtering has the potential for both high deposition rates and low substrate temperatures over large areas. Sputtering in a thermionically supported discharge offers the advantage of high rates and uniform target erosion.

Each year new complex alloys, compounds, and structures are being produced by sputter deposition. The use of sputtering for thin-film production appears to be limited only by the engineer's imagination. Sputtering need not be restricted to making thin films; it can also be used for producing thick films and freestanding structures.

Sputter deposition takes place in a complex environment. The adhesion and the properties of the films are closely linked to the many inherent process variables. Continued success in this field depends on thoroughly understanding what takes place at the growing surface and adequately characterizing the product. Fortunately, two well-established industries have grown parallel to the sputtering apparatus business. Vacuum system components are continually being improved with the addition of better pumps, traps, gauges, residual gas analyzers, and gas-metering equipment. A completely new industry offering surface analysis equipment has revolutionized the characterization and understanding of film growth.

Sputtered films are now routinely characterized by many different methods including AES, ESCA, EDS, SIMS, ISS, SNMS, RBS, STM, and NRA. The increased analytical power of these aids will hasten development as well as broaden the applications of this versatile coating process. This is especially true in areas involving more complex compounds. Here sputtering has a clear advantage over simpler coating techniques. The phenomenal growth in the published literature on sputtered films is a reliable measure of the importance of this coating technique. Witness the surge in the number of papers on thin-film magnetic and high-temperature superconductors alone.

Acknowledgements. We gratefully acknowledge the encouragement and advice of H.R. Gardner, R.W. Stewart, and R.W. Moss. We are deeply indebted to G. Wehner and K. Wittmaack for critically reading the manuscript and for their many suggestions. This work was supported by the U.S. Department of Energy Office of Basic Energy Sciences (contract DE-AC06-76RLO 1830).

References

7.1   W.R. Grove: Phil. Trans. Roy. Soc. London **142**, 87 (1852)
7.2   J. Plücker: Pogg. Ann. **103**, 88 (1858)
7.3   A.W. Wright: Am. J. Sci. Arts **13**, 49 (1877)
7.4   G.K. Wehner, G.S. Anderson: In *Handbook of Thin Film Technology*, ed. by L.I. Maissel, R. Glang (McGraw-Hill, New York 1970), Chap. 3
7.5   L.M. Reiber, J. Lantaires: Le Vide **168** (1973)
7.6   W.D. Westwood: In *Prog. Surf. Sci.* **7**, ed. by S.G. Davison (Pergamon Press, Oxford 1976) p. 71
7.7   R. Glang: In *Handbook of Thin Film Technology*, ed. by L.I. Maissel, R. Glang (McGraw-Hill, New York 1970) p. 3

7.8   C.F. Powell: In *Vapor Deposition*, ed. by C.F. Powell, J.H. Oxley, J.M. Blocher, Jr. (Wiley, New York 1966) Chap. 9, p. 249

7.9   A.R. Reinberg: In *Annual Review of Material Science*, Vol. 9, ed. by R.A. Huggins (Annual Reviews, Inc., Palo Alto 1979) p. 341

7.10   J.R. Salem, F.O. Sequeda, J. Duran, W.Y. Lee: J. Vac. Sci. Technol. A **4**, 369 (1986)

7.11   G.K. Wehner, G.S. Anderson: In *Handbook of Thin Film Technology*, ed. by L.I. Maissel, R. Glang (McGraw-Hill, New York 1970) Chap. 3, p. 1

7.12   L.M. Reiber: Revue Technique Thompson-CSF **4** (2) (1972)

7.13   E. Groshart: Metal Finishing **71**, 44 (1973)

7.14   E.D. McClanahan, N. Laegreid, R. Busch, E.N. Greenwell, R.W. Moss, J.W. Patten, M.A. Bayne: Government-Industry Workshop on Alternatives for Cadmium Electroplating in Metal Finishing, Gaithersburg, Maryland (1977)

7.15   B.J. Shaw, R.P. Miller: U.S. Patent, 3,918,100 (1975)

7.16   M.A. Bayne, R.W. Moss, E.D. McClanahan: Thin Solid Films **54**, 327 (1978) **63**, 137 (1979)

7.17   M. Kitabatake, K. Wasa: J. Vac. Sci. Technol. A **6**, 1793 (1988)

7.18   G.K. Wehner, Y.H. Kim: Appl. Phys. Lett. **52**, 1187 (1987)

7.19   R. Behrisch (ed.): *Sputtering by Particle Bombardment I*, Topics Appl. Phys., Vol. 47 (Springer, Berlin, Heidelberg 1982)

7.20   R. Behrisch (ed.): *Sputtering by Particle Bombardment II*, Topics Appl. Phys., Vol. 52 (Springer, Berlin, Heidelberg 1983)

7.21   B. Chapman: *Glow Discharge Processes – Sputtering and Plasma Etching* (Wiley, New York 1980)

7.22   P.D. Townsend, J.C. Kelly, N.E.W. Hartley: *Ion Implantation, Sputtering, and Their Applications* (Academic, London, 1976)

7.23   R.V. Stuart: *Vacuum Technology, Thin Films, and Sputtering: An Introduction* (Academic, New York 1983)

7.24   H.D. Hagstrum: J. Vac. Sci. Technol. **12**, 7 (1975)

7.25   H.H. Andersen, H.L. Bay: In *Sputtering by Particle Bombardment I*, Topics Appl. Phys., Vol. 47, ed. by R. Behrisch (Springer, Berlin, Heidelberg 1982) p. 165

7.26   N. Laegreid, G.K. Wehner: J. Appl. Phys. **32**, 356 (1961)

7.27   H. Oechsner: Z. Physik **261**, 37 (1973)

7.28   G.K. Wehner, D. Rosenberg: J. Appl. Phys. **31**, 177 (1960)

7.29   B.M. Gurmin, Y.A. Rhyzon, I.I. Shkarban: Bull. Acad. Sci. USSR, Phys. Ser. (USA) **33**, 752 (1969)

7.30   H.H. Andersen, B. Stenum, T. Sorensen, H.J. Whitlow: Nucl. Instrum. and Methods in Physics Research B **6**, 459 (1985)

7.31   J.B. Bindell, T.C. Tisone: Thin Solid Films **23**, 31 (1974)

7.32   J. Dembowski, H. Oechsner, Y. Yamamura: Nucl. Instrum. and Methods in Physics Research B **18**, 464 (1987)

7.33   T. Motohiro, Y. Taga: Surf. Sci. **118**, 66 (1982)

7.34   K. Meyer, I.K. Schuler, C.M. Falco: J. Appl. Phys. **50**, 5803 (1981)

7.35   P. Sigmund: Nucl. Instrum and Methods, B **18**, 375 (1987)

7.36   G. Betz, G.K. Wehner: In *Sputtering by Particle Bombardment II*, Topics , Appl. Phys., Vol. 52, ed. by R. Behrisch (Springer, Berlin, Heidelberg 1983), p. 16

7.37   L.I. Maissel: In *Handbook of Thin Film Technology*, ed. by L.I. Maissel, R. Glang (McGraw-Hill, NewYork 1970) Chap. 4, p. 39

7.38   R.R. Olson, M.E. King, G.K. Wehner: J. Appl. Phys. **50**, 3677 (1979)

7.39   P. Sigmund: In *Sputtering by Particle Bombardment I*, Topics Appl. Phys., Vol. 47, ed. by R. Behrisch (Springer, Berlin, Heidelberg 1982) p. 15

7.40   J. Bohdansky, H. Linder, E. Hechtl, A.P. Martinelli, J. Roth: Nucl. Instrum. and Methods in Physics Research B **18**, 509 (1987)

7.41   G.K. Wehner: Phys. Rev. **102**, 690 (1956)

7.42   J.J. Cuomo, S.M. Rossnagel: Nucl. Instrum. and Methods in Physics Research B **19/20**, 963 (1987)

7.43   E. Krikorian, R.J. Sneed: J. Appl. Phys. **37**, 3665 (1966)
7.44   S.D. Dahlgren: Met. Trans. **1**, 3095 (1970)
7.45   W.T. Pawlewicz and R. Busch: Thin Solid Films **63**, 251 (1979)
7.46   R. Harrop, P.J. Harrop: Thin Solid Films **3**, 109 (1969)
7.47   T. Spalvins: Thin Solid Films **53**, 285 (1978)
7.48   S. Flügge (ed.): *Encyclopedia of Physics*, Vol. XXI (Springer, Berlin, Göttingen, Heidelberg 1956)
7.49   S. Flügge (ed.): *Encyclopedia of Physics*, Vol. XXII (Springer, Berlin, Göttingen, Heidelberg 1956)
7.50   W.D. Westwood: J. Vac. Sci. Technol. **15**, 1 (1978)
7.51   L.I. Maissel: In *Physics of Thin Films*, ed. by G. Hass and R.E. Thun (Academic, New York 1966) p. 92
7.52   F. Paschen: Wied. Ann. **37**, 69 (1889)
7.53   R.G. Johanson, W.G. Carruthers: J. Vac. Sci. Technol. A **4**, 550 (1986)
7.54   E. Kay: Trans. Materials Research Corporation Conf. and School, Pebble Beach, California (1969) p. 1
7.55   H.C. Cooke, C.W. Covington, J.F. Libsch: Trans. Met. Soc. AIME **236**, 314 (1966)
7.56   V. Hoffman: Electron, Pack. and Prod. **13** (11) 81 (1973)
7.57   S.D. Dahlgren, D.M. Kroeger: J. Appl. Phys. **44**, 5595 (1973)
7.58   L.T. Lamont, Jr.: Solid State Technol. **22**, 107 (1979)
7.59   L.I. Maissel: U.S. Patent 3,294,661 (1966)
7.60   L.I. Maissel, J.H. Vaughn: Vacuum **13**, 421 (1963)
7.61   R.D. Bland: Sandia Tech. Memo SC-TM-71-0526 (1971)
7.62   H. Shimizu, M. Ono, K. Nakayama: J. Appl. Phys. **46**, 460 (1975)
7.63   H.F. Winters, E.E. Horne, E.E. Donaldson: J. Chem. Phys. **41**, 2766 (1964)
7.64   E.V. Kornelsen: Can. J. Phys. **42**, 364 (1964)
7.65   G. Carter: In *Ion Bombardment of Solids* (Elsevier, New York 1968) p. 185
7.66   J.L. Vossen, W. Kern (eds.): *Thin Film Processes* (Academic, New York 1978) p. 54
7.67   J.M.E. Harper, J.J. Cuomo, R.J. Gambino, H.R. Kaufman: Nucl. Instrum. and Methods in Physics Research B **7/8**, 886 (1985)
7.68   B.L. Flur: Proc. Intern. Mag. Conf., 2.4-1 Washington, D.C. (1965)
7.69   A.J. Griest, B.L. Flur: J. Appl. Phys. **38**, 1431 (1967)
7.70   D.M. Mattox: SLA-73-0619, Sandia National Laboratory, Albuquerque, New Mexico (1973)
7.71   P.R. Segatto: J. Vac. Sci. Technol. **6**, 368 (1969)
7.72   H.J. Spitzer: Presented at the 9th Annual Symposium, New Mexico Chapter of the Am. Vac. Soc., Albuquerque, New Mexico (1973)
7.73   E.L. Hollar, F.N. Rebarchik, D.M. Mattox: J. Electrochem. Soc. **117**, 1461 (1970)
7.74   J.W. Patten: Ph.D. Thesis, Washington State University (1975)
7.75   S.D. Dahlgren: J. Appl. Phys. **41**, 5004 (1970)
7.76   C. Deshpandey, L. Holland: Thin Solid Films **96**, 265 (1982)
7.77   R.W. Knoll, E.R. Bradley: Mat. Res. Soc. Symp. Proc. **30**, 235 (1984)
7.78   T. Hata, K. Toriyama, J. Kawahara, M. Ozaki: Thin Solid Film **108**, 325 (1983)
7.79   D. Buhling, L. Michalowsky: Le Vide **185**, 185 (1976)
7.80   K.Y. Ahn, M. Wittmer, C.Y. Ting: Thin Solid Films **107**, 45 (1983)
7.81   W.D. Sproul: Thin Solid Films **107**, 141 (1983)
7.82   S. Alertovitz, J.A. Woollam, L. Kammerdiner, H-L. Luo, C. Martin: U.S. NASA Tech. Memo, NASA TM X-73620 (1977)
7.83   T. Spalvins: Thin Solid Films **96**, 17 (1982)
7.84   W.T. Pawlewicz, P.M. Martin, D.D. Hays, I.B. Mann: In *Optical Thin Films*, Proceedings of SPIE (The International Society of Optical Engineering), ed. by R.I. Seddon, **325**, 105 (1982)
7.85   R. Kelly, N.Q. Lam: Radiat. Eff. **19**, 39 (1973)
7.86   L. Holland: *The Vacuum Deposition of Thin Films* (Wiley, New York 1956) p. 434
7.87   P.J. Burkhardt, L.V. Gregor: Trans. 19th Natl. Vac. Symp., Am. Vac. Soc. (1967), p. 31
7.88   Bomchil, F. Buiquez, A. Monfret, S. Galzin: Thin Solid Films **47**, 235 (1977)

7.89   T. Abe, T. Yamashina: Thin Solid Films **30**, 19 (1975)
7.90   F. Shinoki, A. Itoh: J. Appl. Phys. **46**, 3381 (1975)
7.91   J. Hrbek: Thin Solid Films **42**, 185 (1977)
7.92   B. Goranchev, V. Orlinov, V. Popova: Thin Solid Films **33**, 173 (1976)
7.93   N. Schwartz: In *Trans. 10th Natl. Vac. Symp.*, Am. Vac. Soc. (Macmillan, New York 1963)
7.94   E. Hollands, D.S. Campbell: J. Mater. Sci. **3**, 544 (1968)
7.95   H.F. Winters, D.L. Raimondi, D.E. Horne: J. Appl. Phys. **40**, 2996 (1969)
7.96   S. Schiller, U. Heisig, K. Steinfelder, J. Strümpfel: Thin Solid Films **63**, 369 (1979)
7.97   O. Fiedler, B. Schöneich, G. Reisse, H.J. Erler: Wiss. Z. d. Techn. Hochsch. Karl-Marx-Stadt, **12**, 483 (1970)
7.98   D. Brzezinska: In Works of the Industrial Institute of Electronics, **12**(3) 95 (1971)
7.99   T.I. Putner, G.N. Jackson: Sprechsaal Keram Glass. Email. Silikate **105**, 20 (1972)
7.100 J.W. Nickerson, R. Moseson: In *Trans. 3rd Intern. Vac. Cong.*, ed. by H. Adams (Pergamon Press, New York 1967) p. 625
7.101 E.C. Muly, A.J. Aronson: J. Vac. Sci. Tech. **6**, 128 (1969)
7.102 S.M. Rossnagel: J. Vac. Sci. Technol. A **6**, 19 (1988)
7.103 F.E. Terman: *Electronic and Radio Engineering*, 4th ed. (McGraw-Hill, Toronto 1955) p. 173
7.104 N. Laegreid, G.K. Wehner, B. Meckel: J. Appl. Phys. **30**, 374 (1959)
7.105 E.D. McClanahan, R.W. Moss: U.S. Patent 4,046,666 (1977)
7.106 R.W. Moss, M.D. Merz: J. Vac. Sci. Technol. A **3**, 2694 (1985)
7.107 J.D. Cobine: *Gaseous Conductors* (Dover, New York 1958) p. 109
7.108 D.A. Wright: Proc. Inst. Elec. Engrs. **100**, 125 (1953)
7.109 E.D. McClanahan: Presented at the Am. Vac. Soc. Thin Film Division Topical Symposium on Sputtering, San Diego, California (April 6–8, 1984)
7.110 R.V. Stuart, G.K. Wehner, G.S. Anderson: J. Appl. Phys. **40**, 803 (1969)
7.111 G.S. Anderson, W.N. Mayer, G.K. Wehner: J. Appl. Phys. **33**, 2991 (1962)
7.112 D.P. Davidse, L.I. Maissel: J. Appl. Phys. **37**, 574 (1966)
7.113 P.D. Davidse, L.I. Maissel: In *Trans. 3rd Intern. Vac. Cong.*, ed. by H. Adam (Pergamon, New York 1967) p. 651
7.114 R.B. McDowell: SCP and Solid State Techn. **15**, 23 (1969)
7.115 H. Norstrom: Vacuum **29**, 341 (1979)
7.116 G. Siegle: MRV Metallpraxis/Oberflächentechnik, p. 247 (1972)
7.117 J.J. Bessot: Thin Solid Films **32**, 19 (1976)
7.118 G. Gorinas, Cit-Alcatel: French Patent 74 18999 (1974)
7.119 F.M. Penning: U.S. Patent 2,146,025 (1939)
7.120 E. Kay: J. Appl. Phys. **34**, 760 (1963)
7.121 L.I. Maissel: In *Handbook of Thin Film Technology*, ed. by L.I. Maissel, R. Glang (McGraw-Hill, New York 1970) Chap. 4, p. 9
7.122 C.F. Weston: In *Cold Cathode Glow Discharge Tubes* (ILIFFE Books, London 1968) Chap. 3, p. 108
7.123 J.A. Thornton: J. Vac. Sci. Technol. **15**, 171 (1978)
7.124 J.A. Thornton, A.S. Penfold: In *Thin Film Processes*, ed. by J.L. Vossen, W. Kern (Academic, New York 1978) p. 77
7.125 F.M. Penning, J.H.A. Moubis: Proc. Koninkl. Ned. Akad. Wetenschap. **43**, 41 (1940)
7.126 F.M. Penning: Physica **3**, 873 (1936)
7.127 U. Heisig, K. Goedicke, S. Schiller: Proc. 7th Intern. Symp. Electron and Ion Beam Sci. and Tech., Washington D.C. 1976 (Electrochem. Soc., Princeton, New Jersey 1976) p. 129
7.128 N. Hosakawa, T. Tsukada, T. Misumi: J. Vac. Sci. Technol. **14**, 143 (1977)
7.129 A.S. Penfold, J.A. Thornton: U.S. Patents 4,041,353, 3,995,187, 4,030,996, 4,031,424, (1977); U.S. Patent 3,884,793 (1975); A.S. Penfold: U.S. Patent 3,919,678 (1975)
7.130 K.I. Kirov, N.A. Ivanov, E.D. Atanasova, G.M. Minchev: Vacuum **26**, 237 (1976)
7.131 F.R. Arcidiacono: Proc. 27th Electron. Components Conf. (IEEE, New York 1977) p. 232
7.132 K. Wasa, S. Hayakawa: Rev. Sci. Instrum. **40**, 693 (1969)
7.133 J.A. Thornton, J. Tabock, D.W. Hoffman: Thin Solid Films **64**, 111 (1979)

7.134 D.W. Hoffman: Thin Solid Films **107**, 353 (1983)
7.135 R.L. Cormia, P.S. McLeod, N.K. Tsujimoto: In *Proc. 6th Intern. Conf. on Electron and Ion Beam Sci. and Tech.*, ed. by R. Bakish (Electrochem. Soc. Inc., Princeton 1974) p. 248
7.136 S. Schiller, U. Heisig, K. Goedicke: Thin Solid Films **40**, 327 (1977)
7.137 S. Schiller, G. Beister, E. Buedke, H.J. Becker, H. Schicht: Thin Solid Films **96**, 113 (1982)
7.138 R.K. Waits: In *Thin Film Processes*, ed. by J.C. Vossen, W. Kern (Academic, New York 1978) p. 131
7.139 Sloan Tech. Corp. Brochure, T-133[A]-5M-677
7.140 P.S. McLeod: U.S. Patent 3,956,093 (1976)

List of Symbols

A	Ion bombarded area
A	Electron affinity
A_{BM}	Preexponantial factor of Born-Mayer potential
A_0	Data acquisition area
A_d	Empirical constant
δA	Fractional damaged area
A'	Rate constant in Auger neutralization
a'	Inverse distance
a_{BM}	Characteristic distance of Born-Mayer potential
B	Magnetic field strength
B_i	Conversion factor
b	Empirical exponent
b_i	Characteristic factor in AES
c	Velocity of light
c_i, c_M	Concentration (atomic fraction) of species i or M
c_{cc}, c_{ad}	Apparent concentration due to cross-contamination and residual gas adsorption, respectively
D	Distance between atoms in a crystal direction
D	Diffusion constant
D_s	Sample-electrode spacing
d	Drift length or distance from sample
E_0	Energy of primary ion or electron
E_n	Energy deposition into nuclear motion
E_d	Threshold displacement energy
E_{dp}	Mean displacement energy
E_f	Focusing energy
E_p	Most probable energy
E_i^r	Energy of (back-)scattered ion
E_R	Energy of reflected ions
E_s, E_i	Energy of sputtered ions
δE^Σ	Total energy loss to surrounding lattice
e	Elementary charge
F_{ij}	Isotope fractionation ratio
F_n	Nuclear energy deposition per unit depth
f	Undamaged fraction of bombarded area
f_i	Fraction of atoms in bonding state i
g	Instrumental factor

g_i	Multiplicity
$g(E_i)$	Recoil spectrum
H	Hamiltonian
H_{12}	Transition matrix element
H_e	Auger yield
h	Planck constant
I	Ionization potential
I_0	Primary ion or electron current
I_i	Intensity due to species i
$\hat{I}$	Peak intensity
I_i^e	AES intensity due to species i
I_i^r	ISS or RBS intensity due to species i
I_{true}	True intensity
I_{app}	Apparent intensity
IP	Ionization potential
J	Primary ion flux
J_R	Reflected ion flux
J_{rg}	Residual gas flux
J_x'	Differential primary ion flux, dJ/dx
j	Current density
K	Sticking coefficient of sputtered atoms
K	Calibration factor in TOF mass spectrometry
K	Secondary ion yield enhancement factor per oxygen neighbor
K_i	Kinematic scattering factor
k_c, k_s	Proportionality constants
k_B	Boltzmann constant
L	Radius of ring of atoms
L	Effective escape depth of electrons
L_0	Escape depth of electrons
L_s	Mean depth of origin of sputtered atoms
M	Mass of atom or molecule
$M_{1,2}$	Mass of incident ion or target atom
m	Exponent
N, N_S	Number density of atoms in a sample
$N(E)$	Energy distribution of sputtered atoms
N^A	Areal density of atoms
n	Exponent
n_0	Number of incident primary particles
n_i	Sputtered atom flux
p	Momentum of an atom
p	Pressure
$p^{*,q}$	Probability for a particle to be emitted in excitation state * or ionization state q, respectively
p_i	Normalized mixing profile
p_i^s	Mean value of internal profile at the surface
p'	Numerical factor

p''	Geometrical filling factor
q	charge or charge state
R	Distance of closest approach between colliding atoms
R	Bending radius or radius of hemispherical electrode
R_c	Target-collector distance
R_c	Crossing distance
R_e	Electron backscattering yield
R_d	Mean relocation distance
R^+	Positive ion fraction
r	Radius of circular bombarded area
$r_a(t)$	Time dependent position of atom
r_i	Relative component sputtering yield
T	Energy transferred in a binary collision
T_m	Maximum energy transferred in a binary collision
$T(\theta_e, \phi_e)$	Transformation function
T_c	Collector temperature
T_s	Characteristic parameter in secondary ion emission
t	Time of bombardment
$U(r)$	Interatomic potential
U_{des}	Desorption energy
U_0, U_s	Surface binding energy
$V(r)$	Electronic interaction potential, interatomic potential
V_g	Breakdown voltage
V_0	Ion source terminal voltage
V_t	Target bias
V_a	Acceleration voltage
δV_a	Deficit in acceleration voltage
V'	Volume above the bombarded sample
v	Particle velocity
v_0	Secondary ion velocity
$v_n, v_\perp$	Velocity component normal to the surface
v_i	Erosion velocity of surface
v_s	Scan speed
w_s	Width of image field
x, y	Cartesian coordinates in the plane of the surface
Y	Matrix sputtering yield
dY/dE	Energy distribution of sputtered particles
$dY/d\Omega$	Angular distribution of sputtered particles
Y_i, Y_M	Partial sputtering yield of element i or M
Y_i^c	Component sputtering yield
$y_i(z)$	Normalized depth-differential partial sputtering yield of element i
Z_+, Z_0	Partition function or multiplicity
z	Depth from original surface
$\dot{z}$	Erosion rate
z_i^0	Marker location
$z_{1/e}$	Characteristic depth

z_r	Reference depth
z'	Depth from sputtered surface
z_a^*	Mean depth of altered layer
Δz	Width of broadened profile or profile shift
α	Inclination angle of sputtered slope
α, β	Angles of the facets on a surface
β	Inclination of grain boundary
β_i	Detector efficiency
γ	Fitting parameter in secondary ion emission
γ_{st}	Sticking coefficient
$\Delta(z)$	Width of atomic level
δ_s	Image displacement due to time-of-flight effects
ε	Stopping cross section
$[\varepsilon]$	Stopping cross section factor
ε	Electron energy level in a solid
ε_F	Fermi energy
$\varepsilon_0, \varepsilon_n, \varepsilon_p$	Characteristic energy parameters in secondary ion emission
ε_{cc}	Cross-contamination efficiency
ζ_a	Broadening parameter equivalent to the decay length
η	Transmission and detection efficiency of spectrometer
θ	Scattering angle
θ_{in}	Angle of incidence (to surface normal)
θ_{out}	Angle of emission (to surface normal)
θ_0	Orientation of the axis of primary ion beam propagation
θ_r	Recoil angle in a binary collision
$\theta_{1/2}$	Half width of sputtered angular distribution
θ_s	Fractional surface coverage
Λ	Focusing parameter
λ	Decay length
λ	Mean free path between collisions in a solid
λ_0	Wavelength of laser light
λ_{jk}	Wavelength of light for transition $j \rightarrow k$
μ	Mass ratio M_i/M_1
ν	Exponent
$\nu, \Delta\nu$	Frequency, frequency change
ξ	Reduced depth
$\dot{\xi}$	Reduced erosion rate
$\varrho(\varepsilon)$	Density of states
$\bar{\sigma}$	Damage cross section
σ	Scattering cross section
σ_d	Cross section for displacement collisions
σ_i^e	Cross section for electron impact ionization
τ	Lifetime of excited state
τ_d	Time-of-flight from sample to detector
τ_{ex}	Time of secondary ion extraction
τ_{dr}	Drift time

τ_s	Line scan period
Φ_0	Primary ion fluence
Φ, Φ_e	Work function
ϕ''	Recoil angle in a binary collision
χ	Fraction of neutrals present in postionization volume
χ_0	Freezing distance
ψ	Phase difference of light
$\Psi_k(t)$	Molecular wavefunction
Ω	Solid angle of detection
ω_e	Auger line width

Author Index

In this index the first numbers refer to the pages and the numbers in brackets refer to the relevant references

Subject Index

R. Behrisch (Ed.)

Sputtering by Particle Bombardment I

Physical Sputtering of Single-Element Solids

With contributions by numerous experts

1981. (Topics in Applied Physics, Vol. 47)
Second edition in preparation

Contents: *R. Behrisch:* Introduction and Overview. –
P. Sigmund: Sputtering by Ion Bombardment: Theoretical
Concepts. – *M. T. Robinson:* Theoretical Aspects of Mono-
crystal Sputtering. – *H. H. Anderson, H. Bay:* Sputtering Yield
Measurements. – *H. E. Roosendaal:* Sputtering Yields of Single
Crystalline Targets.

The erosion of a solid surface due to bombardment with energetic ions is a phenomenon widely observed and applied in
today's experimental physics and technology. Within the last
15 years considerable progress has been achieved in measuring
sputtering yields and in getting a physical understanding
of the physics of the sputtering process.
This volume – the first in a three-part
series – covers the sputtering of single-
element solids. The book can be used
as an introduction to sputtering phenomena and their possible applications as
well as a reference for information and
data widely distributed in the literature.

Springer-Verlag
Berlin
Heidelberg
New York
London
Paris
Tokyo
Hong Kong
Barcelona
Budapest

R. Behrisch (Ed.)

Sputtering by Particle Bombardment II

Sputtering of Alloys and Compounds, Electron and Neutron Sputtering, Surface Topography

With contributions by numerous experts

1983. (Topics in Applied Physics, Vol. 52)
Second edition in preparation.

Contents: Introduction and Overview. – Sputtering of Multicomponent Materials. – Chemical Sputtering. – Sputtering by Electrons and Photons. – Sputtering of Solids with Neutrons. – Heavy Ion Sputtering Induced Surface Topography Development. – Development of Surface Topography Due to Gas Ion Implantation. – List of Symbols. – Author Index. – Subject Index.

"Sputtering by Particle Bombardment I" covered physical sputtering in single element solids, which applies predominantly for sputtering of metals with noble gas ions. In this volume several additional effects which contribute to the sputtering process are described. Its coverage includes: sputtering of multicomponent targets and the related compositional charges; chemical effects due to ion implantation; sputtering with electrons and photons (mostly ionic crystals); neutron sputtering; and details about the surface structures which develop due to implantation and surface erosion at particle bombardment.

Springer-Verlag
Berlin
Heidelberg
New York
London
Paris
Tokyo
Hong Kong
Barcelona
Budapest

Topics in Applied Physics   Founded by Helmut K. V. Lotsch